AF358828

Temporal Resolution, a Key Factor in Environmental Risk Assessment

Temporal Resolution, a Key Factor in Environmental Risk Assessment

Editors

Adrian Ursu
Cristian Constantin Stoleriu

Basel • Beijing • Wuhan • Barcelona • Belgrade • Novi Sad • Cluj • Manchester

Editors

Adrian Ursu
University "Alexandru Ioan
Cuza"
Iași
Romania

Cristian Constantin Stoleriu
Alexandru Ioan Cuza
University of Iași
Iași
Romania

Editorial Office
MDPI AG
Grosspeteranlage 5
4052 Basel, Switzerland

This is a reprint of articles from the Special Issue published online in the open access journal *Remote Sensing* (ISSN 2072-4292) (available at: https://www.mdpi.com/journal/remotesensing/special_issues/Temporal_Resolution_Environmental_Risk_Assessment).

For citation purposes, cite each article independently as indicated on the article page online and as indicated below:

Lastname, A.A.; Lastname, B.B. Article Title. *Journal Name* **Year**, *Volume Number*, Page Range.

ISBN 978-3-7258-1887-7 (Hbk)
ISBN 978-3-7258-1888-4 (PDF)
doi.org/10.3390/books978-3-7258-1888-4

Contents

About the Editors

Adrian Ursu

Adrian Ursu, PhD., is an associate professor in the Department of Geography at the Faculty of Geography and Geology, Alexandru Ioan Cuza University, Iasi, Romania. His scientific interests are environmental science, remote sensing, GIS, nature conservation, environmental geography, land use change, risk assessment, and heritage. He has participated in international projects such as Corine Land Cover 2000 and Corine Land Cover 2006 for Romania, funded by the European Environment Agency, and other national research projects concerning risk assessment, sustainable development, and land degradation, GIS for Natura 2000 sites and noise pollution modeling in urban areas. He is Editor-in-Chief of the journal Present Environment and Sustainable Development (ESCI -Clarivate), and he was the Guest Editor of two Special Issues in Remote Sensing (Q1 IF) and Land (Q2 IF). Since 2019, he has been the coordinator of the International Conference 'Present Environment and Sustainable Development', which held its XIXth edition in 2024. He is also of the coordinators of the International Symposium on Geographical Information Systems, which started in 1991 and is organized in partnership with four other academic institutions from Romania and the Republic of Moldova. He has published over 38 scientific papers indexed in Web of Science with 189 citations.

Cristian Constantin Stoleriu

Cristian Constantin Stoleriu, PhD., is an associate professor in the Department of Geography at the Faculty of Geography and Geology, Alexandru Ioan Cuza University, from Iasi, Romania. He is also responsible for the master's specialization in Geomatics within the institution. His main research topics include 1D and 2D flood modeling, spatial analyses of land-use dynamics in protected areas and their adjacent zones, the use of drones' products and LiDAR data for spatial analysis, and ecological changes induced by human impact. He has co-authored approximately 40 scientific articles published in ISI -indexed journals and is involved in research projects, both in academia and geospatial studies related to protected areas with economic agents.

Preface

Assessing the impact of natural or anthropogenic events on the environment and society, in order to better understand the consequences and provide appropriate solutions, is a cross-cutting issue that encompasses various research areas related to the Earth's surface, landscapess and environmental change. It involves the use of spatial data, remote sensing, and geographic information systems (GIS) to study natural processes, human impacts, and ecological vulnerability. It is concerned with understanding and mitigating environmental risks, with a focus on assessing and managing risks associated with natural hazards, urbanization, climate change, and ecological change.

Researchers aim to understand the complex interactions between natural systems and human activities. They try to predict future environmental changes and their consequences. They aim to develop strategies for adapting to environmental challenges. The purpose of these studies is to inform policy, planning, and sustainable practices. By understanding environmental dynamics, we can better protect ecosystems, communities, and infrastructure. For example, temporal resolution is a key factor in environmental risk assessment. Researchers seek to fill gaps in understanding, enhance resilience, and improve environmental governance. This scientific work is aimed at fellow researchers, policy makers, urban planners, and environmental practitioners. Students and educators interested in spatial science and environmental management will also benefit from these findings.

The specialists engaged in investigating ecological vulnerability, flood risk, the territorial impact of urban development, glacier shrinkage, drought, fire detection, sand deposition dynamics, and related matters originate from Romania, China, India, and Vietnam. These research areas are of significance for their respective countries, yet they also present approaches and solutions that can be applied in other parts of the world.

Research studies in this reprint received funding from diverse sources, including government grants, institutional support, interdisciplinary research initiatives, and doctoral/postdoctoral programs. Furthermore, some articles emerged from the dedication of passionate specialists who explored the mechanisms driving structural and functional transformations in both natural and human-made systems.

Adrian Ursu and Cristian Constantin Stoleriu
Editors

 *remote sensing*

Article

Normalized Sand Index for Identification of Bare Sand Areas in Temperate Climates Using Landsat Images, Application to the South of Romania

Cristian Vasilică Secu, Cristian Constantin Stoleriu *, Cristian Dan Lesenciuc and Adrian Ursu

Department of Geography, Faculty of Geography and Geology, University "Alexandru Ioan Cuza", 700505 Iassy, Romania
* Correspondence: cristian.stoleriu@uaic.ro

Citation: Secu, C.V.; Stoleriu, C.C.; Lesenciuc, C.D.; Ursu, A. Normalized Sand Index for Identification of Bare Sand Areas in Temperate Climates Using Landsat Images, Application to the South of Romania. *Remote Sens.* **2022**, *14*, 3802. https://doi.org/10.3390/rs14153802

Academic Editor: Parth Sarathi Roy

Received: 23 June 2022
Accepted: 31 July 2022
Published: 7 August 2022

Publisher's Note: MDPI stays neutral with regard to jurisdictional claims in published maps and institutional affiliations.

Abstract: The expansion of bare sand surfaces indicates a tendency towards desertfication in certain periods as a result of the improper agricultural use of sand soils and of the significant changes in the climate in the past 30 years. The Normalised Sand Index (NSI) is a new index used to identify bare sand areas and their spatio-temporal evolution in SW Romania. Landsat scenes (1988, 2001, 2019), spectral and soil texture analysis (36 samples), covariates (e.g., soil map), and field observations allowed for the validation of the results. The performance of the NSI was compared with indices from the sand index family (e.g., Normalized Differential Sand Areas Index) and supervised classifications (e.g., Maximum Likelihood Classification) based on 47 random control square areas for which the soil texture is known. A statistical analysis of the NSI showed 23.6% (27,310.14 hectares) of bare sands in 1988, followed by an accelerated increase to 47.2% (54,737.73 hectares) in 2001 because of economic and land-use changes, and a lower increase by 2019, which reached 52.5% (60,852.42 hectares) due to reforestation programs. Compared to the NSI, the bare sand areas obtained with the tested indicator were almost 20% higher. The traditional classification shows smaller areas of bare sands but uses a higher complexity of land use classes, while the producer accuracy values are lower than those of the NSI. The new index has achieved a correct spatial delimitation of soils in the interdune-dune and major riverbed-interfluvial areas, but it is limited to the transition Arenosols-Chernozems by humus content and agrotechnical works. The new spectral index favours bare sand monitoring and is a fast and inexpensive method of observing the desertification trend of temperate sandy agroecosystems in the context of climate change.

Keywords: sand index normalized; arenosols; landsat; spectral indices; land use

1. Introduction

The issue of climate change and its effects on natural and man-made environments is of increasing concern to decision-makers around the world. In 2022, several countries of the world adopted the Glasgow Climate Pact (COP26, United Nations Framework Convention on Climate Change) to try to respond to the climate emergency by switching away from fossil fuels [1].

In this context, it should be mentioned that desertification is a topic of utmost importance at the global level; however, traditionally, desertification refers to the extension of desert ecosystems to neighbouring areas due to the reduction of rainfall or due to the displacement of dunes by wind.

In the case of Romania, the mere appearance of sand is interpreted by environmental activists as the result of desertification, but things are much more complex than they seem. In this article, we intend to trace the changes that have taken place over 30 years in the Oltenia area of southern Romania, where there were traditionally shifting sands that had been stabilized by land improvement works during the communist period.

After the fall of the communist system, the changes that occurred in the area had different effects on the local ecosystems.

In Europe, sands of aeolian origin are the most commonly found and widespread, forming the so-called European Sand Belt, while continental sands have multiple sources [2]. The sands of the Oltenia Plain are alluvium transported by the Danube or its tributaries from the Carpathians and the Balkan mountains [3].

Over the last half century, landforms, soils, and land use have undergone several stages of transformation.

In 1970, the development of the Sadova–Corabia irrigation system (>74,000 ha) began, which included the southern part of the study area, and one of the subprojects was to establish modern plantations (e.g., peaches, apricots, cherries, grapes) [4].

Part of the dunes were levelled and excavated for irrigation canals, and the sandy material was redistributed to the interdune areas. The sands were covered with organic material, and the agricultural plots were delimited by forest buffer strips at larger distances for permanent crops and smaller distances for arable land.

The post-communist legislative framework was not very favourable for the maintenance of the irrigation system, plot sizes, and crop types. The irrigation system was an excessive consumer of energy as water was pumped upstream, but during the communist period, little attention was paid to the cost of energy [5]. The administrative transfer of the irrigation system from one entity to another (from national authorities to private associations) due to high energy and maintenance costs, coupled with the lack of subsidies in the following years, led to the deterioration of the irrigation system, especially after 1991 [6]. Following the retrocession of agricultural land [7], the farming areas decreased (5–10 ha), which is also one of the reasons for the abandonment of the irrigation system [8].

The Psamosols of the Oltenia Plain have undergone several stages of deforestation and reforestation. At the end of the 19th century, the sand dunes were stabilized by acacia plantations (about 9000 ha), which were cleared during the development of the irrigation system, thus leading to the reactivation of the dunes [9]. After 1970, however, protective forestry barriers (8–10 m wide, at a distance of 288 m) were set up perpendicular to the prevailing wind direction, some of which were illegally cleared after 1989 [10]. The National Rural Development Programme (2014–2020), funded by the Romanian Government, is a prerequisite for increasing the reforested areas in the Oltenia Plain.

The legislative and economic context has led to changes in the crop structure. Thus, the permanent crops grown on sandy soils (vines, peaches, mulberries) during the period of the irrigation system were replaced by traditional crops (cereals).

The present study aimed at observing how these changes were reflected in the agricultural spatial planning that was shown in the satellite images by using the classical sandy surface mapping indices and proposing a new index.

The problem of desertification in the Oltenia Plain was assessed using remote sensing indices for vegetation (NDVI, MSAVI2) [11] and via the strong correlation between the ground surface temperature obtained from satellite images and the surface temperature from meteorological stations in the area [12]. The intensification of climate aridization, which has been emphasized by future climate scenarios (e.g., RCP4.5), indicates that in the plain and plateau areas outside of the Carpathian Mountains, the desertification process can be significant and long-lasting [13].

Desertification is a dynamic process; therefore, it is necessary to monitor the dynamics at certain time intervals [14]. Spectral indices are used to detect, map, and monitor land degradation. Many studies use sand indices [15–17], such as the bare sand index [18,19] or its derivatives [20], which belong to the spectral index family that uses covariates and field surveys to capture the dynamics of sandy areas from medium-resolution satellite imagery.

The mapping and inventory of bare sand provide the baseline data needed in order to prioritize areas and prepare action plans for combating desertification.

The objectives of the study are as follows:

a. To propose a new index (Normalized Sand Index—NSI) that is able to capture bare sand areas and scattered vegetation covering the sandy soils;
b. To evaluate of sandy surfaces based on literature indicators and supervised classifications (Maximum Likelihood Classification—MLK, Support Vector Machine—SVM);
c. To perform a comparative analysis of the accuracy of the NSI with indicators in the literature and with the supervised classifications (MLK, SVM);
d. To assess the spatio-temporal dynamics of sands (1988–2019) based on satellite images (Landsat sensor).

In remote sensing, most spectral indices can detect dunes and mobile sands in arid and semi-arid climates using spectral images such as visible (VIS), near infrared (NIR), and shortwave infrared (SWIR) by applying normalization [15,17,21,22] (Table 1). Other indices use band 1 (Coastal aerosol) and partial normalization [19] or employ a surface reflectivity analysis using red and near-infrared bands and correlation with fractional vegetation cover [23,24]. Different from single index modelling, desertification is quantified by combining an active surface property (e.g., albedo) with vegetation and soil indices (e.g., soil-adjusted vegetation index) in point-to-point and point-to-line models [25].

Table 1. Indices used for mapping bare sand areas.

Index	Bands Math	Feature Extraction, Sand Value/Climate	Satellite	References
Normalized Differential Sand Dune Index (NDSDI)	$NDSDI = \frac{R-SWIR2}{R+SWIR2}$	Sand, <0/dry	Landsat 5 TM, Landsat 7 ETM	[15]
Normalized Differential Sand AreasIndex (NDSAI)	$NDSAI = \frac{SWIR1-R}{SWIR1+R}$	Sand, <0/dry or humid	Landsat 5 TM, Landsat 7 ETM, Landsat 8 OLI	[17]
Normalized Difference Enhanced Sand Index (NDESI)	$NDESI = \left(\frac{b4-b2}{b4+b2}\right) + \left(\frac{b7-b6}{b7+b6}\right)$	−2, +2/arid	Sentinel 2 and Landsat 8 OLI	[21]
Sand differential emissivity index (SDEI)	$SDEI = \frac{b13-b12}{b13+b12}$	1 to 0.28/extremely arid	Aster	[22]

The indices for sand use simple mathematical equations [21] that are easy to apply, and if the validation is supported by the analysis of properties with low variability over time (e.g., soil texture versus humus content), the results are good. There is no best method to map bare sands. In many papers, data normalization is used [15,17,26], while in others, partial normalization is preferred [19,27] as a way to differentiate sand from soils. The disadvantage of indices that detect sand is their exclusive use in arid climates [21,26], thus making the results obtained in temperate climates not adequate [27].

2. Study Area

The study area is bounded to the west and south by the slope that transitions to the floodplains of the Jiu and Danube rivers (Figure 1). The eastern boundary is characterized by a change in soil type and texture, i.e., the transition from Arenosols to Chernozems. Towards the east, a buffer of 2000 m was generated based on the soil sampling points so that the sandy boundary could be correctly identified.

Sand deposits in the study area are located on the left side of the Jiu River, taking the form of an elongated strip in the north–south direction (about 70 km) and with a narrower width in the north (4–6 km), but reaching 28–30 km in the southern sector (Dăbuleni). These are closely related to the paleogeographic evolution of the Jiu and Danube rivers. In the Pleistocene and Holocene, the course of the Danube moved southwards, leaving a well-developed system of up to six river terraces on the left side. The terrace deposits consist of alternating sand, clay, and loess deposits.

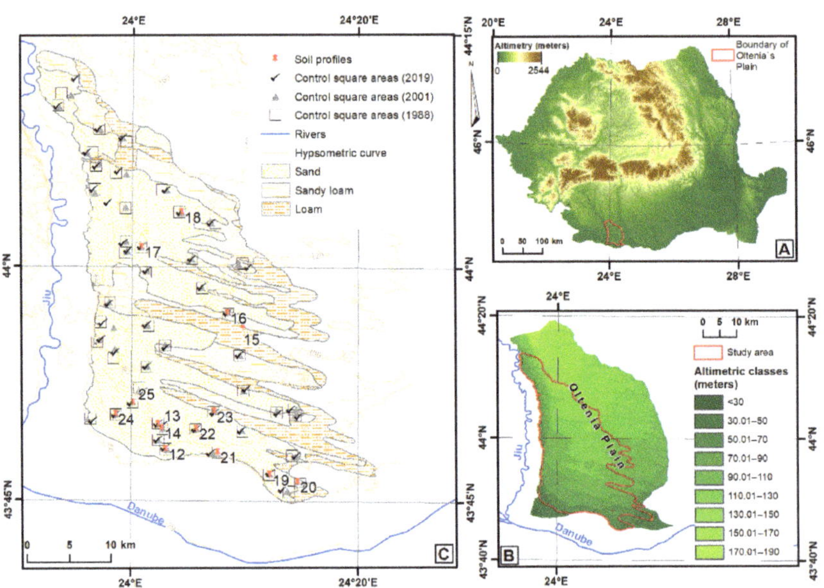

Figure 1. Soil texture and location of soil profiles and control square areas ((**A**)—the geographical location of the Oltenia's Plain within Romania, (**B**)—the geographical location of the study area within the Oltenia's Plain, (**C**)—the numbers represent soil profiles extracted from scientific literature).

Morphologically, two types of dunes can be distinguished. There are old, stabilized, or semi-stabilized dunes that have a symmetrical profile. The interval between two rows of dunes varies between 100 and 500 m, and their height does not exceed 8–15 m. Towards the north, the gap between the dunes increases with distance from the Danube while the height decreases. There are recent, mobile, or semi-stabilized dunes that have an asymmetrical morphology, with a less steep slope towards the wind and steeper slopes on the opposite side. They may occur in the form of barrens that are very sparsely covered by vegetation, and they consist of quartz sands and additional minerals such as mica [3,28].

The climate of the Oltenia Plain is temperate-continental, but spring time is the most critical period of deflation because the soil has not yet been covered by vegetation, and the most severe droughts affect agricultural crops [29], therefore increasing the sandy areas that are not covered with vegetation for a longer period of the year.

The dominant wind direction is NW-SE, with an average speed ranging from 3.2 m·s^{-1} from the NW to 3.8 m·s^{-1} from the W, but with speeds that can exceed 15–17 m·s^{-1} becoming more frequent in recent years [30]. The most active and harmful form of wind erosion occurs in the steppe and partly wooded-steppe zones, where wind speeds exceed 4 m·s^{-1} [31].

Soil distribution and texture were taken from the Map of Romanian Soils (1:200,000), and the names were adapted according to the International System for Soil Classification (WRB) [32].

Soils are represented by Arenosols, which vary spatially according to the configuration of the land. Sandy deposits are favourable for the formation of Arenosols [33], but mollic arenosols occupy the interdune areas, while Eutric Arenosols are dominated on dune tops [34].

In the sector related to the irrigation system (Figure 2), the sandy soils have been anthropogenically modified by levelling, resulting in Arenosols being displaced and transported. A particular characteristic of the sandy soils of the Oltenia Plain is the very high proportion of coarse sand (60–80%) and the very low proportion of clay (2–19%) [35].

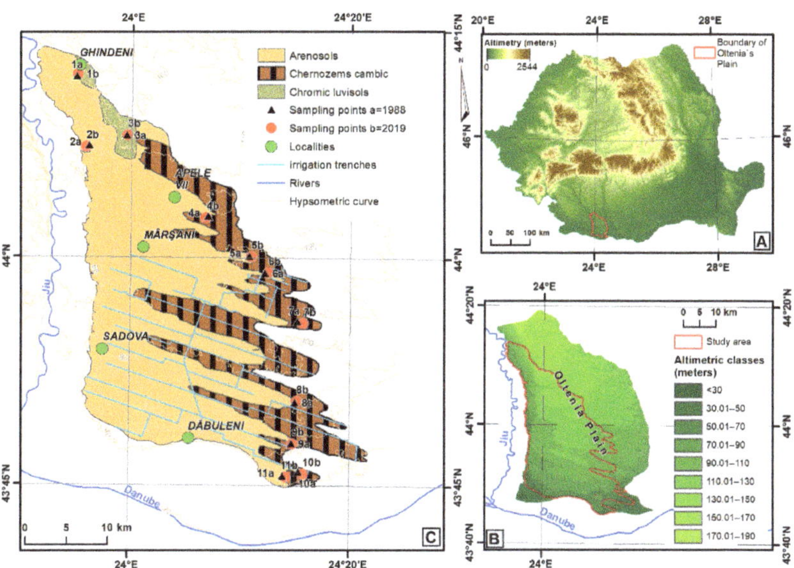

Figure 2. Soils map and sampling points ((**A**)—the geographical location of the Oltenia's Plain within Romania, (**B**)—the geographical location of the study area within the Oltenia's Plain, (**C**)—the numbers represent soil profiles sampled from the field).

Towards the east, the sand deposits thin out, intercalating into Cambic Chernozems, while in the northeast, the transition is better expressed towards Chromic Luvisols (Figure 2).

The texture on the surface horizon is sandy in the west and changes to sandy loam towards the east and northeast. In the south-eastern half, the sandy-loam texture sometimes forms bands to the south of the sandy-textured soils (Arenosols) as a consequence of the dominant wind direction (VNV-ESE) (Figure 1).

Land use for the first hierarchical level in Corine Land Cover (2018) [19] includes: agricultural land, forests and semi-natural areas, water bodies, and artificial surfaces. Agricultural land nomenclature has been adapted to local characteristics, and the following classes have been identified: arable land and pastures on sandy soils (APSS), arable land-pastures on soils with different texture (APSDT), autumn crops (AC), and permanent crops (PC). Forests and semi-natural areas include: compact forests (CF), scattered forests (SF), and sands (S).

Arable land on sandy soils (Arenosols) is partially covered by vegetation, and the plots are rectangular in shape. Arable land and grassland on soils with other textures have low reflectance values due to humus content on the surface horizon, and the plot size is larger than in the previous case. Autumn crops are found on both soil types. The moisture of the Arenosols influences plant density so that the identification of sandy textured soils can be blurred in rainy years. Permanent crops (PC) are represented by vineyards and orchards, which are traditional crops on sandy soils.

Compact forests (CF) are represented by even-aged acacia plantations, and scattered forests (SF) are older acacia plantations where sparse trees form mixtures with grassy vegetation and scrub. Detecting and analysing the spatial dynamics of scattered forests can be an indicator of areas at risk of waste. Bare sand areas (S) are frequently encountered in the SE of the territory, within modified relief. Water bodies include two categories: small lakes (L) resulting from the damming of streams and lakes of semi-permanent character, located between dunes.

3. Materials and Methods

The study methodology involves both the use of remote sensing data such as satellite images and the use of existing GIS databases such as soil maps, topographic maps, Corine Land Cover, etc., in addition to field data processed in the laboratory. A brief methodological workflow is presented in Figure 3.

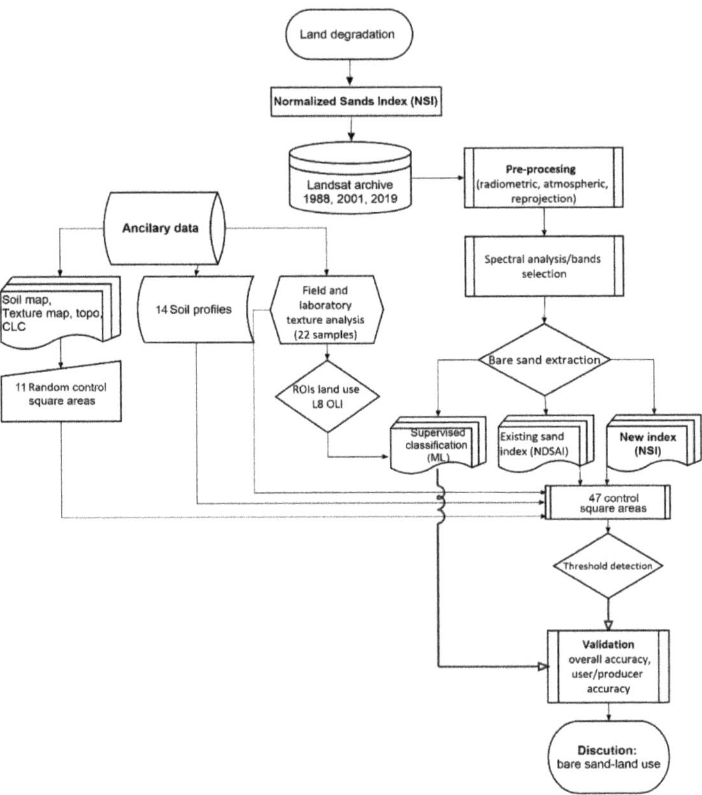

Figure 3. Methodological flowchart.

Three methods were used for mapping areas with uncovered or poorly covered sand: (a) the introduction of a new index, called the Normalized Sands Index (NSI), (b) the use of spectral indices from the literature that are able to map sandy and bare soils (Table 1), and (c) the use of a traditional pixel-based image classification method (e.g., Maximum Likelihood image) to facilitate the interpretation of sand bars in relation to land use. Finally, the performance of NSI is validated by statistical analysis.

3.1. Data Types

Due to the fact that the study of land use evolution on sandy soils in Oltenia spans a period of 30 years, we decided to use ***Landsat satellite images*** as these are characterized by a high temporal consistency; images from the communist period (1988), the first agricultural developments, and recent images (2019) are available with comparable spatial and radiometric characteristics. In this study, we used the highest quality data available for Landsat scenes (Tier 1), which are suitable data for time-series processing analysis [36] and are consistently georegistered within a ≤12 m radial root mean square error, making them suitable for time-series pixel-level analysis [37]. The study area is entirely included in path 184, row 29 for all Landsat scenes, and therefore spatial enhancement is not necessary.

Multispectral Landsat TM5, Landsat 7 ETM, and Landsat 8 OLI data were used for the remote sensing of surface soil attributes, and each row/path date is shown in Table 2. The scenes were obtained from the United States Geological Survey (USGS) and downloaded from the USGS Earth Explorer online platform; all scenes are included in the Level 1 precision terrain (L1TP).

Table 2. Landsat scenes used in the analysis.

Satellite	Row	Path	Date of Acquisition
Landsat 5 TM	29	184	27 January 1988
Landsat 7 ETM	29	184	7 February 2001
Landsat 8 OLI	29	184	17 February 2019

The selection of spectral data was based on several factors that take into account the characteristics of the objects on the images and the variation of the climate during the year. Firstly, it was ensured that the entire study area was free of cloud, haze, and fog. The selection of scenes and time intervals took into account the economic changes outlined in the introduction and the criteria for processing Landsat images. The selection of spectral bands in sandy soil areas for the multi-temporal analysis of land degradation dynamics took into account annual climatic conditions in semi-arid climates, with images from both high rainfall months and months of maximum vegetation growth [38] and low rainfall [39]. Unlike in the drylands, in the temperate zone, the use of images from the same season and the successive years was not possible due to cloud cover.

The choice of images from the winter season took into account the following aspects:

a. Identification of the maximum uncovered sand areas during the year;
b. Avoiding confusion between the light (yellow) colour of the sand and the straw resulting from grain harvesting;
c. In autumn, plant residues are burnt and dark coloured surfaces can be confused with the soft horizon of Chernozems, while smoke influences the quality of the images;
d. In winter, evapotranspiration is lower than in summer, therefore image quality will be less affected by the volume of vapour in the atmosphere. Winter evapotranspiration dynamics, analysed over a long period of time, was more stable compared to other seasons [40].

Soil reflectance is affected by the presence of vegetation, water bodies, irrigation, and other land features. In this study, an attempt was made to obtain the highest possible image quality (with the maximum number of pixels) of the bare soil in the research area. To get the least vegetation cover, the satellite images were acquired in winter (January–February), when the soil was not covered by snow.

The *field study* looked at two aspects. A total of 11 pairs of soil samples (1a,b to 11a,b) (Figure 1) were taken from the eastern extremity of sandy soils (a) to show the maximum distribution of Arenosols and the transition mode to other soils, and (b) to identify land use types in the vicinity of the sampling points that validate the Landsat 8 OLI (2019) scene classification and NSI accuracy.

Field observations have allowed for a differentiation of the soils, based on the texture of the surface horizon, into two classes: sandy (sandy and sandy loam) and soils with a different texture (loam). For each point (Figure 1), soil texture was estimated by testing the sandy texture between fingers, according to the FAO flowchart [41], with the relative deviation from laboratory analysis being lower (3.8%) for sand [42]. The coordinates of the sampling points were acquired with a RTK UNISTRONG G975 geodetic GPS.

Soil samples weighing 500 g were collected from the surface soil horizon. After drying in the laboratory for 14 days, the colour of each soil sample was determined with the Munsell Soil Colour Chart (2000) [43]. After sample preparation [44], soil texture was determined using Analysette 22NanoTec (Fritsch GmbH, Idar-Oberstein, Germany). Then, the sand (2–0.02 mm), silt (0.02–0.002 mm), and clay (<0.002 mm) were classified in accordance with the Romanian soil Taxonomy [45], which is compatible with the texture used by the International Society of Soil Science. This approach allowed for a comparison of the texture results of this study with those in the literature. In order to validate the NSI results, the weight of sand on the surface horizon from 14 soil profiles (12 to 25) specified in the literature was used (Figure 1) [34,35].

3.2. Data Processing

The Landsat scenes were *pre-processed (radiometric calibration, atmospheric correction)* and subsequently spectrally analysed and processed.

The Landsat scenes were radiometrically corrected [46], thus obtaining surface reflectance, and a Quick atmospheric correction was then applied; all images were reprojected in Stereographic 1970 (EPSG 31700) using the ENVI (Exelis Visual Information Solutions) software.

In many studies, soil properties were investigated based on the visible (VIS), near infrared (NIR), and shortwave infrared (SWIR) spectral range [47]. Landsat bands are known for particular applications: band 3 (discriminant of soil from vegetation), band 5 (soil–rocks), band 7 (geology) [48]; the red band (Landsat 5 TM) was used to discriminate bare soil from other land [9].

The multispectral digital classification of remote sensing data requires the *selection of only the bands containing the most information*. The high value of the Optimal Index Factor (OIF) indicates that the bands contain a lot of information (e.g., high standard deviation) with little duplication (e.g., low correlation between the bands) [49].

ILWIS 3.8 Academic software and the Optimal Index Factor (OIF) algorithm were used to select the bands. For each scene, two combination possibilities were identified based on the highest Index Highest Ranking (IHR) values: bands 4,5,7 (1527.26) and bands 3–5 (1282.48) for Landsat 5 TM; bands 4,5,7 (1597.26) and bands 3,5,7 (1406.53) for Landsat 7 ETM; bands 5–7 (2982.32) and bands 4,5,7 (2982.32) with bands 3,5,7 (2449.33) for Landsat 8 OLI.

Prior to *soil sampling*, the eastern boundary of the sandy textured soils for scene Landsat 5 TM (27 January 1988) was estimated by transforming the synthetic colour image (Transformation module in Envi 5.3), which enabled us to locate the soil sampling points, noted on the map from the 1st to the 11th (Figure 1). The pair of each point was identified in the field based on texture and noted from 1b to 11b (Figure 2). Using ArcGIS 10.2 (Environmental Systems Research Institute, Redlands, CA, USA), the twenty-two points were generated in the eastern part of the sand area, which were used for sampling and land-use validation. In the second step, the position of the points was maintained or modified according to the texture estimation in the field.

Using ArcGIS 10.2, the *representative area for each land use type was digitized*, resulting in a different number of ROIs. When the landscape of a study area is complex-heterogeneous, selecting sufficient samples becomes difficult [50]. Auxiliary data were used to extract training areas associated with each land cover type. Thus, certain land use types were extracted from the topographic map (1984) and used for the classification of the Landsat 5 TM scene (1988), and the CLC (2000) [51] was used as a covariate to identify certain land use types (e.g., vineyards and orchards) for scene Landsat 7 ETM (2001). Only a few land cover types could be identified on the basis of topographic map information (lakes, vineyards and orchards, forests), while others could be inferred from plot shape (e.g., agricultural land on sandy soils is rectangular in shape and separated by tree belts).

Within localities, different pixels' reflectance values are determined by the diversity of material types and agricultural practices (e.g., addition of natural fertilizers, burning of vegetation). Under these conditions, the small areas within the localities induce a high level of noise in the identification of some classes. In the study area, there were only rural settlements whose area had not changed considerably during the study period. Based on the above considerations, the perimeter of the localities was extracted from Landsat 8 OLI (2019), which represents the unique mask for the whole set of analysed images.

The ***Jeffries–Matusita (JM) separability criterion*** is a common ***statistical index*** that allows for the selection of a suitable feature subset associated with high class separability [52]. A spectral curve for each training area associated with one land use was directly determined by the software ENVI 5.3. For increased sand separability, we used a set of non-overlapping areas for each Landsat scene. This allowed us to retrieve the average spectral patterns of bare sand and compare them with other land-use types (Figure 4). In such a comparison, the spectral curve could be used to indicate the spatial cover of sands with the other land components (e.g., lakes). In the case of sandy soils, some studies show that spectral reflectance in Landsat bands increase rapidly to B4, with a peak in B5, especially in the absence of opaque minerals [19,53].

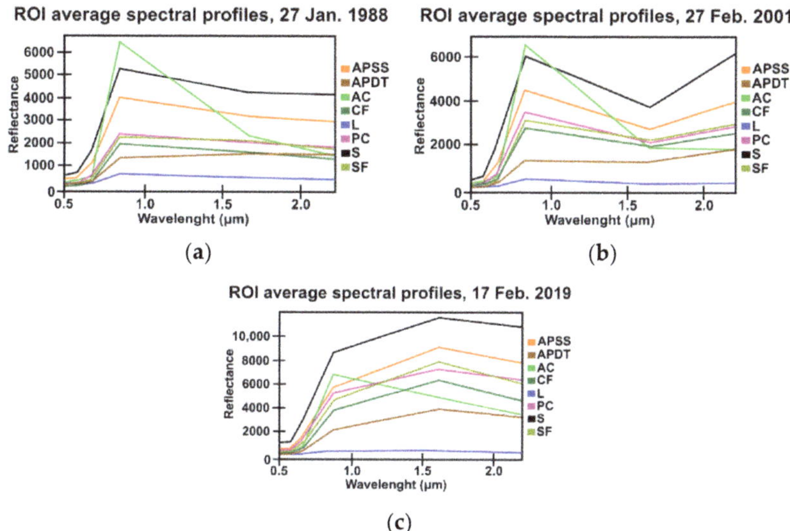

Figure 4. AC = autumn crops, APSDT = arable lands-pastures on soils with a different texture, APSS = arable lands-pastures on sandy soils, CF = compact forests, L = lakes, PC = permanent crops, S = sands, SF = scattered forests.

Based on the JM-transformed divergence algorithm, which was applied to eight classes of land use, twenty-eight pairs with different values were obtained. The separability of the selected training sites for all classes was examined by computing their spectral separability in the ENVI 5.3. software.

3.3. Normalized Sand Index (NSI)

To date, the most commonly used indices for assessing sand surfaces, based on pixel information from satellite images, are the bare sand indices (Table 3).

Table 3. Indices used to extract sandy surfaces.

Index	Band Math	Feature Extraction, Sand Value/Climate	References
Normalized Differential Sand Dune Index (NDSDI)	$NDSDI = \frac{R-SWIR2}{R+SWIR2}$	Sand, <0/dry	[15]
Normalized Differential Sand Areas Index (NDSAI)	$NDSAI = \frac{SWIR1-R}{SWIR1+R}$	Sand, <0/dry or humid	[17]

Two traditional supervised classifications were tested: Maximum Likelihood Classification (MLK) and Support Vector Machine (SVM); however, MLK was used as it is one of the most popular and widely used conventional classifications, thanks to its robustness [54] and its high performance for regional scale land cover mapping [55].

In order to get better results for our geographical area, we proposed a new index—*Normalized Sand Index (NSI)*, which is calculated based on the Red (R), Green (G), and Short-wave Infrared (SWIR1) bands (Equation (1)). In order to minimize the differences induced by atmospheric and moisture conditions between images collected in different time periods, the sand index values were max–min normalized, resulting in the linear transformation of the data in the range 0–1, which preserves the relationships among the original data values. The application of normalization is a widely used method that restricts a large range of values to a small range (e.g., 0–1). Under these conditions, we can compare images from different years or results from different indices, applied to the same type of surface (e.g., sand) [56–58].

$$NSI = (G + R)/(\log (SWIR1)) \tag{1}$$

NSI is an index that can be used in the temperate zone, where bare sands and soils have different moisture values. For example, in a major riverbed, soil moisture is higher than on the slopes, and the soil is covered with vegetation. NSI applied to satellite imagery from 2019 maps these details better than NDSAI. Under these conditions, NSI is able to map sandy surfaces without being influenced by moisture. NSI does not generate confusion between light-coloured concrete irrigation channels or other surfaces (e.g., roads) and sand bars.

In remote sensing, some software (e.g., ENVI 5.3) uses traditional transformations (Principal Component Analyses, Tassaled Cap) in order to highlight certain properties of land surface objects, but also the dynamics of some phenomena (e.g., desertification) [59,60].

The mapping of some soil properties (e.g., salinization, sodicity) can be improved by transforming variables (1/log), and the spectral reflectance increases with the value of electro-conductivity (e.g., EC) [61]. Bare sand as well as salinized soils show high reflectance values. The distribution of DN values for SWIR 1 (2001) analysed in SPSS 10 does not clearly show the presence of sandy textured soils. Following the SWIR1 log transformation, an increase in reflectance values was observed for sandy textured soils, with a clear clustering in the range of 7.5–8.5. Cartographically, sandy soils are better highlighted and will have higher values than soils with other textures.

The *classification accuracy check* followed two components:

(a) Assessment of the accuracy of the NSI classification against the selected spectral indicator (NDSAI);

(b) Assessment of the ability of traditional classification (TL) to map bare sand surfaces.

The priority was to identify the threshold at which the NSI value and the tested indicators show the presence of sand. For this purpose, 47 polygons of 3×3 pixel sides were constructed in Arc GIS 10.8; these will be referred to as control square areas hereafter. By observing each control square area, it was found that it may also contain pixels indicating sand with sparse vegetation cover. Using only pure pixels to assess accuracy would result in unduly high values for overall accuracy [62]; therefore, decreasing the size of the control square areas would not be justified.

Since NSI tests for the existence of uncovered sands, each control square area was assigned to sandy textured soils (Arenosols) based on the soil map (Figure 1). For some of the 1–3 control square areas, there was the uncertainty of not belonging to Arenosols as they were placed at the transition to soils with other textures.

The minimum threshold value at which pixels were assigned to sand, as estimated by the NSI index based on satellite images, was extracted by the statistical calculation of the 9 pixels integrated in the averaging. The threshold value was identified from the histogram of the 47 control square areas by summing the mean $\pm$ standard deviation [63].

The minimum threshold value for NSI varied from year to year as follows: 0.38 in 1988, 0.24 in 2001, and 0.19 in 2019.

3.4. Accuracy of Traditional Classifications

The confusion matrix from the ENVI 5.3. software was the basis for the evaluation of the MLK classification performance, and the correlation matrix for the NSI and NDSAI indicators was generated in Arc GIS 10.8. Due to the large time difference between the 1988 and 2001 scenes, a confusion matrix using ground truth image was preferred. For the scene of 17 February 2019, ground truth ROIs were generated based on observations in the vicinity of the sampling points (1a,b–11a,b) and texture on the surface horizon [34,44].

The classification accuracy for multi-temporal studies and soil properties was evaluated based on overall accuracy and the Kappa coefficient [64], or the user's and producer's accuracy per class [65].

4. Results and Discussion

The assessment of bare sands or soils with high sand content is really difficult in temperate areas because the precipitation excess—compared to the average value in desert areas—creates favourable conditions for the growth of grassy vegetation. Natural and cultivated vegetation is closely related to the consumption habits of the community.

The different configurations of the spectral curves in SWIR 1 and SWIR 2 in 2019 as compared to 1988 and 2001 may be a consequence of the change in land use and thus the number of pixels allocated to each training area (Table 4).

Table 4. Pixels count for each training areas.

Land Use	Pixels Count for Each Land Use		
	27 January 1988	7 February 2001	17 February 2019
Autumn crops (AC)	7702	7730	4274
Arable lands-pastures on sandy soils (APSS)	3724	3213	2167
Arable lands-pastures on different soil textures (APSDT)	8651	7688	6104
Permanent crops (PC)	2995	1298	778
Compact forests (CF)	933	809	1619
Scattered forests (SF)	2123	4013	2140
Sands (S)	464	627	635
Lakes (L)	120	112	146

For the distance, Jeffries–Matusita (JM) values higher than 1.9 were interpreted as very good, and separability values between 1.5 and 1.9 indicated good separability [66].

Low separability values may have been the result of the combined types of vegetation (such as cultivated, herbaceous, bushes) and soil surface characteristics (e.g., colour). In the case of low separability, two classes indicating the same object may be merged into one. The weak values of the JM results obtained for the CF-SF pair are not a major problem because the two forest types can form a single class, but this segregation was preferred for observing the spatial dynamics of forest plantations (Figure 5).

The band selection was based on the spectral behaviour of sand surfaces in the range of visible bands (Green and Red) and for other land uses in the Short-wave Infrared band (SWIR 1). In the studied area, the sand showed a continuous increase in the reflectance of the Green (b2, Landsat 5 TM, Landsat 7 ETM 0.53–0.59 μm, b3 Landsat 8 OLI = 0.53–0.59 μm) and Red (b3, Landsat 5 TM, Landsat 7 ETM 0.63–0.69 μm, b4 Landsat 8 OLI = 0.64–0.67 μm) bands, but with a variation for SWIR1 (Figure 4). The reflectance of the SWIR1 bands (b5, Landsat 5 TM, Landsat 7 ETM 1.55–1.75 µm, b6 Landsat 8 OLI = 1.57–1.65 µm) accentuated the difference between soil moisture content and vegetation.

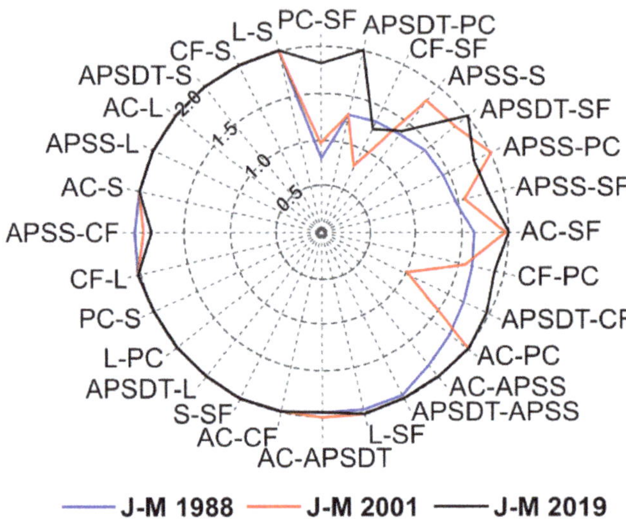

Figure 5. Jeffries–Matusita values for each land use type.

From the category of indices used for bare sand and sand dunes, those that generated normalized values were preferred (Table 4), such that the results are comparable with those obtained by the indicator proposed in this study (NSI). Two indicators were tested to identify sandy surfaces: NDSAI [17] and NDSDI [16]. However, NDSAI was selected because its spatial distribution was closer to NSI, and because it could capture sandy areas in wet climates [17]. The dynamics of non-vegetated sandy areas are associated with the desertification process. One of the methods used to monitor bare sands is the use of spectral indices (e.g., Bare soil index, Normalized Differential Sand Dune Index, Normalized Differential Sand Areas) [15,17,20]. Some of these monitor sandy surfaces in arid climates, while the Normalized Differential Sand Areas (NDSAI) uses Red and SWIR1 bins that are sensitive to soil moisture [17]. As the study area is located in the temperate zone, the snow sands received variable amounts of water. Of the tested indicators, NDSAI best estimated the areas with sand that lacked vegetation, including wet areas where sand rapidly loses moisture, and was compared with the indicator proposed in this study (NSI).

Compared to the uncovered sandy areas identified by NSI and shown above, NDSAI captures larger areas at about 20% of the total number of pixels. Thus, in 1988, the uncovered sand area was 45.5% (586,408 pixels), it reached 57.4% (739,699 pixels) in 2001, and it continued to increase to 71.6% (922,397 pixels) in 2019. Climatically, February 2019 was considered the warmest month, with an average of 2.7 °C, and it was excessively rainy; the last decade, however, has seen cooling air masses and wind intensities reaching 80 km h^{-1} [67]. Thus, as the wind velocity increases, plant roots are exposed, and plants are damaged by grain bombardment or by being covered with sand [68]. These rapid changes in weather can cause the unvegetated sand surfaces to manifest in a pulsating manner from year to year, and even within the same season.

The comparative analysis of the user/producer accuracy matrix shows the close values of the two indicators, but NSI stands out with higher values for producer accuracy in Arenosols only in 2001 (Table 5). For the other soil types (e.g., Cernozems), which also have a sandy texture, the user accuracy results obtained with NSI are better, as a spatially correct delimitation of soils in the interdune-dune and major riverbed-interfluvial areas was obtained.

Table 5. NSI and NDSAI calibration accuracy analysis for Arenosols.

Index	NSI 1988				NDSAI 1988			
	NSI (1)	NSI (2)	Total (User)	User Accuracy (%)	NDSAI (1)	NDSAI (2)	Total (User)	User Accuracy (%)
Arenosol (1)	42	8	50	84	44	9	53	83
Other soil classes (2)	3	5	8	62.5	1	4	5	80
Total (Producer)	45	13	58	0	45	13	58	0
Producer accuracy (%)	93.3	38.5	0	81.0	97.8	30.8	0	82.8
Overall accuracy for arenosols (%)				81.4				82
	NSI 2001				NDSAI 2001			
Arenosol (1)	44	8	52	84.6	43	9	52	82.7
Other soil classes (2)	1	5	6	83.3	2	4	6	66.7
Total (Producer)	45	13	58	0	45	13	58	0
Producer accuracy (%)	97.8	38.5	0	84.5	95.6	30.8	0	81
Overall accuracy for arenosols (%)				84.5				97
	NSI 2019				NDSAI 2019			
Arenosol (1)	43	8	51	84.3	44	10	54	81.5
Other soil classes (2)	2	5	7	74.4	1	3	4	75
Total (Producer)	45	13	58	0	45	13	58	0
Producer accuracy (%)	95.6	38.5	0	85.8	97.8	23.1	0	81
Overall accuracy for arenosols (%)				82.4				81

The overall accuracy values obtained with NDSAI for the study territory are comparable to that determined by Sahar et al. [17], but NSI has a lower deviation from the three-year average (82.7%) as compared to NDSAI. If NDSAI is the best sandy index for mapping most of the sandy surfaces [17], NSI highlights the bare sand surfaces in a temperate zone territory characterized by a higher complexity of soil types and cultivated species, but also by anthropogenic influence on the environment.

The statistical processing of the NSI data shows that the area of bare sand continuously increased at an accelerated rate from 1988 (23.6% representing 303,446 pixels) to 2001 (47.2%, representing 608,197 pixels), and then moderately until 2019 (52.5%, representing 676,138 pixels).

The higher frequency of dry years in the 1983–2002 period [44] may be one of the causes of the expansion of uncovered sand areas. This is compounded by intervals of extreme temperatures, such as those in June and July 2000, which exceeded 43 °C in the southern part of the territory and had heavy rainfall recorded on a single day [69]; these increased the area of uncovered sand due to erosion. It is possible that the NSI performance shown by producer accuracy was also influenced by the climatic characteristics of the year 2000.

The classification maps produced from the implementation of the MLK classifier is illustrated in Figure 6, and the accuracy statistics are summarized in Table 6.

Although climatic conditions are not very favourable for revegetation during dry years, which have also been very frequent after 2001 [30,70], natural revegetation through the abandonment of agricultural land [71] and national revegetation programs slowed the pace of sand expansion until 2019.

Overall accuracy (Oa) and the kappa coefficient (k) obtained for MLK in the three years were close (1988-L5, Oa = 86.77%, k = 0.83%; 2001-L7, Oa = 84.79%, k = 0.81%; 2019-L8, Oa = 85.21%, k = 0.82), but it is recommended that the kappa coefficient be discarded, which is not an adequate index for describing classification accuracy [63]; alternatively, user's and producer's accuracy per classes was used.

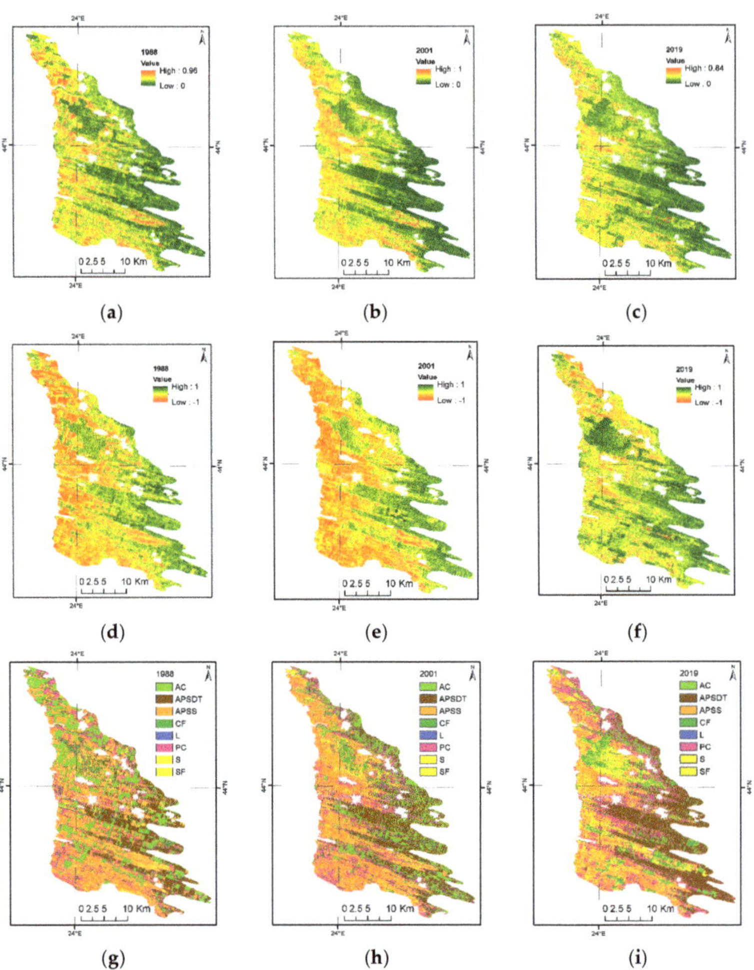

Figure 6. NSI (**a–c**), NDSAI (**d–f**), and MLK (**g–i**); blank areas represent the settlements which were excluded from the analysis.

Table 6. User's and producer's accuracy by classes.

	MLK 1988				MLK 2001				MLK 2019			
Land Use	Prod. Acc.	User Acc.	Prod. Acc.	User Acc.	Pro. Acc.	User Acc.	Prod. Acc.	User Acc.	Prod. Acc.	User Acc.	Prod. Acc.	User Acc.
	(%)	(%)	Pixels	Pixels	(%)	(%)	Pixels	Pixels	(%)	(%)	Pixels	Pixels
AC	98.87	95.64	263/266	263/275	93.24	92.34	193/207	193/209	100	99.38	641/641	641/645
APSS	95.68	86.46	332/347	332/384	92.27	85.42	334/362	334/391	62.99	93.27	291/462	291/312
APSDT	92.35	88.95	169/183	169/190	94.01	90.23	157/167	157/174	100	87.77	165/165	165/188
PC	79.82	77.39	178/223	178/230	81.72	80	228/279	228/285	62.07	31.03	36/58	36/116
CF	80.88	85.94	55/68	55/64	75.7	76.42	81/107	81/106	79.29	91.79	425/536	425/463
SF	58.08	80.83	97/167	97/120	61.54	79.28	88/143	88/111	85.96	57.83	251/292	251/434
S	66.67	94.74	18/27	18/19	52.17	92.31	12/23	12/13	92.65	94.19	227/245	227/241
L	75	100	3/4	3/3	0	0	0/1	0/0	100	100	56/56	56/56

For the sand class, user accuracy was considerably higher for the three years (Table 6), and this could be explained by the constant shape of the spectral signature in the green and red bands. The homogeneous spectral signature associated with high building density surfaces resulted in higher values for user accuracy than producer accuracy [34], and sand indicated higher homogeneity than the other land use classes in the studied case.

Some Chernozems do not meet the diagnostic criterion for colour to be classified as a mollic horizon (e.g., samples 4a,4b,5a) (Table 7), being less than five within dry material [32]. This feature suggests the idea of the sands extending eastward from the active areas (e.g., east of Stefan cel Mare locality).

Table 7. Soil properties and land use.

Soil Sample Code	Latitude	Longitude	Fine Sand [a]	Coarse Sand [b]	Sand	Land Use/Vegetation	Soils	Soil Colour [c]
1a	44°19′91″	23°91′59″	24.9	62.1	87	Forest	Arenosols	10YR8/8
1b	44°20′16″	23°91′82″	71.2	22.1	93.3	Arable		10YR6/4
2a	44°12′31″	23°93′63″	49,8	36.2	93	Herbaceous plants	Arenosols	10YR6/4
2b	44°12′16″	23°93′17″	12.6	87.1	86	Herbaceous plants		10YR7/6
3a	44°86′72″	23°59′34″	29.8	59.4	89.2	Herbaceous plants (grassland)	Luvisols	10YR6/3
3b	44°83′93″	23°59′39″	49.5	41	90.5	Herbaceous plants (grassland)		10YR6/3
4a	44°24′64″	24°71′31″	19.2	70.5	89.7	Arable	Chernozems	10YR6/3 [c]
4b	44°24′10″	24°65′21″	13.8	79.6	93.4	Arable		10YR6/4 [c]
5a	44°01′30″	24°10′49″	17.3	75.6	92.9	Arable	Chernozems	10YR6/3 [c]
5b	44°01′38″	24°10′49″	14.9	79	93.9	Arable		10YR5/3
6a	43°59′14″	24°12″	17.6	75.8	93.4	Arable	Chernozems	10YR5/3
6b	43°98′41″	24°20′91″	8.3	87	95.3	Arable		10YR4/3
7a	43°55′49″	24°15′21″	14.8	77.7	92.5	Vineyard	Chernozems	10YR4/2
7b	43°92′85″	24°26′41″	19.9	72.4	92.3	Arable		10YR5/3
8a	43°84′46″	24°25′56″	20.1	72.9	93.0	Arable	Arenosols	10YR4/3
8b	43°84′49″	24°25′66″	19.8	71.9	91.7	Arable		10YR5/4
9a	43°79′66″	24°2′46″	52.4	33.4	85.8	Herbaceous plants	Arenosols	10YR5/2
9b	43°79′64″	24°24′78″	25.4	62.9	88.3	Arable	Cherozems	10YR4/2
10a	43°76′56″	24°25′77″	0.7	85	85.7	Arable	Arenosols	10YR4/4
10b	43°76′34″	24°26′4″	57.6	38	95.6	Arable	Cherozems	10YR4/3
11a	43°75′95″	24°23′46″	23.2	72.4	95.6	Abandoned vineyard	Arenosols	10YR4/4
11b	43°75′88″	24°23′79″	25.1	62.4	87.5	Abandoned vineyard		10YR4/3

Fine sand [a] = 0.02–0.2 mm, coarse sand [b] = 0.2–2 mm, 10YR6/3 [c] = colour criteria not met for mollic horizon.

There are no significant differences between the texture values in the sampling points (Table 7) and those on the surface horizon of the soils in the literature (average sand in samples = 91.5%, in profiles = 89.14%; StdDev in samples = 3.52, in profiles = 4.35) (Figure 7). A detailed study of the soils of the Oltenia Plain revealed the existence of several soil types (e.g., Cambisols) included in the Arenosols area [35], but also some particular properties of the latter [34]. The maximum and minimum texture values express both the pedological diversity and anthropic influence on soils (e.g., stripping/covering).

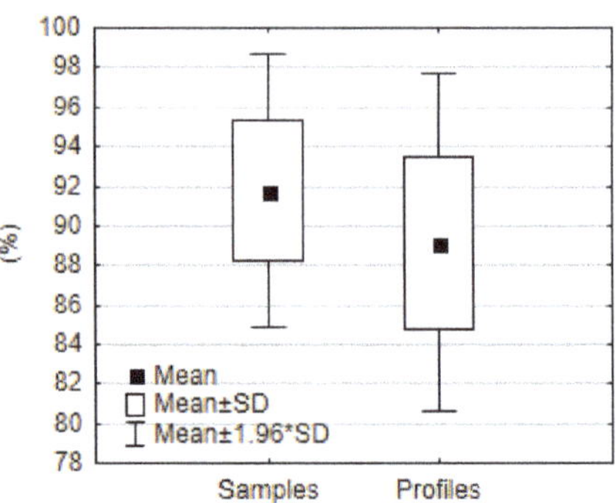

Figure 7. Sand content in the samples and in the literature profiles (*—refers to the multiplication sign).

The statistical analysis of land use classes obtained from MLK did not reflect the increase in uncovered sand areas in 2019 as compared to 2001, and APSS was observed to be declining. The lower exposure of soils to deflation through the decrease in the area of autumn crops and the areal extension of compact forests in the period of 2001–2019 explains the slower evolution of uncovered sandy areas (Table 8). In addition to these, there was also the abandonment of some agricultural land due to the low productivity of sandy soils (e.g., Sadova) [71]. The values for S and APSS resulting from the MLK classification (Table 8) indirectly validated the NSI, which reflects a moderate increase in the last interval.

Table 8. Pixels count and percent for each land use (according to MLK).

	27 January 1988		7 February 2001		17 February 2019	
Land Use	**MLK**		**MLK**		**MLK**	
	Pixel Count	**%**	**Pixel Count**	**%**	**Pixel Count**	**%**
AC	266,683	8.14	207,204	6.33	132,014	4.02
APSS	346,531	10.58	362,030	11.06	265,433	8.1
APSDT	184,225	5.62	168,077	5.13	265,382	8.1
PC	224,362	6.85	278,978	8.52	208,548	6.36
CF	68,577	2.09	106,872	3.26	141,542	4.32
SF	167,130	5.10	142,871	4.36	256,403	7.82
S	26,718	0.81	22,530	0.68	20,179	0.61
L	3920	0.11	1278	0.03	1214	0.03

NSI emphasizes certain peculiarities in the distribution of unvegetated sands according to land use. The rectangular shape of the unvegetated areas in the north-west corresponds to large agricultural land from the period of planned agriculture (1988) (Figure 6a). The high NSI values from the same area in 2001 are due to excessive land parcelling (Figure 6b) and poor soil cover by autumn crops (Figure 6h), which are factors favouring the reactivation of sands. In response to land degradation, the area of compact forests was extended by acacia plantations in the north-west of the territory (e.g., near Marsani) [72] (Figure 6c,i).

After 1989, the surfaces occupied by vineyards or by orchards decreased abruptly, as opposed to the surfaces used as pastures and agricultural land; in the latter case, an increase could be observed [73–75]. The deforestation of vine plots to make way for new plantations damaged the fragile equilibrium of sandy soils. In the Oltenia region during the period of 2007–2018, through the conversion/restructuring programs, there was a significant increase in the areas planted with vines destined to produce quality wines [76]. The disappearance of some traditional crops on Arenosols (Figure 8a) in the southern part of the territory is evidenced by NSI in the rectangular shape of the unvegetated plots, (Figure 8d) but with smaller dimensions as compared to the agricultural lands in the north-west.

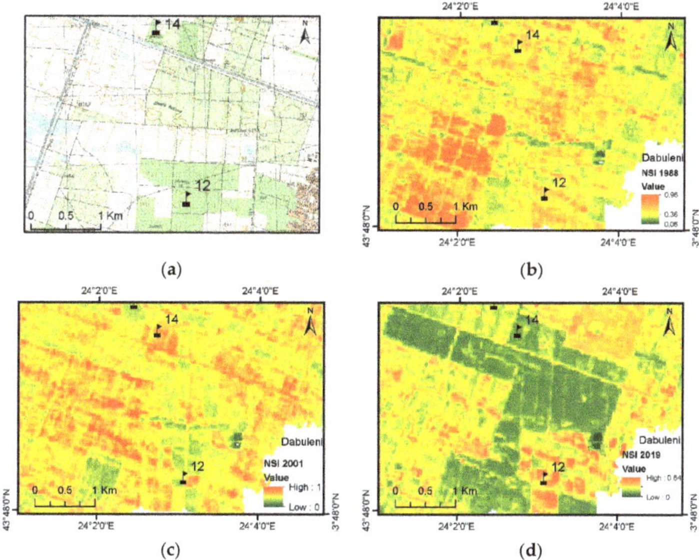

Figure 8. Orchards (**a**) replaced by farmland and compact forest in the vicinity of profiles 12 and 14 (**b–d**).

With the levelling of the relief, the lakes between the dunes (Figure 8a) disappeared, leading to the expansion of the uncovered sandy areas (Figure 8b). The mitigation of hot areas was achieved by the expansion of compact forests (Figure 8d, Table 8), with the rate of reforestation in SW Oltenia being higher than the rate of deforestation [77], at least in some periods.

In the central and southern part, linear dunes are more clearly expressed in the relief than in the north, their characteristics being parallelism, regular spacing, and partial vegetation [78]. The dune peaks, in the shape of an arrow, intersect with the rectangular surface of the farmland, penetrating the Chernozems area (Figure 9a). The change in colour of the A horizon over a distance of several meters marks the transition from Arenosols to Chernozems (Table 7). The agricultural use of dune soils keeps the peak active, but agricultural work and wildlife in Chernozems limit the correct estimate of peak dynamics (Figure 9d).

In the areas where the dunes were modelled, the unaltered sand brought to the surface had high reflectivity, as was the NSI value for 1988 (Figure 10a). Over time, the alteration and accumulation of organic matter attenuated the NSI value (2001), and through land cover, bare sands became less identifiable (Figure 10c).

Most often, the boundary between two different soil types is gradual, but while one property (texture) may be similar for both types, another differentiates them significantly. For example, Arenosols and the other soils (Chernozems and Luvisols) have a high proportion of sand, while the humus content is low in Arenosols (<1–2%) and high for Chernozems (>5%) [34].

Soil is a dynamic system, and profile development through the aeolian input of material is a component of progressive pedogenesis in which current processes and land use are able to assimilate the addition of material from the surface. The aeolian input of material can be assimilated with regressive pedogenesis if the material prevents horizon differentiation or profile development at depth [79]. The transport and accumulation of sand on the surface horizon of Chernozems (CZ) near Arenosols influences the spectral reflectance and therefore the NSI value at different time intervals (Figure 10d).

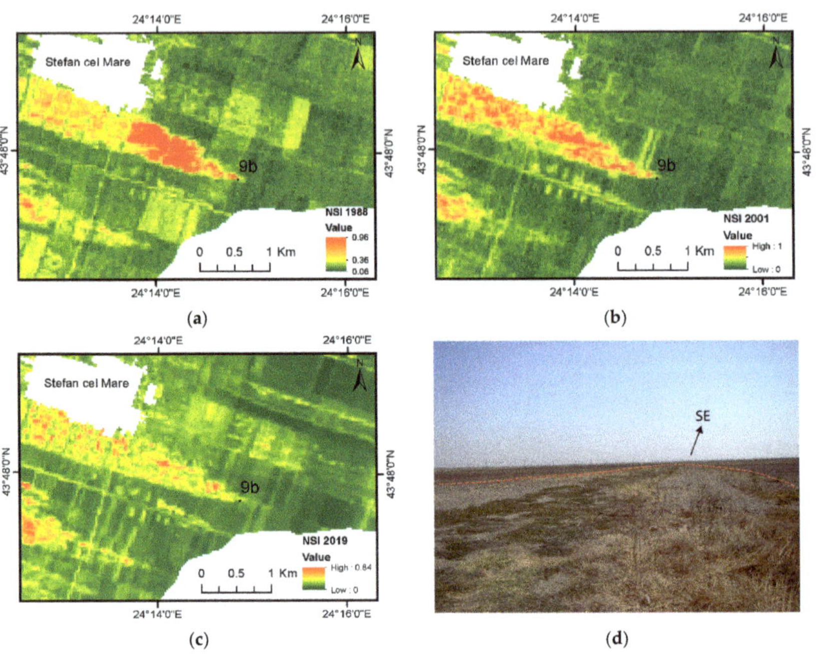

Figure 9. Evolution of sands evidenced by NSI in the vicinity of point 9b in 1988 (**a**), 2001 (**b**), and 2019 (**c**); (**d**) photo of a sample point location (2019).

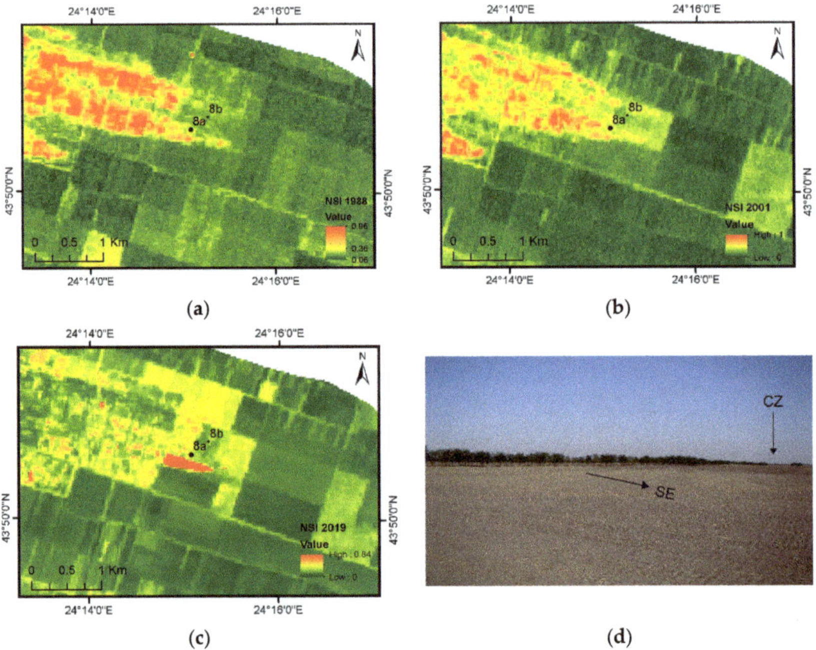

Figure 10. Evolution of sands evidenced by NSI in the vicinity of points 8a and 8b in 1988 (**a**), 2001 (**b**), and 2019 (**c,d**).

5. Conclusions

In this study, the evolution of bare sand surfaces was assessed using a new Normalized Sand Index (NSI) indicator based on a Landsat medium-resolution image series from 1988, 2001, and 2019.

The existence of bare sandy areas is conditioned by anthropogenic factors, in particular land use and the variability of climatic conditions. The development of the irrigation system in the south-eastern part of the territory and the large plots in the north, which lack erosion protection from the period of planned agriculture before 1989, have kept the bare sand areas active. After this date, the decrease in irrigated areas, the abandonment of traditional crops on sandy soil, and the decline in forested areas led to an increase in bare sand surfaces. Sand areas increased by 2019, but at a slower pace due to national reforestation programmes and the natural revegetation of abandoned agricultural land.

Statistical analysis of the NSI showed the presence of bare sands at 23.6% (27,310.14 hectares) in 1988, followed by an accelerated increase to 47.2% (54,737.73 hectares) in 2001, and a smaller increase to 52.5% (60,852.42 hectares) by 2019.

Compared to NSI, the bare sand areas detected with the tested indicator were almost 20% higher. The traditional classification (MLK) showed smaller areas of bare sand but used a higher complexity of land use classes; the producer accuracy values were also lower.

Bare sand surfaces in temperate climates can increase or decrease from one year to the next, depending on climate conditions and even during the same season. The uncovered sandy area in 2001 may have been a consequence of the previous year's drought, but the use of images from the succeeding years was limited by cloud cover in the temperate climates.

The spatial distribution of bare sands is the result of a complex of economic, social, and political factors. The deforestation of forest vegetation and the collapse of the Sadova–Corabia irrigation system (1970), followed by its abandonment, the clearing of forestry fences, and excessive land subdivision following the change in legislation (1989) influenced the distribution of bare sands. Two areas with bare sands persisted during the analysis period: one in the NW, where sand protection measures were lacking, and another in the SE, in the area of levelled dunes. The national reforestation programmes of recent years (2019) have decreased the areas with bare sands, especially in the south.

NSI accurately captures sandy areas, but it is limited to the transition to Chernozems by humus content and agrotechnical works.

This study will continue in the future because the area studied is vulnerable to climate change.

Author Contributions: All authors equally contributed to the article. Conceptualization, C.V.S., C.C.S., C.D.L. and A.U.; field work and methodology, C.D.L. and C.V.S.; laboratory analyses, C.V.S.; data validation, A.U. and C.C.S.; writing—original draft preparation, C.V.S., C.C.S. and C.D.L.; writing—review and editing, A.U., C.V.S. and C.C.S. All authors have read and agreed to the published version of the manuscript.

Funding: The analysis of soil samples from this research paper was funded by Operational Program Competitiveness 2014–2020, Axis 1, under POC/448/1/1 research infrastructure projects for public R&D institutions/Sections F 2018, through the Research Centre with Integrated Techniques for Atmospheric Aerosol Investigation in Romania (RECENT AIR) project, under grant agreement MySMIS no. 127324, Contract number 322/04.09.2020. The publishing cost of this paper was assumed by the authors.

Data Availability Statement: Not applicable.

Acknowledgments: This project received technical support from the Department of Geography, Faculty of Geography and Geology, "Alexandru Ioan Cuza" University of Iassy, Romania who offered us full access to Remote sensing and/GIS laboratories and to the Pedology Laboratory.

Conflicts of Interest: The authors declare no conflict of interest. The funders had no role in the design of the study; in the collection, analyses, or interpretation of data; in the writing of the manuscript, or in the decision to publish the results.

References

1. UN Climate Change Conference (COP26) at the SEC-Glasgow 2021. Available online: https://ukcop26.org (accessed on 10 April 2022).
2. Bertran, P.; Bosq, M.; Borderie, Q.; Coussot, C.; Coutard, S.; Deschodt, L.; Franc, O.; Gardère, P.; Liard, M.; Wuscher, P. Revised Map of European Aeolian Deposits Derived from Soil Texture Data. *Quat. Sci. Rev.* **2021**, *266*, 107085. [CrossRef]
3. Coteţ, P. *Oltenia Plain. Geomorphological Study (with Special Reference to the Quaternary)*; Scientific Publisher: Bucharest, Romania, 1957.
4. Document of International Bank for Reconstruction and Development. *Romania–Sadova–Corabia Agricultural Credit Project P-1555-RO (English)*; World Bank Group: Washington, DC, USA, 1975.
5. Rusu, M.; Simion, G. Farm Structure Adjustments under the Irrigation Systems Rehabilitation in the Southern Plain of Romania: A First Step towards Sustainable Development. *Carpathian J. Earth Environ. Sci.* **2015**, *10*, 91–100.
6. Pravalie, R. Aspects Regarding Spatial and Temporal Dynamic of Irrigated Agricultural Areas from Southern Oltenia in the Last Two Decades. *Present Environ. Sustain. Dev.* **2013**, *7*, 133–143.
7. The Government of Romania. *Low 18*; The Government of Romania, Official Monitor: Bucharest, Romania, 1991.
8. Vorovencii, I. Applying the Change Vector Analysis Technique to Assess the Desertification Risk in the South-West of Romania in the Period 1984–2011. *Environ. Monit. Assess.* **2017**, *189*, 524. [CrossRef]
9. Nuta, S. Structural and functional characteristics of the forest curtains for the protection of the agricultural field in the south of Oltenia. *Ann. For. Res.* **2005**, *48*, 161–169.
10. Achim, E.; Manea, G.; Vijulie, I.; Cocos, O.; Tirla, L. Ecological Reconstruction of the Plain Areas Prone to Climate Aridity through Forest Protection Belts. Case Study: Dabuleni Town, Oltenia Plain, Romania. *Procedia Environ. Sci.* **2012**, *14*, 154–163. [CrossRef]
11. Pravalie, R.; Sirodoev, I.; Peptenatu, D. Changes in the Forest Ecosystems in Areas Impacted by Aridization in South-Western Romania. *J. Environ. Health Sci. Eng.* **2014**, *12*, 2. [CrossRef]
12. Rosca, F.C.; Harpa, G.V.; Croitoru, A.E.; Herbel, I.; Imbroane, A.M.; Burada, D.C. The Impact of Climatic and Non-Climatic Factors on Land Surface Temperature in Southwestern Romania. *Theor. Appl. Clim.* **2017**, *130*, 775–790. [CrossRef]
13. Irimia, L.M.; Patriche, C.V.; LeRoux, R.; Quenol, H.; Tissot, C.; Sfica, L. Projections of climate suitability for wine production for the cotnari wine region (Romania). *Present Environ. Sustain. Dev.* **2019**, *13*, 5–18. [CrossRef]
14. Dharumarajan, S.; Bishop, T.F.A.; Hegde, R.; Singh, S.K. Desertification Vulnerability Index-an Effective Approach to Assess Desertification Processes: A Case Study in Anantapur District, Andhra Pradesh, India. *Land Degrad. Dev.* **2018**, *29*, 150–161. [CrossRef]
15. Fadhil Al-Quraishi, A.M. Land Degradation Detection Using Geo-Information Technology for Some Sites in Iraq. *Al-Nahrain. J. Sci.* **2009**, *12*, 94–108. [CrossRef]
16. Fadhil Al-Quraishi, A.M. Sand Dunes Monitoring Using Remote Sensing and GIS Techniques for Some Sites in Iraq. In *PIAGENG 2013: INTELLIGENT Information, Control, and Communication Technology for Agricultural Engineering*; SPIE: Bellingham, WA, USA, 2013.
17. Sahar, A.A.; Rasheed, M.J.; Uaid, D.A.A.-H.; Jasimm, A.A. Mapping Sandy Areas and Their Changes Using Remote Sensing. A Case Study at North-East Al-Muthanna Province, South of Iraq. *Rev. Teledetec.* **2021**, *58*, 39–52. [CrossRef]
18. Wentzel, K. Determination of the Overall Soil Erosion Potential in the Nsikazi District (Mpumalanga Province, South Africa) Using Remote Sensing and GIS. *Can. J. Remote Sens.* **2002**, *22*, 322–327. [CrossRef]
19. Zhao, H.; Chen, X.; Zhang, Z.; Zhou, Y. Exploring an Efficient Sandy Barren Index for Rapid Mapping of Sandy Barren Land from Landsat TM/OLI Images. *Int. J. Appl. Earth Obs. Geoinf.* **2019**, *80*, 38–46. [CrossRef]
20. Afrasinei, G.M.; Melis, M.T.; Arras, C.; Pistis, M.; Buttau, C.; Ghiglieri, G. Spatiotemporal and Spectral Analysis of Sand Encroachment Dynamics in Southern Tunisia. *Eur. J. Remote Sens.* **2018**, *51*, 352–374. [CrossRef]
21. Marzouki, A.; Dridri, A. Normalized Difference Enhanced Sand Index for desert sand dunes detection using Sentinel-2 and Landsat 8 OLI data, application to the north of Figuig, Morocco. *J. Arid. Environ.* **2022**, *198*, 104693. [CrossRef]
22. Chen, S.; Ren, H.; Liu, R.; Tao, Y.; Zheng, Y.; Liu, H. Mapping Sandy Land Using the New Sand Differential Emissivity Index From Thermal Infrared Emissivity Data. *IEEE Trans. Geosci. Remote Sens.* **2021**, *59*, 5464. [CrossRef]
23. Wang, X.; Song, J.; Xiao, Z.; Wang, J.; Hu, F. Desertification in the Mu Us Sandy Land in China: Response to climate change and human activity from 2000 to 2020. *Geogr. Sustain.* **2022**, *3*, 177. [CrossRef]
24. Yang, Z.; Gao, X.; Lei, J.; Meng, X.; Zhou, N. Analysis of spatiotemporal changes and driving factors of desertification in the Africa Sahel. *CATENA* **2022**, *213*, 106213. [CrossRef]
25. Guo, B.; Wei, C.; Yu, Y.; Liu, Y.; Li, J.; Meng, C.; Cai, Y. The dominant influencing factors of desertification changes in the source region of Yellow River: Climate change or human activity? *Sci. Total Environ.* **2022**, *813*, 152512. [CrossRef]
26. Pan, X.; Zhu, X.; Yang, Y.; Cao, C.; Zhang, X.; Shan, L. Applicability of downscaling land surface temperature by using Normalized Difference Sand Index. *Sci. Rep.* **2018**, *8*, 9530. [CrossRef] [PubMed]
27. Rasul, A.; Baltzer, H.; Faqe Ibrahim, G.R.; Hameed, H.M.; Wheeler, J.; Adamu, B.; Ibrahim, S.; Najmaddin, P.M. Applying Built-Up and Bare-Soil Indices from Landsat 8 to Cities in Dry Climates. *Land* **2018**, *7*, 81. [CrossRef]
28. Simulescu, D. Geographical Study of Sandy Lands in the Romanati Plain. Ph.D. Thesis, Romanian Academy, Institute of Geography, Bucharest, Romania, 2019.
29. Angearu, C.-V.; Ontel, I.; Boldeanu, G.; Mihailescu, D.; Nertan, A.; Craciunescu, V.; Catana, S.; Irimescu, A. Multi-Temporal Analysis and Trends of the Drought Based on MODIS Data in Agricultural Areas, Romania. *Remote Sens.* **2020**, *12*, 3940. [CrossRef]

30. Bercea, I.; Dinucă, N.C. Considerations on Zoning and Micro-Zoning of the Dolj County Area for Potential Forest Vegetation in the Context of Anthropic Changes in Forest Lands and Climatic Changes. *Ann. Univ. Craiova-Agric. Montanol. Cadastre Ser.* **2018**, *48*, 18–34.
31. Dudiak, N.; Pichura, V.; Potravka, L.; Stratichuk, N. Environmental and economic effects of water and deflation destruction of steppe soil in Ukraine. *J. Water Land Dev.* **2021**, *50*, 10. [CrossRef]
32. WRB. *World Reference Base for Soil Resources 2014, International Soil Classification System for Naming Soils and Creating Legend for Soil Map*; FAO: Rome, Italy, 2014.
33. Grigoraş, C.; Boengiu, S.; Vlăduţ, A.; Grigoraş, E.N.; Avram, S. *Romania's Soils*; Universitaria: Craiova, Romania, 2008; Volume 2.
34. Ignat, P.; Gherghina, A.; Vrînceanu, A.; Anghel, A. Assesment of Degradation Processes and Limitative Factors Concerning the Arenosols from Dăbuleni–Romania. *Geogr. Forum Stud. Res. Geogr. Environ. Prot.* **2009**, *8*, 64–71.
35. Stănilă, A.L.; Simota, C.C.; Dumitru, M. Contributions to the Knowledge of Sandy Soils from Oltenia Plain. *Rev. Chim.* **2020**, *71*, 192–200. [CrossRef]
36. Landsat 8 Data Users Handbook | U.S. Geological Survey. Available online: https://www.usgs.gov/landsat-missions/landsat-8-data-users-handbook (accessed on 17 February 2022).
37. Young, N.E.; Anderson, R.S.; Chignell, S.M.; Vorster, A.G.; Lawrence, R.; Evangelista, P.H. A Survival Guide to Landsat Preprocessing. *Ecology* **2017**, *98*, 920–932. [CrossRef]
38. Guo, Q.; Fu, B.; Shi, P.; Cudahy, T.; Zhang, J.; Xu, H. Satellite Monitoring the Spatial-Temporal Dynamics of Desertification in Response to Climate Change and Human Activities across the Ordos Plateau, China. *Remote Sens.* **2017**, *9*, 525. [CrossRef]
39. Salih, A.A.M.; Ganawa, E.T.; Elmahl, A.A. Spectral Mixture Analysis (SMA)-Change Vector Analysis (CVA) Methods for Monitoring-Mapping Land Degradation/Desertification in Arid-Semiarid Areas (Sudan), Using Landsat Imagery. *Egypt. J. Remote Sens. Space Sci.* **2017**, *20*, 22–29. [CrossRef]
40. Pravalie, R. Analysis of temperature, precipitation and potential evapotranspiration trends in southern Oltenia in the context of climate change. *Geogr. Tech.* **2014**, *9*, 68.
41. FAO. *Guidelines for Soil Description*, 4th ed.; Food and Agriculture Organization of the United Nations: Rome, Italy, 2006.
42. Vos, C.; Axel, D.; Prietz, R.; Heidkamp, A.; Freibauer, A. Field-based soil-texture estimates could replace laboratory analysis. *Geoderma* **2016**, *267*, 215–219. [CrossRef]
43. Munsell Color Co., Inc. *Revised Washable Edition*; GretagMacbeth: New Windsor, NY, USA, 2000.
44. Gee, G.; Or, D. Particle-Size Analysis. In *Methods of Soil Analysis*; Physical Methods; Soil Science Society of America: Madison, WI, USA, 2002; pp. 255–293.
45. Florea, N.; Munteanu, I. *Romanian System of Soil Taxonomy*, 2nd ed.; Sitech: Craiova, Romania, 2012.
46. Chander, G.; Markham, B.L.; Helder, D.L. Summary of Current Radiometric Calibration Coefficients for Landsat MSS, TM, ETM+,-EO-1 ALI Sensors. *Remote Sens. Environ.* **2009**, *113*, 893–903. [CrossRef]
47. Diek, S.; Fornallaz, F.; Schaepman, M.E.; Rogier De Jong. Barest Pixel Composite for Agricultural Areas Using Landsat Time Series. *Remote Sens.* **2017**, *9*, 1245. [CrossRef]
48. Boettinger, J.L.; Ramsy, R.D.; Bodily, J.M.; Cole, N.J.; Kienast-Brown, S.; Nield, S.J.; Saunder, A.M.; Stum, A.K. Landsat Spectral Data for Digital Soil Mapping. In *Digital Soil Mapping with Limited Data*; Springer: Berlin/Heidelberg, Germany, 2008; pp. 193–202.
49. Patel, N.; Kaushal, B. Improvement of User's Accuracy through Classification of Principal Component Images and Stacked Temporal Images. *Geo-Spat. Inf. Sci.* **2010**, *13*, 243–248. [CrossRef]
50. Lu, D.; Weng, Q. A Survey of Image Classification Methods and Techniques for Improving Classification Performance. *Int. J. Remote Sens.* **2007**, *28*, 823–870. [CrossRef]
51. Bossard, M.; Feranec, J.; Otahel, J. *CORINE Land Cover Technical Guide—Addendum 2000*; Technical Report No 40; European Environmental Agency: Copenhagen, Denmark, 2000; p. 105.
52. Dabboor, M.; Howell, S.; Shokr, M.; Yackel, J. The Jeffries–Matusita Distance for the Case of Complex Wishart Distribution as a Separability Criterion for Fully Polarimetric SAR Data. *Int. J. Remote Sens.* **2014**, *35*, 6859–6873. [CrossRef]
53. Fongaro, C.; Demattê, J.; Rizzo, R.; Lucas Safanelli, J.; Mendes, W.; Dotto, A.; Vicente, L.; Franceschini, M.; Ustin, S. Improvement of Clay and Sand Quantification Based on a Novel Approach with a Focus on Multispectral Satellite Images. *Remote Sens.* **2018**, *10*, 1555. [CrossRef]
54. Bindel, M.; Hese, S.; Berger, C.; Schmullius, C. Feature Selection from High Resolution Remote Sensing Data for Biotope Mapping. Int. Arch. Photogramm. *Remote Sens. Spat. Inf. Sci.* **2011**, *38*, 39–44. [CrossRef]
55. Gong, P.; Wang, J.; Yu, L.; Zhao, Y.; Zhao, Y.; Liang, L.; Niu, Z.; Huang, X.; Fu, H.; Liu, S.; et al. Finer Resolution Observation and Monitoring of Global Land Cover: First Mapping Results with Landsat TM and ETM+ Data. *Int. J. Remote Sens.* **2013**, *34*, 2607–2654. [CrossRef]
56. Wu, W.; Zucca, C.; Karam, F.; Liu, G. Enhancing the Performance of Regional Land Cover Mapping. *Int. J. Appl. Earth Obs. Geoinf.* **2016**, *52*, 422–432. [CrossRef]
57. Curell, G.; Dowman, A. *Essential Mathematics and Statistics for Science*, 2nd ed.; Wiley-Blackwell: Hoboken, NJ, USA, 2009; 416p, ISBN 9780470694480.
58. Reimann, R.C.; Filzmoster, P.; Garrett, R.G.; Dutter, R. *Statistical Data Analysis Explained: Applied Environmental Statistics*; John Wiley and Sons: Hoboken, NJ, USA, 2008; 343p. [CrossRef]
59. Weiss, N.A. *Introductory Statistics*, 9th ed.; Pearson Education, Inc.: London, UK, 2012; 912p, ISBN 9780321691224.

60. Zhang, J.; Liu, M.; Liu, X.; Luo, W.; Wu, L.; Zhu, L. Spectral analysis of seasonal rock and vegetation changes for detecting karst rocky desertification in southwest China. *Int. J. Appl. Earth Obs. Geoinf.* **2021**, *100*, 102337. [CrossRef]
61. Zachentta, A.; Bitelli, G.; Karnieli, A. Monitoring desertification by remote sensing using the Tasselled Cap transform for long-term change detection. *Nat. Hazards* **2016**, *83*, 223. [CrossRef]
62. Yu, H.; Liu, M.; Du, B.; Wang, Z.; Hu, L.; Zhang, B. Mapping Soil Salinity/Sodicity by using Landsat OLI Imagery and PLSR Algorithm over Semiarid West Jilin Province, China. *Sensors* **2018**, *18*, 1048. [CrossRef]
63. Shao, G.; Tang, L.; Liao, J. Overselling Overall Map Accuracy Misinforms about Research Reliability. *Landsc. Ecol.* **2019**, *34*, 2487–2492. [CrossRef]
64. Fung, T.; LeDrew, E. The Determination of Optimal Threshold Levels for Change Detection Using Various Accuracy Indices. *Photogramm. Eng. Remote Sens.* **1988**, *54*, 1449–1454.
65. Kantakumar, L.N.; Neelamsetti, P. Multi-Temporal Land Use Classification Using Hybrid Approach. *Egypt. J. Remote Sens. Space Sci.* **2015**, *18*, 289–295. [CrossRef]
66. Foody, G.M. Explaining the Unsuitability of the Kappa Coefficient in the Assessment and Comparison of the Accuracy of Thematic Maps Obtained by Image Classification. *Remote Sens. Environ.* **2020**, *239*, 111630. [CrossRef]
67. Marinică, A.F.; Marinică, I.; Chimisliu, C. Climatic Variability in Southwestern Romania in the Context of Climate Changes during the Winter of 2018–2019. *Stud. Commun. Nat. Sci. Olten. Mus. Craiova* **2019**, *35*, 169–172.
68. Tsoar, H. 11.21 Critical Environments: Sand Dunes and Climate Change. In *Treatise on Geomorphology*; Elsevier: Amsterdam, The Netherlands, 2013; pp. 414–427. [CrossRef]
69. Marinica, I.; Chimisliu, C. Climatic Changes on Regional Plan in Oltenia and Their Effects on the Biosphere. *Stud. Commun. Nat. Sci. Olten. Mus. Craiova* **2008**, *24*, 221–229.
70. Alexandru, D.; Mateescu, E.; Tudor, R.; Leonard, I. Analysis of Agroclimatic Resources in Romania in the Current and Foreseeable Climate Change—Concept and Methodology of Approaching. *Agron. Ser. Sci. Res.* **2019**, *62*, 221–229.
71. Dumitrașcu, M.; Mocanu, I.; Mitrică, B.; Dragotă, C.; Grigorescu, I.; Dumitrică, C. The Assessment of Socio-Economic Vulnerability to Drought in Southern Romania (Oltenia Plain). *Int. J. Disaster Risk Reduct.* **2018**, *27*, 142–154. [CrossRef]
72. Enescu, C.M. Sandy Soils from Oltenia and Carei Plains: A Problem or an Opportunity to Increase the Forest Fund in Romania? *Sci. Pap. Ser. Manag. Econ. Eng. Agric. Rural Dev.* **2019**, *19*, 203–206.
73. Simulescu, D. The Impact of Human Activities on the Environment in the Romanați Plain (Romania), during the Postcommunist Era. *Forum Geogr.* **2018**, *17*, 123–134. [CrossRef]
74. Petrișor, A.I.; Petrișor, L.E. 2006-2012 land cover and use changes in Romania—An overall assessment based on Corine data. *Present Environ. Sustain. Dev.* **2017**, *11*, 119–127. [CrossRef]
75. Ursu, A.; Stoleriu, C.C.; Ion, C.; Jitariu, V.; Enea, A. Romanian Natura 2000 Network: Evaluation of the Threats and Pressures through the Corine Land Cover Dataset. *Remote Sens.* **2020**, *12*, 2075. [CrossRef]
76. Vladu, C.E. Reconversion/Restructuring of Vineyard Plantings in Oltenia in the Period 2007–2018 with the Access of European Funds. *Ann. Univ. Craiova-Agric. Montanology Cadastre Ser.* **2019**, *49*, 395–399.
77. Andronache, I.; Fensholt, R.; Ahammer, H.; Ciobotaru, A.-M.; Pintilii, R.-D.; Peptenatu, D.; Drăghici, C.-C.; Diaconu, D.; Radulović, M.; Pulighe, G.; et al. Assessment of Textural Differentiations in Forest Resources in Romania Using Fractal Analysis. *Forests* **2017**, *8*, 54. [CrossRef]
78. Lancaster, N. Dune Morphology and Dynamics. In *Geomorphology of Desert Environments*; Springer: Dordrecht, The Netherlands, 2009.
79. Eger, P.; Almold, P.C.; Condron, L.M. Upbuilding Pedogenesis under Active Loess Deposition in a Super-Humid, Temperate Climate—Quantification of Deposition Rates, Soil Chemistry and Pedogenic Thresholds. *Geoderma* **2012**, *189–190*, 491–501. [CrossRef]

Article

Spatio-Temporal Analysis of Ecological Vulnerability and Driving Factor Analysis in the Dongjiang River Basin, China, in the Recent 20 Years

Jiao Wu [1], Zhijun Zhang [2,3], Qinjie He [1] and Guorui Ma [1,*]

[1] State Key Laboratory of Information Engineering in Surveying, Mapping and Remote Sensing, Wuhan University, Wuhan 430079, China; wujiaors@whu.edu.cn (J.W.); 2018206190073@whu.edu.cn (Q.H.)

[2] Xining Center of Natural Resources Comprehensive Survey, China Geological Survey, Xining 810000, China; zhangzhijun@mail.cgs.gov.cn

[3] Ministry of Education Key Laboratory of Geological Survey and Evaluation, China University of Geosciences (Wuhan), Wuhan 430074, China

* Correspondence: mgr@whu.edu.cn

Abstract: The global ecological environment faces many challenges. Landsat thematic mapper time-series, digital elevation models, meteorology, soil types, net primary production data, socio-economic data, and auxiliary data were collected in order to construct a comprehensive evaluation system for ecological vulnerability (EV) using multi-source remote sensing data. EV was divided into five vulnerability levels: potential I, slight II, mild III, moderate IV, and severe V. Then, we analyzed and explored the spatio-temporal patterns and driving mechanisms of EV in the region over the past 20 years. Our research results showed that, from 2001 to 2019, the DRB was generally characterized as being in the severe vulnerability class, with higher upstream and downstream EV classes and a certain amount of reduction in the midstream EV classes. Moreover, EV in the DRB continues to decrease. The spatio-temporal EV patterns in the DRB were significantly influenced by the relative humidity, average annual temperature, and vegetation cover over the past 20 years. Our work can provide a basis for decision-making and technical support for ecosystem protection, ecological restoration, and ecological management in the DRB.

Keywords: ecological vulnerability; driving mechanisms; remote sensing; Dongjiang River Basin

Citation: Wu, J.; Zhang, Z.; He, Q.; Ma, G. Spatio-Temporal Analysis of Ecological Vulnerability and Driving Factor Analysis in the Dongjiang River Basin, China, in the Recent 20 Years. *Remote Sens.* **2021**, *13*, 4636. https://doi.org/10.3390/rs13224636

Academic Editors: Adrian Ursu and Cristian Constantin Stoleriu

Received: 11 October 2021
Accepted: 15 November 2021
Published: 17 November 2021

Publisher's Note: MDPI stays neutral with regard to jurisdictional claims in published maps and institutional affiliations.

1. Introduction

Since the middle of the 20th century, the intensification of human activities has led to frequent climate-change-related disasters [1–4] and the intensification of ecological and environmental crises [5]. These scenarios pose serious challenges to the ecosystems on which humans depend for their survival. Among them, the issue of ecological vulnerability (EV) is particularly prominent. EV was first introduced into ecological theory by Clements as an "ecological staggering zone" [6,7] and was further discussed at the seventh SCOPE (Scientific Committee of Environmental Problems) Conference in 1989 [6,8,9]. The ability of a system to resist external environmental change and to be disturbed and recover itself is defined as EV [10–13]. EV is determined by a combination of internal and external vulnerability [14,15]. Internal vulnerability usually stems from the structure of the ecosystem itself and is mainly influenced by natural conditions such as topography and climate. External vulnerability is influenced by human activities [16]. Exploring regional EV is important for ecological change and socio-economic development.

Scholars around the world have conducted relevant studies on the spatio–temporal evolution patterns of EV and its driving mechanisms [2,13,17–23]. Most of these EV studies are focused on China [21] and can be based on individual cities or regions [24], e.g., on the northwestern part of the Songnun Plain [17], the Tibetan Plateau [18], the Loess Plateau [19], the southwestern karst mountains [10], agricultural and pastoral areas [22], coal mining areas [2], the southern Shaanxi region [25], islands [26], coastal wetlands [27]; there are also studies that focus on the Bangladesh–China–India–Myanmar economic corridor [21]. Recently, the ecology of red loam hilly areas in southern China has become a focus of attention for many scholars [28].

Dongjiang is the third largest water system in the Pearl River basin and belongs to the South China basin, which is characterized as having a typical tropical–subtropical climate. The Dongjiang River Basin (DRB) is located in the red-loamy hilly region of southern China, which, as a pioneering region for reform and opening up, has seen a rapid economic development over the past 40 years. China's 13th Five-Year Plan supports the construction of an open and innovative transformation of the Pearl River Delta region, namely the Guangdong–Hong Kong–Macao Greater Bay Area. Meanwhile, the DRB is an important drinking water source for the Pearl River Delta and Hong Kong and an essential focus of the sustainable development strategy in the Pearl River Delta region. The socio-economic development of the basin and changes in the natural environment affect the quality of the ecological environment and the sustainable use of resources in the basin [29]. With accelerated urbanization, dramatic climate change, and increased disturbance from human activities, the quality of the ecological environment in the DRB faces unprecedented threats, and thus the DRB has become an ideal area for analyzing EV. Hu [24] used the AHP method to evaluate the EV of Weifang city, China; however, the study used limited the quantitative data and was heavily dependent on qualitative inputs. These factors led to unconvincing results, especially when there were many factors, and it was difficult to precisely determine the weights. Wu [30] used the fuzzy hierarchical analysis method to evaluate the EV of the Yellow River Delta; however, there were drawbacks related to the subjective factor weights and large computational effort. Principal component analysis (PCA) is one of the most commonly used methods for model evaluation, as it can reduce data complexity, identify the most important multiple features, and save significant computational resources, and is easily implemented on a computer. In this paper, PCA was used to model and analyze the time-series EV in the DRB.

Factors commonly used in EV assessment can be divided into two categories: those reflecting human activities, such as population density, GDP per capita, etc.; and ecological and natural conditions, such as slope, temperature, precipitation, vegetative cover, etc. As a result of the complexity of human–nature interactions, there is currently no uniform standard for the selection of factors for EV assessment models [31]. Moreover, there is the problem of regional adaptability in EV assessment models [22]. The ecological environment in the red soil hilly areas of southern China is fragile [32]; furthermore, there is a lack of an EV zoning system. In addition, many of the current studies on EV are analyzed at a regional scale and are less specific to county EV evaluations. Therefore, there is a need to develop a reliable methodology to assess EV in the DRB.

Our work established a dynamic evaluation scheme for monitoring the spatio-temporal EV changes in the DRB and calculated the dynamic EV weights in 2001–2019 using PCA. Finally, we analyzed the spatio-temporal evolution of EV in the DRB and its driving mechanisms over the past 20 years. The research results and methods provide a theoretical basis and technical support for the protection and restoration of the environment within the DRB. The flowchart for this work is shown in Figure 1.

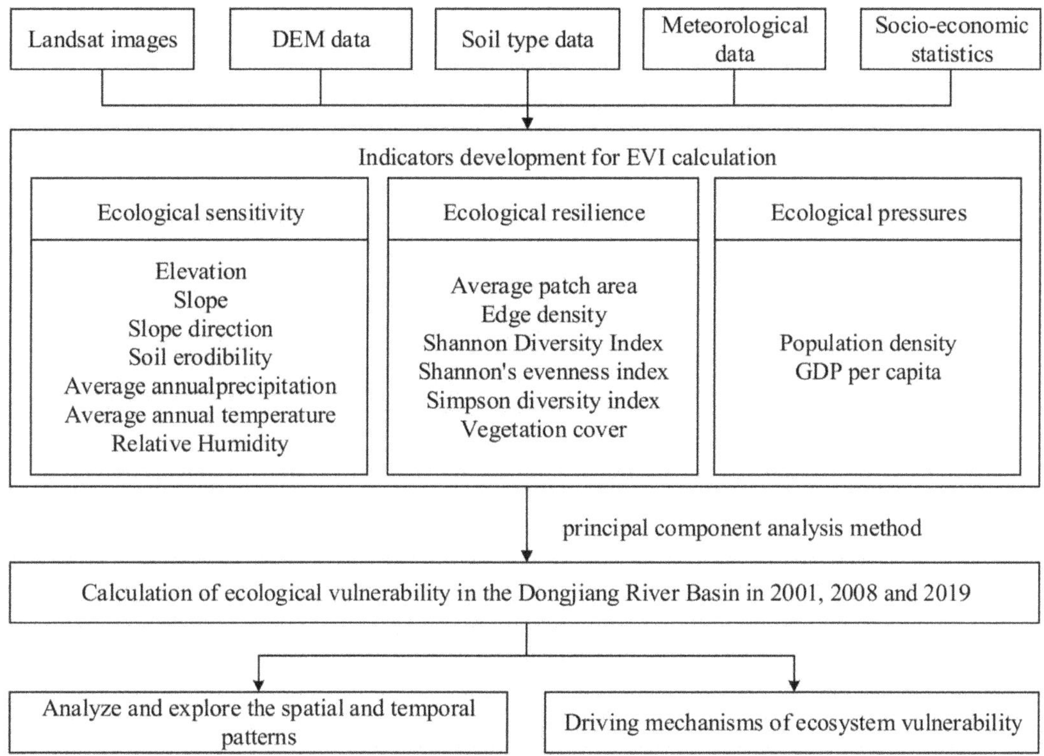

Figure 1. Stepwise procedure for spatio-temporal EVI modeling.

2. Materials and Methods

2.1. Study Area

The DRB (Figure 2) is connected to Meishan in eastern Guangdong; to its west, it is bordered by Shaoguan and Qingyuan in northern Guangdong; the South China Sea and Hong Kong lie to the south, and the Dongjiang source area of southern Ganzhou lies to the north. The geographical location of the DRB is 113°52′–115°52′ E and 22°38′–25°14′ N, with hilly mountains in the center and north. The DRB is a valuable water resource for social development within the basin, providing water to Hong Kong for production, living, and ecology, and to more than 40 million people who live in the urban clusters along the basin. The middle and lower reaches of the DRB are economically developed, and the total GDP of the regions in Guangdong, China, that are supported by Dongjiang water was RMB 4.17 trillion in 2019, accounting for approximately 39% of the total GDP of Guangdong Province.

2.2. Evaluation Factors and Data Sources

We studied the "China Guangdong Environmental Protection Department on the Comprehensive Water Environment Improvement Programme of the Dongjiang River Basin (2015–2020)", combined with related studies [33–35], and considered the availability of relevant assessment factors. We selected factors from topography, climate, landscape, vegetation, and socio-economic aspects to construct an EV evaluation system. The data sources are shown in Table 1. There were some differences as regards the data sources and spatial accuracies of the factors. In the Arc GIS10.1 software platform, the WGS 1984 coordinate system and Mercator projection were uniformly used to ensure acceptable spatial coincidences for the factors. Moreover, all of these were unified to a raster image element size of 30 × 30 m using the ArcGIS10.2 platform to accommodate subsequent spatial calculations. The data sources are shown in Table 1.

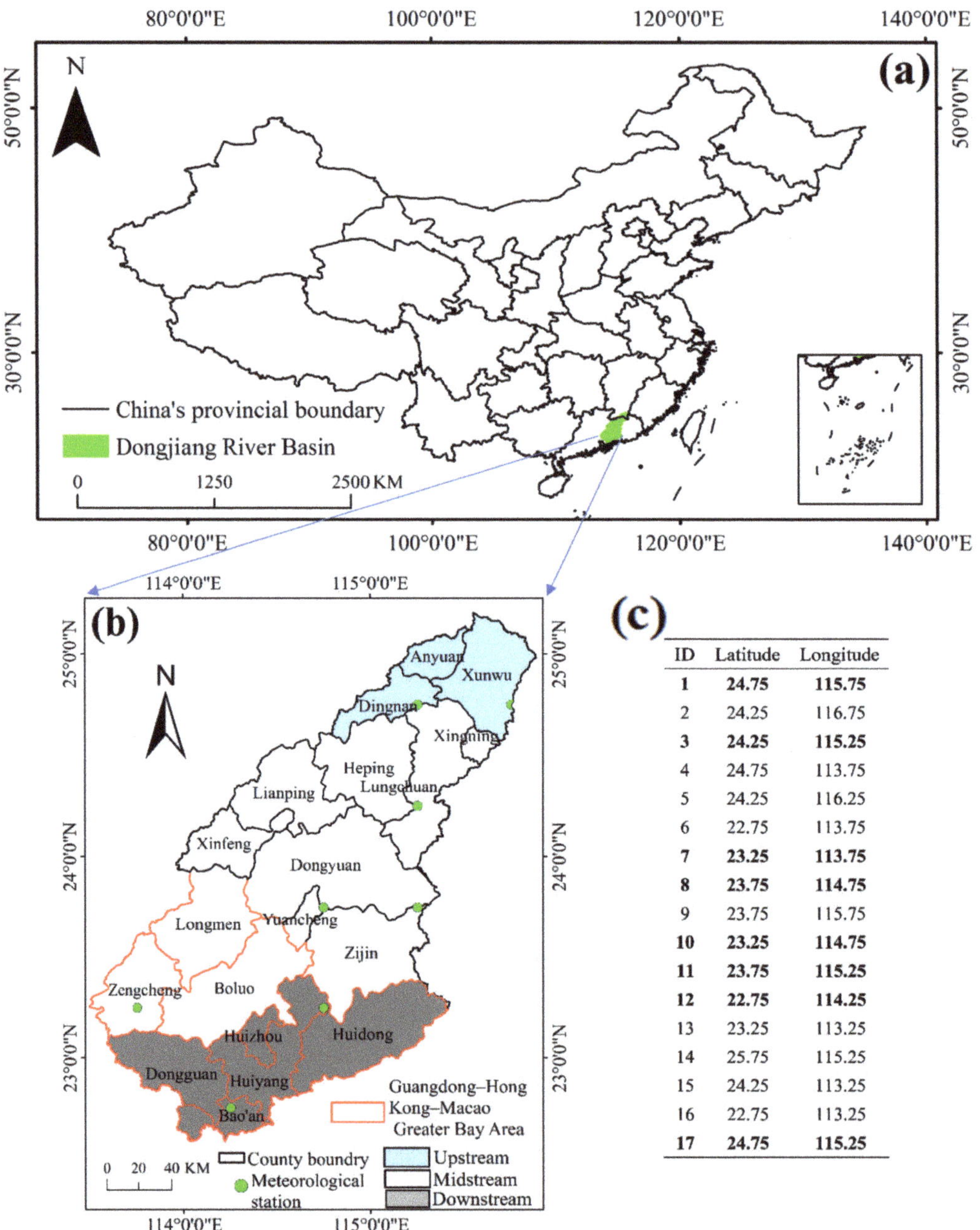

ID	Latitude	Longitude
1	**24.75**	**115.75**
2	24.25	116.75
3	**24.25**	**115.25**
4	24.75	113.75
5	24.25	116.25
6	22.75	113.75
7	**23.25**	**113.75**
8	**23.75**	**114.75**
9	23.75	115.75
10	**23.25**	**114.75**
11	**23.75**	**115.25**
12	**22.75**	**114.25**
13	23.25	113.25
14	25.75	115.25
15	24.25	113.25
16	22.75	113.25
17	**24.75**	**115.25**

Figure 2. Study area. (**a**) Location of the study area in China; (**b**) The extent of the DRB; (**c**) The geographical location of meteorological stations. (The meteorological stations marked in bold are located within the DBR, and the meteorological stations not marked in bold are located outside the DBR).

Table 1. Data sources and characteristics.

Name of Data	Data Production Unit	Data Source Website	Resolution	Processing Method
Landsat remote sensing satellite data	USGS	https://earthexplorer.usgs.gov/	30 m	Spatial analysis
GDEMV2 Elevation Data	Geospatial Data Cloud	http://www.gscloud.cn/	30 m	Spatial analysis
GDP per capita	Statistical Yearbook of Jiangxi and Guangdong Province, China	–	–	Statistical analysis
1:4 million Chinese soil type data	National Earth System Science Data Sharing Platform	http://www.geodata.cn/	–	Spatial analysis
Population density	WorldPOP dataset	https://www.worldpop.org/	100 m	Spatial analysis
Meteorological data	China Weather Data website	http://data.cma.cn/	–	Spatial analysis

Note: We last accessed the link above on 16 November 2020.

Topographic factors included elevation, slope, slope orientation, soil erosion, and other factors. The GDEMV2 elevation data, which have a 30 m spatial resolution covering 29 scenes in Jiangxi and 31 scenes in Guangdong, were mainly obtained from the Geospatial Data Cloud website (http://www.gscloud.cn/). Data such as elevation, slope, and slope direction were calculated from the DEM data [29]. Soil erosion is one of the major factors related to soil degradation worldwide [36–38]. The widely used soil loss equation model [39] was used to calculate soil erosion in the DRB. These data were resampled to 30 m in ArcGIS.

Climate factors included average annual precipitation, average annual temperature, and relative humidity. These data were mainly obtained from the China Meteorological Data website (http://data.cma.cn/), which catalogues 17 meteorological stations. The inverse distance weighting method is a reliable method for spatial distribution that takes full account of the geographical links between factors. The above data were obtained by interpolation with the inverse distance weight method in ArcGIS 10.2.

Landscape factors included the mean patch area, boundary density, Shannon diversity index, Shannon evenness index, and Simpson diversity index, from Landsat remote sensing satellite data from the USGS website. Landsat-8 maintained a basic consistency with Landsat 1–7 in terms of spatial resolution and spectral characteristics. The satellite has a total of 11 bands, i.e., bands 1–7, and 9 bands have a spatial resolution of 30 m; band 8 is a panchromatic band with a 15 m resolution; and bands 10 and 11 have a spatial resolution of 100 m; the satellite can achieve global coverage once every 16 days [40]. For the 12 scenes, Landsat satellite image data containing few clouds were selected and are detailed in Appendix A Table A1. The calculation of landscape pattern indices, such as mean patch area [41], boundary density [41], Shannon diversity index [42], Shannon evenness index [42], and Simpson diversity index [42], was performed in Fragstats 4.2 as described at https://www.umass.edu/landeco/research/fragstats/documents/fragstats.help.4.2.pdf.

Vegetation factors, which comprise vegetative cover, were derived from Landsat remote sensing satellite data, as described in the previous subsection. The vegetative cover data were calculated using the linear spectral mixture model (LSMM) from Landsat remote sensing satellite data [43].

Socio-economic statistics included population density and GDP per capita. Population density data were sourced from the WorldPOP dataset, and GDP per capita data were sourced from the 'Statistical Yearbook of Jiangxi Province' [44] and the 'Statistical Yearbook of Guangdong Province' [45]. Population density and GDP per capita were selected to calculate the impact of socio-economic activities on the EV of the watershed. The above data were obtained using the inverse distance weighting method in ArcGIS 10.2.

2.3. The Principal Component Analysis

The main objective of this study was to establish a comprehensive factor system based on PCA theory for assessing EV in DRB. PCA is a method for converting existing variables into a small number of summary factors that can best represent the most original information [46]. We first calculate the contribution rate, then look at the value of the main factor contribution rate to determine how many main factors there are, and the cumulative contribution rate of the main factors is considered to be satisfactory at 90% or more [47]. This assessment method includes a series of processing steps, e.g., the selection of relevant factors, standardization, the determination of the variance of common factors, the calculation of weights, and finally EV assessment.

2.3.1. The PCA Structure of EV Based on the PSR (Ecological Pressure Ecological Sensitivity, Ecological Resilience)

EV is the combined result of PSR [25,29,48,49]. Both natural and social factors can constrain the development of EV. Therefore, EV can also affect the economic development of a region, bringing about a series of developmental problems such as poor economic activity, slow development of modern industries, and low per capita education [13]. By reviewing the relevant information on EV in the DRB and assessing the relevant factor systems selected by researchers for the southern hilly mountains [50–55], it was found that a causal relationship between man and nature can be established with the PSR model. The general aim of this study was to establish a comprehensive PSR evaluation system based on PCA theory, and our analysis and evaluation system contains a target level, a criterion level, and a factor level. Both human activities and natural conditions affect EV, and so the evaluation factor system should include these two factors [56,57]. We developed a system analysis model consisting of three levels and 15 factors (Table A2 in the Appendix A). We consulted relevant experts from the Guangdong Dongjiang River Basin Authority to understand the current situation in the basin.

Ecological response factors are the most direct characteristics expressed by the long-term interactions of various factors within an ecosystem [29]. In the vicinity of the DRB, the topography is fragmented, it has varying heights of terrain, there are hilly mountains in the center and north, and deltas, lowlands, and coastal plains in the south; the DRB has a subtropical monsoon climate, with an average annual temperature of 21 °C and an average annual precipitation of 1750 mm, which is uneven and mainly concentrated in April–September [34]. Ecological sensitivity factors can be divided into topographic and climatic factors: topographic factors are selected for elevation, slope, slope orientation, and soil erosion, which directly contribute to increased soil erosion problems [58]; climatic factors include mean annual precipitation, average annual temperature, and relative humidity. The DRB is cloudy and rainy. The average annual temperature and precipitation are key climatic factors affecting the amount of vegetative production. Moreover, relative humidity can reflect vegetation transpiration, so it is crucial for ecological protection [59].

Ecological state factors have a role in protecting ecosystems and can be divided into landscape factors and vegetation factors. The mean patch area (mean patch area, area_mn) can reflect landscape heterogeneity. It can, on the one hand, constrain the minimum patches of the image landscape and, on the other hand, reflect the degree of fragmentation of the landscape [41]. The boundary density (edge density, ED) is an important factor for analyzing patch shape, as it indicates the extent to which the landscape is fragmented, i.e., the higher the value, the more the boundary is fragmented, the more dispersed the layout, and the more compact the patches [41]. The Shannon diversity index (SHDI) reflects the landscape heterogeneity, primarily used to assess the contribution of rare patches to information, i.e., the richer the land use, the greater the fragmentation in the landscape, the higher the value, and the less damage to the landscape [42]. The Shannon evenness index (SHEI) indicates the maximum likelihood of a given landscape, i.e., the richer the landscape, the healthier and more stable its ecosystem [42]. The Simpson diversity index reflects the heterogeneity of the community, and the higher its value, the more stable the ecosystem [34].

Ecological pressure factors are mainly associated with human activities that constrain the health of the ecosystem to a certain extent. The DRB is located in a humid zone. Its watershed is important to Guangdong, Hong Kong, and Macau, which include the more economically active cities of Shenzhen, Guangzhou, Dongguan, Huizhou, and Hong

Kong, the core city of the Greater Bay Area. In recent years, ecological and environmental pollution and degradation have posed serious challenges as the population has increased, and urbanization and industrial upgrading have accelerated [33]. Socio-economic activities constantly influence the magnitude of EV within the watershed area; therefore, human activities have a huge impact on the evolution of the ecological environment [60,61]. Population density increase can lead to excessive resource consumption, and increases in GDP per capita mean that more resources need to be consumed to achieve economic productivity, which constrains the health of regional ecosystems to some extent [62]. It follows that many anthropogenic activities can directly lead to environmental pollution and other serious consequences, such as the degradation and depletion of natural resources.

2.3.2. Weight Calculation Based on the PCA

Given that the scale criteria and inter-factor attribute interval values are different for each factor, standardization is used to maintain the same scale and criteria across factors in order to facilitate the subsequent modeling analysis [25]. All the assessment factors were controlled to be within [0, 1], i.e., values closer to 1 representing stronger vulnerability and values closer to 0 denoting weaker vulnerability. Increasing positive factors brings about increased vulnerability, and consequently, ecological damage becomes more frequent and/or even more serious. Increasing negative factors partly alleviates EV and improves the ecological environment. The formulae for the positive and negative correlation factors calculated by normalizing the above data are as follows:

$$\text{Positive: } m_i^{'} = (m_i - m_{imin})/(m_{imax} - m_{imin}) \tag{1}$$

$$\text{Negative: } m_i^{'} = 1 - (m_i - m_{imin})/(m_{imax} - m_{imin}) \tag{2}$$

where $m_i^{'}$ is the value of the first factor after standard processing; m_i is the initial value of the i-th factor; m_{imin} is the minimum of the i-th factor in the layer; and m_{imax} indicates the maximum of the i-th factor in the layer.

The weights were calculated using PCA [56]. Furthermore, the weight of each principal component α_i was calculated using the following formula:

$$\alpha_i = \lambda_i / \sum_{i=1}^{m} \lambda_i \tag{3}$$

where λ_i denotes the variation degree of the i-th principal component. The PCA results were obtained using ArcGIS10.2, and the common factor variance of each evaluation factor was calculated from the factor matrix of the results.

$$H_j = \sum_{j=1}^{m} \lambda_{jk}^2 \ (j = 1,2,\ldots,9; \ k = 1,2,\ldots,m) \tag{4}$$

where j is the number of identified evaluation factor factors, k is the number of principal components, and m is the total number of principal components. Equations (3) and (4) were used to standardize the H_j (common factor variances) of the evaluation factors. Finally, the weights of each factor were obtained using Formula (5).

$$W_j = H_j / \sum_{j=1}^{15} H_j \ (j = 1,2,\ldots,15) \tag{5}$$

where W_j denotes the weights of the j-th factor.

2.3.3. Ecological Vulnerability Model Calculation

The original evaluation factors were standardized using the extreme difference method, as was described in the previous subsection. We calculated the weighting coefficients of the 15 evaluation factors in the DRB. Moreover, we were able to construct a comprehensive EV evaluation model, which included the state of natural resources and socio-economic development in the DRB. Formula (6) is able to represent the EV of the DRB through the comprehensive analysis of multiple evaluation factors, and thus the values of the comprehensive evaluation factors were derived to calculate the EVI of the DRB.

$$EVI_i = \sum\nolimits_{i=1}^{15} p_i * W_i \tag{6}$$

where EVI_i is the EV index of the i-th image raster, between $[0, 1]$; p_i is the value obtained after normalization of the i-th factor of the raster image; and W_i is the weighting factor of a factor for EV. The weighting values for the 3 years are shown in Table A2.

2.3.4. Threshold Definition Based on Net Primary Production

NPP (net primary production), which indicates the total amount of organic dry matter produced by green plants per unit time and unit area, can be used to measure ecological changes [10]. We used GEE (Google Earth Engine) to download annual NPP data with a pixel resolution of 500 m (m) from MOD17A3H V6 (2001, 2008, and 2019). These were introduced to assist in determining the EV thresholds for different classes within each time series.

Current studies generally define EV thresholds randomly [18,24,63]. We used the method for classifying EV described by Guo Bing [10] et al. To avoid randomness in the definition of vulnerability (EV) thresholds, we introduced NPP data to assist in the determination of EV thresholds for the 3 years (2001, 2008, and 2019), thus to some extent ensuring EV comparability in the same region over time. The main steps were as follows: (1) we divided the three NPP data periods (2001, 2008, 2019) into four classes using the equidistance method; (2) we combined the three NPP data periods to derive the EV values of the corresponding periods; (3) the EV values were reverse-ordered to obtain the EV thresholds of different classes in the three periods. The level was set at five levels: potential, slight, mild, moderate, and severe (Table 2). The potential level indicates a stable, fully functional ecosystem with a high sensitivity to external disturbance, a high self-recovery rate, and abundant vegetative cover. Slight-level ecosystems are relatively stable, fully functional, with a low sensitivity to external disturbances, a weak self-recovery rate, and good vegetative cover. The mild level indicates an ecosystem that is marginally sensitive to external disturbances but generally stable. Moderate-level ecosystems are relatively unstable, functionally deficient, find it difficult to recover from damage, are highly sensitive to external disturbances, and have poor vegetative cover. Severe-level ecosystems are extremely unstable, severely degraded, find it extremely difficult or even impossible to recover from damage, are highly sensitive to external disturbance, and have poor vegetative cover.

Table 2. Table of EV levels in the DRB in different periods.

Vulnerability Level	2001	2008	2019	NPP
Potential	<0.47	<0.41	<0.38	-
Slight	0.47–0.51	0.41–0.48	0.38–0.42	0.75
Mild	0.51–0.52	0.48–0.50	0.42–0.44	0.5
Moderate	0.52–0.59	0.50–0.53	0.44–0.46	0.25
Severe	>0.59	>0.53	>0.46	-

2.4. Geodetector

Geodetector is a statistical method for detecting spatial differentiation and revealing the factors that influence it [64]. The four detectors of the Geodetector are factor detection, interaction detection, risk zone detection, and ecological detection. The spatial distribution of EV varies significantly and is influenced by a combination of factors. We use the Geodetector to perform factor detection and interaction detection, i.e., to calculate the drivers of variation in the spatial extent of EV affecting DRB, and then to infer the interaction between two variables.

(1) Factor detector: It detects the spatial heterogeneity of EV change Y and the explanatory power of different factors X on EV change Y. Measured by the q-value, the expression is [65]:

$$q = 1 - \sum_{h=1}^{L} N_h \sigma_h^2 / N\sigma^2 \tag{7}$$

where $h = 1, \ldots, L$, L is the stratification of variable Y or factor X, i.e., classification or zoning; N_h and N are the number of cells in stratum h and the whole area, respectively; σ_h^2 and σ^2 are the variance of stratum h and the whole area of Y values, respectively. q has a range of [0, 1], with larger values of q indicating a stronger explanation of changes in EV Y by the independent variable X and vice versa.

(2) Interaction detector: analyzes the possible causal relationships between different influencing factors, i.e., whether the combined effect of different factors enhances the explanatory power of EV. In the evaluation process, we first calculate the q-values of Y for each of the two factors: q(X1) and q(X2); calculate the q-value of Y when the two layers are tangent: q(X1∩X2); and compare q(X1), q(X2), and q(X1∩X2). The relationship is detailed in the Table 3 [64].

Table 3. Interaction relationship.

Criterion	Interaction
q(X1∩X2) < Min(q(X1), q X2))	Non-linear weakening
Min(q(X1), q(X2)) < q(X1∩X2) < Max(q(X1), q(X2))	Single-factor non-linear attenuation
q(X1∩X2) > Max(q(X1), q(X2))	Two-factor enhancement
q(X1∩X2) = q(X1) + q(X2)	Independent
q(X1∩X2) > q(X1) + q(X2)	Non-linear enhancement

3. Results

3.1. Temporal Evolution Characteristics of Ecological Vulnerability

On the basis of the establishment of the PSR EV model in the previous section, the spatial distribution of the EV index in the study area was calculated using Equation (6). The grading map (Figure 3) and the percentage of area in EV levels (Table 4) were calculated.

This research reclassified the EV values in 2001, 2008, and 2019 in the DRB, and then obtained three periods of vulnerability level results in this area, which are shown in Figure 4. According to the level results, the research also counted the coverage and area proportion of each classification, which are shown in Table 5. In 2001, 2008, and 2019, the EV index of the DRB ranged from 0.15 to 0.82, with an annual average of 0.51, as shown in Figure 5. According to the analysis of statistics in Table 6, the proportion of severe (V) level in the DBR was the highest in all three periods, increasing and then decreasing, indicating that EV in the DBR deteriorated and then improved. The proportion of moderate (IV) level was the lowest in all periods. The proportion of mild (III) level decreased year by year. The proportion of slight (II) level increased and then decreased, and the proportion of potential (I) level followed the opposite trend. At the macro level, the overall fragility of the eco-environment in the DRB decreased from north to south.

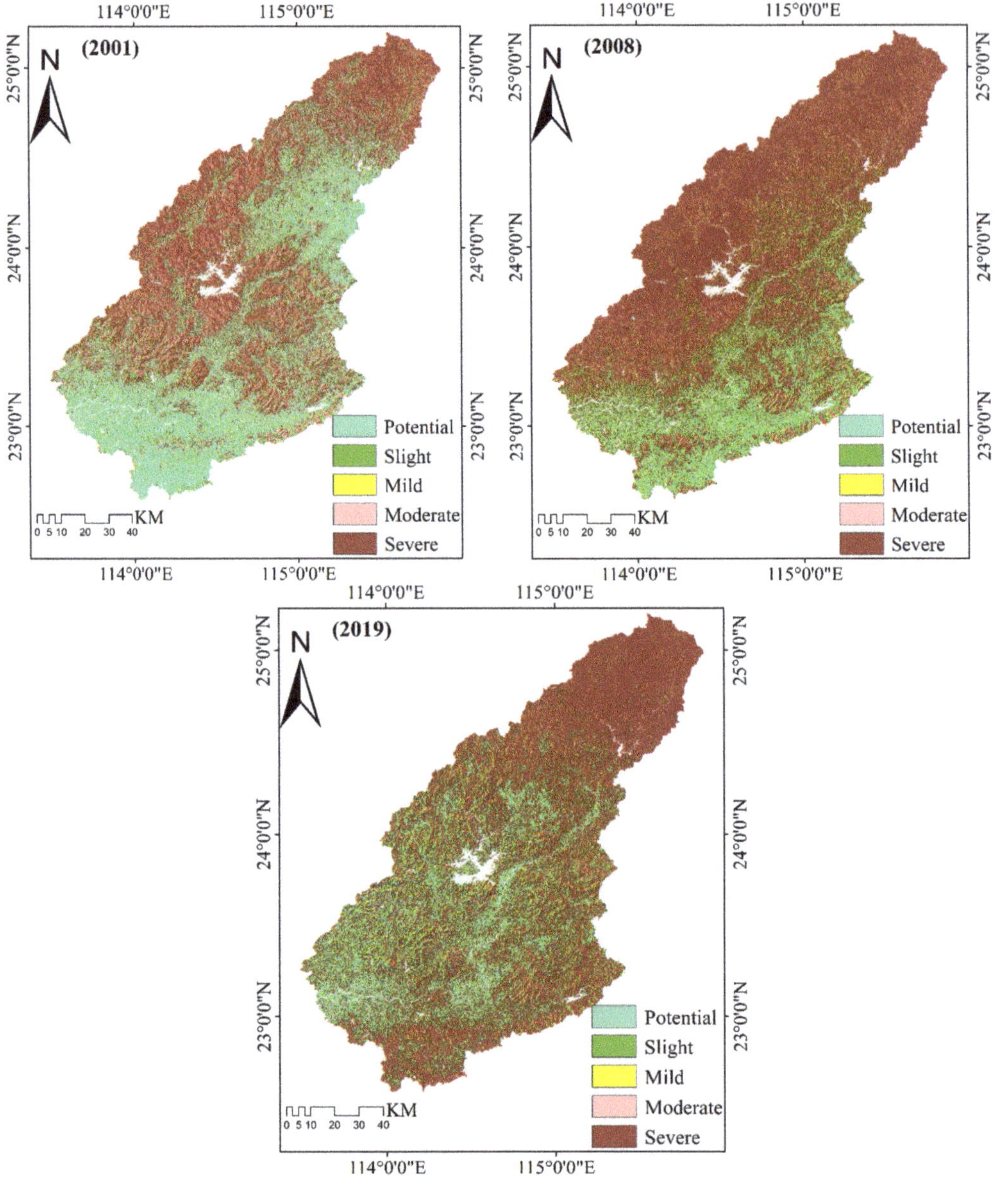

Figure 3. Distribution of EV levels in the DRB, 2001–2019.

Table 4. Statistics on the percentage of area at each EV level in the DRB (2001, 2008, 2019).

Level	Vulnerability Level	2001	2008	2019
		Percentage of the Total Area (%)	Percentage of the Total Area (%)	Percentage of the Total Area (%)
I	Potential	23	8	17
II	Slight	14	17	15
III	Mild	4	7	9
IV	Moderate	29	13	9
V	Severe	30	55	50

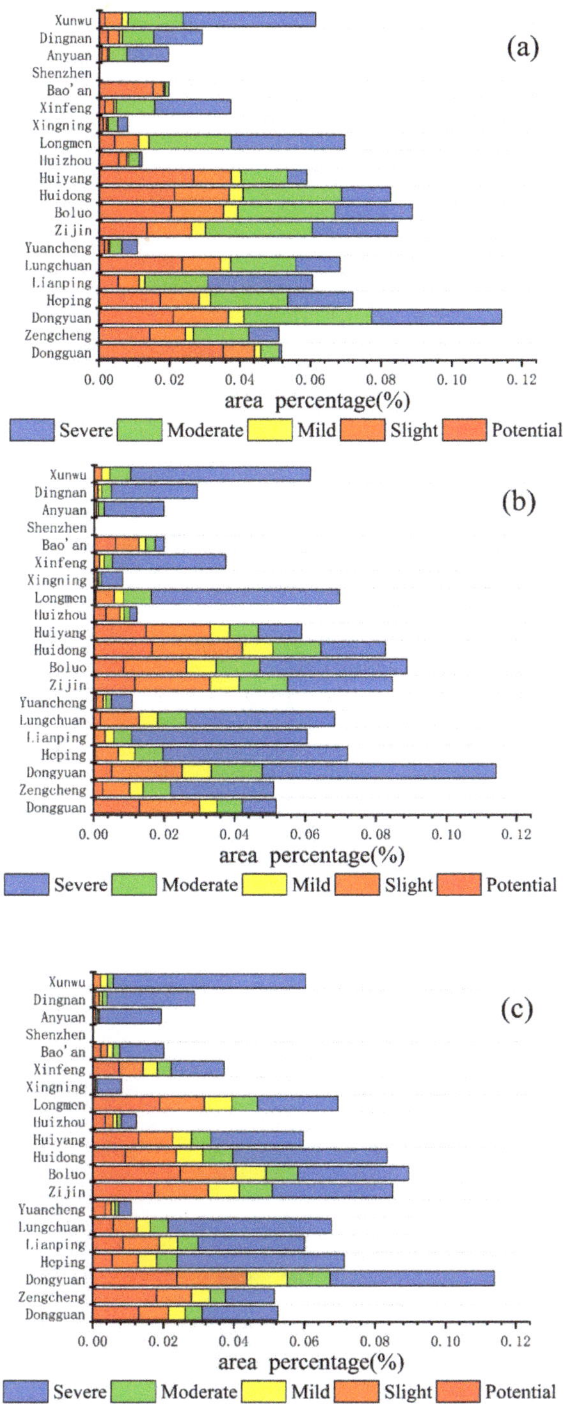

Figure 4. Scale map of vulnerability areas at different levels (**a**) 2001; (**b**) 2008; (**c**) 2019.

Table 5. Statistics of q-values for ecological vulnerability factor detection.

Name of the Factor	2001			2008			2019		
	q	q Ranking	p	q	q Ranking	p	q	q Ranking	p
Soil erosion(X4)	0.109	10	0	0.066	10	0	0.037	12	0
Area_mn(X8)	0.057	15	0	0.046	14	0	0.080	9	0
Slope orientation(X3)	0.250	5	0	0.259	2	0	0.002	15	0
Vegetation cover(X13)	0.280	3	0	0.208	4	0	0.102	4	0
Elevation(X1)	0.258	4	0	0.207	5	0	0.049	11	0
Boundary density (ed) (X9)	0.057	14	0	0.049	13	0	0.084	5	0
GDP per capita(X15)	0.132	9	0	0.089	9	0	0.148	3	0
Average annual precipitation(X5)	0.245	6	0	0.145	6	0	0.076	10	0
Population density(X14)	0.172	7	0	0.133	7	0	0.023	13	0
Average annual temperature(X6)	0.284	2	0	0.226	3	0	0.235	1	0
Shannon Diversity Index (SHDI) (X10)	0.059	11	0	0.050	11	0	0.083	8	0
Shannon's evenness index (SHEI) (X11)	0.058	12	0	0.009	15	0	0.083	7	0
Relative Humidity(X7)	0.325	1	0	0.329	1	0	0.209	2	0
Simpson diversity index (SIDI) (X12)	0.058	13	0	0.050	12	0	0.084	6	0
Slope(X2)	0.144	8	0	0.090	8	0	0.015	14	0

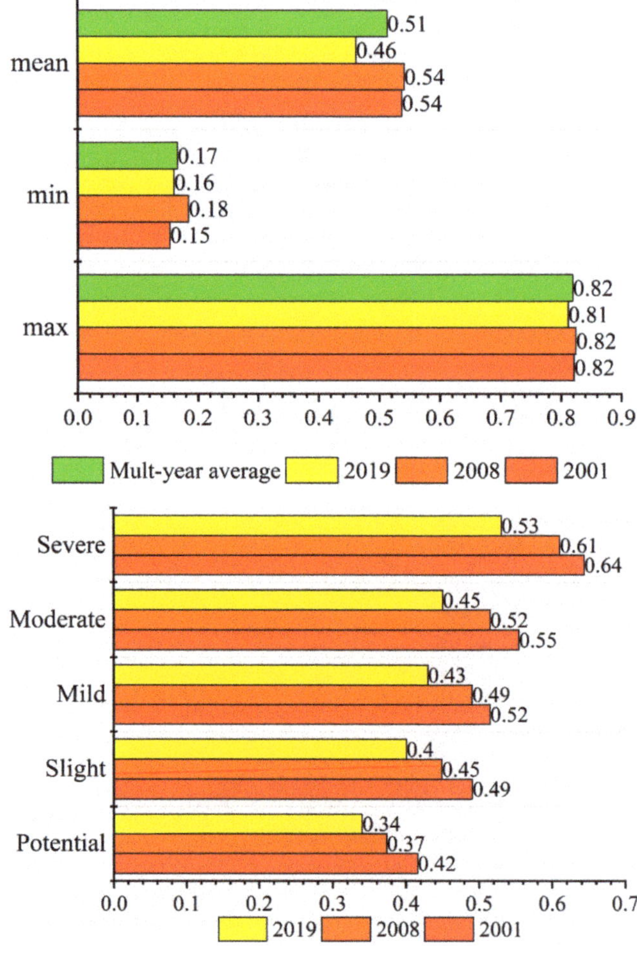

Figure 5. Histogram of minimum, maximum, and mean values based on EV in DRB.

Table 6. Interaction results for selected factors of ecological vulnerability in the DRB.

2001		2008		2019	
X3/X7 *	0.564	X3/X7 *	0.570	X13/X6 *	0.351
X3/X6 **	0.535	X13/X7 *	0.505	X13/X7 *	0.321
X3/X13 *	0.528	X3/X6 **	0.476	X9/X6 *	0.311
X13/X7 *	0.517	X3/X13 *	0.471	X6/X12 *	0.311
X3/X1 *	0.506	X3/X1 *	0.464	X6/X11 *	0.310
X3/X5 **	0.502	X4/X7 **	0.442	X6/X10 *	0.310
X13/X6 *	0.481	X13/X6 *	0.434	X8/X6 *	0.307
X13/X5 *	0.454	X3/X5 **	0.402	X7/X12 *	0.291
X4/X7 **	0.447	X15/X7 *	0.394	X11/X7 *	0.291
X3/X14 **	0.426	X3/X14 **	0.394	X9/X7 *	0.291
X3/X2 **	0.405	X10/X7 *	0.394	X10/X7 *	0.290
X4/X6 **	0.403	X7/X12 *	0.394	X8/X7 *	0.287
X1/X7 *	0.397	X9/X7 *	0.393	X4/X6 **	0.284
X5/X7 *	0.396	X8/X7 *	0.388	X5/X6 *	0.280
X1/X6 *	0.393	X13/X5 *	0.379	X5/X7 *	0.280
X15/X7 *	0.388	X5/X7 *	0.379	X13/X6 *	0.270
X6/X7 *	0.385	X1/X7 *	0.368	X15/X6 *	0.265
X3/X15 **	0.383	X6/X7 *	0.367	X4/X7 **	0.257
X13/X15 *	0.382	X15/X6 *	0.354	X15/X7 *	0.255
X10/X7 *	0.380	X3/X15 **	0.354	X14/X6 *	0.253

(* represents a two-factor enhanced interaction, ** represents a non-linear enhanced interaction).

3.2. Change of Ecological Vulnerability Grade Index

We calculated the spatial distribution of the EV change intensity in 2001, 2008, and 2019. Then, we differentiated the vulnerability maps for the three years (2001, 2008, and 2019), and the spatio-temporal changes in EV in the DRB were further analyzed. On the basis of the histogram distribution and related spatio-temporal map changes, the four stages of EV change intensity (CI) were classified as follows: decreasing intensity ($CI \leq -0.20$); mildly decreasing ($-0.20 < CI \leq -0.1$); stable ($-0.1 < CI \leq 0$); mildly increasing ($0 < CI \leq 0.1$); and intensity increase ($CI > 0.1$). We used the ArcGIS zonal statistics tool to calculate the average of each factor for each district and county and combined it with Figure 6. The results show that, from 2001 to 2008 (Figure 6A), the DRB was dominated by two intensities, i.e., mild increase and stable, with mild increase occupying more than half and stable mainly distributed in Huidong, Zijin, Huizhou, Yuancheng, Dongyuan, and Boluo. From 2008 to 2019 (Figure 6B), mild decrease was mainly distributed in Zengcheng, Boluo, Longmen, Lianping, Dongyuan, and Yuancheng, etc., and mild increase dominated in Shenzhen, Dongguan, Huiyang, Huizhou, and Huidong in the southern DRB. The quality of the environment has improved over the last 20 years in the DRB, with a trend towards increased EV in a few areas, these being more obvious is the southernmost part of the DRB (Figure 6C).

3.3. Analysis of the Drivering Factors of Ecological Vulnerability

We used the ArcGIS 10.2 software to create 30 m $\times$ 30 m grid points and extracted 18,249 sample points. We relied on the aforementioned sample points to extract the values of each dependent and independent variable for the quantitative analysis and evaluation of the driving force factors. EV was selected as the dependent variable, and the 15 EV factors were selected as the independent variables. After standardizing them, the factors for each period (2001, 2008, and 2019) were transformed from numerical quantities to typological quantities using the natural breakpoint method at five levels. Finally, we used Geodetector to analyze the impact of each factor on EV. Factor detectors were used to calculate the influence of each factor on EV. As can be seen from Table 5, the *p*-value of each factor was 0, which indicates that the explanatory power of the three period factors on EV was sufficient. This also proves that the selection of topography, climate, vegetation, landscape, and economic factors to calculate the EV of the DRB was feasible.

In general, the influence of the DRB factors on EV was basically stable. From these, the top two q-values were relative humidity and average annual temperature, with mean q-values of 0.288 and 0.248, respectively, indicating that the EV of the DRB was most influenced by relative humidity and average annual temperature. This is mainly due to the fact that relative humidity and average annual temperature are the dominant factors affecting evapotranspiration in most parts of China [65]. The DRB is located in the humid subtropical monsoon climate of southern China, with abundant but unevenly distributed rainfall; the average annual temperature is 20–22 °C, with little average annual temperature variation [66]. Evapotranspiration affects regional climate change and vegetative growth, and thus the EV of the DRB [67].

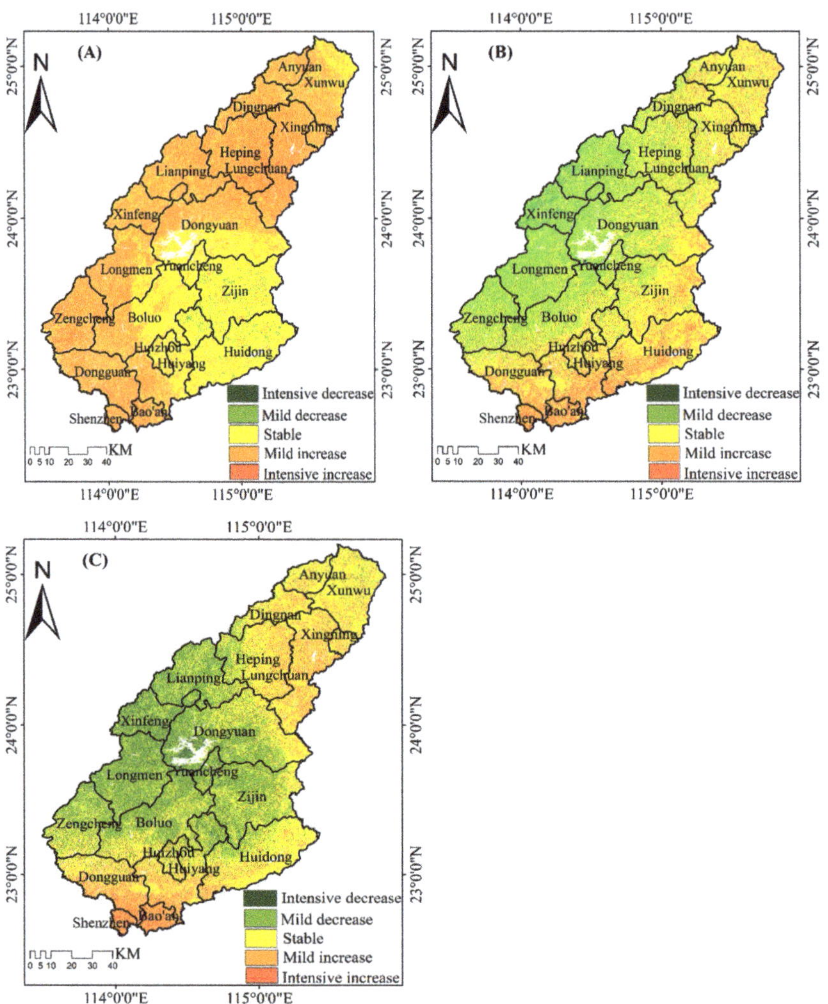

Figure 6. Dynamic changes in EV in DRB. (**A**) 2008–2001; (**B**) 2019–2008; (**C**) 2019–2001.

From 2001 to 2008, the q-values in the 3–6 range were mainly vegetative cover, elevation, and precipitation. They exhibit a decreasing trend, indicating that the influence of these three factors on the EV was gradually weakening. This is mainly because the DRB is located in the humid region of southern China, with undulating terrain and dense vegetation; changes in the ecological environment are largely influenced by vegetative cover, elevation, and precipitation. The q-values in 2019 were mainly for GDP per capita, which is very different from the rankings in 2001 and 2008. As the DRB has undergone a rapid developmental process in recent years, in general, the q-value of GDP per capita exhibits an upward trend, which indicates a gradual rise in the impact of social activities on EV.

Interaction detection was used to assess whether the explanatory power of EV was enhanced when the two factors acted together. Overall (Table 6, Figure 7), the values of the EV factor interactions were greater than the maximum values of the individual factors, thus indicating that the effects of the factors on EV were not independent of each other, but occurred synergistically. As shown in the table, from 2001 to 2008, all the factors were bi-factorially enhanced, except for slope orientation, which was non-linearly enhanced with precipitation, average annual temperature, population density, and GDP per capita. In 2019, soil erosion was non-linearly enhanced with average annual temperature and relative humidity, and each was bi-factorially enhanced. From 2001 to 2019, the DRB interaction detection results were generally more stable, with relative humidity exhibiting the highest interaction with other factors. Moreover, the interaction of the relative humidity with the other factors was much higher than the number of interactions between two of the other factors that occurred. The interaction between topography, climate, vegetation, landscape, and economic factors was stronger than the interaction within each element, and the interaction between natural and social factors was stronger than the interaction within each factor, indicating that the EV of the DRB was the result of the combined effect of all factors.

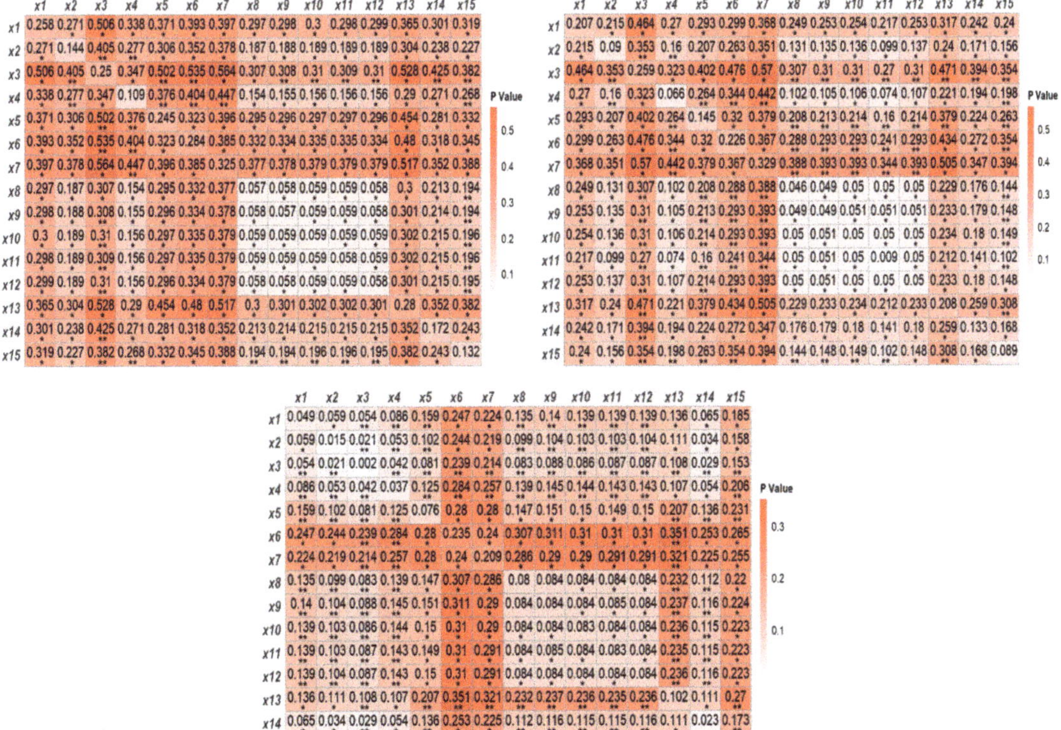

2001:

	x1	x2	x3	x4	x5	x6	x7	x8	x9	x10	x11	x12	x13	x14	x15
x1	0.258	0.271	0.506	0.338	0.371	0.393	0.397	0.297	0.298	0.3	0.298	0.299	0.365	0.301	0.319
x2	0.271	0.144	0.405	0.277	0.306	0.352	0.378	0.187	0.188	0.189	0.189	0.189	0.304	0.238	0.227
x3	0.506	0.405	0.25	0.347	0.502	0.535	0.564	0.307	0.308	0.31	0.309	0.31	0.528	0.425	0.382
x4	0.338	0.277	0.347	0.109	0.376	0.404	0.447	0.154	0.155	0.156	0.156	0.156	0.29	0.271	0.268
x5	0.371	0.306	0.502	0.376	0.245	0.323	0.396	0.295	0.296	0.297	0.297	0.296	0.454	0.281	0.332
x6	0.393	0.352	0.535	0.404	0.323	0.284	0.385	0.332	0.334	0.335	0.335	0.334	0.48	0.318	0.345
x7	0.397	0.378	0.564	0.447	0.396	0.385	0.325	0.377	0.378	0.379	0.379	0.379	0.517	0.352	0.388
x8	0.297	0.187	0.307	0.154	0.295	0.332	0.377	0.057	0.058	0.059	0.059	0.058	0.3	0.213	0.194
x9	0.298	0.188	0.308	0.155	0.296	0.334	0.378	0.058	0.057	0.059	0.059	0.058	0.301	0.214	0.194
x10	0.3	0.189	0.31	0.156	0.297	0.335	0.379	0.059	0.059	0.059	0.059	0.059	0.302	0.215	0.196
x11	0.298	0.189	0.309	0.156	0.297	0.335	0.379	0.059	0.059	0.059	0.058	0.059	0.302	0.215	0.196
x12	0.299	0.189	0.31	0.156	0.296	0.334	0.379	0.058	0.058	0.059	0.059	0.058	0.301	0.215	0.195
x13	0.365	0.304	0.528	0.29	0.454	0.48	0.517	0.3	0.301	0.302	0.302	0.301	0.28	0.352	0.382
x14	0.301	0.238	0.425	0.271	0.281	0.318	0.352	0.213	0.214	0.215	0.215	0.215	0.352	0.172	0.243
x15	0.319	0.227	0.382	0.268	0.332	0.345	0.388	0.194	0.194	0.196	0.196	0.195	0.382	0.243	0.132

2008:

	x1	x2	x3	x4	x5	x6	x7	x8	x9	x10	x11	x12	x13	x14	x15
x1	0.207	0.215	0.464	0.27	0.293	0.299	0.368	0.249	0.253	0.254	0.217	0.253	0.317	0.242	0.24
x2	0.215	0.09	0.353	0.16	0.207	0.263	0.351	0.131	0.135	0.136	0.099	0.137	0.24	0.171	0.156
x3	0.464	0.353	0.259	0.323	0.402	0.476	0.57	0.307	0.31	0.31	0.27	0.31	0.471	0.394	0.354
x4	0.27	0.16	0.323	0.066	0.264	0.344	0.442	0.102	0.105	0.106	0.074	0.107	0.221	0.194	0.198
x5	0.293	0.207	0.402	0.264	0.145	0.32	0.379	0.208	0.213	0.214	0.16	0.214	0.379	0.224	0.263
x6	0.299	0.263	0.476	0.344	0.32	0.226	0.367	0.288	0.293	0.293	0.241	0.293	0.434	0.272	0.354
x7	0.368	0.351	0.57	0.442	0.379	0.367	0.329	0.388	0.393	0.393	0.344	0.393	0.505	0.347	0.394
x8	0.249	0.131	0.307	0.102	0.208	0.288	0.388	0.046	0.049	0.05	0.05	0.05	0.229	0.176	0.144
x9	0.253	0.135	0.31	0.105	0.213	0.293	0.393	0.049	0.049	0.051	0.051	0.051	0.233	0.179	0.148
x10	0.254	0.136	0.31	0.106	0.214	0.293	0.393	0.05	0.051	0.05	0.05	0.05	0.234	0.18	0.149
x11	0.217	0.099	0.27	0.074	0.16	0.241	0.344	0.05	0.051	0.05	0.009	0.05	0.212	0.141	0.102
x12	0.253	0.137	0.31	0.107	0.214	0.293	0.393	0.05	0.051	0.05	0.05	0.05	0.233	0.18	0.148
x13	0.317	0.24	0.471	0.221	0.379	0.434	0.505	0.229	0.233	0.234	0.212	0.233	0.208	0.259	0.308
x14	0.242	0.171	0.394	0.194	0.224	0.272	0.347	0.176	0.179	0.18	0.141	0.18	0.259	0.133	0.168
x15	0.24	0.156	0.354	0.198	0.263	0.354	0.394	0.144	0.148	0.149	0.102	0.148	0.308	0.168	0.089

2019:

	x1	x2	x3	x4	x5	x6	x7	x8	x9	x10	x11	x12	x13	x14	x15
x1	0.049	0.059	0.054	0.086	0.159	0.247	0.224	0.135	0.14	0.139	0.139	0.139	0.136	0.065	0.185
x2	0.059	0.015	0.021	0.053	0.102	0.244	0.219	0.099	0.104	0.103	0.103	0.104	0.111	0.034	0.158
x3	0.054	0.021	0.002	0.042	0.081	0.239	0.214	0.083	0.088	0.086	0.087	0.087	0.108	0.029	0.153
x4	0.086	0.053	0.042	0.037	0.125	0.284	0.257	0.139	0.145	0.144	0.143	0.143	0.107	0.054	0.206
x5	0.159	0.102	0.081	0.125	0.076	0.28	0.28	0.147	0.151	0.15	0.149	0.15	0.207	0.136	0.231
x6	0.247	0.244	0.239	0.284	0.28	0.235	0.24	0.307	0.311	0.31	0.31	0.31	0.351	0.253	0.265
x7	0.224	0.219	0.214	0.257	0.28	0.24	0.209	0.286	0.29	0.29	0.291	0.291	0.321	0.225	0.255
x8	0.135	0.099	0.083	0.139	0.147	0.307	0.286	0.08	0.084	0.084	0.084	0.084	0.232	0.112	0.22
x9	0.14	0.104	0.088	0.145	0.151	0.311	0.29	0.084	0.084	0.084	0.085	0.084	0.237	0.116	0.224
x10	0.139	0.103	0.086	0.144	0.15	0.31	0.29	0.084	0.084	0.083	0.084	0.084	0.236	0.115	0.223
x11	0.139	0.103	0.087	0.143	0.149	0.31	0.291	0.084	0.085	0.084	0.083	0.084	0.235	0.115	0.223
x12	0.139	0.104	0.087	0.143	0.15	0.31	0.291	0.084	0.084	0.084	0.084	0.084	0.236	0.116	0.223
x13	0.136	0.111	0.108	0.107	0.207	0.351	0.321	0.232	0.237	0.236	0.235	0.236	0.102	0.111	0.27
x14	0.065	0.034	0.029	0.054	0.136	0.253	0.225	0.112	0.116	0.115	0.115	0.116	0.111	0.023	0.173
x15	0.185	0.158	0.153	0.206	0.231	0.265	0.255	0.22	0.224	0.223	0.223	0.223	0.27	0.173	0.148

(P Value legend: 0.1 – 0.5)

Figure 7. The explanatory power of interaction between factors in 2001, 2008, 2019. (* represents a two-factor enhanced interaction, ** represents a non-linear enhanced interaction, *p* value represents the interaction detection value of the two factors).

4. Conclusions and Discussions

4.1. Discussions

We took the DRB as the study area and constructed factor models to analyze the current situation through regional ecological problems. On the basis of the statistical data, such as multi-temporal remote sensing data and multi-year meteorological station data, this study used socio-economic statistics to characterize anthropogenic disturbance factors and analyze the characteristics of EV. As a result of the high level of human activity in the DRB, its ecosystem is under a certain amount of development pressure. In this study, we also conducted a scientific analysis of the manifestation factors of EV from 2001–2019 in terms of ecological pressure, ecological sensitivity, and ecological resilience, and screened out various practical and typical integrated factors. In addition, vegetation inversion, soil erosion factors, landscape pattern factors, and spatial interpolation were constructed as evaluation factors based on three time periods from 2001–2019. Finally, the DRB EV evaluation index system was established.

PCA generates principal components that are independent of each other after the transformation of the original factors, which can reduce the workload of factor selection. The weights of the principal components, i.e., the contribution rate, contain the proportion of the information from the original data to the total information, and thus, this is considered an objective and reasonable assessment. Therefore, we used PCA to evaluate the EV of the DRB. In addition, using the interpolation method, the overall trend state index, and the comprehensive EV index, with the method of dynamically determining the EV rating of NPP data used by Guo Bing [10], we finally analyzed the change in the EV rating in the area, the overall EV trend, and the quality state of each district in the DRB for three periods of vulnerability from 2001 to 2019.

The factors influencing the EV in the DRB were analyzed with the help of the Geodetector software developed by Wang Jinsong [64] et al. in order to obtain objective and scientific results. This can also be confirmed by the research results of Guo Zeqing [56] et al. and Liu Jiaru [68] et al. In this study, Geodetector was used to analyze the magnitude of interactions between various factors in the EV model for each period and the anthropogenic drivers of EV in the DRB.

On the basis of a PSR model with 15 evaluation factors, we analyzed the spatial distribution of EV in the DRB over three periods and, using a geographic probe, carried out a quantitative statistical analysis of the factors driving EV. The results show that EV is decreasing from 2001 to 2019, with a stronger explanatory power for relative humidity, mean annual temperature, and vegetation cover as the main driving factors of EV in the DRB in the 3 years. In 2019, the explanatory power of GDP per capita increases, showing that human activities further influence EV in the DRB. The relative humidity and mean annual temperature did not change much [66]. Moreover, the woodland cover of the DRB increased from 2001 to 2019 (Figure 8), showing that woodland ecosystems are the most functional and structurally complex natural ecosystems of terrestrial ecosystems. Thus, they have a decisive influence on the terrestrial ecosystem environment and play an important role in safeguarding the ecosystem [69,70]. They are not only capable of providing production and living resources for humans, they also have a variety of ecological service functions, such as water containment, carbon sequestration, oxygen release, and maintaining regional ecological balance, which are important for maintaining the environment and promoting a sustainable urban development [71]. This coincides with a consistent decrease in EV. Furthermore, in the interaction detection results, the study found that the interaction between two driving factors had a greater impact on EV in the DRB as compared to a single driving factor (Table 5, Table 6). In both driver indicators, the slope orientation (X3)/relative humidity (X7), slope orientation (X3)/average annual temperature (X6), and slope orientation (X3)/vegetation cover (X13) interactions had the strongest impact on EV in the DBR, in 2001; In 2008, the strongest interaction factors were slope orientation (X3)/relative humidity (X7), vegetation cover (X13)/relative humidity (X7), and slope orientation (X3)/average annual temperature (X6). The interaction factors of vegetation cover (X13)/average annual temperature (X6), vegetation cover (X13)/relative humidity (X7), and (X9)/average annual temperature (X6) were strongest in 2019 (Table 6). Ecosystems are complex systems where some factors interact with each other. For example, the transpiration of the vegetation increases relative humidity, and changes in vegetation cover affect relative humidity [59]. Jiang [72] showed that vegetation has the ability to change the climate of an ecosystem. Moreover, we found that vegetation cover had a strong EV explanatory power in all three years (Table 5). Limpid waters and lush mountains are invaluable assets. Therefore, we believe that planned afforestation is a worthwhile project that may influence (improve) EV.

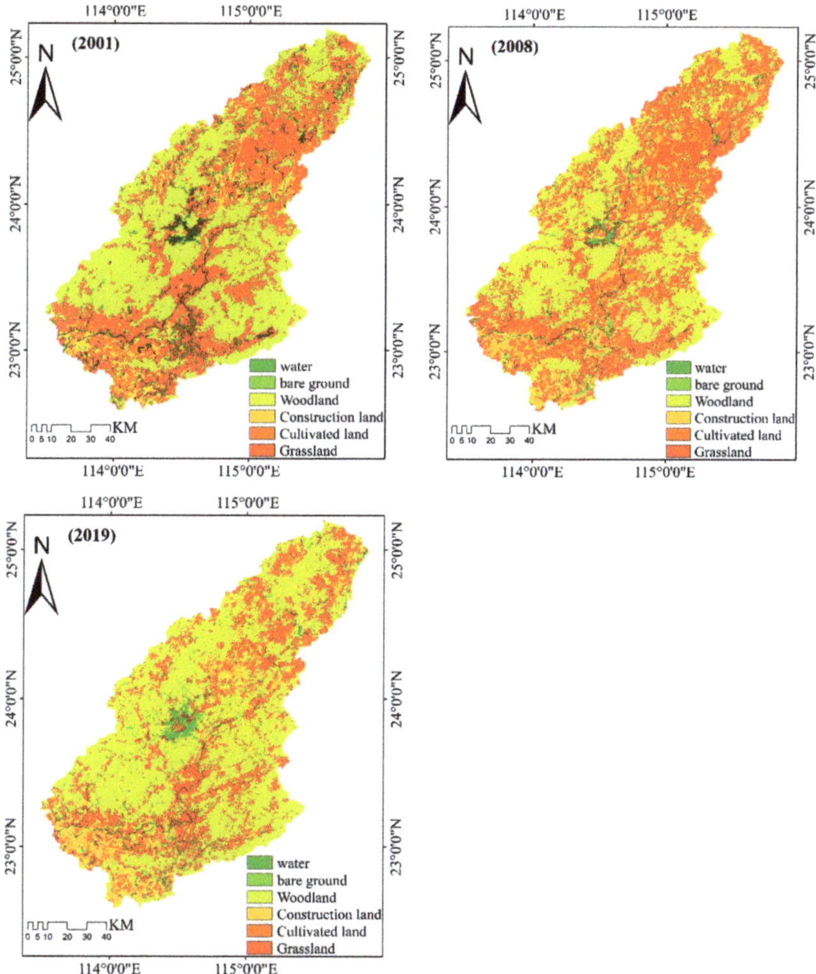

Figure 8. Land use map of DRB (2001, 2008, 2019).

However, the evolution of the spatio-temporal patterns of EV in the DRB from 2001 to 2019 was also influenced by many other uncontrollable natural factors and multi-level human socio-economic factors. Importantly, future studies need to cover a wide range of scientific fields and thus analyze the evolution of EV in the DRB in all its complexity. To this end, more factors must be considered in future studies.

4.2. Conclusions

The DRB, which is an ecologically fragile area in South China, has an important strategic position in terms of resources, environment, and ecology. In view of the unique geographical characteristics and the limitations of relevant studies, we selected 15 well-structured and statistically significant factors, such as topography, climate, landscape, and socio-economy, to analyze the EV status of the DRB. Furthermore, we analyzed its spatio-temporal changes based on the PCA analysis, and detected the driving factors using geographic probes. The following conclusions were reached:

(1)   The PCA method can objectively and reasonably calculate the changes in each factor in the process of assigning weights to vulnerability factors in multi-temporal studies. The method can better reflect the change process of each factor in the EV system, and has good applicability in the southern red soil hilly ecosystem of China.

(2)   NPP data can be associated with the assessment of the health of land surface ecosystems, and the EV level thresholds in different periods can be obtained with the aid of NPP data calculation, which is important for the analysis of EV in different years.

(3)   Over the past 20 years, the overall EV intensity in the DRB can be characterized by a mild decrease, while the upstream and downstream EV intensity in the DRB can be characterized by a mild increase. The midstream exhibited a mild decrease. From 2001 to 2019, the mean EV value gradually decreased. From 2001 to 2008, The area of EV intensity for mild increase is much larger than stable. From 2008 to 2019, EV intensity is more widely distributed in areas of mild decrease than mild increase.

(4)   During 2001–2019, the spatio-temporal pattern of EV in the DRB was significantly affected by the relative humidity, average annual temperature, and vegetation cover.

Author Contributions: J.W. designed experiments; J.W. and G.M. wrote the manuscripts; Z.Z., Q.H. and J.W. processed the experimental data; Z.Z. and J.W. did the methodology; G.M. supervised the manuscript; J.W., Z.Z., Q.H. and G.M. did reviewed and edited the manuscript; all authors contributed to the interpretation of the results and the writing of the paper. All authors have read and agreed to the published version of the manuscript.

Funding: The State Key Research and Development Plan (2018YFB10046) and This research was financially supported by China Geological Survey Project (DD20191016).

Institutional Review Board Statement: Not applicable.

Informed Consent Statement: Not applicable.

Data Availability Statement: The data that support the findings of this study are available from the author upon reasonable request.

Conflicts of Interest: The authors declare no conflict of interest.

Appendix A

Table A1. Landsat image identifier, acquisition time, and type characteristics.

Image Identifier	Acquisition Time	Sensor Type	Track Number (ROW/PATH)	Sun Elevation/(°)	Solar Azimuth/(°)
LT05_L1TP_121043_20011121_20161209_01_T1	21/11/2001		43/121	39.428	149.338
LT05_L1TP_121044_20011121_20161209_01_T1	21/11/2001	TM	44/122	40.544	148.454
LT05_L1TP_122043_20011230_20161209_01_T1	30/12/2001		43/121	34.59	147.199
LT05_L1TP_122044_20081201_20161028_01_T1	30/12/2001		44/122	35.663	146.415
LT05_L1TP_121043_20081210_20161028_01_T1	10/12/2008		43/121	36.556	150.666
LT05_L1TP_121044_20081210_20161028_01_T1	10/12/2008	TM	44/122	37.69	149.878
LT05_L1TP_122043_20081201_20161028_01_T1	1/12/2008		43/121	37.903	150.895
LT05_L1TP_122044_20081201_20161028_01_T1	1/12/2008		44/122	39.041	150.076
LC08_L1TP_121043_20191123_20191203_01_T1	23/11/2019		43/121	41.304	155.234
LC08_L1TP_121044_20191107_20191115_01_T1	7/11/2019	OLI	44/122	46.471	152.692
LC08_L1TP_122043_20191114_20191202_01_T1	14/11/2019		43/121	43.436	154.605
LC08_L1TP_122044_20191114_20191202_01_T1	14/11/2019		44/122	44.634	153.701

Table A2. Data sources and evaluation indicators.

Target Layer	Criterion Layer	Indicator Layer	Name of Data	Positive and Negative	Weight of 2001	Weight of 2008	Weight of 2019
Ecological Response	Terrain indicators	Elevation(X1)	GDEMV2 elevation data	Negative	0.03598	0.01263	0.00885
		Slope(X2)		Positive	0.0185	0.00718	0.00407
		Slope orientation(X3)		Positive	0.24991	0.24989	0.2499
		soil erosion(X4)	1:4 million Chinese soil type data	Positive	0.00021	0.00017	0.00008
		Average annual precipitation(X5)		Negative	0.08956	0.09075	0.20449
		Average annual temperature(X6)	Meteorological Data	Negative	0.04795	0.05347	0.10973
		Relative Humidity(X7)		Positive	0.17487	0.20386	0.11149
Ecological State	Landscape indicators	Mean Patch area (Area_mn) (X8)	Landsat remote sensing satellite data	Negative	0.00059	0.00003	0.00012
		Boundary density (ed) (X9)		Positive	0.00107	0.00042	0.00071
		Shannon Diversity Index (SHDI) (X10)		Positive	0.00059	0.00039	0.00058
		Shannon's evenness index (SHEI) (X11)		Negative	0.00001	0.00003	0.00012
		Simpson diversity index (SIDI) (X12)		Negative	0.00001	0.00003	0.00012
	Vegetation indicators	Vegetation cover(X13)		Negative	0.24466	0.24056	0.24606
Ecological Pressure	Social indicators	Population density(X14)	Population density	Positive	0.00015	0.00009	0.00011
		GDP per capita(X15)	GDP density	Positive	0.13595	0.1405	0.06357

References

1. Depietri, Y. The social–ecological dimension of vulnerability and risk to natural hazards. *Sustain. Sci.* **2020**, *15*, 587–604. [CrossRef]
2. Lv, X.; Xiao, W.; Zhao, Y.; Zhang, W.; Li, S.; Sun, H. Drivers of spatio-temporal ecological vulnerability in an arid, coal mining region in Western China. *Ecol. Indic.* **2019**, *106*, 105475. [CrossRef]
3. Bouwer, L.M. Have disaster losses increased due to anthropogenic climate change? *Bull. Am. Meteorol. Soc.* **2011**, *92*, 39–46. [CrossRef]
4. Gu, D.; Gerland, P.; Pelletier, F.; Cohen, B. Risks of exposure and vulnerability to natural disasters at the city level: A global overview. *United Nations Tech. Pap.* **2015**, *2*, 1–40.
5. Pachauri, R.K.; Allen, M.R.; Barros, V.R.; Broome, J.; Cramer, W.; Christ, R.; Church, J.A.; Clarke, L.; Dahe, Q.; Dasgupta, P.; et al. Climate Change 2014: Synthesis Report. In *Contribution of Working Groups I, II and III to the Fifth Assessment Report of the Intergovernmental Panel on Climate Change*; IPCC: Geneva, Switzerland, 2014.
6. Niu, W.Y. Improvement of ablers model with regard to searching of geographical space. *Chin. Sci. Bull.* **1989**, *34*, 155–157.
7. Clements, F.E. *Research Methods in Ecology*; University Publishing Company: Lincoln, Nebraska, 1905.
8. Liao, X.; Li, W.; Hou, J. Application of GIS based ecological vulnerability evaluation in environmental impact assessment of master plan of coal mining area. *Procedia Environ. Sci.* **2013**, *18*, 271–276. [CrossRef]
9. Nelson, R.; Kokic, P.; Crimp, S.; Martin, P.; Meinke, H.; Howden, S.M.; Voil, P.D.; Nidumolu, U. The vulnerability of Australian rural communities to climate variability and change: Part II—Integrating impacts with adaptive capacity. *Environ. Sci. Policy* **2010**, *13*, 18–27. [CrossRef]
10. Guo, B.; Zang, W.; Luo, W. Spatial-temporal shifts of ecological vulnerability of Karst Mountain ecosystem-impacts of global change and anthropogenic interference. *Sci. Total Environ.* **2020**, *741*, 140256. [CrossRef]
11. Keyes, A.A.; McLaughlin, J.P.; Barner, A.K.; Dee, L.E. An ecological network approach to predict ecosystem service vulnerability to species losses. *Nat. Commun.* **2021**, *12*, 1586. [CrossRef]
12. Chang, H.; Pallathadka, A.; Sauer, J.; Grimm, N.B.; Herreros-Cantis, P. Assessment of urban flood vulnerability using the social-ecological-technological systems framework in six US cities. *Sustain. Cities Soc.* **2021**, *68*, 102786. [CrossRef]
13. Boori, M.S.; Choudhary, K.; Paringer, R.; Kupriyanov, A. Spatiotemporal ecological vulnerability analysis with statistical correlation based on satellite remote sensing in Samara, Russia. *J. Environ. Manag.* **2021**, *285*, 112138. [CrossRef]
14. Mafi-Gholami, D.; Pirasteh, S.; Ellison, J.C.; Jaafari, A. Fuzzy-based vulnerability assessment of coupled social-ecological systems to multiple environmental hazards and climate change. *J. Environ. Manag.* **2021**, *299*, 113573. [CrossRef]
15. Turner, B.L.; Kasperson, R.E.; Matson, P.A.; McCarthy, J.J.; Corell, R.W.; Christensen, L.; Eckley, N.; Kasperson, J.X.; Luers, A.; Martello, M.L.; et al. A framework for vulnerability analysis in sustainability science. *Proc. Natl. Acad. Sci. USA* **2003**, *100*, 8074–8079. [CrossRef]
16. Wang, B.; Ding, M.; Guan, Q.; Ai, J. Gridded assessment of eco-environmental vulnerability in Nanchang city. *Acta Ecol. Sin.* **2019**, *39*, 5460–5472.
17. Yang, Y.; Song, G. Human disturbance changes based on spatiotemporal heterogeneity of regional ecological vulnerability: A case study of Qiqihaer city, northwestern Songnen Plain, China. *J. Clean. Prod.* **2021**, *291*, 125262. [CrossRef]
18. Xia, M.; Jia, K.; Zhao, W.; Liu, S. Spatio-temporal changes of ecological vulnerability across the Qinghai-Tibetan Plateau. *Ecol. Indic.* **2021**, *123*, 107274. [CrossRef]
19. Li, Q.; Shi, X.; Wu, Q. Effects of protection and restoration on reducing ecological vulnerability. *Sci. Total Environ.* **2021**, *761*, 143180. [CrossRef]

20. Tang, Q.; Wang, J.; Jing, Z. Tempo-spatial changes of ecological vulnerability in resource-based urban based on genetic projection pursuit model. *Ecol. Indic.* **2021**, *121*, 107059. [CrossRef]
21. Jin, Y.; Li, A.; Bian, J.; Nan, X.; Lei, G.; Muhammand, K. Spatiotemporal analysis of ecological vulnerability along Bangladesh-China-India-Myanmar economic corridor through a grid level prototype model. *Ecol. Indic.* **2021**, *120*, 106933. [CrossRef]
22. Dai, X.; Gao, Y.; He, X.; Liu, T.; Jiang, B.; Shao, H.; Yao, Y. Spatial-temporal pattern evolution and driving force analysis of ecological environment vulnerability in Panzhihua City. *Environ. Sci. Pollut. Res.* **2021**, *28*, 7151–7166. [CrossRef]
23. Huang, M.; Zhong, Y.; Feng, S.; Zhang, J. Spatial and temporal characteristics and drivers of landscape ecological vulnerability in the Chaohu Lake Basin since 1970s. *Lake Sci.* **2020**, *32*, 977–988.
24. Hu, X.; Ma, C.; Huang, P.; Guo, X. Ecological vulnerability assessment based on AHP-PSR method and analysis of its single parameter sensitivity and spatial autocorrelation for ecological protection–A case of Weifang City, China. *Ecol. Indic.* **2021**, *125*, 107464. [CrossRef]
25. Kang, H.; Tao, W.; Chang, Y.; Zhang, Y.; Li, X.; Chen, P. A feasible method for the division of ecological vulnerability and its driving forces in Southern Shaanxi. *J. Clean. Prod.* **2018**, *205*, 619–628. [CrossRef]
26. Xie, Z.; Li, X.; Chi, Y.; Jiang, D.; Chen, S. Ecosystem service value decreases more rapidly under the dual pressures of land use change and ecological vulnerability: A case study in Zhujiajian Island. *Ocean Coast. Manag.* **2021**, *201*, 105493. [CrossRef]
27. Shi, H.; Lu, J.; Zheng, W.; Sun, J.; Ding, D. Evaluation system of coastal wetland ecological vulnerability under the synergetic influence of land and sea: A case study in the Yellow River Delta, China. *Mar. Pollut. Bull.* **2020**, *161*, 111735. [CrossRef]
28. Liu, M.; Liu, X.; Wu, L.; Tang, Y.; Li, Y.; Zhang, Y.; Ye, L.; Zhang, B. Establishing forest resilience indicators in the hilly red soil region of southern China from vegetation greenness and landscape metrics using dense Landsat time series. *Ecol. Indic.* **2021**, *121*, 106985. [CrossRef]
29. Xue, L.; Jing, W.; Zhang, L.; Wei, G.; Zhou, B. Spatiotemporal analysis of ecological vulnerability and management in the Tarim River Basin, China. *Sci. Total Environ.* **2019**, *649*, 876–888. [CrossRef]
30. Wu, C.; Liu, G.; Huang, C.; Liu, Q.; Guan, X. Ecological vulnerability assessment based on fuzzy analytical method and analytic hierarchy process in Yellow River Delta. *Int. J. Environ. Res. Public Health* **2018**, *15*, 855. [CrossRef] [PubMed]
31. Nguyen, K.A.; Liou, Y.A. Global mapping of eco-environmental vulnerability from human and nature disturbances. *Sci. Total Environ.* **2019**, *664*, 995–1004. [CrossRef] [PubMed]
32. Wang, Y. *Effects of Different Vegetation Restoration Patterns on Soil Reactive Organic Carbon in the Antaibao Mining Area*; China University of Geosciences: Beijing, China, 2015.
33. He, Y.; Guo, H.; Tan, Q.; Pan, W.; Chen, S. Identification of significant water problems in the Dongjiang River Basin of Guangdong Province. *Water Resour. Conserv.* **2021**, *37*, 16–21.
34. Lv, L.; Gao, X.Q.; Liu, Q.; Jiang, Y. Effects of landscape patterns on nitrogen and phosphorus export in the Dongjiang River Basin. *J. Ecol.* **2021**, *41*, 1758–1765.
35. Lv, L.T.; Zhang, J.; Peng, Q.Z.; Ren, F.P.; Jiang, Y. Analysis of landscape pattern evolution and prediction of changes in the Dongjiang River Basin. *J. Ecol.* **2019**, *39*, 6850–6859.
36. Jiang, Q.; Zhou, P.; Liao, C.; Liu, Y.; Liu, F. Spatial pattern of soil erodibility factor (K) as affected by ecological restoration in a typical degraded watershed of central China. *Sci. Total Environ.* **2020**, *749*, 141609. [CrossRef]
37. Amundson, R.; Berhe, A.A.; Hopmans, J.W.; Olson, C.; Sztein, A.E.; Sparks, D.L. Soil and human security in the 21st century. *Science* **2015**, *348*. [CrossRef]
38. Wang, B.; Zheng, F.; Römkens, M.J.M.; Darboux, F. Soil erodibility for water erosion: A perspective and Chinese experiences. *Geomorphology* **2013**, *187*, 1–10. [CrossRef]
39. Williams, J.R. The erosion-productivity impact calculator (EPIC) model: A case history. Philosophical Transactions of the Royal Society of London. *Ser. B Biol. Sci.* **1990**, *329*, 421–428.
40. Diao, C.; Wang, L. Landsat time series-based multiyear spectral angle clustering (MSAC) model to monitor the inter-annual leaf senescence of exotic saltcedar. *Remote Sens. Environ.* **2018**, *209*, 581–593. [CrossRef]
41. Mao, X.P.; Diao, J.J.; Fan, J.H.; Lui, Y.Y.; Xu, N.G.; Wang, Z.; Li, M.S. Analysis and prediction of landscape dynamics in the forest-grass mosaic zone of the Daxingan Mountains, Inner Mongolia. *J. Ecol.* **2021**, 1–12.
42. Xue, S.-S.; Gao, F.; He, B.; Yan, Z.-G. Analysis of landscape patterns and driving forces in the Wulungu River basin from 1989–2017. *Ecol. Sci.* **2021**, *40*, 33–41.
43. Cui, T.-X.; Gong, Z.-N.; Zhao, W.-J.; Zhao, Y.-L.; Lin, C. Extraction method of wetland vegetation cover under different end element models: An example from the Beijing Wild Duck Lake Wetland Nature Reserve. *J. Ecol.* **2013**, *33*, 1160–1171.
44. National Bureau of Statistics. *Statistical Yearbook of Jiangxi Province*; China Statistics Press: Beijing, China, 2002 2009 2020.
45. National Bureau of Statistics. *Statistical Yearbook of Guangdong Province*; China Statistics Press: Beijing, China, 2002 2009 2020.
46. Johnson, R.A.; Wichern, D.W. *Practical Multivariate Statistical Analysis*; Tsinghua University Press: Beijing, China, 2001.
47. Zou, T.; Yoshino, K. Environmental vulnerability evaluation using a spatial principal components approach in the Daxing'anling region, China. *Ecol. Indic.* **2017**, *78*, 405–415. [CrossRef]
48. Siegel, K.J.; Cabral, R.B.; McHenry, J.; Ojea, E.; Owashi, B.; Lester, S.l. Sovereign states in the Caribbean have lower social-ecological vulnerability to coral bleaching than overseas territories. *Proc. R. Soc. B* **2019**, *286*, 20182365. [CrossRef]

49. Guo, B.; Fan, Y.; Yang, F.; Jiang, L.; Yang, W.; Chen, S.; Gong, R.; Liang, T. Quantitative assessment model of ecological vulnerability of the Silk Road Economic Belt, China, utilizing remote sensing based on the partition-integration concept. *Geomat. Nat. Hazards Risk.* **2019**, *10*, 1346–1366. [CrossRef]
50. Zhou, S.; Tian, Y.; Liu, L. Adaptability of agricultural ecosystems in the hilly areas in Southern China a case study in Hengyang Basin. Adaptability of Agricultural Ecosystems in the Hilly Areas in Southern China: A Case Study in Hengyang Basin. *Acta Ecol. Sin.* **2015**, *35*, 1991–2002.
51. Dai, E.; Li, S.; Wu, Z.; Yan, H.; Zhao, D. Spatial pattern of net primary productivity and its relationship with climatic factors in Hilly Red Soil Region of southern China: A case study in Taihe county, Jiangxi province. *Geogr. Res.* **2015**, *34*, 1222–1234.
52. Yao, X.; Yu, K.; Liu, J.; Yang, S.; He, P.; Deng, Y.; Yu, X.; Chen, Z. Spatial and temporal changes of the ecological vulnerability in a serious soil erosion area, Southern China. *Chin. J. Appl. Ecol.* **2016**, *27*, 735–745.
53. Chen, Z.; Yao, X.; Yu, K.; Liu, J. Evolutionary Relation between Ecological Vulnerability and Soil Erosion in the Typical Reddish Soil Region of Southern China. *J. Southwest For. Univ. (Nat. Sci.)* **2017**, *37*, 82–90.
54. Tian, Y.; Liu, P.; Zheng, W. Vulnerability assessment and analysis of hilly area in Southern China: A case study in the Hengyang Basin. *Geogr. Res.* **2005**, *24*, 843–852.
55. Fan, S.; Guo, Y.; Qiu, L.; Jiang, C.; Huang, Y. Analyzing the effects of land cover change on urban ecological vulnerability in the central districts of Fuzhou city. *J. Fujian Norm. Univ. (Nat. Sci. Ed.)* **2018**, *34*, 92–98.
56. Guo, Z.C.; Wei, W.; Pang, S.F.; Zhen, Y.; Zhou, J.; Xie, B. Spatio-temporal evolution and motivation analysis of ecological vulnerability in arid inland river basin based on SPCA and remote sensing index: A case study on the Shiyang River Basin. *Acta Ecol. Sin.* **2019**, *39*, 2558–2572.
57. Yao, K.; Zhou, B.; Li, X.; He, L.; Li, Y. Evaluation of Ecological Environment Vulnerability in the Upper-Middle Reaches of Dadu River Basin Based on AHP-PCA Entropy Weight Model. *Res. Soil Water Conserv.* **2019**, *26*, 265–271.
58. Xie, Y.; Tu, X.; Wu, H.; Zhou, W.; Huang, B. Evolution of drought levels and impacts of main factors in the Dongjiang River basin. *J. Nat. Disasters* **2020**, *29*, 69–82.
59. Shi, X.; Wu, M.; Zhang, N. Characteristics of water use efficiency of typical terrestrial ecosystems in China and its response. *Trans. Chin. Soc. Agric. Eng.* **2020**, *36*, 152–159.
60. Abd El-Hamid, H.T.; Caiyong, W.; Hafiz, M.A.; Mustafa, E.K. Effects of land use/land cover and climatic change on the ecosystem of North Ningxia, China. *Arab. J. Geosci.* **2020**, *13*, 1099. [CrossRef]
61. Chen, J.; Luo, M.; Ma, R.; Zhou, H.; Zou, S.; Gan, Y. Nitrate distribution under the influence of seasonal hydrodynamic changes and human activities in Huixian karst wetland, South China. *J. Contam. Hydrol.* **2020**, *234*, 103700. [CrossRef]
62. Guo, B.; Kong, W.; Jiang, L.; Fan, Y. Analysis of spatial and temporal changes and its driving mechanism of ecological vulnerability of alpine ecosystem in Qinghai Tibet Plateau. *Ecol. Sci.* **2018**, *37*, 96–106.
63. Jiang, Y.; Li, R.; Shi, Y.; Guo, L. Natural and Political Determinants of Ecological Vulnerability in the Qinghai–Tibet Plateau: A Case Study of Shannan, China. *ISPRS Int. J. Geo-Inf.* **2021**, *10*, 327. [CrossRef]
64. Wang, J.; Xu, C. Geodetector:principle and prospective. *Acta Geogr. Sin.* **2017**, *72*, 116–134.
65. Zhou, B.; Li, F.; Xiao, H.; Hu, A.; Yan, L. Characteristics and climate explanation of spatial distribution and temporal variation of potential evapotranspiration in Headwaters of the Three Rivers. *J. Nat. Resour.* **2014**, *29*, 2068–2077.
66. Lin, K.; He, Y.; Lei, X.; Chen, X. Climate change and its impact on runoff during 1956–2009 in Dongjiang basin. *Ecol. Environ. Sci.* **2011**, *20*, 1783–1787.
67. Liang, S.; Bai, R.; Chen, X.; Chen, J.; Fan, W.; He, T.; Jia, K.; Jiang, B.; Jiang, L.; Jiao, Z.; et al. Review of China's land surface quantitative remote sensing development in 2019. *J. Remote Sens.* **2020**, *24*, 618–671.
68. Liu, J.; Zhao, J.; Shen, S.; Zhao, Y. Ecological vulnerability assessment of Qilian Mountains region based on SRP conceptual model. *Arid Land Geogr.* **2020**, *43*, 1573–1582.
69. Canadell, J.G.; Raupach, M.R. Managing forests for climate change mitigation. *Science* **2008**, *320*, 1456–1457. [CrossRef]
70. Gibson, L.; Lee, T.M.; Koh, L.P.; Brook, B.W.; Gardner, T.A.; Barlow, J.; Peres, C.A.; Bradshaw, C.J.; Laurance, W.F.; Lovejoy, T.E.; et al. Primary forests are irreplaceable for sustaining tropical biodiversity. *Nature* **2011**, *478*, 378–381. [CrossRef]
71. Taye, F.A.; Folkersen, M.V.; Fleming, C.M.; Buckwell, A.; Mackey, B.; Diwakar, K.C.; Le, D.; Hasan, S.; Saint Ange, C. The economic values of global forest ecosystem services: A meta-analysis. *Ecol. Econ.* **2021**, *189*, 107145. [CrossRef]
72. Jiang, P.; Ding, W.; Yuan, Y.; Ye, W.; Mu, Y. Interannual variability of vegetation sensitivity to climate in China. *J. Environ. Manag.* **2022**, *301*, 113768. [CrossRef]

remote sensing

Article

Managing Flood Hazard in a Complex Cross-Border Region Using Sentinel-1 SAR and Sentinel-2 Optical Data: A Case Study from Prut River Basin (NE Romania)

Cătălin I. Cîmpianu [1], Alin Mihu-Pintilie [2,*], Cristian C. Stoleriu [1], Andrei Urzică [1] and Elena Huțanu [1]

[1] Department of Geography, Faculty of Geography and Geology, "Alexandru Ioan Cuza" University of Iași, Bd. Carol I 20A, 700505 Iași, Romania; catalin.cimpianu90@gmail.com (C.I.C.); cristian.stoleriu@uaic.ro (C.C.S.); urzica.andrei94@gmail.com (A.U.); hutanu.elena@yahoo.com (E.H.)

[2] Department of Exact and Natural Sciences, Institute of Interdisciplinary Research, "Alexandru Ioan Cuza" University of Iași, St. Lascăr Catargi 54, 700107 Iași, Romania

* Correspondence: mihu.pintilie.alin@gmail.com or alin.mihu.pintilie@mail.uaic.ro; Tel.: +40-741-912-245

Citation: Cîmpianu, C.I.; Mihu -Pintilie, A.; Stoleriu, C.C.; Urzică, A.; Huțanu, E. Managing Flood Hazard in a Complex Cross-Border Region Using Sentinel-1 SAR and Sentinel-2 Optical Data: A Case Study from Prut River Basin (NE Romania). *Remote Sens.* **2021**, *13*, 4934. https://doi.org/ 10.3390/rs13234934

Academic Editor: Chiara Corbari

Received: 11 November 2021
Accepted: 2 December 2021
Published: 4 December 2021

Publisher's Note: MDPI stays neutral with regard to jurisdictional claims in published maps and institutional affiliations.

Abstract: In this study, an alternative solution for flood risk management in complex cross-border regions is presented. In these cases, due to different flood risk management legislative approaches, there is a lack of joint cooperation between the involved countries. As a main consequence, LiDAR-derived digital elevation models and accurate flood hazard maps obtained by means of hydrological and hydraulic modeling are missing or are incomplete. This is also the case for the Prut River, which acts as a natural boundary between European Union (EU) member Romania and non-EU countries Ukraine and Republic of Moldova. Here, flood hazard maps were developed under the European Floods Directive (2007/60/EC) only for the Romanian territory and only for the 1% exceeding probability (respectively floods that can occur once every 100 years). For this reason, in order to improve the flood hazard management in the area and consider all cross-border territories, a fully remote sensing approach was considered. Using open-source SAR Sentinel-1 and Sentinel-2 data characterized by an improved temporal resolution, we managed to capture the maximum spatial extent of a flood event that took place in the aforementioned river sector (middle Prut River course) during the 24 and 27 June 2020. Moreover, by means of flood frequency analysis, the development of a transboundary flood hazard map with an assigned probability, specific to the maximum flow rate recorded during the event, was realized.

Keywords: Sentinel-1 SAR and Sentinel-2 optical data; GIS; flood hazard map; cross-border area; temporal resolution

1. Introduction

When it comes to floods, Romania is among the most affected countries [1–5]. Starting at the beginning of 21st century, Romania witnessed flood events at increasingly high levels and frequencies, with floods occurring every two years (2005, 2006, 2008, 2010, 2013 and 2014) [4,6–9]. The negative consequences associated with the 40 major flood events recorded in Romania in the last 20 years (235,105 people affected, 10,731 people left homeless, 243 deaths, €2.6 billion worth of damage) ranks Romania as the second place in Europe after Russian Federation in terms of flood events incidence and people killed [10]. Many of these catastrophic flood events experienced by Romania in the last 20 years took place in the north-eastern part of the country, an area crossed by the catchments of Siret and Prut Rivers, some of the largest river basins in Romania [11]. In this area, historical flow rates for the entire Romanian territory were recorded: 4650 m^3/s in the Siret River in 2005—the maximum value ever recorded, and 4240 m^3/s on the Prut River in 2008—the second maximum value recorded for Romania [4]. The initial communicated flow rate for Prut River in 2008 was 7140 m^3/s, which would have been the maximum ever recorded for Romania, but the Stânca-Costeşti Reservoir located on the Prut River decreased the discharge value at Stânca-Costeşti hydrometric station to 1260 m^3/s [12,13].

With the accession of Romania to the European Union (EU) in 2007, the country was forced to align its flood risk management activities in compliance with the European Floods Directive (EFD) 2007/60/EC specifications [3,14]. Consequently, flood hazard maps (FHM), flood risk maps (FRM) and flood risk management plans (FRMP) were developed for all areas with potentially significant flood risk, identified within the country's main hydrographic basins and by the district administrations [6]. Although the process of flood hazard assessment and production of flood hazard maps in Romania was carried out in concordance with current scientific standards and state of the art technology, comprising mainly of advanced hydrological and hydraulic modeling along with LiDAR digital elevation models (DEMs) and collaboration with other EU members (e.g., Danube Flood Risk Project, Monitoring of extreme flood events in Romania and Hungary using Earth Observation data Project) [15], a series of issues regarding the elaboration of these products along the transboundary Prut River (natural border between EU country Romania and non-EU countries Ukraine and Republic of Moldova) can be addressed [9,16]. Here, even though Prut River was responsible for catastrophic flood events that affected all three countries [4,12,17,18], the flood hazard was quantified by using simplified Geographic Information System (GIS) procedures that involved the reconstruction of the water levels produced at certain events, only for the Romanian territory and only for the 1% probability. Moreover, according to the EU Member State Report: RO—Romania on Assessment of Flood Hazard and Flood Risk Maps prepared in December 2014, no transnational cooperation project regarding the development of flood hazard and risk maps (FHRM) was initiated or implemented.

This aspect was favored by a cumulus of obstacles which consisted of the overlay of the Prut River with the European Union eastern border and the existence of no harmonized legislation concerning flood risk management [19]. Due to these issues, even if the EFD regulates the collaboration between EU member states and third-party countries in terms of effective flood prevention and mitigation, in practice, a cross-boundary cooperation it is difficult to implement [20]. In this particular case, one of the main disadvantages is represented by the fact that terrestrial laser scans and acquisition of LiDAR digital elevation models (essential data for hydraulic modeling techniques and elaboration of probabilistic flood hazard maps) cannot be performed at large scales. The existence of LiDAR data only for the Romanian territory is not able to offer an ample perspective regarding the flood hazard in the area along Prut River, the flood hazard maps developed here may underestimate the flood impact. However, despite all the mentioned impediments, a joint operational program between Romania, Ukraine and Republic of Moldova was implemented and completed in 2017. Following this project, called EASTAVERT PROJECT "The prevention and protection against floods in the upper Siret and Prut River Basins, through the implementation of a modern monitoring system with automatic stations", significant progress has been made in terms of flood hazard and risk mapping within the transboundary area of Prut River.

Usually, the development of flood hazard maps involves hydrodynamic models, based on a series of data that describe statistical river discharge or rainfall values and high precision DEMs (e.g., LiDAR), along with remotely sensed maps of flood extents for calibration and validation purposes [8,21–24]. In the absence of long-term hydrometric monitoring data and high precision DEMs, flood hazard assessment and flood hazard maps can be realized on the basis of remote sensing data [25–27]. Flood probability maps can be obtained by correlating satellite images acquired regularly throughout the stages of a flood event, and flow rates values recorded at hydrometric stations located preferably upstream of the area affected by the flood [28]. In this context, one emergent solution to manage the flood hazard in complex cross-border regions could rely on the use of synthetic aperture radar (SAR) imagery [29,30]. Due to its active radar which operates day and night in all weather capabilities, this type of data is invaluable for uninterrupted monitoring and mapping of flood events [31–34]. Flood extent maps derived from SAR data can support the elaboration of flood hazard maps and therefore specific flood risk management measures and activities [35]. Until recently, the use of SAR data in similar applications was rather limited, due to the high costs and the complexity of their interpretation by non-expert users [36]. The launch of European Space Agency (ESA) Copernicus Program and its particular SAR mission Sentinel-1 in April 2014 brought up new opportunities in terms of near real-time flood disaster monitoring, flood extent mapping and emergency management [37–39]. Given the new improved capabilities in terms of global coverage, free availability, 3–6 days revisit time and 10 m spatial resolution, Sentinel-1 SAR imagery proved its efficiency in numerous disaster-related studies that focused on similar topics [40–44]. Moreover, the practice of using Sentinel-1 SAR data in combination with cloud-free ESA Sentinel-2 optical imagery (when available) can lead to more acurate flood hazard and flood risk analysis products, such as spatio-temporal inundation maps [45].

This paper aims to delineate the maximum inundation extent of a flood event that took place over a four day period (from 24 to 27 June 2020) in a complex cross-border region of the Prut River, which acts as a natural boundary between EU country Romania and non-EU countries Ukraine and Republic of Moldova. The main objectives consist of: (1) the exploration of Sentinel-1 SAR and Sentinel-2 optical data capabilities to capture all flood development stages, considering the image availability and repetitiveness (temporal resolution); (2) the development of quick and efficient methodologies to delineate flooded areas within Sentinel-1 SAR and Sentinel-2 optical images; (3) the development of a flood hazard map with an assigned probability for the entire study area which can offer an integrated perspective without any cross-border related limitations.

2. Study Area

The Prut River basin (total surface: 27,500 km^2; total hydrological network length: 11,000 km) is located in the northeastern sector of Danube basin and overlaps with the territory of three countries: Ukraine (28%–7700 km^2); Romania (39%–10,725 km^2); and Republic of Moldova (33%–9075 km^2) [4,18,46] (Figure 1a). It springs from the Ukrainian Wooded Carpathians at an altitude of 1580 m a.s.l. and flows into the Danube River near the Romanian village of Giurgiulești at an altitude of 2 m a.s.l. [12,18]. As the last major tributary of Danube River, it represents the natural border between Ukraine and Romania (31 km border length) and between Romania and Republic of Moldova (711 km border length) [3,4,9,12].

From a morphometric point of view, the Prut River has three sectors:

- Upper watercourse sector: from springs to Cernăuți where the river flows out of the mountain region. This river sector presents itself as a typical mountain river, with a narrow and deep valley, medium slopes (4.5°) and steep banks. The riverbed has a width of 50–70 m, depth 0.5–1.5 m and rate flow of 1.0–1.5 m/s [13];
- Middle watercourse sector: from Cernăuți to Ungheni (plain sector) with a floodplain width of 5–6 km with low banks and small slopes (0.1–0.2°). The riverbed has a width of 50–85 m, a depth of 2–3 m and a flow rate of 0.6–1.0 m/s [13];
- Lower watercourse sector: between Ungheni and the confluence with Danube River. The lower course is characterized by small slopes (0.08–0.1°), wide floodplain (10–12 km) and low flow rate (0.5–0.8 m/s). The river has a width of 60–100 m and a depth of 2–4 m [13].

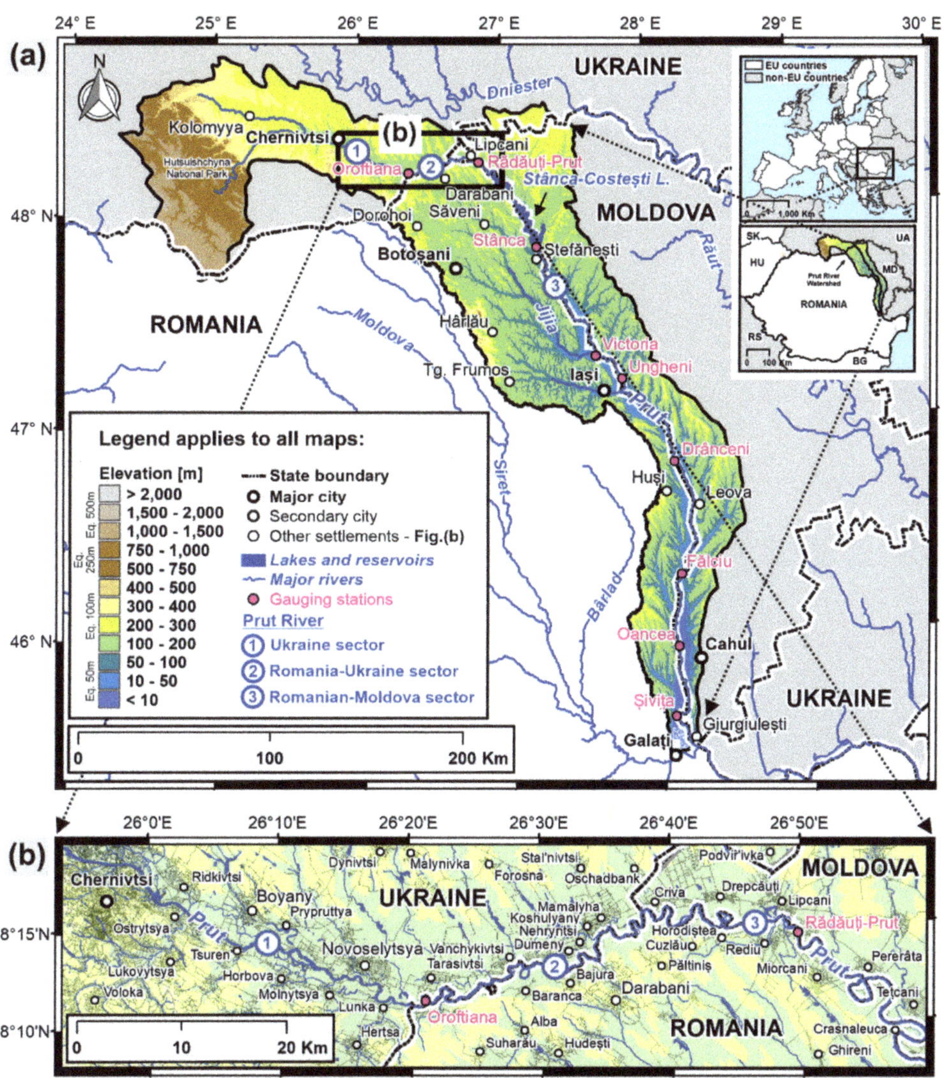

Figure 1. Geographical location of the (**a**) Prut River basin and (**b**) the study area found between Ukraine, Romania and the Republic of Moldova.

Due to the lack of a legislation to ensure a transparent cooperation between these three countries, we obtained hydrological data just for the Romanian sector of Prut River (the common border with Ukraine and Republic of Moldova) [3] (Figure 1b). Therefore, within the Romanian territory, Prut River has nine gauging stations from the entry of Prut River to Prut-Danube confluence as follows: Oroftiana (included only water level measurements), Rădăuți-Prut, Stânca Aval, Ungheni, Prisăcani, Drânceni, Fălciu, Oancea and Șivița [3,4,12] (Table 1). The first installed gauging station was at Ungheni (1914) and the newest at Șivița (1978) [12]. On Ukrainian territory, one gauging station was installed in Cernăuți city that included water level measurements (as in the case of Oroftiana gauging station) [12].

Table 1. The main characteristics of the Prut River gauging stations located on the Romanian territory (see Figure 1a).

Gauge Station	Year of Inauguration	Elevation (m a.s.l.)	Latitude	Longitude	Max. Water Level Recorded (cm)	Max. Flow Rate Recorded (m³/s)	Date of Max. Flow Rate
Oroftiana [1]	1976	123.47	48°11′12″	26°21′04″	876	-	-
Rădăuți-Prut	1976	101.87	48°14′55″	26°48′14″	1130	4240	28 July 2008
Stânca Aval	1978	62.00	47°47′00″	27°16′00″	512	1050	31 July 2008
Ungheni	1914	31.41	47°11′04″	27°48′28″	654	796	8–10 July 2010
Prisăcani	1976	28.08	47°05′19″	27°53′38″	622	900	9–10 July 2010
Drânceni	1915	18.65	46°48′45″	28°08′04″	718	736	17–18 July 2010
Fălciu	1927	10.04	46°18′52″	28°09′13″	650	722	19 July 2010
Oancea	1928	6.30	45°53′37″	28°03′04″	622	757	24 April 1979
Șivița [2]	1978	1.66	45°37′10″	28°05′23″	-	-	-

[1] Oroftiana gauge station was designed for measuring only the water level. [2] The Prut flow rate and water level at Șivița gauge station were directly influenced by Danube waters.

3. Database and Methodology

The Sentinel-1 SAR and Sentinel 2 data were obtained for the upper area within Cernăuți–Ungheni sector, from Cernăuți city (Ukraine) to Cotu Miculinți village (Romania) —Tețcani village (Republic of Moldova) (Figure 2). Along this river sector are located 26 Ukrainian settlements (of which two cities) with 323,788 inhabitants, 11 Romanian settlements (of which one city) with 20,059 inhabitants and seven Moldavian settlements (one city) with 13,406 inhabitants. The water level and flow rate for this sector was monitored by two of nine gauging stations: Oroftiana and Rădăuți-Prut gauging station (Figure 1b, Table 1).

3.1. Sentinel-1 SAR and Sentinel-2 Data

Synthetic Aperture Radar (SAR) is an imaging active radar system mounted on a moving platform that operates independently of Sun illumination and cloud coverage [47]. Such systems make use of an antenna to generate, transmit and collect electromagnetic pulses and backscattered echoes, respectively, in order to acquire information about Earth's features and thus create SAR images [48]. ESA Sentinel-1 mission consists of a constellation of two polar-orbiting satellites (Sentinel-1A and Sentinel-1B), equipped with a C-band (wavelengths [λ] = 5.6 cm) active SAR sensor, which facilitates the acquisition of imagery regardless of day, night or weather conditions every 6 days or even 1–3 days for areas such as Europe [49–51]. The data were acquired under different configurations in four different modes: Stripmap (SM), Interferometric Wide swath (IW), Extra-Wide swath (EW), Wave (WV); and distributed freely at various product levels (Level-0, Level-1, and Level-2), each with specific resolutions, polarizations (single polarization HH or VV and dual polarization HH+HV or VV+VH) and extents [39,52].

Level-1 data are the generally available products intended for most data users and applications. Level-1 products are distributed as Single Look Complex (SLC) and Ground Range Detected (GRD) [53,54]. SLC data consist of focused complex imagery that uses the full C signal bandwidth, having amplitude and phase information preserved [55]. At the same time, an SLC dataset provides the highest possible spatial resolution [56,57]. Level-1 Ground Range Detected (GRD) comprises of focused SAR data that has been detected, multi-looked and projected to ground range using an Earth ellipsoid model, such as WGS84, storing only amplitude (pixel intensity) without any remaining phase information [44]. The information recorded in both Level-1 product types lies in the measured echoes of the backscattered signal at the C-band wavelength [58]. The backscattering coefficient, or sigma nought (σ^0), is the measure of the incident microwave radiation scattered by the radiated terrain. The amplitude and phase of the backscattered signal depends on the land surface properties of terrain elements and their electromagnetic characteristics (e.g., materials) [59,60]. One of the main advantages of using SAR images in flood mapping applications is that they show a high contrast between ground and water surfaces [61]. Smooth open water or flooded areas act as ideal specular reflectors and are characterized by low SAR backscatter values [43,51]. Therefore, in SAR imagery, the water covered surfaces appear black [62].

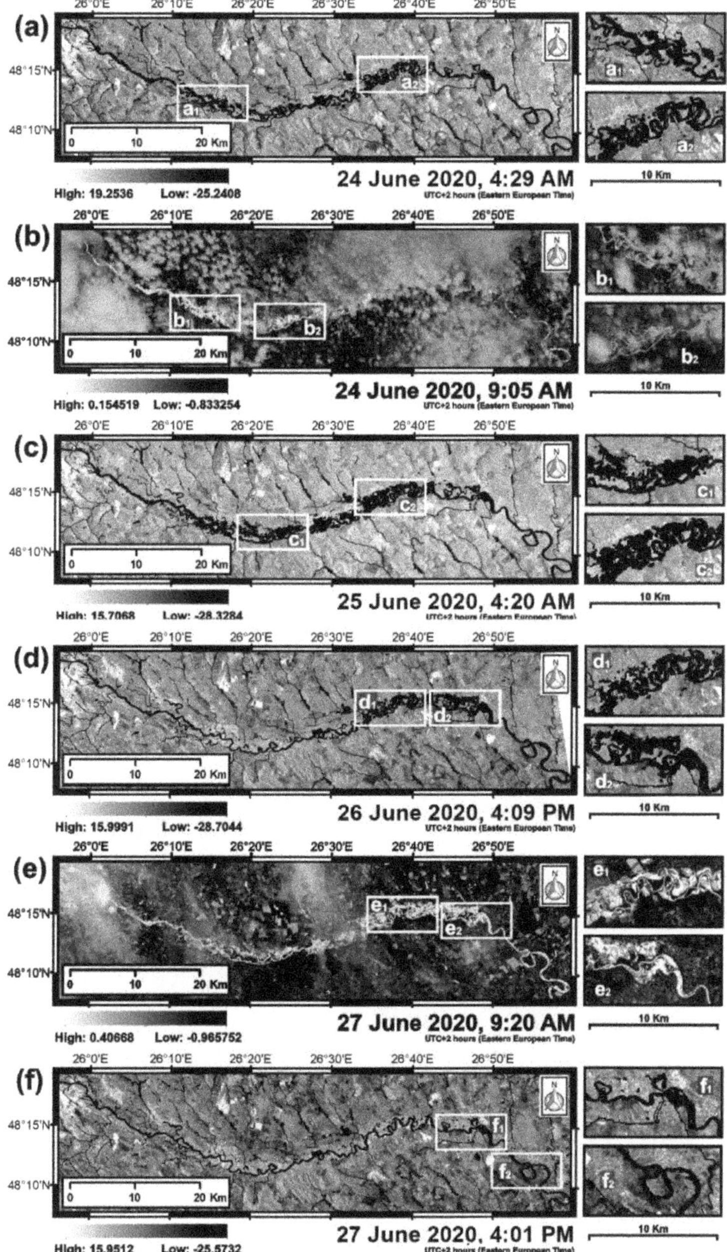

Figure 2. Sentinel-1 SAR (**a,c,d,f**) and Sentinel-2 optical (**b,e**) ground-projected imagery collected over the study area (see Figure 1b) during the flood event that took place between 24 and 27 June 2020.

Sentinel-2 consists of a constelation of two satellites (Sentinel-2A launched in June 2015 and Sentinel-2B launched in March 2017), equipped with a multi-spectral instrument (MSI). The multispectral sensor aquires optical imagery in 13 spectral bands from VIS/NIR to SWIR at a spatial resolution ranging from 10 to 60 m and at a high revisit time (5 days at the equator with two satellites and every 2–3 days at mid-latitudes.) The data are freely available under two distinct product levels (Level-1C and Level-2A) [63,64]. Level-1C data are distributed as Top of the Atmosphere (ToA) reflectance in cartographic geometry (100×100 km^2 orthoimages found in UTM/WGS84 projection). Level-2A data are based on Level-1C data and are atmospherically corrected to Bottom of Atmosphere (BoA) reflectance [65,66].

For the purpose of this study, freely available GRD dual-polarized (VV/VH) Sentinel-1 SAR images acquired by the sensor under interferometric wide-swath (IW) mode (the default acquisition mode over land) at a pixel spacing of 10 m and a swath of 250 km were downloaded using the Copernicus Open Access Hub [42,67–70]. Level-2A Sentinel-2 atmospherically corrected products were downloaded using the QGIS application and the semi-automatic classification plugin [71].

Due to their specific characteristics (especially the improved temporal resolution), the Sentinel-1 and Sentinel-2 imagery obtained managed to depict all development stages of a flood event that took place on a shared Prut River sector located between Ukraine, Romania and Republic of Moldova during 24 to 27 June 2020 (Figure 2). This aspect allowed near real-time flood monitoring as the images were available for download in less than 24 h since acquisition by sensor. A characterization of the images used is listed in Table 2.

Table 2. Characteristics of the Sentinel-1 SAR and Sentinel-2 optical images used in the study (see Figure 2).

ID	Image Identifier	Satellite	Date Acquired/Time	Product Type	Mode	Polarization
A	S1A_IW_GRDH_1SDV_20200624T042904_20200624T042929_033154_03D73C_540B	S1A	2020-06-24/T04:29:04.880Z	GRD	IW	VV VH
B	S2B_MSIL2A_20200624T090559_N0214_R050_T35UMP_20200624T123246	S2B	2020-06-24/T09:05:59.024Z	LEVEL-2A	-	-
C	S1B_IW_GRDH_1SDV_20200625T042007_20200625T042036_022185_02A1B4_C73E	S1B	2020-06-25/T04:20:07.597Z	GRD	IW	VV VH
D	S1B_IW_GRDH_1SDV_20200626T160915_20200626T160940_022207_02A256_2724	S1B	2020-06-26/T16:09:15.797Z	GRD	IW	VV VH
E	S2B_MSIL2A_20200627T092029_N0214_R093_T35UMP_20200627T121756	S2B	2020-06-27/T09:20:29.024Z	LEVEL-2A	-	-
F	S1A_IW_GRDH_1SDV_20200627T160148_20200627T160213_033205_03D8C4_F195	S1A	2020-06-27/T16:01:48.625Z	GRD	IW	VV VH

3.2. SAR and Optical Data Pre-Processing

The GRD Sentinel-1 images downloaded were pre-processed using Sentinel Application Platform (SNAP) software, an ESA open-source architecture [72,73]. The pre-processing steps were based on a workflow proposed by Filipponi [74] and consisted mainly of (Figure 3a): data preparation, area and band subset, image calibration, speckle filtering, terrain correction and data conversion to dB.

In the first step, in order to reduce the processing time and memory use, the data were sub-sampled according to study area limit and polarization. Next, the precise orbit file, thermal and border noise were respectively applied and removed [37,75]. Even if both VH and VV polarized Sentinel-1 data can be employed in flood monitoring and mapping research [59,76], for the purpose of this study the VV polarization was preferred over VH polarization, as it is more suitable for delimiting flood waters within mixed, forest and agricultural land areas [35,50,77].

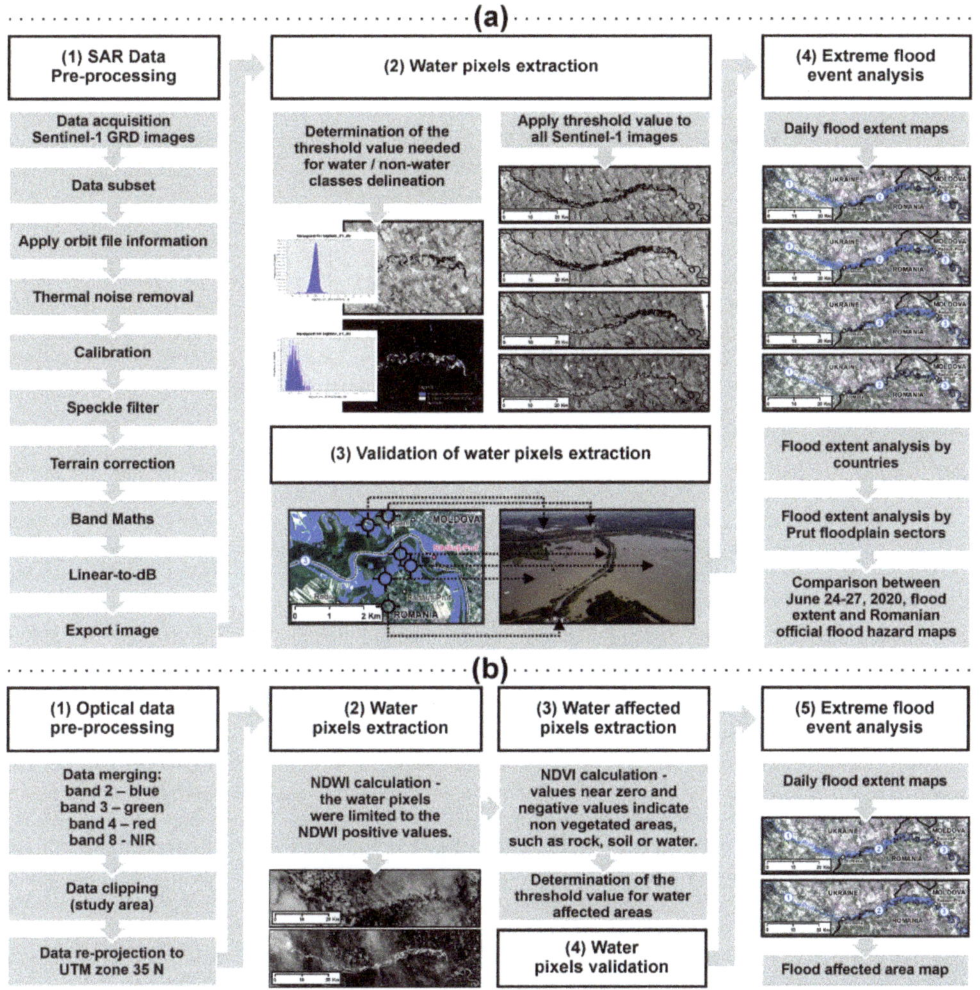

Figure 3. Workflow chart of the study including: (**a**) SAR data processing: step (1) Sentinel-1 SAR data pre-processing; step (2) water pixel extraction for flood extent delineation; step (3) validation of water pixels extraction using vector overlay analysis and aerial images; step (4) flood extent analysis. (**b**) Optical data processing: step (1) Sentinel-2 optical data pre-processing; step (2) water pixels extraction (NDWI calculation); step (3) water affected pixels extraction (NDVI calculation); step (4) water pixels validation (see step a3); step (5) extreme flood event analysis.

Image calibration consisted of the conversion of raw DN values to radiometrically calibrated SAR backscatter or sigma nought values (σ^0) [78]. Speckle-filtering implies a procedure to increase image quality by reducing speckle noise. In this context, many speckle filters based on smoothing windows on a weighted summation of neighboring pixels were developed [77]. In this study, a Lee-Sigma filter with a window kernel size 5×5 pixels was applied [79]. Terrain corrections were performed in order to reduce the distortions determined by the varying viewing angles involved in the image acquisition process [80]. The SAR images underwent range Doppler terrain correction using the Shuttle Radar Topography Mission (SRTM) (1 arc second) DEM as a reference. During this operation, the images were corrected for topographical distortions determined by the side looking SAR mechanism, geolocated and re-projected to Universal Transverse Mercator (UTM) zone 35 N, with a 10 m ground sampling [81].

In order to ensure that all four datasets considered in this study were spatially aligned, an image co-registration was performed. This was carried out using the co-registration tool found in SNAP software and the Sentinel-1 image acquired on 24 June 2020 as a reference. In the last step, the calibrated backscatter coefficient values were converted to dB logarithmic units and exported for further water and non-water pixel delineation operations [76,82].

Level-2A Sentinel-2 imagery was downloaded using the the QGIS application and the semi-automatic classification plugin as Bottom of Atmosphere (BoA) reflectance and its pre-processing consisted of: data merging using only the 10 m spatial resolution spectral bands (band 2—blue, band 3—green, band 4—red and band 8—NIR), study area data clipping and data re-projection to Universal Transverse Mercator (UTM) zone 35 N [63]. To spatially match the Sentinel-2 optical images with the Sentinel-1 SAR images, a co-registration step was also performed using the SNAP application [83] (Figure 3b).

3.3. Water Pixel Extraction from SAR and Optical Data

Various methods have been developed for the detection of water surfaces and thus for flood mapping using radar SAR images [29,84]. Among them, considering water's well-defined backscattering signature, backscatter histogram thresholding is the most widely used [85,86]. The threshold method can be used to differentiate between the water pixels, represented by low values from the no-water pixels characterized by high values within the image histogram [87].

Thresholding implies SAR backscatter intensity histogram analysis and the estimation of water and non-water pixel distribution. This method is based on histogram manipulation and on the distinction of two partially overlapping distributions that correspond to water and non-water pixels [88,89]. The intersection point between wet and dry distributions determines the threshold value needed for the two classes separation (i.e., open water surfaces and non-water surfaces) [62,90]. For Sentinel-1 SAR images considered in the present study, the extraction of the two distinct classes was conducted based on a threshold value which was initially obtained and then applied to all image's histograms (Figure 3).

The threshold value was determined by analyzing the backscatter coefficient values, corresponding to a series of points manually digitized by means of visual interpretation, along the Prut River course and within the most visible flooded areas. Subsequently, for each Sentinel-1 image, a different set of points was acquired in areas that overlap with low intensity dark pixels that represent water. To make sure that the point locations coincide exclusively with water surfaces, a cross-examination with the cloud-free areas of two Sentinel-2 multispectral images dating from 24 to 27 June 2020 was performed. In this way, the extraction of the matching raster pixel values set an interval for the true water areas found within each image. The interval highest values that were equal to -15 dB (Image A), -16 dB (Image B), -17 dB (Image C) and -15 dB (Image D), established the threshold between water and land classes. Consequently, all values below these thresholds were considered representative for the water class and respectively for the flooded areas. Next, a binary geotiff layer showing the water and dry areas was obtained for all four images.

For data integration purposes, a shapefile vector format conversion was also realized. The detected water class consists of flooded areas and permanent water bodies. To highlight only the flooded areas, it was necessary to subtract the permanent water bodies from the initial water class dataset [91]. In this case, the permanent water bodies acquired from OpenStreetMap were used. The final step consisted of the export of four distinct datasets regarding the flooded areas, one dataset for each corresponding day of the flood (Figure 3).

The water pixel extraction from Sentinel-2 optical data was realized by means of Normalized Difference Water Index (NDWI) spectral index calculation and simple thresh-olding. The McFeeters NDWI [92] maximizes the reflectance of the water bodies in the green band and minimizes it in the NIR band [93,94]. NDWI was calculated according to the Equation (1):

$$NDWI = \frac{\rho Green - \rho NIR}{\rho Green + \rho NIR} \tag{1}$$

Considering this multispectral advantage, the NDWI is a simple and effective method to extract water bodies [95]. The NDWI is a dimensionless quantity which varies between -1.0 and 1.0. The positive NDWI values indicate the presence of water [96,97]. In this study, the water pixels were limited to the NDWI positive values. The flooded areas for the two Sentinel-2 datasets were delimited by permanent water bodies' subtraction. The final result consisted of two layers, which illustrate the flooded areas within the cloud-free regions of the Sentinel-2 satellite images.

3.4. Water-Affected Pixel Extraction from Optical Data

The identification of the flooded areas can be realized even after the water has already withdrawn, by analyzing the flood's effects on the soil [98]. Using the Sentinel-2 image acquired on 27 June 2020 (on the last day of the flood), the extraction of the flood affected pixels was possible. By means of Normalized Difference Vegetation Index (NDVI) and simple thresholding, we managed to explore the high contrast between the bare soil flood affected areas and the surrounding high vegetated agricultural land. NDVI was calculated according to the Equation (2):

$$NDVI = \frac{\rho NIR - \rho Red}{\rho NIR + \rho Red} \tag{2}$$

NDVI values range between -1.0 and $+1.0$. The NDVI values near zero and negative values indicate nonvegetated areas, such as rock, soil or water [99]. In this study, the water-affected areas were restricted through visual interpretation of the NDVI product and Sentinel-2 natural color composite image to values lower than 0.35. The threshold value was determined based on a set of points manually digitized in the visible flood-affected areas. As the result also included the areas covered with water, a Sentinel-1/Sentinel-2 combined flooded area from 27 June 2020 and permanent water bodies subtraction was realized. The final result consisted of a layer which depicted all the flood-affected areas from 24 to 27 June 2020.

3.5. Water Pixel Validation

The subtracted flooded areas and water-affected areas underwent a validation process in order to confirm the proposed methodologies and the determined threshold values. The validation for the Sentinel-1 extracted flood areas was realized by vector overlay analysis (Figure 3). Two vector datasets delineating the flooded areas on 26 June 2020 within the study area, one obtained by the proposed methodology and the other one offered by Copernicus Mapping Emergency Management Service (CMEMS-M), EMSR445 activation code product [100], were compared. The analysis revealed an overlay degree of over 95%, which was considered satisfactory for validation purposes of the Sentinel-1 water pixel extraction methodology adopted in the present study. Moreover, an aerial image acquired in the proximity of Rădăuți-Prut cross-border point between Romania and Republic of Moldova, which depicted the same flood event was also used for data visual interpretation and confirmation (Figure 4). The validation for Sentinel-2 flooded areas and water-affected areas was realized by means of visual interpretation.

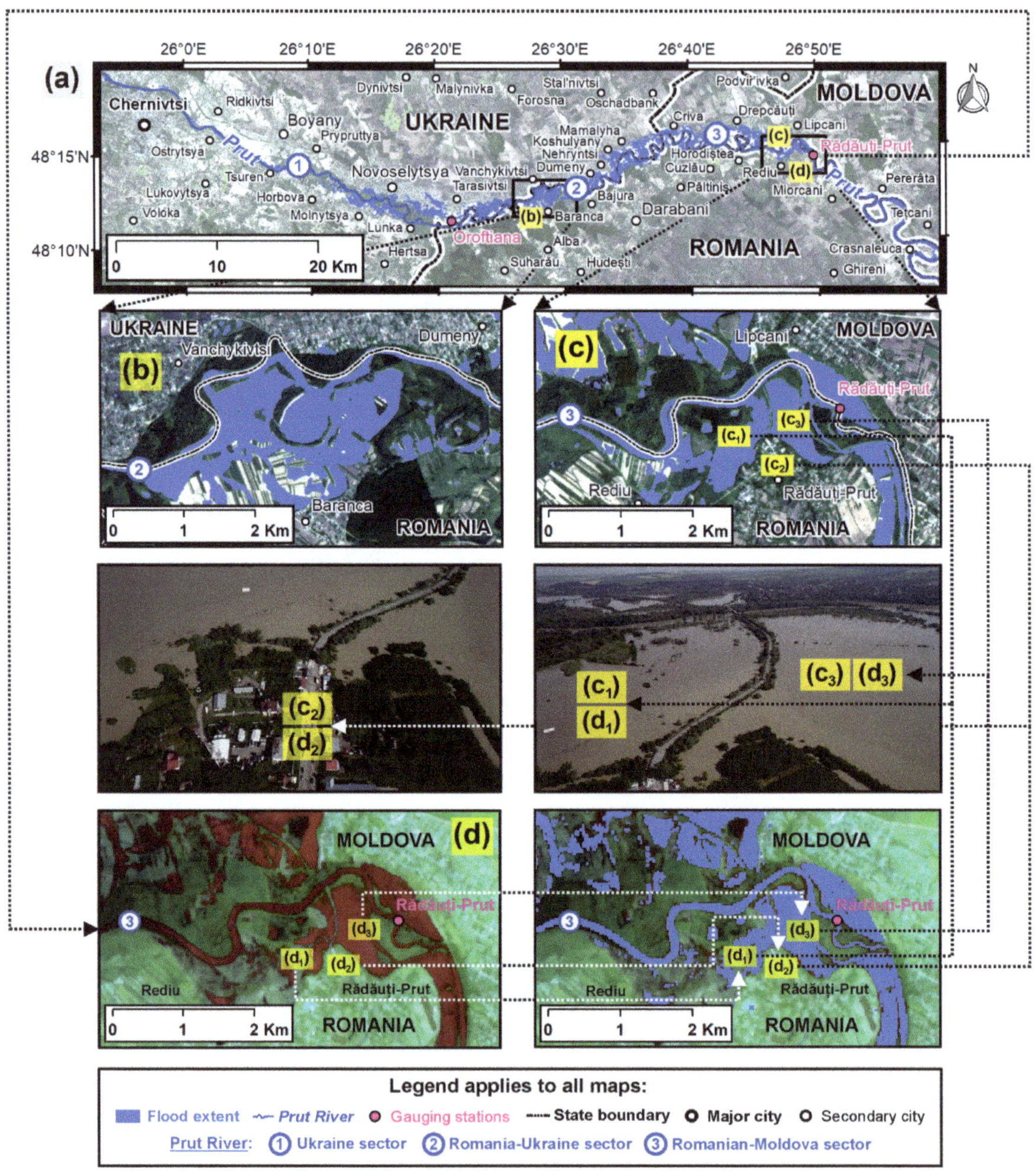

Figure 4. Validation of water pixel extracted using vector, satellite and aerial images overlay analysis: (**a**) 26 June flood extent in the study area, highlighted over (**b**) Baranca (Romania)—Vanchykivtsi (Ukraine) and (**c**) Rădăuți-Prut (Romania)—Lipcani (Republic of Moldova) areas—flooded area extracted from Sentinel-1 SAR images; (**d**) 27 June flood extent highlighted over Rădăuți-Prut (Romania)—Lipcani (Republic of Moldova) area—flooded area extracted from Sentinel-2 optical images.

3.6. Flood Frequency Analysis

In order to estimate the return periods of the flood events, their corresponding probabilities and probable discharges at Rădăuți-Prut gauging station, we considered the Gumbel's distribution method, first introduced in 1941 [101]. Gumbel probability distribution [102] is expressed as Equation (3):

$$X_T = X + K\sigma_x, \tag{3}$$

where, X_T is the probable discharge with a return period of T years (the magnitude of floods); X is the mean flood; σ_x is the standard deviation of the maximum annual flow rates; K is the frequency factor, expressed as Equation (4):

$$K = \frac{Y_T - Y_n}{S_n}, \tag{4}$$

where, Y_n is the expected mean and S_n is the expected standard deviations of reduced extremes found within Gumbel's table, depending on the sample size, and Y_T is the reduced variate, expressed as Equation (5):

$$Y_T = -\left[LnLn\left(\frac{T}{T-1} \right) \right], \tag{5}$$

This is one of the most common distribution functions used to predict flood return periods and their probabilities based on time series discharge data [103,104].

4. Results and Discussion

4.1. Flood Hazard Map

All seven datasets corresponding to the flooded areas and flooded-affected areas were merged into one. In this way, a new flood extent map derived from satellite images was obtained (Figure 5).

As the map developed managed to capture the maximum flood extent, the conversion to a flood hazard map with an assigned probability was taken into consideration. To achieve this, a return period was established for the maximum flow rate of 2965 m^3/s recorded at Rădăuți-Prut gauging station on 26 June 2020 at 10:00 AM (considering the whole flood event). This value was the only one considered representative for the flood wave, as at Oroftiana gauging station this type of data is not available. In order to carry out this statistical operation within the study area, a sample of the maximum annual flow rates recorded at Rădăuți-Prut gauging station in the last 42 years (from 1978 to 2019) was analyzed.

Table 3. Flood frequencies estimated according to Gumbel's distribution method at Rădăuți-Prut gauge station on the Prut River (see Figure 1b).

Return Period (T) in Years	Probability of Occurrence (%)	Reduced Variate (Yt)	Frequency Factor (K)	Computed Flood Discharges (X_T) (m^3/s)
1000	0.1	6.907255	5.5528495	5191.5417
100	1	4.600149	3.539316	3670.2634
33.3	3	3.491366	2.571624	2939.1456
20	5	2.970195	2.116770	2595.4911
10	10	2.250367	1.488538	2120.8451
5	20	1.499939	0.833600	1626.0222

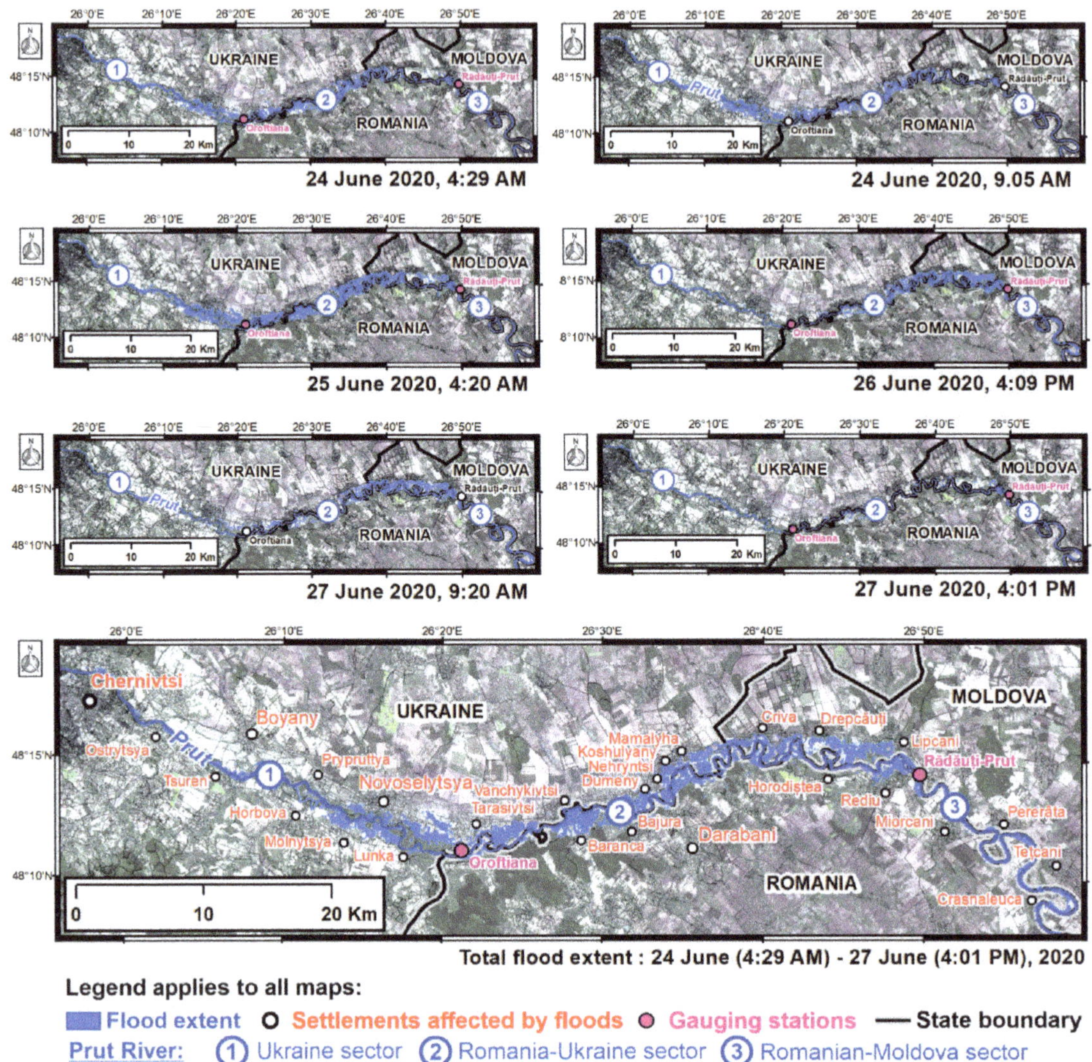

Figure 5. Daily flood extents derived from Sentinel-1 SAR (24 June, 4:29 AM; 25 June, 4:20 AM; 26 June, 4:09 PM; 27 June, 4:01 PM) and Sentinel-2 optical data (24 June, 9:05 AM; 27 June, 9:20 AM), and the total flood extent later converted to a flood hazard map by means of flood frequency analysis (see Table 3).

The flood frequency analysis consisted of: (1) the computation of the mean flood X and standard deviation σ_x considering the maximum annual flow rates (in our case these values were equal to: $X = 996.21$ and $\sigma_x = 755.52$); (2) selection of the corresponding values of Y_n and S_n from Gumbel's table (in our case, given the 42 year sample size, the values were 0.5448 and 1.1458, respectively); (3) depending on the given return periods T, the reduced variate Y_T was calculated using Equation (5); (4) the calculation of the flood frequency factor K using Equation (4); (5) the probable flood discharges corresponding to different return periods and probabilities at Rădăuți-Prut gauging station were computed according to Equation (3). Detailed information about the computational process can be found in Table 3.

Considering the computed flood discharges (Table 3) and the maximum flow rate of 2965 m^3/s registered at Rădăuți-Prut gauging, a probability of occurrence and a return period was determined for the flood event that took place between 24 and 27 June 2020, on the middle course of Prut River. Consequently, this led to the conversion of the flood extent map derived initially from Sentinel-1 data into a flood hazard map with an assigned probability of 2.97% and a return period of 33.67 years (Figure 5).

4.2. Flood Wave Development Stages Based of Sentinel-1 SAR Satellite Images

4.2.1. Flood Status on 24 June 2020

On the first day of the flood, the most affected areas were located within the Ukrainian territory and corresponding Prut River sector (Figure 6). Here, during this day, 1609 ha and 1091 ha, respectively, were covered by water. At the same time, the flood wave started spreading across Romania, flooding 951 ha and the Republic of Moldova, where 303 ha were flooded. Meanwhile, the figures within the Romanian–Ukrainian and Romanian–Moldavian shared river sectors stood at 1003 ha and 772 ha. At 4:00 AM (the approximate time of image acquisition), the flow rate registered at Rădăuți-Prut gauging station was 990 m^3/s (probability of occurrence 32.84%). The maximum value of the day, 1585 m^3/s was recorded at 23:00 PM (probability of occurrence 20.51%) (Table 4).

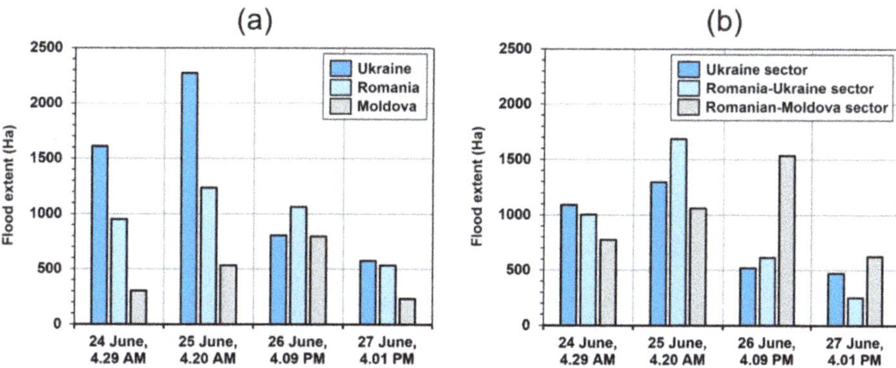

Figure 6. Daily flood status within the study area considering the Sentinel-1 data and (**a**) Ukraine, Romania and Republic of Moldova territories; (**b**) the shared Prut River sectors (see Figure 1b).

Table 4. The maximum flow rates recorded during the flood event at the Rădăuți-Prut gauge station and the corresponding probability of occurrence estimated using the Gumbel's method.

Date	Flow Rate (m^3/s)/Time of Image Acquisition	Probability of Occurrence (%)/Time of Image Acquisition	Maximum Flow Rate (m^3/s) of the Day	Maximum Probability of Occurrence (%) of the Day
24 June 2020	990/4:29 AM	32.84/4:29 AM	1585	20.51
25 June 2020	1705/4:20 AM	19.07/4:20 AM	2610	4.97
26 June 2020	2860/4:09 PM	3.08/4:09 PM	2965	2.97
27 June 2020	2410/4:01 PM	5.38/4:01 PM	-	-

4.2.2. Flood Status on 25 June 2020

On 25 June, the situation escalated and the flood reached its peak. The most affected country was Ukraine with 2275 ha covered by water, followed by Romania with 1234 ha and Republic of Moldova with 532 ha. Regarding the three river sectors considered, the Romanian–Ukrainian was the most affected (1686 ha). On this day, the total area occupied by water reached 4042 ha (Figure 6). The flow rate recorded at 4:00 AM (the approximate time of image acquisition) at Rădăuți-Prut was 1705 m^3/s (probability of occurrence 19.07%). The maximum value of the day, 2610 m^3/s, was recorded at 23:00 PM (probability of occurrence 4.97%) (Table 4).

4.2.3. Flood Status on 26 June 2020

Starting on 26 June, the flood wave diminished over Ukrainian territory and shared Romanian–Ukrainian river sector. The flood that considerably affected Ukraine on 24 and 25 June decreased by half in terms of affected areas and remained constant over Romania. During this day, the flood waters began to cover territories located within the Republic of Moldova and Romania, the shared river sector being the most affected (Figure 6). The flow rate recorded at 4:00 AM (the approximate time of image acquisition) at Rădăuți-Prut was 2860 m^3/s (probability of occurrence 3.08%). The maximum value of the day and also over the entire duration of the flood, 2965 m^3/s, was recorded at 10:00 PM (probability of occurrence 2.97%) (Table 4).

4.2.4. Flood Status on 27 June 2020

On the last day of the flood event, few territories were still covered by water. As the flood wave followed its course, the downstream Romanian–Moldavian sector was the most affected (623 ha) (Figure 6). At 4:00 AM, the flow rate registered at Rădăuți-Prut had a value of 2410 m^3/s (probability of occurrence 5.38%). Starting on 28 June 2020, the flow rates dropped below 1000 m^3/s marking the end of the flood event (Table 4).

4.3. The Sentinel-1 SAR and Sentinel-2 Optical Data: Applicability and Limitations for Flood Assessment

In this study, an alternative solution for flood hazard map development in complex cross-border regions is presented. In this cases, due to different flood risk management legislative approaches, there is a lack of joint cooperation between the involved countries. As a main consequence, LiDAR high resolution digital elevation models and accurate flood hazard maps obtained by means of hydrological and hydraulic modeling are missing or are incomplete. This is also the case of Prut River middle course, which acts as a natural boundary between Romania and non-EU countries Ukraine and Republic of Moldova. Here, flood hazard maps were developed under EFD by the National Administration "Romanian Waters", only for the 1% exceeding probability (floods that can occur once at every 100 years), using simplified Geographic Information System (GIS) procedures. For this reason, in order to improve the flood hazard management in the area, alternative solutions were considered. Consequently, such an alternative could rely on the use of Sentinel-1 SAR and Sentinel-2 optical imagery.

The launch of the Sentinel-1 mission in 2014 has significantly improved flood mapping capabilities mainly due to: the mission active radar sensor (SAR) that captures images both day and night (regardless of weather and its characteristics—clouds, fog, aerosols), high spatial resolution (10 m) and also due to the ability to acquire images at shorter intervals of time, even on consecutive days. Therefore, the mapping of floods that occurred in the past using Sentinel-1 SAR data can represent a viable solution in the attempt to identify areas with high flood potential and in the development of flood hazard maps. Sentinel-2 optical imagery has limited application in flood mapping, due to the presence of clouds and other atmospheric impediments during flood events. However, this type of data can complement the Sentinel-1 SAR data and improve the flood monitoring process. This can be realized through flood area delineation within the image's cloud-free regions and flood-affected areas delineation within post-event images.

Even if the merging of the daily flood extents derived from Sentinel-1 SAR and Sentinel-2 optical imagery led to a map that illustrates a maximum flood extent, which subsequently was considered for conversion to a flood probability map, some limitations regarding the water pixel extraction from Sentinel-1 and Sentinel-2 data needs to be addressed. Among them which is the most challenging is the possibility that in the time interval between the acquisition of the satellite images, some areas may have been flooded and thus missed. This may lead to the underestimation of the flooded areas. To address this challenge and achieve a complete flood characterization, the flood-affected areas were extracted using the 27 June 2020 Sentinel-2 satellite image (acquired in the last day of the flood).

Realizing a flood hazard map with an assigned probability using exclusively satellite imagery is difficult to achieve. This is limited by correlations with statistical data from gauging stations (used to determine the probability of flooding) and by the availability and continuity of satellite images throughout the entire temporal development of the flood event. Regardless of these issues, in an attempt to map a flood event that occurred on the Prut River middle course during 24–27 June 2020, both conditions mentioned above were accomplished and Sentinel-1 SAR and Sentinel-2 optical satellite images managed to capture the daily development stages of the flood event and the flood-affected areas.

In this way, an overview regarding the start date of the flood, the withdrawal rate, the affected areas throughout the flood wave and the maximum extent was obtained. Depending on the surface of the flooded areas within the Sentinel-1 images, we managed to determine that the flood started on Romanian territory on 24 June 2020, reached its peak on 25 June 2020 and experienced a decrease in intensity on 26 June 2020.

Even if the merging of the daily flood extent and flood-affected areas derived from Sentinel-1 SAR and Sentinel-2 optical imagery led to a map that illustrates the maximum flood area, this cannot be considered a flood hazard map as it does not contain any information regarding any exceedance probability. Therefore, following the proposed methodology, a statistical flood frequency analysis based on the maximum flow rate values located at a nearby gauging station—Rădăuți-Prut—was conducted and a flood return period was assigned.

5. Conclusions

One of the most effective methods of reducing the catastrophic impact of floods is the prior scientific assessment of flood hazard, its prediction, modeling and mapping. Flood hazard maps can significantly contribute to reducing the negative effects associated with floods. In recent years, advances in remote sensing and Geographic Information Systems (GIS) have revolutionized flood hazard assessment methodologies, thus facilitating new interdisciplinary approaches. Under certain conditions, remote sensing and time series discharge data can offer an alternative solution for generating flood hazard maps, especially in complex cross-border regions such as in our study area.

Using Sentinel-1 SAR and Sentinel-2 optical data along with GIS, we managed to: (i) capture the maximum extent of a flood event that took place on the middle course of Prut River, a complex cross-border region shared by the EU member Romania and non-EU countries Ukraine and Republic of Moldova, where the flood hazard maps are either incomplete or missing; (ii) develop two simple and efficient water pixel extraction methodologies considering the high contrast between water, water affected areas and land within the Sentinel-1 SAR and Sentinel-2 data; (iii) explore the improved temporal resolution of combined Sentinel-1 SAR and Sentinel-2 optical data and obtain a transboundary flood hazard map with an assigned probability.

This study proved that alternative solutions for flood hazard maps obtained by means of hydrological and hydraulic modeling exist and can be successfully implemented. Considering the legislative and emergency situations issues, determined by the cross-border regions, this kind of methodology can significantly contribute to the improvement of flood risk management and can stimulate international cooperation. Moreover, given the open-source character of Sentinel-1 and Sentinel-2 data and the adaptability of the proposed methodologies, this workflow can be easily extrapolated to other areas where the flood hazard has not been yet regulated and no related information exists.

Author Contributions: Conceptualization, C.I.C. and A.M.-P.; methodology, C.I.C., A.M.-P., C.C.S., A.U. and E.H.; software, C.I.C., A.U. and E.H.; validation, C.I.C., A.M.-P. and C.C.S.; formal analysis, C.I.C., A.M.-P., C.C.S., A.U. and E.H.; investigation, C.I.C. and A.M.-P.; resources, C.I.C., A.M.-P., C.C.S., A.U. and E.H.; data curation, C.I.C., A.M.-P. and C.C.S.; writing—original draft preparation,

C.I.C. and A.M.-P.; writing—review and editing, C.I.C. and A.M.-P.; visualization, C.I.C., A.M.-P., C.C.S., A.U. and E.H.; supervision, A.M.-P. and C.C.S.; project administration, C.I.C.; funding acquisition, C.I.C., A.M.-P., C.C.S., A.U. and E.H. All authors have read and agreed to the published version of the manuscript.

Funding: This work was supported and APC funded by a grant of the Ministry of Research, Innovation and Digitization, CNCS/CCCDI—UEFISCDI, project number: CNFIS-FDI-2021-0501.

Institutional Review Board Statement: Not applicable.

Informed Consent Statement: Not applicable.

Data Availability Statement: Data sharing not applicable.

Acknowledgments: The authors would like to express their gratitude to the employees of the Prut-Bîrlad Water Administration (Iaşi, Romania), particularly to Claudiu Pricop, who kindly provided a significant part of the hydrological data used in the present study. All data were processed in both Geomatics Laboratory of Doctoral School of Geoscience, Department of Geography, Faculty of Geography and Geology, "Alexandru Ioan Cuza" University of Iaşi (UAIC) and the Geoarchaeology Laboratory of Institute for Interdisciplinary Research, Science Research Department, "Alexandru Ioan Cuza" University of Iaşi (UAIC), Romania. This work was financed by the Doctoral School of Geoscience, Department of Geography, Faculty of Geography and Geology, "Alexandru Ioan Cuza" University of Iaşi (UAIC), Romania.

Conflicts of Interest: The authors declare no conflict of interest. The founding sponsors had no role in the design of the study; in the collection, analyses or interpretation of data; in the writing of the manuscript, and in the decision to publish the results.

Abbreviations

The following abbreviations are used in this manuscript:

BoA	Bottom of Atmosphere
CM-EMS	Copernicus Mapping—Emergency Management Service
DEM	Digital Elevation Model
EFD	European Floods Directive
EM-DAT	Emergency Events Database
ESA	European Space Agency—Copernicus Program
EU	European Union
EW	Extra-Wide swath
FHM	Flood Hazard Maps
FRMP	Flood Risk Management Plans
FRM	Flood Risk Maps
GIS	Geographic Information System
GRD	Ground Range Detected
HH or VV	Single polarization
HH+HV or VV+VH	Dual polarization
ICPDR	International Commission for the Protection of the Danube River
IW	Interferometric Wide swath
LiDAR	Light Intensity Detection and Ranging
NDVI	Normalized Difference Vegetation Index
NDWI	Normalized Difference Water Index
SAR	Synthetic Aperture Radar
SLC	Single Look Complex
SNAP	Sentinel Application Platform
SM	Stripmap
SRTM	Shuttle Radar Topography Mission
ToA	Top of the Atmosphere
UTM 35N	Universal Transverse Mercator—zone 35 North
WV	Wave

References

1. Costache, R.; Popa, M.C.; Tien Bui, D.; Diaconu, D.C.; Ciubotaru, N.; Minea, G.; Bao Pham, Q. Spatial predicting of flood potential areas using novel hybridizations of fuzzy decision-making, bivariate statistics, and machine learning. *J. Hydrol.* **2020**, *585*, 124808. [CrossRef]
2. Dumitriu, D. Sediment flux during flood events along the Trotuș River channel: Hydrogeomorphological approach. *J. Soils Sediments* **2020**, *20*, 4083–4102. [CrossRef]
3. Romanescu, G.; Stoleriu, C.C.; Mihu-Pintilie, A. Implementation of EU Water Framework Directive (2000/60/EC) in Romania—European Qualitative Requirements. In *Water Resources Management in Romania*; Negm, A., Romanescu, G., Zeleňáková, M., Eds.; Springer: Cham, Switzerland, 2020; pp. 17–55. [CrossRef]
4. Romanescu, G.; Cîmpianu, C.I.; Mihu-Pintilie, A.; Stoleriu, C.C. Historic flood events in NE Romania (post-1990). *J. Maps* **2017**, *13*, 787–798. [CrossRef]
5. Salit, F.; Zaharia, L.; Beltrando, G. Assessment of the warning system against floods on a rural area: The case of the lower Siret River (Romania). *Nat. Hazards Earth Syst. Sci.* **2013**, *13*, 409–416. [CrossRef]
6. Huţanu, E.; Mihu-Pintilie, A.; Urzică, A.; Paveluc, L.E.; Stoleriu, C.C.; Grozavu, A. Using 1D HEC-RAS Modeling and LiDAR Data to Improve Flood Hazard Maps Accuracy: A Case Study from Jijia Floodplain (NE Romania). *Water* **2020**, *12*, 1624. [CrossRef]
7. Romanescu, G.; Mihu-Pintilie, A.; Stoleriu, C.C.; Carboni, D.; Paveluc, L.; Cîmpianu, C.I. A Comparative Analysis of Exceptional Flood Events in the Context of Heavy Rains in the Summer of 2010: Siret Basin (NE Romania) Case Study. *Water* **2018**, *10*, 216. [CrossRef]
8. Mihu-Pintilie, A.; Cîmpianu, C.I.; Stoleriu, C.C.; Pérez, M.N.; Paveluc, L.E. Using High-Density LiDAR Data and 2D Streamflow Hydraulic Modeling to Improve Urban Flood Hazard Maps: A HEC-RAS Multi-Scenario Approach. *Water* **2019**, *11*, 1832. [CrossRef]
9. Mihu-Pintilie, A.; Nicu, I.C. GIS-based Landform Classification of Eneolithic Archaeological Sites in the Plateau-plain Transition Zone (NE Romania): Habitation Practices vs. Flood Hazard Perception. *Remote Sens.* **2019**, *11*, 915. [CrossRef]
10. EM-DAT (Emergency Events Database). The Emergency Events Database of Université Catholique de Louvain (UCL)—CRED, D. Guha-Sapir, Brussels, Belgium. Available online: https://www.emdat.be (accessed on 25 October 2021).
11. Stoleriu, C.C.; Urzică, A.; Mihu-Pintilie, A. Improving flood risk map accuracy using high-density LiDAR data and the HEC-RAS river analysis system: A case study from north-eastern Romania. *J. Flood Risk Manag.* **2020**, *13*, e12572. [CrossRef]
12. Romanescu, G.; Stoleriu, C.C.; Romanescu, A.-M. Water reservoirs and the risk of accidental flood occurrence. Case study: Stanca–Costesti reservoir and the historical floods of the Prut river in the period July–August 2008, Romania. *Hydrol. Process.* **2011**, *25*, 2056–2070. [CrossRef]
13. Stoleriu, C.C.; Romanescu, G.; Mihu-Pintilie, A. Using single-beam echo-sounder for assessing the silting rate from the largest cross-border reservoir of the Eastern Europe: Stanca-Costesti Lake, Romania and Republic of Moldova. *Carpathian J. Earth Environ. Sci.* **2019**, *14*, 83–94. [CrossRef]
14. Costache, R. Flood Susceptibility Assessment by Using Bivariate Statistics and Machine Learning Models—A Useful Tool for Flood Risk Management. *Water Resour. Manag.* **2019**, *33*, 3239–3256. [CrossRef]
15. Irimescu, A.; Stancalie, G.; Craciunescu, V.; Flueraru, C.; Anderson, E. The Use of Remote Sensing and GIS Techniques in Flood Monitoring and Damage Assessment: A Study Case in Romania. In *Threats to Global Water Security. NATO Science for Peace and Security, Series C: Environmental Security*; Jones, J.A.A., Vardanian, T.G., Hakopian, C., Eds.; Springer: Cham, Switzerland, 2009; pp. 167–177. [CrossRef]
16. Arseni, M.; Rosu, A.; Calmuc, M.; Calmuc, V.A.; Iticescu, C.; Georgescu, L.P. Development of Flood Risk and Hazard Maps for the Lower Course of the Siret River, Romania. *Sustainability* **2020**, *12*, 6588. [CrossRef]
17. Romanescu, G.; Stoleriu, C.C. An inter-basin backwater overflow (the Buhai Brook and the Ezer reservoir on the Jijia River, Romania). *Hydrol. Process.* **2013**, *28*, 3118–3131. [CrossRef]
18. Romanescu, G.; Stoleriu, C.C. Exceptional floods in the Prut basin, Romania, in the context of heavy rains in the summer of 2010. *Nat. Hazards Earth Syst. Sci.* **2017**, *17*, 381–396. [CrossRef]
19. ICPDR (International Commission for the Protection of the Danube River). Flood Action Programme Prut-Siret Sub-Basin. Available online: http://www.icpdr.org/main/activities-projects/flood-action-plans (accessed on 25 October 2021).
20. Kjeldsen, T.R.; Macdonald, N.; Lang, M.; Mediero, L.; Albuquerque, T.; Bogdanowicz, E.; Brázdil, R.; Castellarin, A.; David, V.; Fleig, A.; et al. Documentary evidence of past floods in Europe and their utility in flood frequency estimation. *J. Hydrol.* **2014**, *517*, 963–973. [CrossRef]
21. Ezzine, A.; Saidi, S.; Hermassi, T.; Kammessi, I.; Darragi, F.; Rajhi, H. Flood mapping using hydraulic modeling and Sentinel-1 image: Case study of Medjerda Basin, northern Tunisia. *Egypt. J. Remote Sens. Space Sci.* **2020**, *23*, 303–310. [CrossRef]
22. Horritt, M.S.; Bates, P.D. Evaluation of 1D and 2D numerical models for predicting river flood inundation. *J. Hydrol.* **2002**, *268*, 87–99. [CrossRef]
23. Machado, M.J.; Botero, B.A.; López, J.; Francés, F.; Díez-Herrero, A.; Benito, G. Flood frequency analysis of historical flood data under stationary and non-stationary modelling. *Hydrol. Earth Syst. Sci.* **2015**, *19*, 2561–2576. [CrossRef]
24. Samela, C.; Albano, R.; Sole, A.; Manfreda, S. A GIS tool for cost-effective delineation of flood-prone areas. *Comput. Environ. Urban Syst.* **2018**, *70*, 43–52. [CrossRef]

25. Guerriero, L.; Ruzza, G.; Guadagno, F.M.; Revellino, P. Flood hazard mapping incorporating multiple probability models. *J. Hydrol.* **2020**, *587*, 125020. [CrossRef]
26. Islam, M.M.; Sado, K. Development of flood hazard maps of Bangladesh using NOAA-AVHRR images with GIS. *Hydrol. Sci. J.* **2000**, *45*, 337–355. [CrossRef]
27. Schumann, G.J.-P.; Domeneghetti, A. Exploiting the proliferation of current and future satellite observations of rivers. *Hydrol. Process.* **2016**, *30*, 2891–2896. [CrossRef]
28. Huang, C.; Chen, Y.; Wu, J. Mapping spatio-temporal flood inundation dynamics at large river basin scale using time-series flow data and MODIS imagery. *Int. J. Appl. Earth Obs. Geoinf.* **2014**, *26*, 350–362. [CrossRef]
29. Rahman, M.S.; Di, L. The state of the art of spaceborne remote sensing in flood management. *Nat. Hazards* **2017**, *85*, 1223–1248. [CrossRef]
30. Sanyal, J.; Lu, X.X. Application of remote sensing in flood management with special reference to monsoon Asia: A review. *Nat. Hazards* **2004**, *33*, 283–301. [CrossRef]
31. Borah, S.B.; Sivasankar, T.; Ramya, M.N.S.; Raju, P.L.N. Flood inundation mapping and monitoring in Kaziranga National Park, Assam using Sentinel-1 SAR data. *Environ. Monit. Assess.* **2018**, *190*, 520. [CrossRef] [PubMed]
32. Chowdhury, E.H.; Hassan, Q.K. Use of remote sensing data in comprehending an extremely unusual flooding event over southwest Bangladesh. *Nat. Hazards* **2017**, *88*, 1805–1823. [CrossRef]
33. Hoque, R.; Nakayama, D.; Matsuyama, H.; Matsumoto, J. Flood monitoring, mapping and assessing capabilities using RADARSAT remote sensing, GIS and ground data for Bangladesh. *Nat. Hazards* **2011**, *57*, 525–548. [CrossRef]
34. Schumann, G.; Moller, D.K. Microwave remote sensing of flood inundation. *Phys. Chem. Earth* **2015**, *83–84*, 84–95. [CrossRef]
35. Twele, A.; Cao, W.; Plank, S.; Martinis, S. Sentinel-1- based flood mapping: A fully automated processing chain. *Int. J. Remote Sens.* **2016**, *37*, 2990–3004. [CrossRef]
36. Amitrano, D.; Martino, G.D.; Iodice, A.; Mitidieri, F.; Papa, M.N.; Riccio, D.; Ruello, G. Sentinel-1 for Monitoring Reservoirs: A Performance Analysis. *Remote Sens.* **2014**, *6*, 10676–10693. [CrossRef]
37. Landuyt, L.; Verhoest, N.E.K.; van Coillie, F.M.B. Flood Mapping in Vegetated Areas Using an Unsupervised Clustering Approach on Sentinel-1 and -2 Imagery. *Remote Sens.* **2020**, *12*, 3611. [CrossRef]
38. Malenovský, Z.; Helmut, R.; Cihlar, J.; Schaepman, M.E.; García-Santos, G.; Fernandes, R.; Berger, M. Sentinels for science: Potential of Sentinel-1, -2, and -3 missions for scientific observations of ocean, cryosphere, and land. *Remote Sens. Environ.* **2012**, *120*, 91–101. [CrossRef]
39. Poursanidis, D.; Chrysoulakis, N. Remote Sensing, natural hazards and the contribution of ESA Sentinels missions. *Remote Sens. Appl. Soc. Environ.* **2017**, *6*, 25–38. [CrossRef]
40. Gebremichael, E.; Molthan, A.L.; Bell, J.R.; Schultz, L.A.; Hain, C. Flood Hazard and Risk Assessment of Extreme Weather Events Using Synthetic Aperture Radar and Auxiliary Data: A Case Study. *Remote Sens.* **2020**, *12*, 3588. [CrossRef]
41. Huang, M.; Jin, S. Rapid Flood Mapping and Evaluation with a Supervised Classifier and Change Detection in Shouguang Using Sentinel-1 SAR and Sentinel-2 Optical Data. *Remote Sens.* **2020**, *12*, 2073. [CrossRef]
42. Li, Y.; Martinis, S.; Plank, S.; Ludwig, R. An automatic change detection approach for rapid flood mapping in Sentinel-1 SAR data. *Int. J. Appl. Earth Obs. Geoinf.* **2018**, *73*, 123–135. [CrossRef]
43. Romero, N.A.; Cigna, F.; Tapete, D. ERS-1/2 and Sentinel-1 SAR Data Mining for Flood Hazard and Risk Assessment in Lima, Peru. *Appl. Sci.* **2020**, *10*, 6598. [CrossRef]
44. Tapete, D.; Cigna, F. Poorly known 2018 floods in Bosra UNESCO site and Sergiopolis in Syria unveiled from space using Sentinel-1/2 and COSMO-SkyMed. *Sci. Rep.* **2020**, *10*, 12307. [CrossRef] [PubMed]
45. Konapala, G.; Kumar, S.V.; Ahmad, S.K. Exploring Sentinel-1 and Sentinel-2 diversity for flood inundation mapping using deep learning. *Isprs J. Photogramm. Remote Sens.* **2021**, *180*, 163–173. [CrossRef]
46. Burcea, S.; Cică, R.; Bojariu, R. Radar-derived convective storms' climatology for the Prut River basin: 2003–2017. *Nat. Hazards Earth Syst. Sci.* **2019**, *19*, 1305–1318. [CrossRef]
47. Bamler, R. Principles of Synthetic Aperture Radar. *Surv. Geophys.* **2000**, *21*, 147–157. [CrossRef]
48. Moreira, A.; Prats-Iraola, P.; Younis, M.; Krieger, G.; Hajnsek, I.; Papathanassiou, K.P. A Tutorial on Synthetic Aperture Radar. *IEEE Geosci. Remote Sens. Mag.* **2013**, *1*, 6–43. [CrossRef]
49. Aschbacher, J.; Milagro-Pérez, M.P. The European Earth monitoring (GMES) programme: Status and perspectives. *Remote Sens. Environ.* **2012**, *120*, 3–8. [CrossRef]
50. Clement, M.; Kilsby, C.; Moore, P. Multi-temporal synthetic aperture radar flood mapping using change detection. *J. Flood Risk Manag.* **2018**, *11*, 152–168. [CrossRef]
51. Tsyganskaya, V.; Martinis, S.; Marzahn, P.; Ludwig, R. Detection of Temporary Flooded Vegetation Using Sentinel-1 Time Series Data. *Remote Sens.* **2018**, *10*, 1286. [CrossRef]
52. DeVries, B.; Huang, C.; Armston, J.; Huang, W.; Jones, J.W.; Lange, M.W. Rapid and robust monitoring of flood events using Sentinel-1 and Landsat data on the Google Earth Engine. *Remote Sens. Environ.* **2020**, *240*, 111664. [CrossRef]
53. Chojka, A.; Artiemjew, P.; Rapiński, J. RFI Artefacts Detection in Sentinel-1 Level-1 SLC Data Based On Image Processing Techniques. *Sensors* **2020**, *20*, 2919. [CrossRef]
54. Stasolla, M.; Neyt, X. An Operational Tool for the Automatic Detection and Removal of Border Noise in Sentinel-1 GRD Products. *Sensors* **2018**, *18*, 3454. [CrossRef]

55. Muro, J.; Canty, M.; Conradsen, K.; Hüttich, C.; Nielsen, A.A.; Skriver, H.; Remy, F.; Strauch, A.; Thonfeld, F.; Menz, G. Short-Term Change Detection in Wetlands Using Sentinel-1 Time Series. *Remote Sens.* **2016**, *8*, 795. [CrossRef]
56. Goumehei, E.; Tolpekin, V.; Stein, A.; Yan, W. Surface water body detection in polarimetric SAR data using contextual complex Wishart classification. *Water Resour. Res.* **2019**, *55*, 7047–7059. [CrossRef]
57. Schubert, A.; Miranda, N.; Geudtner, D.; Small, D. Sentinel-1A/B Combined Product Geolocation Accuracy. *Remote Sens.* **2017**, *9*, 607. [CrossRef]
58. Clauss, K.; Ottinger, M.; Leinenkugel, P.; Kuenzer, C. Estimating rice production in the Mekong Delta, Vietnam, utilizing time series of Sentinel-1 SAR data. *Int. J. Appl. Earth Obs. Geoinf.* **2018**, *73*, 574–585. [CrossRef]
59. Carreño Conde, F.; de Mata Muñoz, M. Flood Monitoring Based on the Study of Sentinel-1 SAR Images: The Ebro River Case Study. *Water* **2019**, *11*, 2454. [CrossRef]
60. Li, X.; Zhou, Y.; Gong, P.; Seto, K.C.; Clinton, N. Developing a method to estimate building height from Sentinel-1 data. *Remote Sens. Environ.* **2020**, *240*, 111705. [CrossRef]
61. Perrou, T.; Garioud, A.; Parcharidis, I. Use of Sentinel-1 imagery for flood management in a reservoir-regulated river basin. *Front. Earth Sci.* **2018**, *12*, 506–520. [CrossRef]
62. Huth, J.; Gessner, U.; Klein, I.; Yesou, H.; Lai, X.; Oppelt, N.; Kuenzer, C. Analyzing Water Dynamics Based on Sentinel-1 Time Series—A Study for Dongting Lake Wetlands in China. *Remote Sens.* **2020**, *12*, 1761. [CrossRef]
63. Goffi, A.; Stroppiana, D.; Brivio, P.A.; Bordogna, G.; Boschetti, M. Towards an automated approach to map flooded areas from Sentinel-2 MSI data and soft integration of water spectral features. *Int. J. Appl. Earth Obs. Geoinf.* **2020**, *84*, 101951. [CrossRef]
64. Cordeiro, M.C.R.; Martinez, J.-M.; Peña-Luque, S. Automatic water detection from multidimensional hierarchical clustering for Sentinel-2 images and a comparison with Level 2A processors. *Remote Sens. Environ.* **2021**, *253*, 112209. [CrossRef]
65. Pahlevan, N.; Sarkar, S.; Franz, B.A.; Balasubramanian, S.V.; He, J. Sentinel-2 MultiSpectral Instrument (MSI) data processing for aquatic science applications: Demonstrations and validations. *Remote Sens. Environ.* **2017**, *201*, 47–56. [CrossRef]
66. Szantoi, Z.; Strobl, P. Copernicus Sentinel-2 Calibration and Validation. *Eur. J. Remote Sens.* **2019**, *52*, 253–255. [CrossRef]
67. Huang, W.; DeVries, B.; Huang, C.; Lang, M.W.; Jones, J.W.; Creed, I.F.; Carroll, M.L. Automated Extraction of Surface Water Extent from Sentinel-1 Data. *Remote Sens.* **2018**, *10*, 797. [CrossRef]
68. Manakos, I.; Kordelas, G.A.; Marini, K. Fusion of Sentinel-1 data with Sentinel-2 products to overcome non-favourable atmospheric conditions for the delineation of inundation maps. *Eur. J. Remote Sens.* **2020**, *53*, 53–66. [CrossRef]
69. Pulvirenti, L.; Chini, M.; Pierdicca, N.; Boni, G. Use of SAR Data for Detecting Floodwater in Urban and Agricultural Areas: The Role of the Interferometric Coherence. *IEEE Trans. Geosci. Remote Sens.* **2016**, *54*, 1532–1544. [CrossRef]
70. Singha, M.; Dong, J.; Sarmah, S.; You, N.; Zhou, Y.; Zhang, G.; Doughty, R.; Xiao, X. Identifying floods and flood-affected paddy rice fields in Bangladesh based on Sentinel-1 imagery and Google Earth Engine. *Isprs J. Photogramm. Remote Sens.* **2020**, *166*, 278–293. [CrossRef]
71. Leroux, L.; Congedo, L.; Bellón, B.; Gaetano, R.; Bégué, A. Land Cover Mapping Using Sentinel-2 Images and the Semi-Automatic Classification Plugin: A Northern Burkina Faso Case Study. In *QGIS and Applications in Agriculture and Forest*; Baghdadi, N., Mallet, C., Zribi, M., Eds.; Wiley: Hoboken, NJ, USA, 2018; pp. 119–151. [CrossRef]
72. Clerici, N.; Valbuena Calderón, C.A.; Posada, J.M. Fusion of Sentinel-1A and Sentinel-2A data for land cover mapping: A case study in the lower Magdalena region, Colombia. *J. Maps* **2017**, *13*, 718–726. [CrossRef]
73. Truckenbrodt, J.; Freemantle, T.; Williams, C.; Jones, T.; Small, D.; Dubois, C.; Thiel, C.; Rossi, C.; Syriou, A.; Giuliani, G. Towards Sentinel-1 SAR Analysis-Ready Data: A Best Practices Assessment on Preparing Backscatter Data for the Cube. *Data* **2019**, *4*, 93. [CrossRef]
74. Filipponi, F. Sentinel-1 GRD Preprocessing Workflow. In Proceedings of the 3rd International Electronic Conference on Remote Sensing (ECRS 2019), Online, 22 May–5 June 2019. [CrossRef]
75. Mandal, D.; Kumar, V.; Rathaa, D.; Dey, S.; Bhattacharyaa, A.; Lopez-Sanchez, J.M.; McNairn, H.; Rao, Y.S. Dual polarimetric radar vegetation index for crop growth monitoring using sentinel-1 SAR data. *Remote Sens. Environ.* **2020**, *247*, 111954. [CrossRef]
76. Hu, S.; Qin, J.; Ren, J.; Zhao, H.; Ren, J.; Hong, H. Automatic Extraction of Water Inundation Areas Using Sentinel-1 Data for Large Plain Areas. *Remote Sens.* **2020**, *12*, 243. [CrossRef]
77. Gašparović, M.; Dobrinić, D. Comparative Assessment of Machine Learning Methods for Urban Vegetation Mapping Using Multitemporal Sentinel-1 Imagery. *Remote Sens.* **2020**, *12*, 1952. [CrossRef]
78. Mirsoleimani, H.R.; Sahebi, M.R.; Baghdadi, N.; El Hajj, M. Bare Soil Surface Moisture Retrieval from Sentinel-1 SAR Data Based on the Calibrated IEM and Dubois Models Using Neural Networks. *Sensors* **2019**, *19*, 3209. [CrossRef]
79. Benoudjit, A.; Guida, R. A Novel Fully Automated Mapping of the Flood Extent on SAR Images Using a Supervised Classifier. *Remote Sens.* **2019**, *11*, 779. [CrossRef]
80. Prasad, K.A.; Ottinger, M.; Wei, C.; Leinenkugel, P. Assessment of Coastal Aquaculture for India from Sentinel-1 SAR Time Series. *Remote Sens.* **2019**, *11*, 357. [CrossRef]
81. Ahmad, W.; Kim, D. Estimation of flow in various sizes of streams using the Sentinel-1 Synthetic Aperture Radar (SAR) data in Han River Basin, Korea. *Int. J. Appl. Earth Obs. Geoinf.* **2019**, *83*, 101930. [CrossRef]
82. Liu, Y.; Gong, W.; Xing, Y.; Hu, X.; Gong, J. Estimation of the forest stand mean height and above ground biomass in Northeast China using SAR Sentinel-1B, multispectral Sentinel-2A, and DEM imagery. *Isprs J. Photogramm. Remote Sens.* **2019**, *151*, 277–289. [CrossRef]

83. Ye, Y.; Yang, C.; Zhu, B.; Zhou, L.; He, Y.; Jia, H. Improving Co-Registration for Sentinel-1 SAR and Sentinel-2 Optical Images. *Remote Sens.* **2021**, *13*, 928. [CrossRef]

84. Matgen, P.; Hostache, R.; Schumann, G.; Pfister, L.; Hoffmann, L.; Savenije, H.H.G. Towards an automated SAR-based flood monitoring system: Lessons learned from two case studies. *Phys. Chem. Earth* **2011**, *36*, 241–252. [CrossRef]

85. Chini, M.; Pelich, R.; Pulvirenti, L.; Pierdicca, N.; Hostache, R.; Matgen, P. Sentinel-1 InSAR Coherence to Detect Floodwater in Urban Areas: Houston and Hurricane Harvey as A Test Case. *Remote Sens.* **2019**, *11*, 107. [CrossRef]

86. Wan, L.; Liu, M.; Wang, F.; Zhang, T.; You, H.J. Automatic extraction of flood inundation areas from SAR images: A case study of Jilin, China during the 2017 flood disaster. *Int. J. Remote Sens.* **2019**, *40*, 5050–5077. [CrossRef]

87. Gstaiger, V.; Huth, J.; Gebhardt, S.; Wehrmann, T.; Kuenzer, C. Multi-sensoral and automated derivation of inundated areas using TerraSAR-X and ENVISAT ASAR data. *Int. J. Remote Sens.* **2012**, *33*, 7291–7304. [CrossRef]

88. Liang, J.; Liu, D. A local thresholding approach to flood water delineation using Sentinel-1 SAR imagery. *Isprs J. Photogramm. Remote Sens.* **2020**, *159*, 53–62. [CrossRef]

89. Lu, J.; Giustarini, L.; Xiong, B.; Zhao, L.; Jiang, Y.; Kuang, G. Automated flood detection with improved robustness and efficiency using multitemporal SAR data. *Remote Sens. Lett.* **2014**, *5*, 240–248. [CrossRef]

90. Gan, T.Y.; Zunic, F.; Kuo, C.-C.; Strobl, T. Flood mapping of Danube River at Romania using single and multi-date ERS2-SAR images. *Int. J. Appl. Earth Obs. Geoinf.* **2012**, *18*, 69–81. [CrossRef]

91. Rahman, M.R.; Thakur, P.K. Detecting, mapping and analysing of flood water propagation using synthetic aperture radar (SAR) satellite data and GIS: A case study from the Kendrapara District of Orissa State of India. *Egypt. J. Remote Sens. Space Sci.* **2018**, *21*, S37–S41. [CrossRef]

92. McFeeters, S.K. The use of the Normalized Difference Water Index (NDWI) in the delineation of open water features. *Int. J. Remote Sens.* **1996**, *17*, 1425–1432. [CrossRef]

93. Du, Y.; Zhang, Y.; Ling, F.; Wang, Q.; Li, W.; Li, X. Water Bodies' Mapping from Sentinel-2 Imagery with Modified Normalized Difference Water Index at 10-m Spatial Resolution Produced by Sharpening the SWIR Band. *Remote Sens.* **2016**, *8*, 354. [CrossRef]

94. Yang, X.; Zhao, S.; Qin, X.; Zhao, N.; Liang, L. Mapping of Urban Surface Water Bodies from Sentinel-2 MSI Imagery at 10 m Resolution via NDWI-Based Image Sharpening. *Remote Sens.* **2017**, *9*, 596. [CrossRef]

95. Jiang, W.; Ni, Y.; Pang, Z.; Li, X.; Ju, H.; He, G.; Lv, J.; Yang, K.; Fu, J.; Qin, X. An Effective Water Body Extraction Method with New Water Index for Sentinel-2 Imagery. *Water* **2021**, *13*, 1647. [CrossRef]

96. Bangira, T.; Alfieri, S.M.; Menenti, M.; van Niekerk, A.; Vekerdy, Z. A Spectral Unmixing Method with Ensemble Estimation of Endmembers: Application to Flood Mapping in the Caprivi Floodplain. *Remote Sens.* **2017**, *9*, 1013. [CrossRef]

97. Gargiulo, M.; Dell'Aglio, D.A.G.; Iodice, A.; Riccio, D.; Ruello, G. Integration of Sentinel-1 and Sentinel-2 Data for Land Cover Mapping Using W-Net. *Sensors* **2020**, *20*, 2969. [CrossRef] [PubMed]

98. Notti, D.; Giordan, D.; Caló, F.; Pepe, A.; Zucca, F.; Galve, J.P. Potential and Limitations of Open Satellite Data for Flood Mapping. *Remote Sens.* **2018**, *10*, 1673. [CrossRef]

99. Saravanan, S.; Jegankumar, R.; Selvaraj, A.; Jacinth Jennifer, J.; Parthasarathy, K.S.S. Chapter 20—Utility of landsat data for assessing mangrove degradation in Muthupet Lagoon, South India. *Coast. Zone Manag.* **2019**, *20*, 471–484. [CrossRef]

100. CM-EMS (Copernicus Mapping—Emergency Management Service). EMSR445: Flood in Romania. Available online: https://emergency.copernicus.eu/mapping/list-of-components/EMSR445 (accessed on 25 October 2021).

101. Gumbel, E.J. The return period of flood flows. *Ann. Math. Stat.* **1941**, *12*, 163–190. [CrossRef]

102. Bhat, M.; Alam, A.; Ahmad, B.; Kotlia, B.S.; Farooq, H.; Taloor, A.K.; Ahmad, S. Flood frequency analysis of river Jhelum in Kashmir basin. *Quat. Int.* **2019**, *507*, 288–294. [CrossRef]

103. Farooq, M.; Shafique, M.; Khattak, M.S. Flood frequency analysis of river swat using Log Pearson type 3, Generalized Extreme Value, Normal, and Gumbel Max distribution methods. *Arab. J. Geosci.* **2018**, *11*, 216. [CrossRef]

104. Kumar, R. Flood Frequency Analysis of the Rapti River Basin using Log Pearson Type-III and Gumbel Extreme Value-1 Methods. *J. Geol. Soc. India* **2019**, *94*, 480–484. [CrossRef]

remote sensing

Article

Mapping the Impact of COVID-19 Lockdown on Urban Surface Ecological Status (USES): A Case Study of Kolkata Metropolitan Area (KMA), India

Manob Das [1], Arijit Das [1,*], Paulo Pereira [2] and Asish Mandal [1]

[1] Department of Geography, University of Gour Banga, Malda 732103, West Bengal, India; dasmanob631@gmail.com (M.D.); asishm811@gmail.com (A.M.)
[2] Environmental Management Laboratory, Mykolas Romeris University, LT-08303 Vilnius, Lithuania; paulo@mruni.eu
* Correspondence: arijitdas@ugb.ac.in

Abstract: An urban ecosystem's ecological structure and functions can be assessed through Urban Surface Ecological Status (USES). USES are affected by human activities and environmental processes. The mapping of USESs are crucial for urban environmental sustainability, particularly in developing countries such as India. The COVID-19 pandemic caused unprecedented negative impacts on socio-economic domains; however, there was a reduction in human pressures on the environment. This study aims to assess the effects of lockdown on the USES in the Kolkata Metropolitan Area (KMA), India, during different lockdown phases (phases I, II and III). The land surface temperature (LST), normalized difference vegetation index (NDVI), and wetness and normalized difference soil index (NDSI) were assessed. The USES was developed by combining all of the biophysical parameters using Principal Component Analysis (PCA). The results showed that there was a substantial USES spatial variability in KMA. During lockdown phase III, the USES in fair and poor sustainability areas decreased from 29% (2019) to 24% (2020), and from 33% (2019) to 25% (2020), respectively. Overall, the areas under poor USES decreased from 30% to 25% during lockdown periods. Our results also showed that the USES mean value was 0.49 in 2019but reached 0.34 during the lockdown period (a decrease of more than 30%). The poor USES area was mainly concentrated in built-up areas (with high LST and NDSI), compared to the rural fringe areas of KMA (high NDVI and wetness). The mapping of USES are crucial in different biophysical environmental conditions, and they can be very helpful for the assessment of urban sustainability.

Keywords: ecological structure; urban surface ecological status (USES); remote sensing; Kolkata Metropolitan Area; environmental sustainability

Citation: Das, M.; Das, A.; Pereira, P.; Mandal, A. Mapping the Impact of COVID-19 Lockdown on Urban Surface Ecological Status (USES): A Case Study of Kolkata Metropolitan Area (KMA), India. *Remote Sens.* **2021**, *13*, 4395. https://doi.org/10.3390/rs13214395

Academic Editors: Adrian Ursu and Cristian Constantin Stoleriu

Received: 21 September 2021
Accepted: 28 October 2021
Published: 31 October 2021

1. Introduction

In recent decades, rapid urban expansion and population growth have dramatically impacted ecosystems [1–3], increasing land degradation and reducing human wellbeing [4–6]. This hashad negative impacts on urban inhabitants [7]. One of the most visible impacts of urban expansion is the reduction in green spaces(GS), which is well known to affect life quality [8]. Previous works highlighted that the Urban Surface Ecological Status (USES) was influenced mainly by the surface biophysical components such as greenness, dryness, wetness, and heat. Therefore, changes in land surface characteristics lead to a variation of USES [9,10]. The conversion of a pervious land surface into an impervious surface is one of the most widespread forms of land use/land cover change (LULCC) [11]. LULCCdramatically changes land-atmosphere interactions, such as albedo andandevapotranspiration [4,8,12].

The application of remote sensing is widely used to map USES. Several spectral indices have been used, such as the normalised difference vegetation index (NDVI), normalised difference built-up index (NDBI), normalised difference water index (NDWI), leaf area index (LAI), normalised difference soil index (NDSI), land surface temperature (LST) and/or using the combination of different indices [9,12,13].

The USES changes greatly affect life quality. Therefore, the quantification of the USES is crucial and urgently required, especially in rapidly growing megacities. To our knowledge, this field remains unexplored and can provide key information to developersand policymakers to improve urban planning. Urban areas in India are unplanned andover-populated, severely impacting their ecosystems [14,15]. Therefore, the assessment of the USES could provide essential information to make thosecities more habitable.

India is the second most populous country in the world. It is expected that the population of India will increase bynearly 273 million people by 2050 and overtake China by 2027. The estimated population growth is especially high in urban areas. In 2050, it is expected that India will have the highest urban population in the world [6,16].According to the National Commission on Population (NCP), it was estimated that approximately 38.6% of the total population of India reside in urban areas. The populations of cities such as Mumbai, Delhi, Kolkata, Chennai, and Hyderabad have increased rapidly and are among the world's most populated urban areas [17]. Kolkata is the thirdlargest megacity in India, after Delhi and Mumbai [18,19]. Recently, Kolkata experienced a rapid urban expansion LULCC, resulting in a substantial alteration of the natural and semi-natural areas [20,21]. Previous studies observed a strong urban growth pattern and land use land cover (LULC) dynamics in theKolkata Metropolitan Area (KMA) [22–26]. The rapid urban expansion within the spatial limit ofthe KMA has resulted in a dramatic change in vegetation cover, water bodies, agricultural areas, and wetlands (Table A1). For example, Ghosh and Das [27] performed a study on an East Kolkata Wetland (EKW). They found a substantial decline in the wetland area (reduced by 5%) and vegetation cover (reduced by 28%). According to Sahana et al. [28], in the KMA from 1990 to 2015, vegetation cover, wetland, and agricultural lands declined by about 6.6%, 5.9%, and 26%, respectively, while the area of urban areas increased by 24.5%. It is clear that that there were substantial transformations of LULC change in KMA. However, very few studies assessed the impacts of thesurface ecological status [29–32]. To our knowledge, this is the first work focused on mapping the USES in the KMA. Moreover, this is the first attempt to assess the impact of COVID-19 on the USES of Indian urban areas.

The emergence of the COVID-19 pandemic significantly affected global public health, causing the deaths millions of people [7]. Many measures were adopted by many countries such as India, the USA, France, Italy, and the UK to combat COVID-19 transmission [7]. Considerable measures such as strict transport restrictions, limited emissions from industries, and the closure of hotels and restaurantssignificantly affected the environment [7].The lockdown imposed by the COVID-19 outbreak had detrimental effects on society and the economy [11]. However, the decrease in human activities reduced the environmental pressure at both local and global scales [12,33–36]. It is critical to identify the impact of the lockdown on environmental quality [35,37]. Recently, several studies assessed the lockdown effects on air quality and water quality [33–36]. In India, a total lockdown was implemented on 25 March 2020, and continued until 30 June 2020. Several unlocking phases were implemented. From 1 June 2020, until 30 December 2020, unlocking phase VI was implemented [38,39]. There were strict restrictions including the complete banning of industrial, transportation, and other socio-economic activities [40]. The severe limits on human activities resulted in areduction in environmental pressures. Thus, we hypothesise that the lockdown affected USESs. This study attemptedto map the impact of the COVID-19 lockdown on the USES in the KMA, India. The USES maps were compared with the previous year (2019) to better understand the effect of lockdown on the USES. Thus, this study is essential to understand the impact of lockdown on the ecological status.

2. Materials and Methods

2.1. Study Area

Kolkata Metropolitan Area is the third-largest megacity in Eastern India and is the capital of West Bengal state (Figure 1). It is located in the lower Gangetic plain (LGP), and it is extended between 88° 32° E, and between 23° 01, N to 22° 19 N,withan area of 1851.41 km^2. According to the Census of India (2011), KMA has 14.06 million (7480 person/km^2/). KMA comprises 6 districts, 3 municipal corporations (Kolkata, Howrah, and Chandannager) and 38 municipalities. According to the Koppen classification, KMA has an AW climate type, with a wet climate during the summer and dry climate during the winter. In 2019, KMA had about 40% and 21% built-up and vegetation cover (Figure 1).

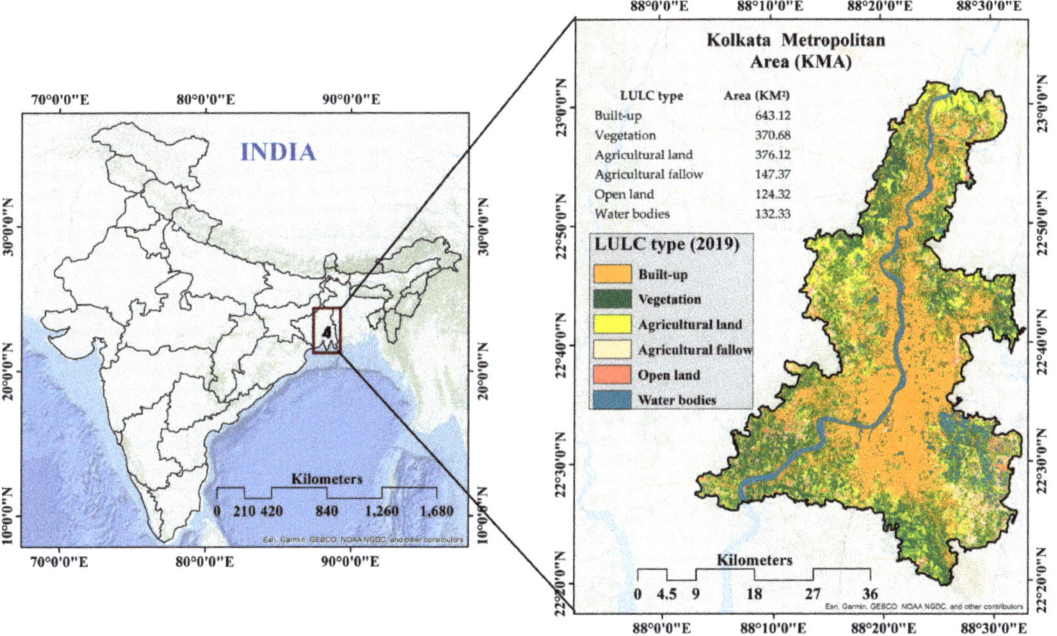

Figure 1. Location and classification of LULC in Kolkata Metropolitan Areain 2019.

2.2. Data Sources

In this study, the satellite images for the assessment of the USES were derived from the United States Geological Survey (USGS) (https://earthexplorer.usgs.gov/accessed on 26 August 2021). Landsat 8 OLI (Operational Land Imager) images were selected in 2019 and 2020 during the different phases of lockdown. Landsat 8 OLI images comprise two sensors, i.e., Operational Land Imager (OLI) and Thermal Infrared Sensor (TIRS), respectively. These sensors have a 30 m resolution (except band 8 of 15m) with nine spectral bands and two thermal bands. In this study, all satellite images were taken into account to develop the USES from the different lockdown phases. These lockdown phases were: pre-lockdown phase (January and February), during lockdown phase (April), and post lockdown phase (November and December), respectively. Two satellite images were selected for the pre-lockdown period, one image for during lockdown, and two images for post lockdown. The satellite images were selected based on the lockdown timeline imposed in India. In India, the full lockdown was imposed from 25 March to 30 June 2020, and after 30 June, unlocking phases (from 1 July 2020) were initiated, and are currently continuing (as of 15 August 2021). The details of the lockdown phases are presented in Table 1.

Table 1. Data used in this study for USES modelling.

Year		Lockdown Phase	Month	Dateof Acquisition	Image ID	Sensor	Resolution (m)	Source
Non Pandemic year	2019	Same period during Prelockdown (Phase I)	January	30 January 2019	LC08_L2SP_138044_20190130_20200829_02_T1	LANDSAT8 OLI (Operational LandImager)	30	USGS (https:// earthexplorer. usgs.gov/) acessed on 26 August 2021
			February	15 February 2019	LC08_L2SP_138044_20190215_20200829_02_T1			
		Same period during lockdown (Phase II)	April	20 April 2019	LC08_L2SP_138044_20190420_20200828_02_T1			
		Same period during after lockdown (Phase III)	November	14 November 2019	LC08_L2SP_138044_20191114_20200825_02_T1			
			December	16 December 2019	LC08_L2SP_138044_20191216_20201023_02_T1			
Pandemic year	2020	Prelockdown (Phase I)	January	17 January 2020	LC08_L2SP_138044_20200117_20200823_02_T1			
			February	18 February 2020	LC08_L2SP_138044_20200218_20200823_02_T1			
		During lockdown (Phase II)	April	6 April 2020	LC08_L2SP_138044_20200406_20200822_02_T1			
		After lockdown (Phase III)	November	16 November 2020	LC08_L2SP_138044_20201116_20210315_02_T1			
			December	18 December 2020	LC08_L2SP_138044_20201218_20210309_02_T1			

2.3. Methods

2.3.1. Spectral Indices and Framework

For the assessment of the environmental quality in KMA, four biophysical parameters were extracted: normalised difference in the vegetation index for greenness, normalised difference in the soil index for dryness, wetness derived from Tasseled cap transformation (TCT) (wetness), and land surface temperature for heat, respectively (Table 2). These spectral indices were previously used for modelling USES [10,12,41,42]. The study flowchart is shown in Figure 2 and was developed to evaluate the impact of COVID-19 amid the lockdown on the USES. In step one, LANDSAT 8 OLI images from 2019 and 2020 were pre-processed and corrected. In step two, the spectral indices related to USES such as NDVI, NDSI, LST, and wetness were calculated on the basis of LANDSAT-8 (OLI) reflective and thermal bands for a different lockdown phase. In the third step, the spectral indices were normalised (PCA was used to assign weight), and finally, USESs were developed for 2019 and 2020 during the different phases of lockdown. NDVI was considered one of the significant indexes, and was widely used to assess and model vegetation [43]. NDSI was also considered as one of the essential parameters to state the surface ecological status [44,45]. LST was considered a significant surface biophysical parameter by which the exchange of thermal energy could be assessed [46,47]. Previous studies documented that LST increased with increasing human activity [48], and spatio-temporal changes of LST influenced climatic conditions. The amount of moisture present in various land surface covers such as built-up, vegetation cover, and bare soil could be estimated through wetness. Tasselled cap transformation (TCT) is a commonly used method to model spatial heterogeneity of wetness status [48]:

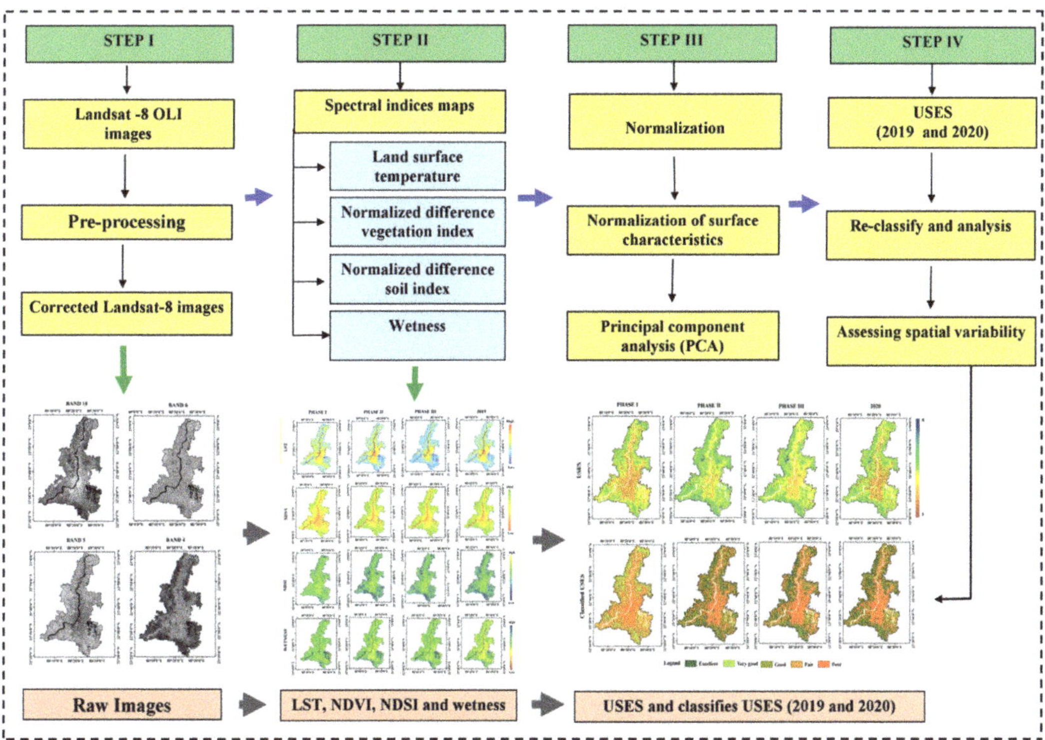

Figure 2. Methodological frameworks for USES mapping of the study.

Table 2. Spectral indices used for USES modeling in this study.

Parameters	Ecological Significance	Equation	Reference
(a) LST	Heat	$LST = T_B / [1 + \{(\lambda * TB/\rho) * \ln\varepsilon\}]$	[46,47]
(b) NDVI	Greenness	$NDVI = \frac{(NIR-\text{Red})}{(NIR+\text{Red})}$	[8,43]
(c) NDSI	Dryness	$NDSI = \frac{\text{SWIR1}-\text{NIR}}{\text{SWIR1}+\text{NIR}}$	[44,45]
(d) Wetnessderivedfrom TCT	Wetness	$0.1115Blue2 - 0.1973Green + 0.3283Red + 0.3407NIR - 0.7117SWIR1 - 0.4559SWIR2$	[48,49]

Where, (a) BT is the brightness temperature and W is the wavelength of the emitted radiance. Thermal band 11 for LANDSAT-8 has error and bias for LST calculation, so thermal band 10 of LANDSAT-8 was considered for LST calculation in this study. (b) NIR and Red are the near-infrared and red bands, (c) SWIR1 is the shortwave infrared bands of satellite imagery, and (d) Tasseled cap transformation (TCT) was calculated for wetness following Baig et al. [48] and Mijani et al. [49].

2.3.2. Modelling Urban Surface Ecological Status (USES) in KMA

The biophysical parameters (greenness, heat, dryness, and wetness indices) were standardised (ranging from 0 to 1) [50]. A principal component analysis (PCA) was applied to combine the assessed indices. The first principal component (PC1) was used for USES analysis of KMA. PCA's application is key to avoiding collinearity problems between the parameters used in this study for USES modelling. Subsequently, USES values were standardised between 0 to 1, where the values close to 0 indicated the best USES (i.e., high values of NDVI, wetness, and low values of LST and NDSI). The values close to 1 show the worst USES (i.e., low NDVI, wetness, and high LST, and NDSI), respectively. The USES values were reclassified into five categories:(a) Excellent (<0.20), (b) Very good (0.20–0.40), (c) Good (0.40–0.60), (d) Fair (0.60–0.80), and (e) Poor (>0.80) [3].

2.3.3. Statistical Analysis

A Mann–Whitney test (M–W test) was applied to identify significant differences between the same lockdown phases of 2019 and 2020. The results of the Mann–Whitney test were carried out at a $p < 0.05$ significance level. The following equation was applied for the Mann–Whitney U test (M–W test):

$$U_1 = n_1 n_2 + \frac{n_1(n_1+1)}{2} - R_1 \qquad U_2 = n_1\, n_2 + \frac{n_2(n_2+1)}{2} - R_1$$

where, n_1 and n_2 are the sample size and R_1 and R_2 represents the sum of ranks, respectively.

3. Results

3.1. Surface Biophysical Parameters

The results showed that the mean LST was 59.48°C in 2020. The highest LST was observed during phase II (64.27°C), followed by phase I (58.15°C),andphase III (56.03°C) in 2020. The LST was relatively lower in 2020 during the entire lockdown phase compared to 2019 during the same period (Figure 3). The results revealed that the average LST was 60.73°C in 2019 (during the same periods of lockdown), with the highest LST recorded from phase II (64.91°C), followed by phase III (59.3°C),and phase I (58°C) (Table 3). As a result, it was observed that there were no significant differences in LST in the different phases of lockdown ($p > 0.05$).

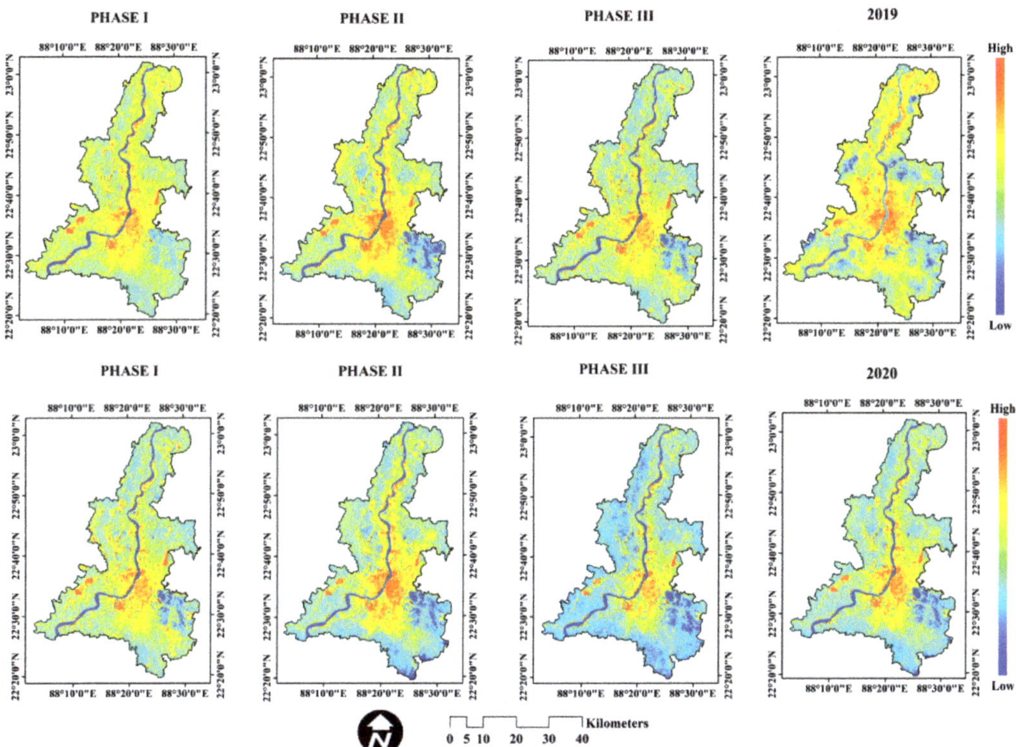

Figure 3. Pattern of LST during 2019 and different phases of lockdown in 2020.

Table 3. Mean (Phase I, II, and III) and coefficient of variation (CV) of surface characteristics of LST (°C), NDVI, NDSI, and wetness in KMA in 2020 (during a different lockdown phase) and 2019 (same periods of lockdown).

Year	Indices	PI	PII	PIII	Mean	CV
2019	LST (°C)	58	64.91	59.3	60.74	0.060
	NDVI	0.14	0.19	0.2	0.18	0.107
	NDSI	0.51	0.26	0.35	0.37	0.703
	Wetness	0.24	0.16	0.22	0.21	0.198
2020	LST (°C)	58.15	64.27	56.03	59.48	0.072
	NDVI	0.16	0.22	0.21	0.20	0.161
	NDSI	0.41	0.22	0.33	0.32	0.298
	Wetness	0.25	0.21	0.22	0.23	0.091

The mean NDVI value was 0.20 in 2020. The highest NDVI was recorded during phase II (0.22), followed by phase III (0.21), and phase I (0.16) in 2020, respectively. The NDVI value slightly increased from 0.18 (2019) to 0.20 (2020). On the other hand, the mean NDVI value during the same period of lockdown in 2019 was 0.18, with the highest NDVI observed during phase III (0.20), followed by phase II (0.19),and phase I (0.14) (Figure 4 and Table 3). From the MW test, it was observed that there were no significant differences in NDVI during the different phases of lockdown in 2019 and 2020 ($p > 0.05$).

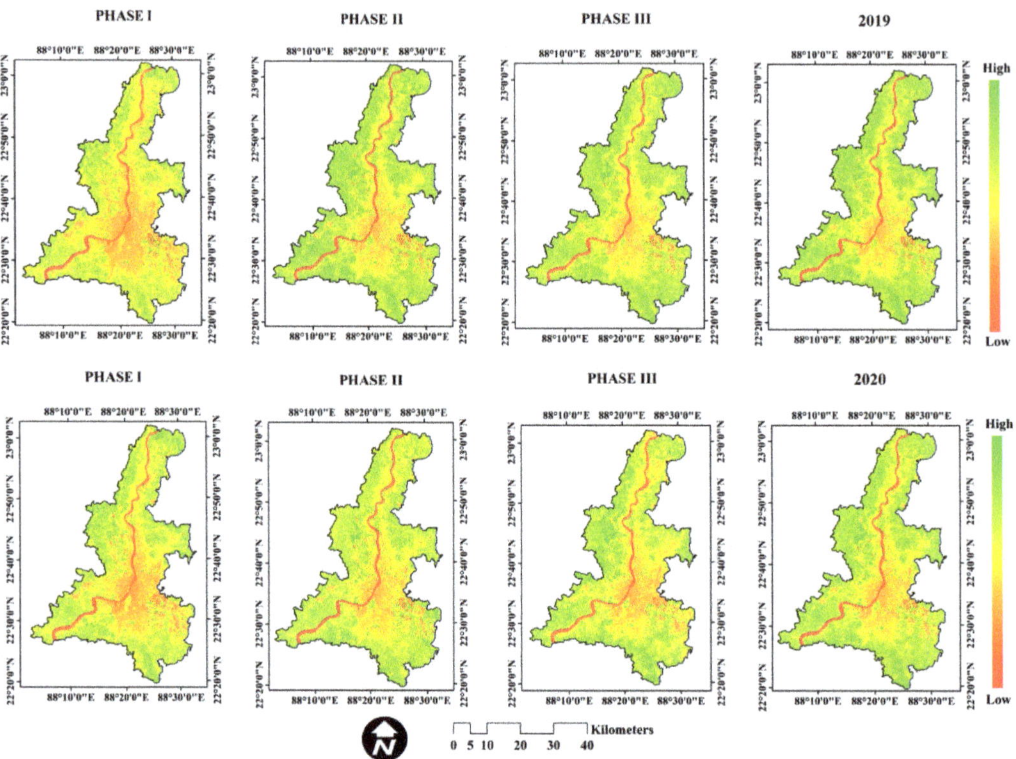

Figure 4. Pattern of NDVI during 2019 and different phases of lockdown in 2020.

There was also substantial variation in NDSI during 2019 and 2020. The mean NDSI value in 2020 was 0.32, with the highest NDSI value recorded in phase I (0.41), followed by phase III (0.33), and phase II (0.22), respectively. In 2019, the NDSI value was higher (0.37) in comparison to 2020 (0.32). In 2019, the highest NDSI value was recorded in phase I (0.56), followed by phase III (0.37), and phase I (0.26), respectively (Figure 5 and Table 3). As per the MW test, it was observed that there were no significant differences in the NDSI value between 2020 and 2019 ($p > 0.05$).

Mean wetness values during lockdown periods slightly increased in comparison to 2019. The result showed that the mean wetness values were 0.23 in 2020 and 0.21 in 2019. During the lockdown periods, the highest wetness value was recorded in phase I (0.25), followed by phase III (0.23), and phase I (0.22) in 2020. In 2019 highest wetness value was observed in phase I (0.24), followed by phase III (0.22), and phase I (0.16), respectively (Figure 6 and Table 3). The MW test result showed no statistically significant differences ($p > 0.05$) in the wetness values during the different phases of lockdown in 2019 and 2020. Thus, from the overall result, it was observed that, though there were variations in the USES parameters in the various phases of lockdown in 2019 and 2020, statistical there were no significant differences.

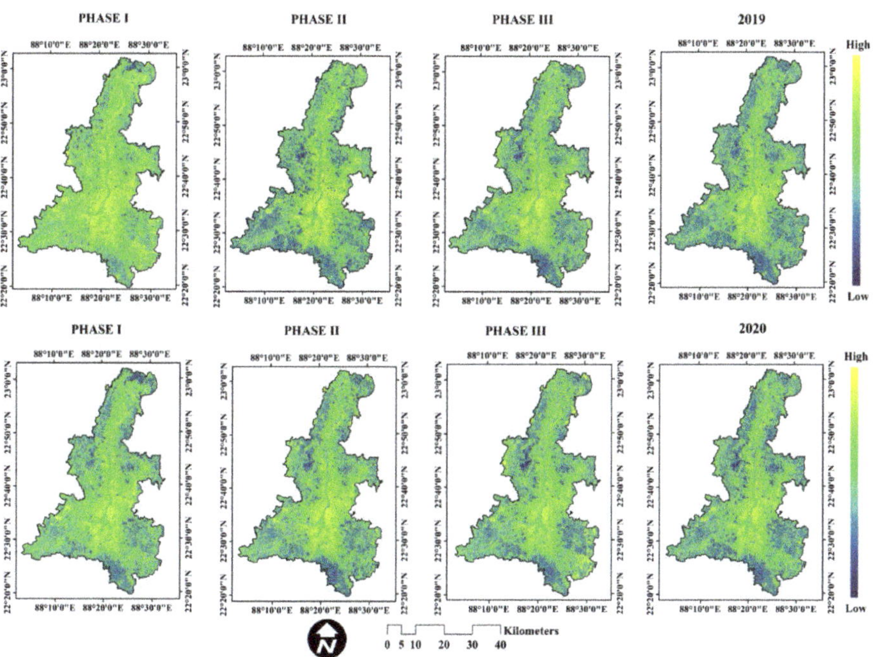

Figure 5. Pattern of NDSI during 2019 and different phases of lockdown in 2020.

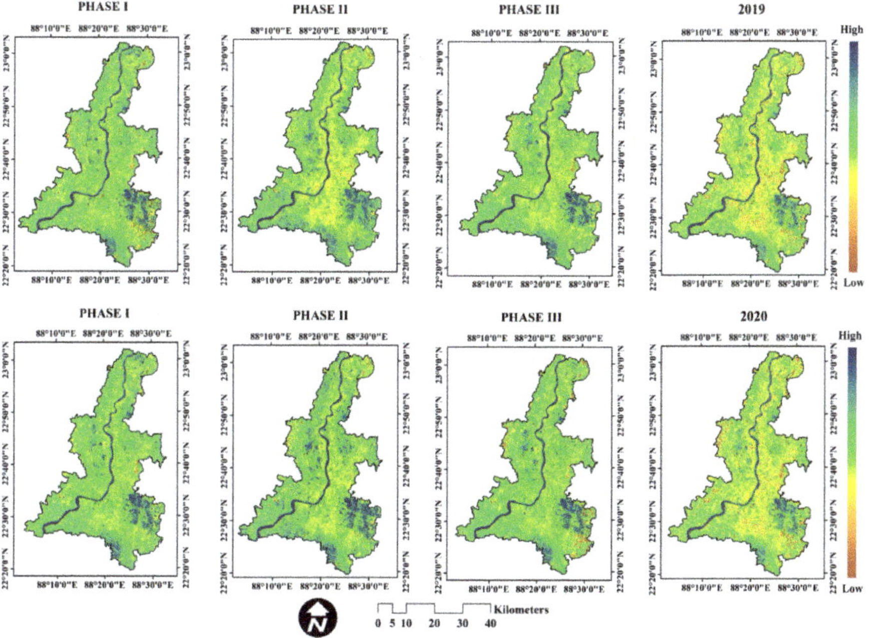

Figure 6. Pattern of wetness during 2019 and different phases of lockdown in 2020.

3.2. USES

The USES was developed for 2019 and 2020 and compared to last year during the same lockdown periods to better understand the impact of lockdown on USES (Table 4). There were slight improvements in the ecological status during lockdown periods (2020) compared to the previous year (2019). For example, the mean value of the USES was 0.49 in 2019 and reached 0.34 during the lockdown periods (decreased by more than 30%). In 2019, the highest USES value was observedin phase I (0.59), followed by phase III (0.48), and phase II (0.41), respectively. Similarly, during the lockdown periods, the highest USES value was observedfrom phase I (0.38), followed by phase III (0.33) and phase II (0.31), respectively. As per comparison, the highest percentage of USES value decrease was recordedin phase I (reduced by 35%) during the lockdown periods, in contrast to 2019, followed by phase III (31%), and phase II (24%). We found important differences in the USES found between the built-up and non-built-up areas of the KMA. The USES value ranged from 0 to 1, where a value close to '0' indicated a relatively better USES and a value close to '1' showed a relatively worse USES, respectively (Figure 7). As per the maps (Figure 7), it was observed that the USES was relatively worsein urban areas (particularly along the Hoogly river) and relatively better in the fringe area of the KMA.

Table 4. Pattern of USES during lockdown periods (2020) and same periods in 2019.

Year	PI	PII	PIII	Mean	CV
2019	0.59	0.41	0.48	0.49	0.184
2020	0.38	0.31	0.33	0.34	0.106

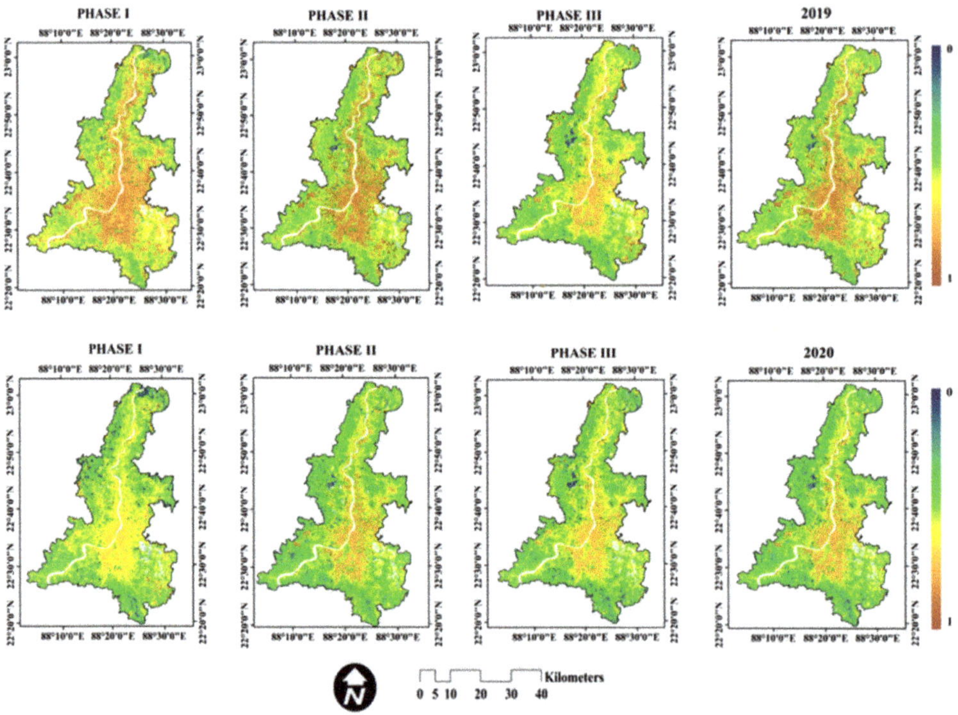

Figure 7. Spatio-temporal patterns of USES in 2019 and different phases of lockdown in 2020.

3.3. USES Spatial Pattern

As per the result of USES, the area under poor USES (>0.80) slightlydecreased during lockdown periods (from 30% to 25%). During lockdown phase III, areasunder fair and poor USES reduced from 29% (2019) to 24% (2020), and 33% (2019) to 25% (2020). Thus, during the entire lockdown phase, the percentage of areas with poor USES reduced from 31% (2019) to 25% (2020). Similarly, during different phases of lockdown, the percentage of the area under the excellent USES increased, with a maximum increase from 16% (2019) to 28% (2020), during lockdown phase II, followed by phase II (from 13% to 15%), respectively. The area under excellent USES during the entire phase of lockdown reached 20% in 2020. Thus, from the overall findings, it was clear that USES improved during the lockdown in the KMA. However, the MW test did not identify significant differences in 2019 and 2020 ($p > 0.05$).

An important variability in the USES was observed in the KMA (Figure 8). Most of the areas along the Hoogly river had a poor and fair USES. The river's eastern bank had a high area covered by poor USES compared to the western bank. More particularly, the urban areas around Kolkata Municipal Corporation were highly characterised by poor USES. Other urban areas, namely Howrah, Baly, Baranagar, Kamarhati, Panihati, Bidhannagar, South Dumdum, North Dumdum, and New Barrackpore were mainly characterised by relatively poor USES (Figure A1). The KMA rural fringe had a good and excellent USES in the different lockdown phases in 2019 and 2020 (Figure 7).

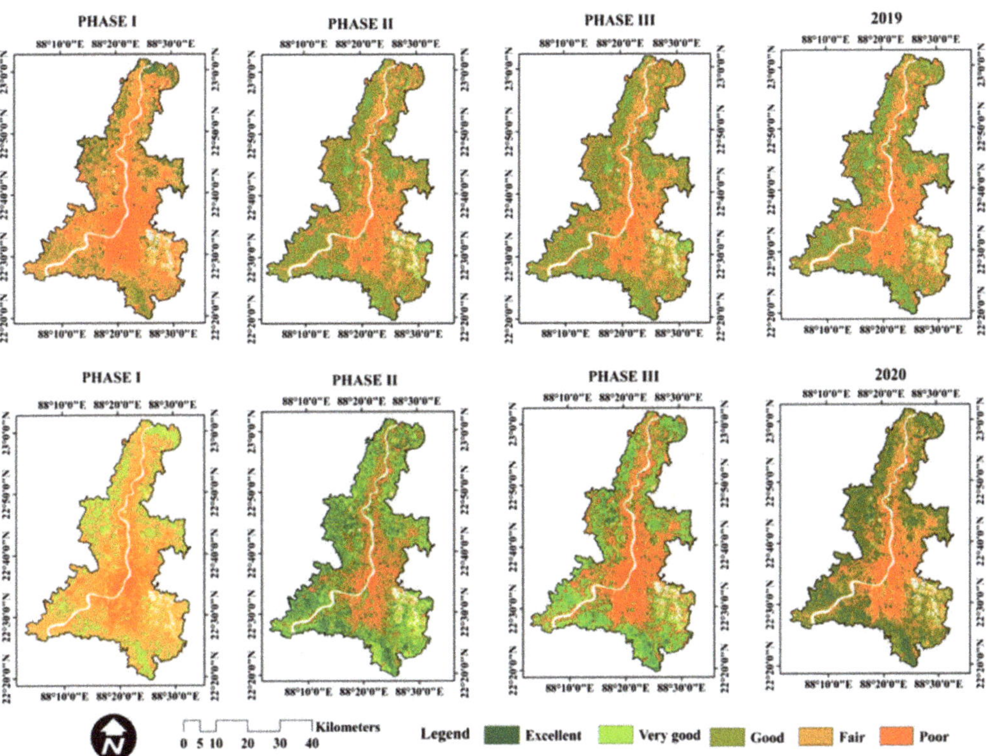

Figure 8. Spatio-temporal patterns of classified USES in 2019 and different phases of lockdown in 2020.

4. Discussion

There was a high spatial variability of the indices measured in KMA. The spatial heterogeneity of the USES existed in the KMA due to micro-level environmental variations and land surface properties [51–53]. The fringe areas of the KMA (rural areas) are characterised by a higher vegetation and water bodies coverage than urban areas. Thus, the variation of landscape configuration is the main cause of the spatial variation of spectral indices and the USES in the KMA. The conversion of natural land covers (such as vegetation cover and water bodies) into built-up areas reduces the surface's greenness and wetness. This increases the surface's dryness, heat, and imperviousness [4,54]. Figures 7 and 8 show that, during all the phases of lockdown in 2020 and the same periods of 2019, the greenness and wetness were higher in the rural fringe areas, and dryness and heat were higher in rural fringe areas than in the urban areas of the KMA. In the urban areas, the impervious surface cover, lack of vegetation cover and evapotranspiration increased the surface heat and dryness [55,56]. In the USES maps, it was observed that the areas with impervious and open land recorded a higher LST and NDSI. In addition to this, the ecological statuses of the surface were largely influenced by anthropogenic activities [14]. Previous studies stated that the USES was highly affected by the functions of biophysical parameters such as the NDVI, normalised water index (NDWI), NDSI, SAVI(Soil-Adjusted Vegetation Index), and LST [3,12]. We observed that areas with a high greenness and wetness and a low LST and NDSI, had a high USES (Figures 5 and 6). The surface areas with vegetation and agricultural lands had a high USES, and built-up and open lands (barren land) had reduced USESs. The leading causes for the high USES in areas with vegetation and agricultural lands were the low surface heat and dryness and relatively higher greenness and wetness. More particularly, the urban areas with impervious surface areas had poor and fair USES in KMA. From the maps, it was observed that, although there were spatio-temporal variations of the USES in the KMA in 2019 and 2020 during different phases of lockdown, no significant differences were observed. Bio-physical conditions largely influenced the USES due to anthropogenic pressures and alterations in the impervious surface configurations [4]. In addition to this, human activities affected the surface cover and altered the thermal capacity, albedo, conductivity, and evapotranspiration, respectively [4,57,58].

Urban environment areas in developing countries have increased rapidly in the past few decades [47,59,60]. Human activities have increased negative environmental impacts, such as the urban heat island effect and USES degradation [35,49,61], leading to decreased wellbeing [3,12,62]. Previous studies highlighted the importance of reducing anthropogenic pressures on the environment to increase the ecological quality of urban areas [63,64]. In this context, the outbreak of the COVID-19 pandemic provides us with a golden opportunity to identify the impact of human activities [39]. The strict restrictions on human mobility, and the closure of industrial activities and other productive activities, resulted in a reduction in pressures on the environment [65]. Thus, the lockdown due to COVID-19 significantly improved environmental quality [7].

There are a very limited number of studies on the impact of lockdown on ecological status [8]. Aside from ecology, a number of studies were performed on the impact of lockdown on air quality, water quality, and noise [34–36]. From the studies, it was documented that the air quality and water quality significantly improved due to the restricted emissions and strict prohibitions on transportation and industrial activities [37,66,67]. This study showed that the USES value was 0.49 in 2019 and reached 0.34 in 2020. This indicates that USES slightly improved during the lockdown periods compared to during the same lockdown periods in 2019. In the KMA, restricted measures were imposed on transport, industrial activities, and human mobility [40]. The slight improvement in USES could be attributed to the limited anthropogenic activities due to strict lockdown. Previous works documented that any ecological conditions were largely influenced by the anthropogenic pressures on the environment [68,69].

4.1. Limitations and Uncertainties of the Work

This study attempted to understand the impact of the COVID-19 lockdown on the USES in the KMA. Thus, the study's findings may be helpful to understand the urban surface ecological status of the Kolkata megacity region and other rapidly growing cities in India. However, a few limitations are identified. This study assessed the impact of lockdown on the USES in 2020 and 2019 during the same periods of lockdown phases. The USESs in 2020 were compared with the previous year (2019) to better understand the impact of the lockdown. Thus, in future studies, the long-term impact of COVID-19 amid lockdown on USES must be considered. Secondly, a total of ten satellite images (five for each year) from three phases of lockdown were used to assess the USES. These images may not be representative of the lockdown phases. Therefore, more images can be taken into consideration for a better analysis of the USES in future. This study used four indices (NDVI, NDSI, wetness and LST) to develop the USES. In future studies, other relevant indices should be taken into account to model the USES. Lastly, the KMA comprises both urban as well as rural areas. The results showed that a lower USES characterised urban areas more than rural areas (located in fringe areas of the KMA). However, this study made no comparison between the urban and rural units to assess the spatial variability of the spectral indices and the USES in the KMA. In future, the regional heterogeneity of the USES must be addressed.

4.2. Implication of Urban Ecological Restoration and Management Policies

From the previous studies, it was well documented that the emergence of the COVID-19 pandemic significantly improved different aspects of the environment, such as air quality [70–72], water quality [73,74], the reduction in noise pollution [74,75], and the improvement of ecosystems [76,77].The unsustainable anthropogenic activities in the urban environments of developing countries alter the urban environmental conditions, for example: the emergence of the urban heat island (UHI) effect, the loss of ecosystem services, the degradation of USES, and increased thermal discomfort conditions [57,61,67]. This study indicates that environmental quality of urban environments can be improved through limiting the human pressures on the environment and ecosystems. The results show that the USES status slightly improved during the lockdown periods compared to the same periods of the previous year (2019).This indicates that the urban ecological status can be improved by reducing and restricting human activities on natural, semi-natural, and artificial ecological landscapes [63,64,78]. During the lockdown periods, strict restrictions were imposed on the human mobility and the use of public spaces (such as parks, gardens), industrial activities and transportation were partially shut down. Thus, the restricted use of green spaces, limited pressures on the landscape, and the partial banning of industrial activities and transportation helped slightly to improve the ecological status in the KMA during the lockdown periods in 2020 as compared to the same periods lockdowns in previous year (2019). The study's findings suggest that the ecological restoration or urban ecological conditions can be enhanced through the restricted human pressures on urban ecology. The outbreak of COVID-19 compelled people to stay away from nature and its benefits for a long time, which caused a tremendous socio-economic burden. Therefore, urban ecological sustainability cannot be achieved by simply enhancing the conditions of a long-term lockdown (due to COVID-19). In a few recent studies, it was observed that there was substantial loss of ecological landscapes such as vegetation cover, water bodies, and the loss of the East Kolkata Wetland (EKW), which resulted in the deteriorations of environmental health [25,30,66,67]. In this context, short-term lockdown can be an alternative tool to achieve urban ecological restoration and ecosystem management.

5. Conclusions

The present study mainly focused on modelling the impact of the COVID-19 lockdown on the USES in the KMA, Eastern India. This was compared with the year 2019 to better understand the effect of lockdown. The spectral indices such as greenness (NDVI), dryness

(NDSI), moisture (wetness), and heat (LST) were used to develop a remotely sensed urban surface ecological status index (RSUSESI). From the results, it was recorded that the greenness (NDVI) and wetness (moisture) conditions slightly increased, and the dryness (NDSI) and heat (LST) slightly decreased, during the lockdown periods. Our findings demonstrated that USES during the lockdown periods improved somewhat in comparison to 2019 during the same periods. The fair to poor USESs were mainly concentrated in urban areas due to impervious surface cover, resulting in a higher heat and dryness, and a lower greenness and wetness, respectively. On the other hand, the rural fringe areas were characterised by excellent to good USESs due to a relatively higher greenness and wetness, and a lower heat and dryness, respectively.

Various aspects of human lives were adversely affected by lockdown in both urban and non-urban environments. The restricted anthropogenic activities in urban areas resulted in the important improvement of USES during the lockdown periods. During the lockdown periods, there were strict restrictions on human mobility, limited access to public space use, and restrictions on transportation and industrial activities. The outbreak of COVID-19 brought about two different sides of the same coin, i.e., the great loss of human lives and the restoration of ecosystems. Thus, from the findings of our study, two notable conclusions may be drawn. Firstly, anthropogenic activities are the main drivers of altering the environment and ecological conditions (directly and indirectly). Secondly, environmental restoration can be achieved (to some extent) through restricted interruptions on the environment. However, the findings of our study suggest that the ecological restoration of the urban areas can be achieved by limiting the anthropogenic activities and pressures on the environment. Notably, in the Indian context, it is essential due to unplanned urban development and the notable deteriorations of ecosystems and their services [66].

Author Contributions: Conceptualization, M.D., A.D., P.P., A.M.; methodology, M.D., A.M., A.D., P.P.; software, M.D., A.M.; formal analysis, M.D., A.D., P.P.; data curation, M.D., A.M.; writing—original draft preparation, M.D., A.D.; writing—review and editing, P.P., A.D., M.D. All authors have read and agreed to the published version of the manuscript.

Funding: This research received no external findings.

Data Availability Statement: The used in this study are available from the corresponding author upon reasonable request.

Acknowledgments: The authors would like to thank USGS for providing data for the study.

Conflicts of Interest: The authors declare no conflict of interest.

Appendix A

Table A1. Previous literature on the alterations of land surface cover in Kolkata Megacity Region.

Study Scale	Reference	Published Year	Study Period	Major Findings
Kolkata Metropolitan Area (KMA)	[66]	2020	2000–2019	Ecosystem Health since 2000 to 2019 declined from 73% to 52% due to mainly rapid built-up areas expansion and loss of vegetation cover.
Kolkata Metropolitan Area (KMA)	[67]	2021	2000–2019	Built-up area increased by about 90% and vegetation cover decreased by about 56% from 2000 to 2019.
Kolkata Metropolitan Area (KMA)	[79]	2020	1990–2020	In the last 30 years (1990 to 2020), cropland area declined by 181 km^2. In core zone (144 municipalities), between 2020 to 2020,built-up areas increased by about 29.37% and wetland and cropland area decreased by 25.66% and 26.43%, respectively.

Table A1. *Cont.*

Study Scale	Reference	Published Year	Study Period	Major Findings
South Kolkata	[80]	2021	2009–2019	Built-up area increased by about 22.11% and vegetation cover decreased by about 5.78%.
East Kolkata Wetland	[81]	2017	2000–2010	Since 2000 to 2011, net loss of wetland was 13.16 km^2 due to built-up growth. 4.76 km2 area of wetland was converted to cropland.
Kolkata Metropolitian Area (KMA)	[82]	2015	2000–2015	Built-up area increased by about 55% and vegetation cover declined by about 25%. Agricultural land decreased(up to 6%) due to built-up expansion.
Pujali Municipality (KMA)	[83]	2017	1980–2015	Built-up area increased by about 25%; vegetation cover and water bodies decreased by about 50%, respectively.
Kolkata Metropolitan Area (KMA)	[84]	2018	1990–2017	Built-up area was increased by about 202% from1990 to 2017 and vegetation cover decreased by about 4%, respectively.
East Kolkata Wetland	[85]	2013	1973–2010	Wetland area reduced by about 26% followed by agricultural land. Built-upareas increased by about 166%.
of Kolkata Urban Agglomeration	[86]	2019	1990–2015	In thelast 25 years, built-up and agricultural land increased by 45% and 62%, respectively. On the other hand, agricultural land and vegetation cover decreased by about 35% and 12%, respectively. Built-up area increased due to conversion of agricultural and open land into built-up area.
Kolkata Municipal Corporation (KMC)	[87]	2021	1980–2018	Low, dense, fragmented built-up areas increased by about 95% and other ecological landscapes significantly decreased, such as vegetation cover (69%), grass land (51%), water bodies (27%), wetland (58%), cropland (56%), respectively.
Howrah Municipal Corporation (HMC)	[88]	2018	1975–2015	In the last 40 years, vegetation cover, agricultural land, water bodies and wetland declined by 14%, 23%, 12% and 10%, respectively. On the other hand, built-up area increased by about 58%.
East Kolkata Wetlands	[89]	2016	1972–2011	Wetland area was reduced by about 28.1 km^2 (decreased by 18%) followed by agricultural land (26%). Wetland decreased due to conversion of wetland intobuilt-up and other land covers.
Kolkata Metropolitan Area	[27]	2019	1991–2017	Vegetation cover and agricultural land decreased by about 16% and 12%, respectively. Moderate dense built-up areas increased by about 23%.
Kolkata and surrounding periphery	[90]	2014	1997–2017	Forests, low vegetation and agricultural land declined by 40%, 8%, and 20%, respectively. Built-up areas increased by 67%.
Kolkata City	[30]	2015	1989–2010	Dense settlement area increased by about 39% and vegetation and wetland vegetation decreased from 178 to 109 km^2 and 34 to 15 km^2.
Kolkata Urban Agglomeration	[28]	2018	1990–2015	Vegetation cover, wetland and agricultural land decreased by about 6.6%, 5.9%, and 26%. Built-up area increased by 24.5%. From 2000 to 2015, 103.7 km^2 agricultural lands were converted into built-up areas.
Kolkata Megacity	[91]	2019	1991–2018	From1991-2018, built-up areas increased by more than 200% and water bodies, dense vegetation and sparse vegetation cover declined by 14%, 47%, and 45%, respectively.

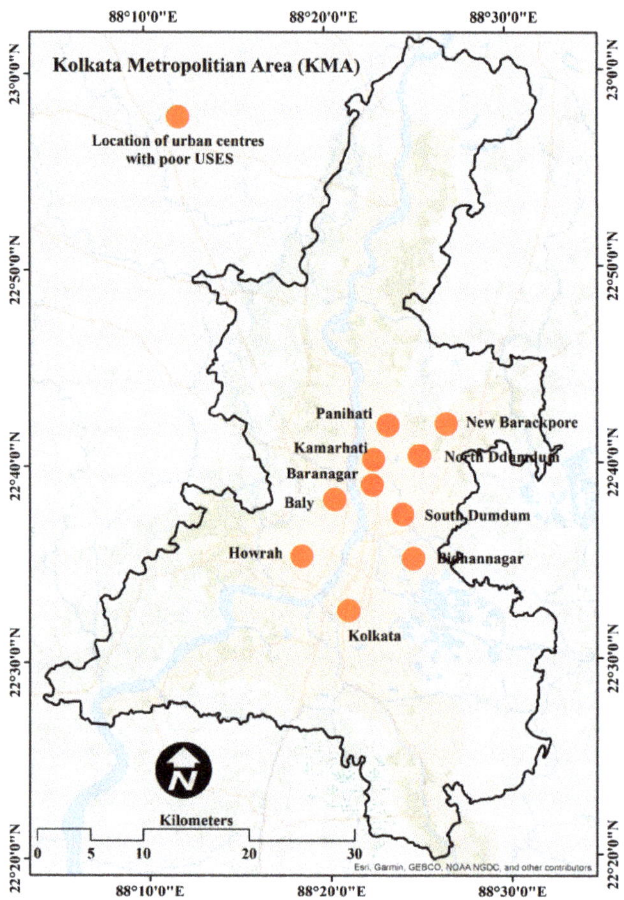

Figure A1. Location of some urban centres with poor USES.

References

1. Lin, T.; Ge, R.; Huang, J.; Zhao, Q.; Lin, J.; Huang, N.; Zhang, G.-Q.; Li, X.; Ye, H.; Yin, K. A quantitative method to assess the ecological indicator system's effectiveness: A case study of the Ecological Province Construction Indicators of China. *Ecol. Indic.* **2016**, *62*, 95–100. [CrossRef]
2. Qureshi, S.; Alavipanah, S.K.; Konyushkova, M.; Mijani, N.; Fathololomi, S.; Firozjaei, M.K.; Homaee, M.; Hamzeh, S.; Kakroodi, A.A. A Remotely Sensed Assessment of Surface Ecological Change over the Gomishan Wetland, Iran. *Remote Sens.* **2020**, *12*, 2989. [CrossRef]
3. Xu, H.; Wang, M.; Shi, T.; Guan, H.; Fang, C.; Lin, Z. Prediction of ecological effects of potential population and impervious surface increases using a remote sensing based ecological index (RSEI). *Ecol. Indic.* **2018**, *93*, 730–740. [CrossRef]
4. Fu, Y.; Li, J.; Weng, Q.; Zheng, Q.; Li, L.; Dai, S.; Guo, B. Characterising the spatial pattern of annual urban growth by using time series Landsat imagery. *Sci. Total Environ.* **2019**, *666*, 274–284. [CrossRef] [PubMed]
5. Jin, K.; Wang, F.; Chen, D.; Liu, H.; Ding, W.; Shi, S. A new global gridded anthropogenic heat flux dataset with high spatial resolution and long-term time series. *Sci. Data* **2019**, *6*, 139. [CrossRef] [PubMed]
6. United Nations, Population Division, Department of Economic and Social Affairs. *World Urbanization Prospects: The 2018 Revision*; United Nations: New York, NY, USA, 2019.
7. Das, A.; Ghosh, S.; Das, K.; Basu, T.; Dutta, I.; Das, M. Living environment matters: Unravelling the spatial clustering of COVID-19 hotspots in Kolkata megacity, India. *Sustain. Cities Soc.* **2021**, *65*, 102577. [CrossRef] [PubMed]
8. Firozjaei, M.K.; Fathololomi, S.; Kiavarz, M.; Arsanjani, J.J.; Homaee, M.; Alavipanah, S.K. Modeling the impact of the COVID-19 lockdowns on urban surface ecological status: A case study of Milan and Wuhan cities. *J. Environ. Manag.* **2021**, *286*, 112236. [CrossRef]

9. Willis, K.S. Remote sensing change detection for ecological monitoring in United States protected areas. *Biol. Conserv.* **2015**, *182*, 233–242. [CrossRef]

10. Xu, H.; Wang, Y.; Guan, H.; Shi, T.; Hu, X. Detecting Ecological Changes with a Remote Sensing Based Ecological Index (RSEI) Produced Time Series and Change Vector Analysis. *Remote Sens.* **2019**, *11*, 2345. [CrossRef]

11. Das, M.; Das, A. Exploring the pattern of outdoor thermal comfort (OTC) in a tropical planning region of eastern India during summer. *Urban Clim.* **2020**, *34*, 100708. [CrossRef]

12. Zhu, D.; Chen, T.; Wang, Z.; Niu, R. Detecting ecological spatial-temporal changes by Remote Sensing Ecological Index with local adaptability. *J. Environ. Manag.* **2021**, *299*, 113655. [CrossRef]

13. Chang, N.-B.; Sun, Z.; Opp, C. Using SPOT-VGT NDVI as a successive ecological indicator for understanding the environmental implications in the Tarim River Basin, China. *J. Appl. Remote Sens.* **2010**, *4*, 043554. [CrossRef]

14. Das, M.; Das, A. Estimation of Ecosystem Services (EESs) loss due to transformation of Local Climatic Zones (LCZs) in Sriniketan-Santiniketan Planning Area (SSPA)West Bengal, India. *Sustain. Cities Soc.* **2019**, *47*, 101474. [CrossRef]

15. Das, M.; Das, A.; Seikh, S.; Pandey, R. Nexus between indigenous ecological knowledge and ecosystem services: A socio-ecological analysis for sustainable ecosystem management. *Environ. Sci. Pollut. Res.* **2021**, 1–18. [CrossRef]

16. Sharma, S.; Nahid, S.; Sharma, M.; Sannigrahi, S.; Anees, M.M.; Sharma, R.; Shekhar, R.; Basu, A.S.; Pilla, F.; Basu, B.; et al. A long-term and comprehensive assessment of urbanization-induced impacts on ecosystem services in the capital city of India. *City Environ. Interact.* **2020**, *7*, 100047. [CrossRef]

17. Bhagat, R.B.; Mohanty, S. Emerging pattern of urbanisation and the contribution of migration in urban growth in India. *Asian Popul. Stud.* **2019**, *5*, 5–20. [CrossRef]

18. Taubenböck, H.; Wegmann, M.; Roth, A.; Mehl, H.; Dech, S. Urbanisation in India–Spatiotemporal analysis using remote sensing data. *Comput. Environ. Urban Syst.* **2009**, *33*, 179–188. [CrossRef]

19. Connor, R. *The United Nations World Water Development Report 2015: Water for a Sustainable World*; UNESCO Publishing: Paris, France, 2015; Volume 1.

20. International Resource Panel, United Nations Environment Programme, Sustainable Consumption, and Production Branch. *Decoupling Natural Resource use and Environmental Impacts from Economic Growth*; UNEP/Earthprint: Nairobi, Kenya, 2011.

21. Ramachandra, T.V.; Bharath, A.H.; Sowmyashree, M.V. Monitoring urbanisation and its implications in a mega city from space: Spatiotemporal patterns and its indicators. *J. Environ. Manag.* **2015**, *148*, 67–81. [CrossRef] [PubMed]

22. Jain, S.; White, M.; Radivojac, P. Estimating the class prior and posterior from noisy positives and unlabeled data. *arXiv* **2016**, arXiv:1606.08561.

23. Bhatta, B. Analysis of urban growth pattern using remote sensing and GIS: A case study of Kolkata, India. *Int. J. Remote Sens.* **2009**, *30*, 4733–4746. [CrossRef]

24. Ramachandra, T.; Bharath, S.; Gupta, N. Modelling landscape dynamics with LST in protected areas of Western Ghats, Karnataka. *J. Environ. Manag.* **2018**, *206*, 1253–1262. [CrossRef]

25. Mondal, S.; Kundu, D.; Haque, S.; Senapati, T.; Ghosh, A.R. Seasonal variation of zooplankton distribution in sewage-fed East Kolkata wetland, West Bengal, India. *Pollut. Res.* **2015**, *34*, 477–787.

26. Kundu, D.; Mondal, S.; Dutta, D.; Haque, S.; Ghosh, A.R. Accumulation and contamination of lead in different trophic levels of food chain in sewage-fed East Kolkata Wetland, West Bengal, India. *Int. J. Env. Tech. Sci.* **2016**, *2*, 61–68.

27. Ghosh, S.; Das, A. Urban expansion induced vulnerability assessment of East Kolkata Wetland using Fuzzy MCDM method. *Remote Sens. Appl. Soc. Environ.* **2019**, *13*, 191–203. [CrossRef]

28. Sahana, M.; Hong, H.; Sajjad, H.; Liu, J.; Zhu, A.-X. Assessing deforestation susceptibility to forest ecosystem in Rudraprayag district, India using fragmentation approach and frequency ratio model. *Sci. Total Environ.* **2018**, *627*, 1264–1275. [CrossRef]

29. Ghosh, S.; Das Chatterjee, N.; Dinda, S. Relation between urban biophysical composition and dynamics of land surface temperature in the Kolkata metropolitan area: A GIS and statistical based analysis for sustainable planning. *Model. Earth Syst. Environ.* **2019**, *5*, 307–329. [CrossRef]

30. Sharma, R.; Chakraborty, A.; Joshi, P.K. Geospatial quantification and analysis of environmental changes in urbanising city of Kolkata (India). *Environ. Monit. Assess.* **2015**, *187*, 4206. [CrossRef]

31. Nimish, G.; Bharath, H.; Lalitha, A. Exploring temperature indices by deriving relationship between land surface temperature and urban landscape. *Remote Sens. Appl. Soc. Environ.* **2020**, *18*, 100299. [CrossRef]

32. Sarkar, U.K.; Mishal, P.; Borah, S.; Karnatak, G.; Chandra, G.; Kumari, S.; Meena, D.K.; Debnath, D.; Yengkokpam, S.; Das, P.; et al. Status, Potential, Prospects, and Issues of Floodplain Wetland Fisheries in India: Synthesis and Review for Sustainable Management. *Rev. Fish. Sci. Aquac.* **2021**, *29*, 1–32. [CrossRef]

33. Chakraborty, I.; Maity, P. COVID-19 outbreak: Migration, effects on society, global environment and prevention. *Sci. Total Environ.* **2020**, *728*, 138882. [CrossRef]

34. Muhammad, S.; Long, X.; Salman, M. COVID-19 pandemic and environmental pollution: A blessing in disguise? *Sci. Total Environ.* **2020**, *728*, 138820. [CrossRef] [PubMed]

35. Nakada, L.Y.K.; Urban, R.C. COVID-19 pandemic: Impacts on the air quality during the partial lockdown in São Paulo state, Brazil. *Sci. Total Environ.* **2020**, *730*, 139087. [CrossRef]

36. Zambrano-Monserrate, M.A.; Ruano, M.A.; Sanchez-Alcalde, L. Indirect effects of COVID-19 on the environment. *Sci. Total Environ.* **2020**, *728*, 138813. [CrossRef] [PubMed]

37. Berman, J.D.; Ebisu, K. Changes in U.S. air pollution during the COVID-19 pandemic. *Sci. Total Environ.* **2020**, *739*, 139864. [CrossRef]

38. Panda, S.; Mallik, C.; Nath, J.; Das, T.; Ramasamy, B. A study on variation of atmospheric pollutants over Bhubaneswar during imposition of nationwide lockdown in India for the COVID-19 pandemic. *Air Qual. Atmos. Health* **2021**, *14*, 97–108. [CrossRef]

39. Das, M.; Das, A.; Sarkar, R.; Saha, S.; Mandal, A. Examining the impact of lockdown (due to COVID-19) on ambient aerosols (PM 2.5): A study on Indo-Gangetic Plain (IGP) Cities, India. *Stoch. Environ. Res. Risk Assess.* **2021**, *35*, 1301–1317. [CrossRef] [PubMed]

40. Bera, B.; Bhattacharjee, S.; Shit, P.K.; Sengupta, N.; Saha, S. Significant impacts of COVID-19 lockdown on urban air pollution in Kolkata (India) and amelioration of environmental health. *Environ. Dev. Sustain.* **2021**, *23*, 6913–6940. [CrossRef]

41. Pettorelli, N.; Vik, J.O.; Mysterud, A.; Gaillard, J.-M.; Tucker, C.J.; Stenseth, N.C. Using the satellite-derived NDVI to assess ecological responses to environmental change. *Trends Ecol. Evol.* **2005**, *20*, 503–510. [CrossRef] [PubMed]

42. Ridd, M.K. Exploring a V-I-S (vegetation-impervious surface-soil) model for urban ecosystem analysis through remote sensing: Comparative anatomy for cities. *Int. J. Remote Sens.* **1995**, *16*, 2165–2185. [CrossRef]

43. Robinson, N.P.; Allred, B.W.; Jones, M.O.; Moreno, A.; Kimball, J.S.; Naugle, D.E.; Erickson, T.A.; Richardson, A.D. A Dynamic Landsat Derived Normalized Difference Vegetation Index (NDVI) Product for the Conterminous United States. *Remote Sens.* **2017**, *9*, 863. [CrossRef]

44. Hu, X.; Xu, H. A new remote sensing index for assessing the spatial heterogeneity in urban ecological quality: A case from Fuzhou City, China. *Ecol. Indic.* **2018**, *89*, 11–21. [CrossRef]

45. Zhu, X.; Wang, X.; Yan, D.; Liu, Z.; Zhou, Y. Analysis of remotely-sensed ecological indexes' influence on urban thermal environment dynamic using an integrated ecological index: A case study of Xi'an, China. *Int. J. Remote Sens.* **2018**, *40*, 3421–3447. [CrossRef]

46. Jiménez-Muñoz, J.C.; Sobrino, J.A.; Skokovic, D.; Mattar, C.; Cristóbal, J. Land Surface Temperature Retrieval Methods from Landsat-8 Thermal Infrared Sensor Data. *IEEE Geosci. Remote Sens. Lett.* **2014**, *11*, 1840–1843. [CrossRef]

47. Weng, Q.; Firozjaei, M.K.; Kiavarz, M.; Alavipanah, S.K.; Hamzeh, S. Normalizing land surface temperature for environmental parameters in mountainous and urban areas of a cold semi-arid climate. *Sci. Total Environ.* **2019**, *650*, 515–529. [CrossRef] [PubMed]

48. Baig, M.H.A.; Zhang, L.; Shuai, T.; Tong, Q. Derivation of a tasselled cap transformation based on Landsat 8 at-satellite reflectance. *Remote Sens. Lett.* **2014**, *5*, 423–431. [CrossRef]

49. Mijani, N.; Alavipanah, S.K.; Firozjaei, M.K.; Arsanjani, J.J.; Hamzeh, S.; Weng, Q. Modeling outdoor thermal comfort using satellite imagery: A principle component analysis-based approach. *Ecol. Indic.* **2020**, *117*, 106555. [CrossRef]

50. Firozjaei, M.K.; Alavipanah, S.K.; Liu, H.; Sedighi, A.; Mijani, N.; Kiavarz, M.; Weng, Q. A PCA–OLS Model for Assessing the Impact of Surface Biophysical Parameters on Land Surface Temperature Variations. *Remote Sens.* **2019**, *11*, 2094. [CrossRef]

51. Sun, Y.; Zhao, S. Spatiotemporal dynamics of urban expansion in 13 cities across the Jing-Jin-Ji Urban Agglomeration from 1978 to 2015. *Ecol. Indic.* **2018**, *87*, 302–313. [CrossRef]

52. Tayyebi, A.; Shafizadeh-Moghadam, H.; Tayyebi, A.H. Analysing long-term spatio-temporal patterns of land surface temperature in response to rapid urbanisation in the megacity of Tehran. *Land Use Polic y* **2018**, *71*, 459–469. [CrossRef]

53. Yu, Z.; Guo, X.; Zeng, Y.; Koga, M.; Vejre, H. Variations in land surface temperature and cooling efficiency of green space in rapid urbanisation: The case of Fuzhou city, China. *Urban For. Urban Green.* **2018**, *29*, 113–121. [CrossRef]

54. Deng, C.; Wu, C. BCI: A biophysical composition index for remote sensing of urban environments. *Remote Sens. Environ.* **2012**, *127*, 247–259. [CrossRef]

55. Firozjaei, M.K.; Kiavarz, M.; Homaee, M.; Arsanjani, J.J.; Alavipanah, S.K. A novel method to quantify urban surface ecological poorness zone: A case study of several European cities. *Sci. Total Environ.* **2021**, *757*, 143755. [CrossRef]

56. Weng, Q.; Firozjaei, M.K.; Sedighi, A.; Kiavarz, M.; Alavipanah, S.K. Statistical analysis of surface urban heat island intensity variations: A case study of Babol city, Iran. *GISci. Remote Sens.* **2019**, *56*, 576–604. [CrossRef]

57. Firozjaei, M.K.; Fathololoumi, S.; Alavipanah, S.K.; Kiavarz, M.; Vaezi, A.R.; Biswas, A. A new approach for modeling near surface temperature lapse rate based on normalised land surface temperature data. *Remote Sens. Environ.* **2020**, *242*, 111746. [CrossRef]

58. Fu, P.; Weng, Q. A time series analysis of urbanisation induced land use and land cover change and its impact on land surface temperature with Landsat imagery. *Remote Sens. Environ.* **2016**, *175*, 205–214. [CrossRef]

59. Firozjaei, M.K.; Weng, Q.; Zhao, C.; Kiavarz, M.; Lu, L.; Alavipanah, S.K. Surface anthropogenic heat islands in six megacities: An assessment based on a triple-source surface energy balance model. *Remote Sens. Environ.* **2020**, *242*, 111751. [CrossRef]

60. Chen, S.; Hu, D.; Wong, M.S.; Ren, H.; Cao, S.; Yu, C.; Ho, H.C. Characterising spatiotemporal dynamics of anthropogenic heat fluxes: A 20-year case study in Beijing–Tianjin–Hebei region in China. *Environ. Pollut.* **2019**, *249*, 923–931. [CrossRef] [PubMed]

61. Chakraborty, S.D.; Kant, Y.; Mitra, D. Assessment of land surface temperature and heat fluxes over Delhi using remote sensing data. *J. Environ. Manag.* **2015**, *148*, 143–152. [CrossRef]

62. Deadman, P.; Brown, R.D.; Gimblett, H.R. Modelling Rural Residential Settlement Patterns with Cellular Automata. *J. Environ. Manag.* **1993**, *37*, 147–160. [CrossRef]

63. Mohajerani, A.; Bakaric, J.; Jeffrey-Bailey, T. The urban heat island effect, its causes, and mitigation, with reference to the thermal properties of asphalt concrete. *J. Environ. Manag.* **2017**, *197*, 522–538. [CrossRef]

64. Gaur, A.; Eichenbaum, M.K.; Simonovic, S.P. Analysis and modelling of surface Urban Heat Island in 20 Canadian cities under climate and land-cover change. *J. Environ. Manag.* **2018**, *206*, 145–157. [CrossRef]
65. Velavan, T.P.; Meyer, C.G. The COVID-19 epidemic. *Trop. Med. Int. Health* **2020**, *25*, 278. [CrossRef]
66. Das, M.; Das, A.; Mandal, A. Research note: Ecosystem Health (EH) assessment of a rapidly urbanising metropolitan city region of eastern India–A study on Kolkata Metropolitan Area. *Landsc. Urban Plan.* **2020**, *204*, 103938. [CrossRef]
67. Das, M.; Das, A.; Pereira, P.; Mandal, A. Exploring the spatio-temporal dynamics of ecosystem health: A study on a rapidly urbanising metropolitan area of Lower Gangetic Plain, India. *Ecol. Indic.* **2021**, *125*, 107584. [CrossRef]
68. Subbaraman, R.; Nathavitharana, R.R.; Mayer, K.H.; Satyanarayana, S.; Chadha, V.K.; Arinaminpathy, N.; Pai, M. Constructing care cascades for active tuberculosis: A strategy for program monitoring and identifying gaps in quality of care. *PLoS Med.* **2019**, *16*, e1002754. [CrossRef] [PubMed]
69. Yang, C.; Zhang, C.; Li, Q.; Liu, H.; Gao, W.; Shi, T.; Wu, G. Rapid urbanisation and policy variation greatly drive ecological quality evolution in Guangdong-Hong Kong-Macau Greater Bay Area of China: A remote sensing perspective. *Ecol. Indic.* **2020**, *115*, 106373. [CrossRef]
70. Singh, R.P.; Chauhan, A. Impact of lockdown on air quality in India during COVID-19 pandemic. *Air Qual. Atmos. Health* **2020**, *13*, 921–928. [CrossRef] [PubMed]
71. Sharma, S.; Zhang, M.; Anshika; Gao, J.; Zhang, H.; Kota, S.H. Effect of restricted emissions during COVID-19 on air quality in India. *Sci. Total Environ.* **2020**, *728*, 138878. [CrossRef] [PubMed]
72. Singhal, S.; Matto, M. COVID-19 Lockdown: A Ventilator for Rivers. 2020. Available online: https://www.downtoearth.org.in/blog/covid-19-lockdown-a-ventilator-for-rivers-70771 (accessed on 16 August 2021).
73. Somani, M.; Srivastava, A.N.; Gummadivalli, S.K.; Sharma, A. Indirect implications of COVID-19 towards sustainable environment: An investigation in Indian context. *Bioresour. Technol. Rep.* **2020**, *11*, 100491. [CrossRef]
74. Terry, C.; Rothendler, M.; Zipf, L.; Dietze, M.C.; Primack, R.B. Effects of the COVID-19 pandemic on noise pollution in three protected areas in metropolitan Boston (USA). *Biol. Conserv.* **2021**, *256*, 109039. [CrossRef]
75. Gandhiok, J.; Ibra, M. COVID-19: Noise Pollution Falls as Lockdown Rings in Sound of Salience. 2020. Available online: https://timesofindia.indiatimes.com/india/covid-19-noise-pollution-falls-as-lockdown-rings-in-sound-of-silence/articleshow/75309318.cms (accessed on 30 August 2021).
76. Kundu, C. Has the Covid-19 Lockdown Returned Dolphins and Swans to ITALIAN Waterways? *The India Today.* 2020. Available online: https://www.indiatoday.in/fact-check/story/has-covid19-lockdown-returned-dolphins-swans-italian-waterways-1658457-2020-03-22 (accessed on 30 August 2021).
77. Rahman, M. Rare Dolphin Sighting as Cox's Bazar Lockdown under COVID-19 Coronavirus. 2020. Available online: https://www.youtube.com/watch?v=gjw8ZllIlbQ (accessed on 30 August 2021).
78. Firozjaei, M.K.; Fathololoumi, S.; Kiavarz, M.; Arsanjani, J.J.; Alavipanah, S.K. Modelling surface heat island intensity according to differences of biophysical characteristics: A case study of Amol city, Iran. *Ecol. Indic.* **2020**, *109*, 105816. [CrossRef]
79. Majumdar, S.; Sivaramakrishnan, L. Mapping of Urban Growth Dynamics in Kolkata Metropolitan Area: A Geospatial Approach. In *Terrorism Revisited*; Springer: Berlin, Germany, 2020; pp. 9–24.
80. Halder, B.; Banik, P.; Bandyopadhyay, J. Mapping and monitoring land dynamic due to urban expansion using geospatial techniques on South Kolkata. *Saf. Extrem. Environ.* **2021**, *3*, 27–42. [CrossRef]
81. Mondal, B.; Dolui, G.; Pramanik, M.; Maity, S.; Biswas, S.S.; Pal, R. Urban expansion and wetland shrinkage estimation using a GIS-based model in the East Kolkata Wetland, India. *Ecol. Indic.* **2017**, *83*, 62–73. [CrossRef]
82. Majumdar, S.; Sivaramakrishnan, L. Monitoring Urban Growth and Land Use Change Detection in The Southern Fringes of Kolkata Metropolitan Area. *Indian Geogr. J.* **2017**, *92*, 168–183.
83. Majumdar, S.; Sivaramakrishnan, L. Patterns of Land Use in and Around Kolkata City: A Spatio-Temporal Analysis. *Indian Cartogr.* **2015**, *35*, 218–223.
84. Shafia, A.; Gaurav, S.; Bharath, H.A. Urban growth modelling using Cellular Automata coupled with land cover indices for Kolkata Metropolitan region. In *IOP Conference Series: Earth and Environmental Science*; IOP Publishing: Bristol, UK, 2018; Volume 169, p. 012090.
85. Parihar, S.M.; Sarkar, S.; Dutta, A.; Sharma, S.; Dutta, T. Characterising wetland dynamics: A post-classification change detection analysis of the East Kolkata Wetlands using open source satellite data. *Geocarto Int.* **2013**, *28*, 273–287. [CrossRef]
86. Rahaman, M.; Dutta, S.; Sahana, M.; Das, D.N. Analysing Urban Sprawl and Spatial Expansion of Kolkata Urban Agglomeration Using Geospatial Approach. In *Applications and Challenges of Geospatial Technology*; Springer: Berlin, Germany, 2019; pp. 205–221.
87. Dinda, S.; Das Chatterjee, N.; Ghosh, S. An integrated simulation approach to the assessment of urban growth pattern and loss in urban green space in Kolkata, India: A GIS-based analysis. *Ecol. Indic.* **2021**, *121*, 107178. [CrossRef]
88. Patra, S.; Sahoo, S.; Mishra, P.; Mahapatra, S.C. Impacts of urbanisation on land use/cover changes and its probable implications on local climate and groundwater level. *J. Urban Manag.* **2018**, *7*, 70–84. [CrossRef]
89. Li, X.; Mitra, C.; Marzen, L.; Yang, Q. Spatial and Temporal Patterns of Wetland Cover Changes in East Kolkata Wetlands, India from 1972 to 2011. *Int. J. Appl. Geospat. Res.* **2016**, *7*, 1–13. [CrossRef]

90. Ghosh, S.; Singh, P.; Kumari, M. *Assessment of Urban Sprawl and Land Use Change Dynamics, Using Remote Sensing Technique: A Study of Kolkata and Surrounding Periphery; WB, India.* 2014. Available online: https://www.researchgate.net/profile/Sukanya-Ghosh-5/publication/321125830_ASSESSMENT_OF_URBAN_SPRAWL_AND_LAND_USE_CHANGE_DYNAMICS_USING_REMOTE_SENSING_TECHNIQUE_A_STUDY_OF_KOLKATA_AND_SURROUNDING_PERIPHERY_WB_INDIA/links/5a1e47e3458515a4c3d1deaa/ASSESSMENT-OF-URBAN-SPRAWL-AND-LAND-USE-CHANGE-DYNAMICS-USING-REMOTE-SENSING-TECHNIQUE-A-STUDY-OF-KOLKATA-AND-SURROUNDING-PERIPHERY-WB-INDIA.pdf (accessed on 16 August 2021).
91. Mandal, J.; Ghosh, N.; Mukhopadhyay, A. Urban growth dynamics and changing land-use land-cover of megacity Kolkata and its environs. *J. Indian Soc. Remote Sens.* **2019**, *47*, 1707–1725. [CrossRef]

remote sensing

Article

Using UAV Survey, High-Density LiDAR Data and Automated Relief Analysis for Habitation Practices Characterization during the Late Bronze Age in NE Romania

Alin Mihu-Pintilie [1,*], Casandra Brașoveanu [1] and Cristian Constantin Stoleriu [2]

[1] Department of Exact and Natural Sciences, Institute of Interdisciplinary Research, Alexandru Ioan Cuza University of Iași (UAIC), St. Lascăr Catargi 54, 700107 Iași, Romania; brasoveanu.casandra@yahoo.com
[2] Department of Geography, Faculty of Geography and Geology, Alexandru Ioan Cuza University of Iași (UAIC), Bd. Carol I 20A, 700505 Iași, Romania; cristian.stoleriu@uaic.ro
* Correspondence: alin.mihu.pintilie@uaic.ro; Tel.: +40-7419-12245

Abstract: The characterization of prehistoric human behavior in terms of habitation practices using GIS cartography methods is an important aspect of any modern geoarchaeological approach. Furthermore, using unmanned aerial vehicle (UAV) surveys to identify archaeological sites with temporal resolution during the spring agro-technical works and automated mapping of the geomorphological features based on LiDAR-derived DEM can provide valuable information about the human–landscape relationships and lead to accurate archaeological and cartographic products. In this study, we applied a GIS-based landform classification method to relief characterization of 362 Late Bronze Age (LBA) settlements belonging to Noua Culture (NC) (cal. 1500/1450-1100 BCE) located in the Jijia catchment (NE Romania). For this purpose, we used an adapted version of Topographic Position Index (TPI) methodology, abbreviated DEV, which consists of: (1) application of standard deviation of TPI for the mean elevation (DEV) around each analyzed LBA site (1000 m buffer zone); (2) classification of the archaeological site's location using six slope position classes (first method), or ten morphological classes by combining the parameters from two small-DEV and large-DEV neighborhood sizes (second method). The results indicate that the populations belonging to Noua Culture preferred to place their settlements on hilltops but close to the steep slope and on the small hills/local ridges in large valleys. From a geoarchaeological perspective, the outcomes indicate a close connection between occupied landform patterns and habitation practices during the Late Bronze Age and contribute to archaeological predictive modelling in the Jijia catchment (NE Romania).

Keywords: UAV survey; LiDAR-derived DEM; TPI and DEV; GIS landform classification; LBA archaeological sites; Jijia catchment; NE Romania

Citation: Mihu-Pintilie, A.; Brașoveanu, C.; Stoleriu, C.C. Using UAV Survey, High-Density LiDAR Data and Automated Relief Analysis for Habitation Practices Characterization during the Late Bronze Age in NE Romania. *Remote Sens.* **2022**, *14*, 2466. https://doi.org/10.3390/rs14102466

Academic Editor: Deodato Tapete

Received: 13 April 2022
Accepted: 18 May 2022
Published: 20 May 2022

Publisher's Note: MDPI stays neutral with regard to jurisdictional claims in published maps and institutional affiliations.

1. Introduction

GIS-based cartography is an essential tool for any modern geoarchaeological approach [1,2] and contributes to archaeological predictive modelling [3] for better protection and management of local or regional cultural heritage [4,5]. Therefore, the GIS-based relief analysis applications well-documented by [6,7] (e.g., combinations of geomorphometric parameters, supervised and unsupervised classification, probabilistic clustering algorithms, double ternary diagram classification and object-oriented image analysis) have proven to be very effective not only in the Earth science disciplines such as geo-pedology [8–11], terrestrial geomorphology [11–15], seafloor mapping [16–18], hydrology [19,20], climatology [21], landscape mapping [22] and landscape ecology [23]; but also in geoarchaeological investigations [24,25]. This statement is supported by the fact that landform characteristics are factors which certainly influenced prehistoric human behavior in terms of habitation practices [25]. In this context, there are many geoarchaeological studies which have focused on the human–environment relationship, testifying to the significant influence of landform features on settlement patterns during the different prehistoric and early historic cultural phases [24–29], among others.

Therefore, even if the automated [24] or semi-automated [6] GIS delineation of small-scale relief features based on Light Detection and Ranging (LiDAR)-derived Digital Elevation Models (DEM's), such as the Topographic Position Index (TPI)-derived methodology abbreviated in this study as DEV after [26], can be considered just an alternative to developing innovative digital geoarchaeological maps [4,30], this method can also solve many other archaeological issues related to human–landscape interactions, especially in areas with high densities of cultural heritage spread across heterogeneous landscapes [24–26]. This is the case for the plateau–plain transition zone of the Jijia catchment (NE Romania), which is characterized by an impressive number of archaeological settlements (>2000 sites) dated throughout the most representative prehistoric periods for eastern Europe (e.g., Neolithic, Chalcolithic, Bronze Age and Iron Age) [25,31,32].

In this framework, due to the fact that each prehistoric cultural phase identified in northeastern Romania left particular habitation traces in the landscape, some of them well-documented, such as Neolithic and Chalcolithic cultures [3,25,31,32], we chose to analyze the spatial patterns and geomorphological characteristics of the settlements belonging to the Late Bronze Age (LBA, chronological framework: cal. 1500/1450-1100 BCE) [33]. This selection was mainly motivated by the relatively low amount of research into LBA sites, and secondly, by the need to complete the existing archaeological database (i.e., National Archaeological Record of Romania) which so far has been lagging for the LBA, in comparison with other prehistoric periods relevant to the region. In this regard, in the present study, we provide the first TPI-based landform classification using the LiDAR-derived DEM, performed on the archaeological sites of Noua Culture (NC), the most representative communities that inhabited the Jijia catchment (NE Romania) during the Bronze Age (BA). The main purposes of this study are (a) to emphasize the role of geomorphological conditions and GIS-based detailed landform mapping as a key tool for investigating the possible influences of landscape features on the spatial evolution of NC settlements and (b) to characterize the preservation status of LBA sites in the current landscape configuration of the Jijia catchment.

2. Study Area

2.1. Geography of the Jijia Catchment

Jijia River drains a territory of 5757 km^2 located in the northeastern part of Romania (watershed centroid: 47°30'N/27°00'E) and is the main tributary (right-bank) of Prut River—a natural border between Romania and the Republic of Moldova [34,35]. The watershed overlaps the lower area of the Moldavian Plateau, also known as the Jijia Plain [25]. The elevation range between 18 m a.s.l. and 584 m a.s.l. (average elevation—150 m a.s.l.) (Figure 1a) induces a relief energy ranging from 0.5–50 m/km^2 (Jijia floodplain) to 50–235 m/km^2 (average relief energy—70.1 m/km^2) (Figure 1b). The highest elevation values (>400 m a.s.l.) indicate the contact area between plateau–plain transition zone in the western (Moldavian Plain–Central Moldavian Plateau) and southern (Moldavian Plain–Suceava Plateau) flanks. The average slope is 5.71°, where the highest declivity (>35°) corresponds to the front of the cuestas, frequently affected by landslides, and the active riverbanks consumed by erosion [25,34,36] (Figure 1c). The climate is temperate continental with a mean annual temperature of 8–10 °C and an average annual precipitation ranging from 460 mm (<150 m a.s.l.) to 670 mm (>500 m a.s.l.) [36].

The general morpho-structure is a monocline with Miocene-Pleistocene dipping strata from north-west to south-east, from Suceava Plateau to the middle Prut Valley [36]. The geological structure consists of a succession of thin layers of limestone and sandstone (2–30 m thick; Lower and Medium Sarmatian deposits) and sands with clays layers (200–300 m thick) and thin layers (2–5 m thick) of limestones and andesitic cinerites (Upper Sarmatian deposits) [37], covered by a loess layer (Pleistocene deposits) with thicknesses between

1.0–2.5 m (most frequent) and 15–30 m (e.g., fluvial terraces, reverse cuesta slopes) [38]. The valleys corresponding to major watercourses (e.g., Jijia, Sitna, Miletin and Bahlui) are characterized by recent alluvial deposits (Holocene period), of which the middle and lower sector of the Jijia floodplain is most developed in the landscape [34,39] (Figure 1d).

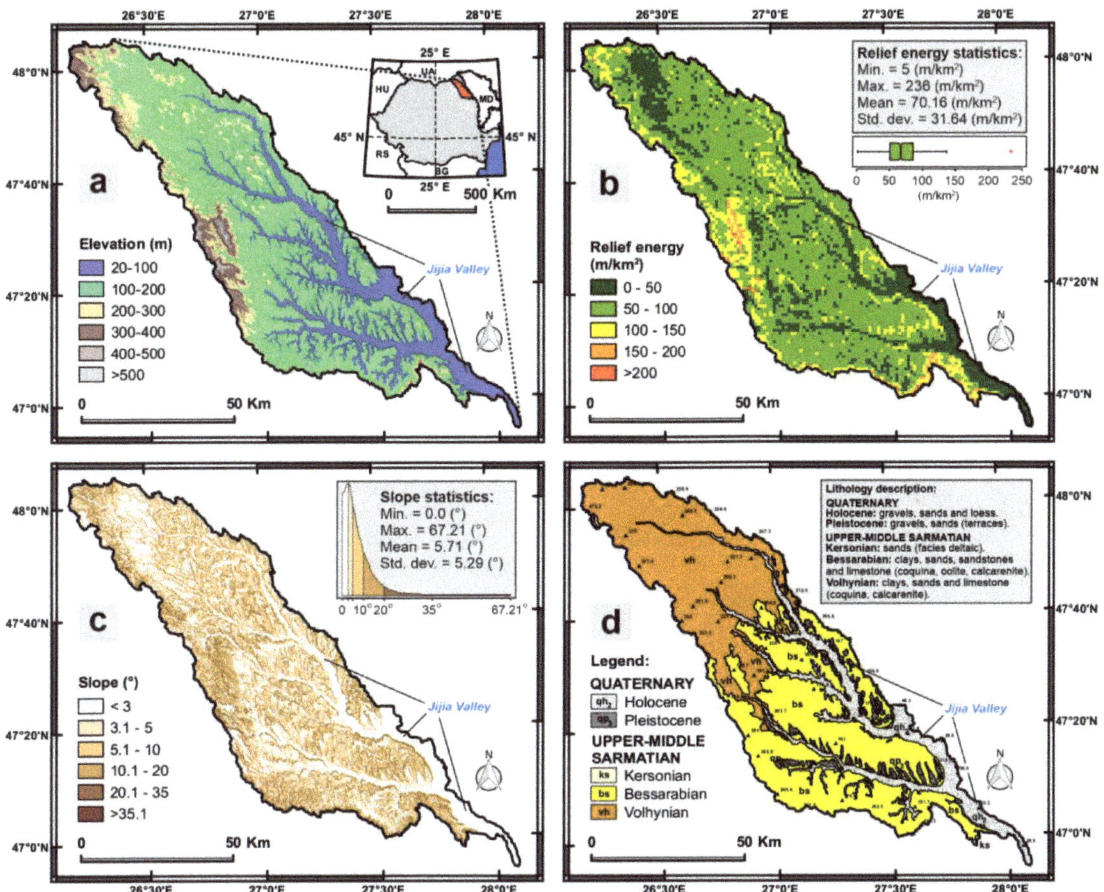

Figure 1. Geographic location of Jijia river basin in northeastern Romania with (**a**) elevation, (**b**) relief energy, (**c**) slope and (**d**) geological sketch maps.

The large-scale landforms which dominate the heterogenous landscape of the Jijia catchment are well-contoured valleys separated by interfluves that are most often of the cuesta type. From a geoarchaeological point of view, the cuesta landforms are the most representative, in terms of prehistoric habitation practices, because they produce two different types of slope–site relationships [25]: (1) the first type consists of the cuesta dip slopes characterized by low roughness—preferred for agro-pastoral practices; and (2) the second type is represented by cuesta scarp slopes, generally affected by deep stream incisions at the base, diffuse and well-defined gully erosion along the slopes and large landslides—preferred for settlement locations due to the dominance in the relief and for the high visibility [25,40]. Generally, this typical morpho-structure along with the open slopes and local ridges in the major floodplains are the main small-scale landforms used by prehistoric populations for placing the settlements. However, more details on the geological, geomorphological, hydrological and climatological aspects can be found in other various studies related to the geoarchaeological context of the studied area [31,37,40–42].

2.2. Archaeological Context: Late Bronze Age—Noua Culture

From a cultural point of view, the end of the Bronze Age (BA) in Romania is characterized by the emergence of two new important cultural complexes, namely, Zimnicea-Plovdiv and Noua–Sabatinovka–Coslogeni. The latter was documented, so far, in eastern Romania and Transylvania, and Transcarpathian Ukraine and Republic of Moldova, reaching the middle and upper Dniester [43–45]. The main characteristics of these human groups is represented by the presence of the so-called ashmounds (grey spots, visible on aerial photographs and on site, with diameters of 20–30 m and small elevations) that sparked interest amongst specialists since the end of the 19th and beginning of the 20th centuries [46,47] (Figure 2).

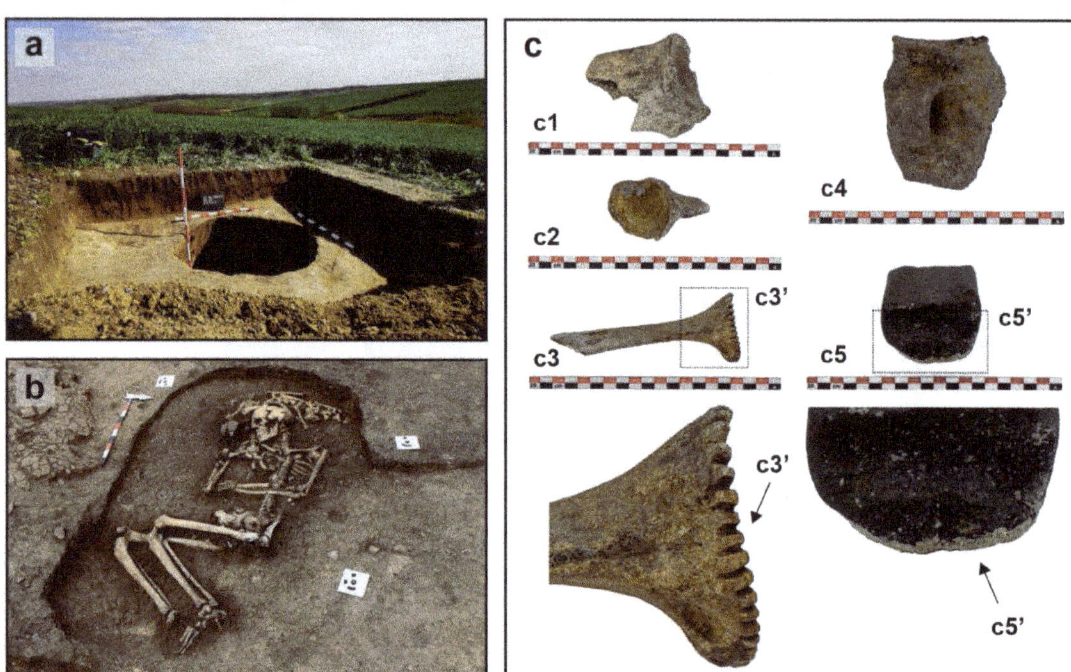

Figure 2. Examples of archaeological remains belonging to Noua Culture (LBA) in the Jijia catchment (NE Romania): (**a**) inside of an LBA site from the Jijia lowland—an excavated ashmound which was visible on aerial photographs and on site; (**b**) a tomb belonging to Noua Culture discovered during the excavation of a Chalcolithic settlement from Jijia-Siret upland; (**c**) LBA tools and ceramic fragments collected from various locations in the Jijia basin: c1, c2 and c3—crenated *scapulae* used as tools for processing animal skin; c4—a ceramic fragment of a cooking pot; c5—a stone-axe.

Until recently, ashmounds were mostly considered linked to possible remains of burnt dwellings or hearths [43,44,48–50] (Figure 2a) and sacred/cultic areas of the settlements [51–54] (Figure 2b). Lately, due to the interdisciplinary studies conducted on such structures from Republic of Moldova [55], a more plausible explanation was issued, specifically the one of household pits, whose organic content led to physicochemical changes in the soil, producing the ash-like color, representing evidence of the type of work practiced, namely, cattle shepherding. The latter is well proven by the high number of animal osteological remains (mostly belonging to cattle or sheep), some of which were processed and converted in different types of tools, present in all the settlements specific to LBA (Figure 2c). Besides the presence of ashmounds and numerous animal bones, the most important cultural markers for Noua Culture (NC) communities are the double-handled *kantharoi* (present especially in funerary contexts); the curved knives made of flint (*krummesser*); crenated *scapulae,* used for processing animal skin; and *tupik* sickles, made out of large animals' mandibles and utilized in agriculture [48].

3. Data and Methods

3.1. Archaeological Data: LBA Sites Inventory

In order to identify the relationship between the LBA settlements and the geomorphological features of the occupied landscape, a geo-referenced database has been generated, firstly in Google Earth Pro 7.1.5., and exported in ArcGIS 10.3 as shapefile data (Table S1). The cartographic and aerial image products used for validation of sites location consisted of old maps; military topographic maps (1:25,000 scale); orthophotoplans existing on portals, such as Inis Viewer and Atlas Explorer; and LiDAR-derived DEM [56] and aerial images collected during the field work. Additionally, the LBA sites inventory was compiled by consulting the archaeological monographs [43,44,57,58] and archaeological repertories existing for the counties Iași and Botoșani [59–62], whose territories overlap, partially, the Jijia catchment. In addition to cartographic documentation, due to the fact that NC settlements are characterized, in most cases, by the presence of the so-called ashmounds, we were able to identify 70 new sites during the field surveys using UAV technology [63–65] (Figure 3). To this end, we used a drone (Phantom 4 Pro v.2) in order to obtain aerial photographs. Two techniques were used: oblique photography and vertical photography, using the mission planner available in the DJI Pilot application. For the vertical one, we used a 70% overlap between each photo. The flight altitude used was between 70 and 100 m. The photos were later imported in Agisoft Metashape in order to obtained large ortorectified images [64].

Overall, regardless of the inventory method used, the site mapping was done only for the discoveries that consisted, among others, of remains from dwellings of certain settlements. Therefore, we have compiled a database consisting of more than 400 certain Bronze Age sites (BA-chronological framework: ca. 2900/2800–1100 BCE), from which 362 belong to the LBA period (out of which 195 present ashmounds) of the Noua Culture (ca. 1500/1450–1100 BCE) (Figure 4a). It should be noted that, besides the LBA settlements, we also identified and mapped the funerary contexts, the hoards and other types of discoveries (isolated or uncertain) belonging to the same chronological interval, but these will be analyzed in the future. In this framework, the last step of the archaeological database construction was to generate a 1000 m buffer zone for each selected LBA site used for automated relief analysis and landform classification purposes (Figure 4b; Table S1).

3.2. Elevation Data: LiDAR-Derived DEM

The elevation data were obtained from National Administration "Romanian Waters"—Prut-Bîrlad Water Administration (NARW-PBWA) which, through the implementation of SMIS-CSNR number 17945: *Works to reduce the flood risk in the Prut–Bîrlad River Basin* (PBRB) (2013) [56], managed to scan the entire northeastern territory of Romania using high-density airborne LiDAR technology, a popular remote sensing method used for measuring the exact length of an object on the Earth's surface [25,34,35]. Broadly, the LiDAR method uses light in the form of a pulsed laser to measure ranges (variable distances) of the Earth. These light pulses—combined with other data recorded by the airborne system—generate precise, three-dimensional information about surface characteristics [25]. Therefore, the elevation data used in this work consisted of more than 700 raster files generated based on raw ground point elevation data collected at spatial density between 4 point/m^2 (out of built-up areas) and 16 point/m^2 (built-up areas).

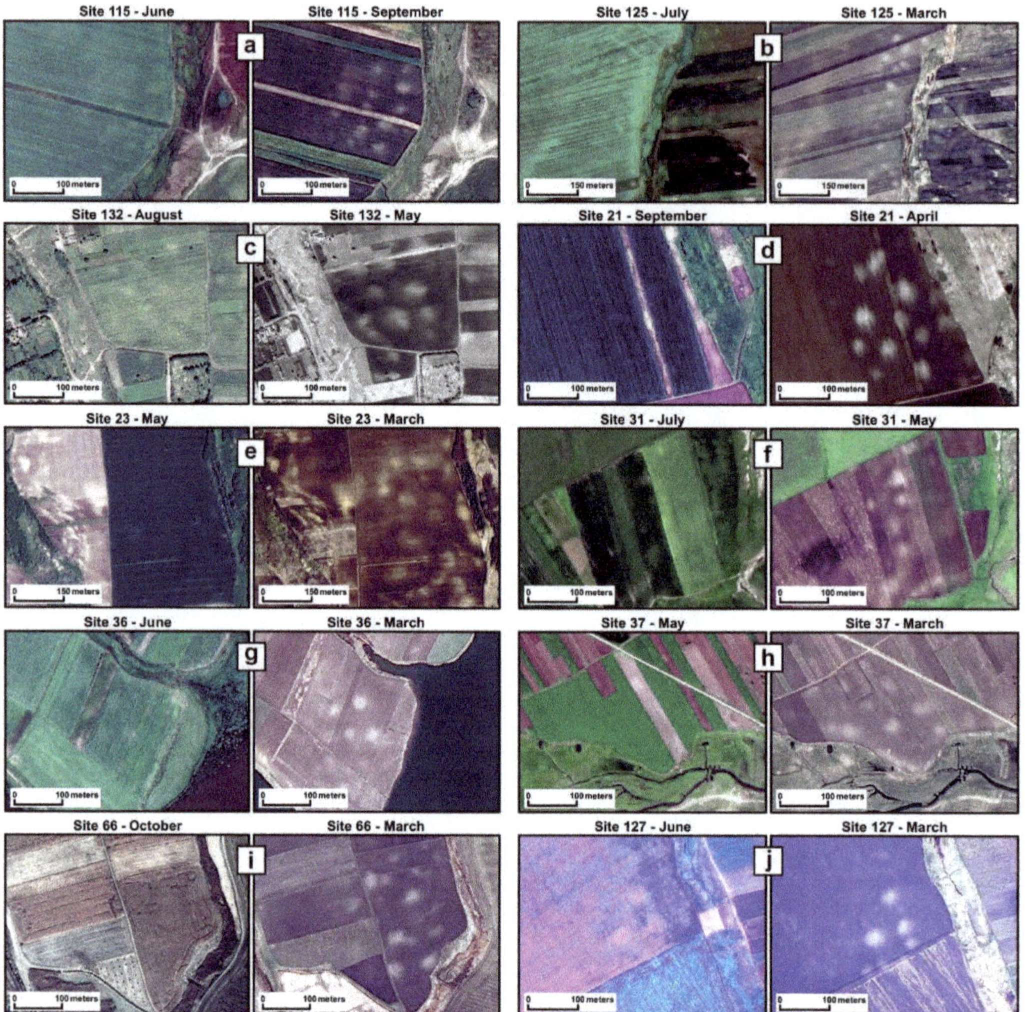

Figure 3. Aerial images (left—during the vegetation season; right—during the spring agro-technical works) with unexcavated LBA sites belonging to Noua Culture in the Jijia catchment: (**a**); Site 115—Dumești, Holmului Hill; (**b**) Site 125—Erbiceni, Spinoasei Hill/Valea Lungă; (**c**) Site 132—Fântânele, Cimitirul Ortodox; (**d**) Site 21—Aroneanu, Șapte Oameni; (**e**) Site 23—Dorobanț, La Chișcă; (**f**) Site 31—Boureni, Popa Mort Hill; (**g**) Site 36—Valea Oilor, Mădârjești Pond; (**h**) Site 37-Valea Oilor, North of the village; (**i**) Site 66—Ceplenița, Ion Clacă Hill; (**j**) Site 127—Spinoasa, Drumul Poștei Hill; the so-called ashmounds (circular gray spots visible on the field) with temporal resolution during the spring agro-technical works (right image of each example).

The LiDAR-derived DEM used for the relief analysis within the buffer zones (1000 m radius) around each LBA site identified in the Jijia catchment was achieved by spatially processing the raster files, and following these steps: (1) the raster's were generated in grid formats through Inverse Distance Weighting (IDW) interpolation [66,67], with 1 m cell sizes; (2) the resulting small-scale DEMs were filtered using flow direction, sink and fill tools, to reduce the errors generated by merging the .tiff files [68,69]; (3) the slope raster was generated using Spatial Analyst Tools via ArcGIS 10.3; (4) the delineation of landform units was performed using an adapted TPI-landform classification tool from Relief Analysis Toolbox for ArcGIS abbreviated by [26] with DEV [8,24,25].

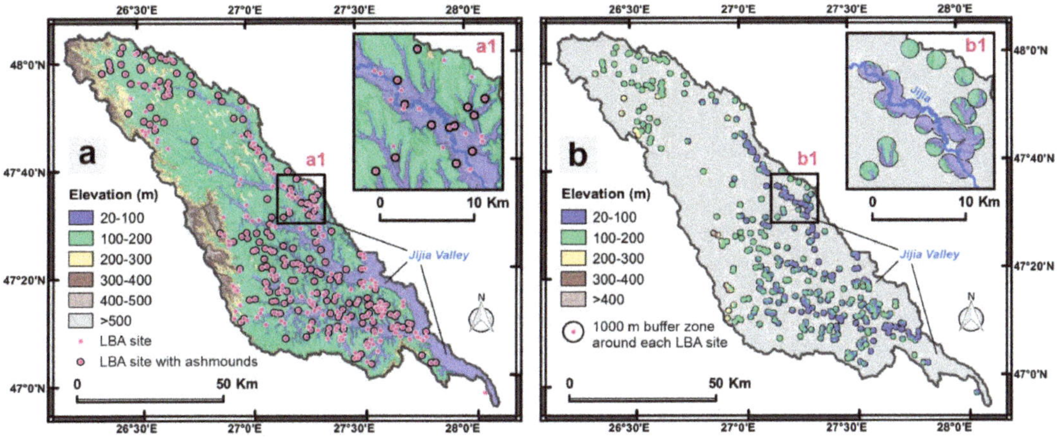

Figure 4. Noua Culture (NC) sites distribution in the Jijia River basin (see Table S1) used for habitation practices characteristics during the Late Bronze Age: (**a**) 362 LBA settlements, of which 167 sites are without ashmounds and 195 sites are with ashmounds (see Figure 3); (**b**) 1000 m buffer zone around each LBA sites used for automated relief analysis and landform classification.

3.3. Automated Relief Analysis: TPI-Based DEV

The GIS algorithm used for landform classification of LBA site locations was based on the Topographic Position Index (TPI) tool developed by Weiss, A. D. [70] and implemented as an ESRI ArcView 3.x. extension by Jenness J. [71]. The TPI tool calculates the difference between elevations at the central point z_0 (Equation (1)), which in our case was the central point of each LBA site, and the average elevation $\bar{z}$ (Equation (2)) around it within a known radius R [25]: R = 100 m, R = 300 m, R = 600 m, R = 1200 m and R = 2000 m around the LBA sites selected for this study. After this process, the positive values of TPI indicate that the central point is located higher ($z_0 > \bar{z}$) than its average surroundings and the negative values of TPI ($z_0 < \bar{z}$) indicate the opposite.

$$TPI = z_0 - \bar{z} \tag{1}$$

$$\bar{z} = \frac{1}{n_R} \sum_{i \in R} z_i \tag{2}$$

Based on the initial TPI algorithm, De Reu J. and his collaborators [72] developed a new computational equation (abbreviated DEV) which uses the TPI calculation and the standard deviation (SD) of the surrounding elevation of z_0 (Equation (3)). According to the authors [72], DEV improves the results because it measures the TPI as a fraction of local relief normalized to local surface roughness (Equation (4)) [25,26]. However, even if both TPI and DEV tools are frequently used in the landscape archaeology studies [24–26,72], we considered it most appropriate to use DEV instead of TPI due to the higher potential accuracy of landform classification in the study area, in the catchment area of Jijia River.

$$DEV = \frac{z_0 - \bar{z}}{SD} \tag{3}$$

$$DEV = \sqrt{\frac{1}{n_R - 1} \sum_{i=1}^{} (z_i - \bar{z})^2} \tag{4}$$

The next step was to divide the landscape into the six discrete slope position classes described in Table 1 (first method) or the ten morphological classes described in Table 2 (second method), both methods using the GIS-based methodology adapted to TPI-derived DEV [70,71]. Therefore, the TPI-based DEV data processing involves classification of the topographic surface into a complex landscape feature by combining the parameters from two small-DEV (e.g., R = 300 m) and large-DEV (e.g., R = 1200 m) neighborhood sizes by using the TPI-landforms class tool from the Relief Analysis Toolbox [8]. As this method has been applied before in a previous study for another prehistoric interval, for the Chalcolithic cultures Precucuteni and Cucuteni [25], we already know that the TPI-based DEV results highlight all landform types [70,71] occupied by the prehistoric settlements in the heterogenous landscape of northeastern Romania. The landform classification accuracy has been verified based on two methods: first, by visual interpretation of morphological limits using the UAV imagery for large-scale landforms (e.g., valleys and hills) and the LiDAR database for small-scale morphological features (e.g., deeply incised streams and small hills in plains); secondly, by comparing the TPI-based DEV results with the specific geomorphological features of LBA site locations identified and characterized during the successive field surveys.

Table 1. Description and abbreviation of slope position classes used for habitation practices' characterization during the LBA in the Jijia catchment, obtained based on TPI-based DEV calculated for various neighborhood sizes and terrain slope (first method).

Slope Position Classes Description [1]	DEV Threshold	Landform Classes Abbreviation
Ridge	$TPI > 1\,SD$	Sp6
Upper slope	$0.5\,SD < TPI \leq 1\,SD$	Sp5
Middle slope	$-0.5\,SD < TPI \leq 0.5\,SD$	Sp4
Flat area	$-0.5\,SD < TPI \leq 0.5\,SD$	Sp3
Lower slope	$-1\,SD < TPI \leq -0.5\,SD$	Sp2
Valley	$TPI \leq -1\,SD$	Sp1

[1] Slope position classes adapted after [70,71] for specific landscape characteristics of the study area.

Table 2. Descriptions and abbreviations of landform classes used for habitation practices characterization during the LBA in the Jijia catchment, obtained based on combined small-DEV and large-DEV neighborhood sizes (second method).

Landform Classes Description [1]	Small-DEV Neighborhood Size	Large-DEV Neighborhood Size	Landform Classes Abbreviation
Hill tops, high ridges	$Z_0 > SD$	$Z_0 > SD$	L10
Middle slope ridges, small hills in plains	$Z_0 > SD$	$0 \leq Z_0 \leq SD$	L9
Local ridges/hills in valley	$Z_0 > SD$	$Z_0 < -SD$	L8
Upper slopes	$-SD \leq Z_0 \leq SD$	$Z_0 > SD$	L7
Open slopes (>5°)	$-SD \leq Z_0 \leq SD$	$0 \leq Z_0 \leq SD$	L6
Plains, flat areas (<5°)	$-SD \leq Z_0 \leq SD$	$-SD \leq Z_0 < 0$	L5
U-shaped valleys	$-SD \leq Z_0 \leq SD$	$Z_0 < -SD$	L4
Upland drainage, headwaters	$Z_0 < -SD$	$Z_0 > SD$	L3
Middle slope drainage, shallow valley	$Z_0 < -SD$	$0 \leq Z_0 \leq SD$	L2
Deeply incised streams	$Z_0 < -SD$	$-SD \leq Z_0 < 0$	L1

[1] Landform classes description after [70,71] for specific landscape characteristics of the study area.

4. Results

4.1. Landform Classification Accuracy and Optimal Neighborhood Sizes Combination

The descriptive statistics of LBA archaeological sites placement classified into six slope position classes, using the first method (see Table 1) for four candidate radii (R = 100 m; R = 300 m; R = 600 m; R = 1200 m), are shown in Table 3 and Figure 5. The descriptive statistics of LBA archaeological sites placement, classified into ten landform classes (second method, see Table 2) for the four combined versions of small TPI-based DEV and large

TPI-based DEV neighborhood sizes (DEV 100 and DEV 300 m; DEV 300 and DEV 1200 m; DEV 300 and DEV 2000 m; DEV 600 and DEV 2000 m), are shown in Table 4 and Figure 6.

Table 3. Number of LBA archaeological sites occurring over six slope position classes in the Jijia catchment (see Figure 5).

[1] Slope Position Classes Description	DEV Threshold	Landform Code	[2] R = 100 m	[2] R = 300 m	[2] R = 600 m	[2] R = 1200 m
Sp6: Ridge	TPI > 1 SD	Sp6	86	153	126	107
Sp5: Upper slope	0.5 SD < TPI ≤ 1 SD	Sp5	42	22	7	4
Sp4: Middle slope	−0.5 SD < TPI ≤ 0.5 SD	Sp4	102	22	16	7
Sp3: Flat area	−0.5 SD < TPI ≤ 0.5 SD	Sp3	66	27	7	5
Sp2: Lower slope	−1 SD < TPI ≤ −0.5 SD	Sp2	30	23	7	6
Sp1: Valley	TPI ≤ −1 SD	Sp1	36	115	199	233

[1] Slope position classes adapted after [70,71] for specific landscape characteristics of the study area; [2] R—radius value around z_0.

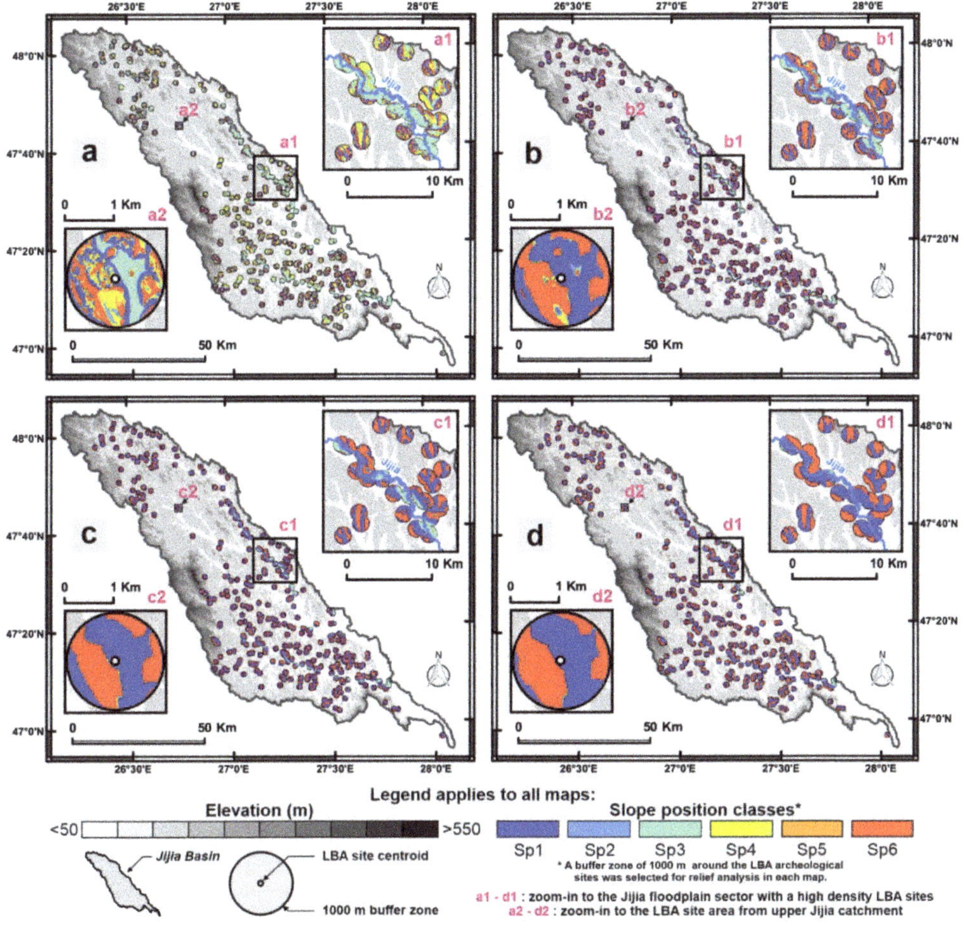

Figure 5. Slope position classification based on TPI-derived DEV of the LBA archaeological sites (Noua Culture) in the Jijia catchment, with six morphological classes (see Table 1); the relief analysis was provided for four TPI-derived DEV neighborhood sizes: (**a**) 100 m, (**b**) 300 m, (**c**) 600 m and (**d**) 1200 m.

Table 4. Number of LBA archaeological sites occurring over ten specific landform classes in the Jijia catchment (see Figure 6).

[1] Landform Classes Description	Small-DEV Neighborhood Size	Large-DEV Neighborhood Size	Combined [2] Small-DEV and [3] Large-DEV			
			100 m and 600 m	300 m and 1200 m	300 m and 2000 m	600 m and 2000 m
L10: Hill tops, high ridges	$Z_0 > SD$	$Z_0 > SD$	50	77	81	85
L9: Middle slope ridges, small hills in plains	$Z_0 > SD$	$0 \leq Z_0 \leq SD$	3	4	5	5
L8: Local ridges/hills in valley	$Z_0 > SD$	$Z_0 < -SD$	33	72	67	36
L7: Upper slopes	$-SD \leq Z_0 \leq SD$	$Z_0 > SD$	71	20	12	8
L6: Open slopes (>5°)	$-SD \leq Z_0 \leq SD$	$0 \leq Z_0 \leq SD$	22	2	5	2
L5: Plains, flat areas (<5°)	$-SD \leq Z_0 \leq SD$	$-SD \leq Z_0 < 0$	11	3	9	3
L4: U-shaped valleys	$-SD \leq Z_0 \leq SD$	$Z_0 < -SD$	136	69	68	24
L3: Upland drainage, headwaters	$Z_0 < -SD$	$Z_0 > SD$	5	20	14	24
L2: Middle slope drainage, shallow valley	$Z_0 < -SD$	$0 \leq Z_0 \leq SD$	1	4	3	3
L1: Deeply incised streams	$Z_0 < -SD$	$-SD \leq Z_0 < 0$	30	91	98	172

[1] Landform classes description after [70,71] for specific landscape characteristics of the study area; [2] R—small radius value around z_0 (100, 300 and 600 m); [3] R—large radius value around z_0 (600 m, 1200 and 2000 m).

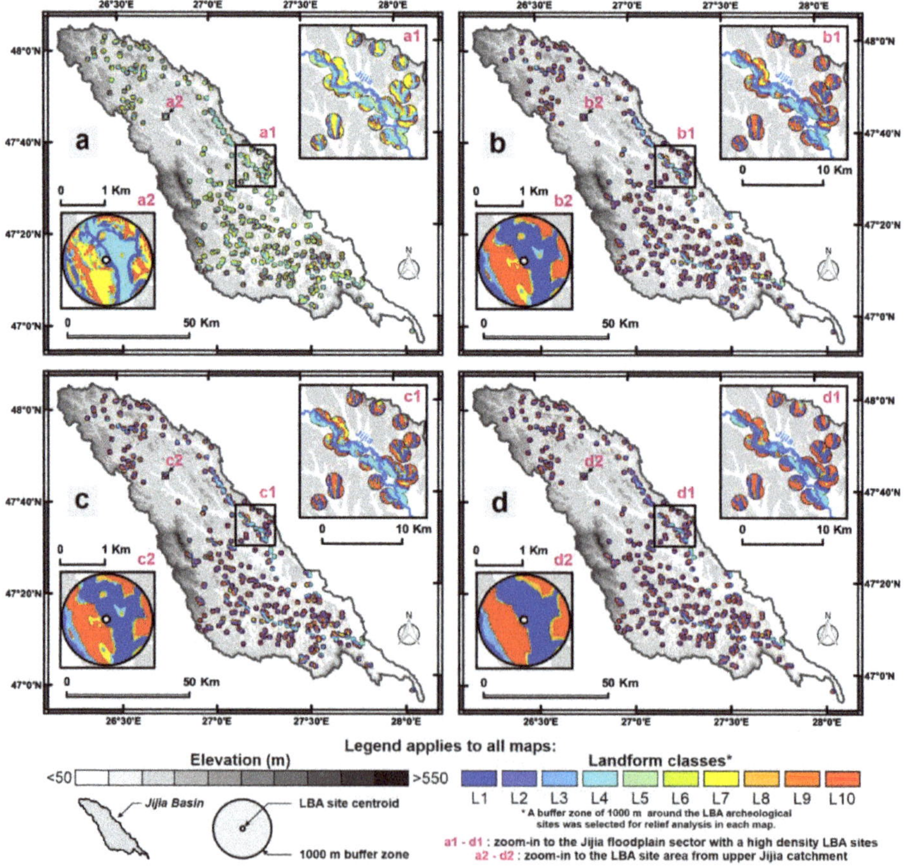

Figure 6. Landform classification based on TPI-derived DEV of the LBA archaeological sites (Noua Culture) in the Jijia catchment, with ten landform types (see Table 2); the relief analysis was provided for the combined of two TPI-derived DEV neighborhood sizes (**a**) 100 and 600 m, (**b**) 300 and 1200 m, (**c**) 300 and 2000 m and (**d**) 600 and 2000 m.

The TPI-based DEV landform classification accuracy has been assessed using the visual interpretation of UAV aerial imagery collected during the field work and by comparing the automated relief analysis results of LBA sites' locations with the specific morphological features of more than 200 prehistoric site locations provided by previous geomorphological and archaeological studies [31,32,37]. Furthermore, based on a previous study [25], in which we applied a similar automated relief classification within the heterogenous landscape of NE Romania (plateau–plain transition zone of Moldavian Plain), we were able to use, in this case also, the same TPI-based DEV thresholds and various neighborhood sizes.

The optimal neighborhood size combinations identified in the previous study [25], which are also applicable in this work, indicate that for the landform classification in six slope position classes (first method), the 300 m radius value around z_0 are the most appropriate for the rest of our approach (Figure 5). We support this statement due to the fact that R = 300 m has a low density of patches and discriminates the various features with less fragmentation, in comparison with R = 100 m, and also does not generalize the relief features such as R = 600 and R = 1200 m (Table 3). For the second relief analysis method in which we classified the landscape into ten morphological classes, the results of small (R = 300 m) and large (R = 1200 m) combined neighborhood sizes highlight the most accurate, dominant landform types occupied by LBA settlements (Figure 6). Other neighborhood size combinations emphasize the U-shaped valleys and discriminate the deeply incised streams (e.g., small-R = 100 m and large-R = 600 m) or upper slopes and headwaters classes (e.g., small-R = 300 m and large-R = 2000 m), or as in cases of small-R = 600 m and large- R = 2000 m, reduce the classes of landform features from ten landform classes to just two dominant relief features, hilltops/high ridges and deeply incised streams (Table 4). Therefore, the geoarchaeological interpretation of LBA site placement per landform classes was achieved based only on this optimal neighborhood size combination.

4.2. Slope Position Classification of LBA Sites' Locations

According to the slope position classification resulting from R = 300 m of TPI-based DEV and LiDAR-derived slope combination, 38.12% of LBA settlements were placed on the concave landforms, 115 sites in valleys (Sp1) and 23 sites on lower slopes (Sp2); 7.45% of LBA sites were located on the flat areas (Sp3) (27 sites); and over 54% of the LBA settlements were placed on the convex landforms (44 sites on the middle and upper slopes—Sp4/Sp5; 153 sites on the ridges—Sp6). The preference manifested by the LBA communities for placing their settlements on the top of cuesta (42.2% sites located on the ridges—Sp6) can be interpreted, firstly, as a necessity to provide defense for at least one or two sides of the settlement, as in the case of Cucuteni Culture [25]; and secondly, as the necessity to have a wide perspective on the valleys. In this context, the settlements which occupied the concave landforms, such as valleys (Sp1) and lower slopes (Sp2) (>140 LBA sites), can be interpreted as agro-pastoral locations during the vegetation season. This affirmation is supported by the fact that the LBA sites which have in their structures the so-called ashmounds (195 sites) occupied a relatively lower number of top/summits/ridges landforms, compared with the ones that do not present the mentioned structures (167 sites). However, the locations of LBA sites on the tops of the hills but near the steep slopes and/or inside of the large valleys indicate human behavior in close connection with agro-pastoral activities.

4.3. Landform Classification of LBA Sites' Locations

According to the TPI-based DEV landform classification using small-DEV 300 m and large-DEV 1200 m combined neighborhood sizes, 47.8% of LBA sites are located on convex landforms: 77 on hilltops, high ridges (L10); 4 sites on middle slope ridges, small hills on plains (L9); 72 on local ridges/hills in valleys (L8); and 20 on upper slopes (L7). Next, 1.38% of the LBA settlements were located on the flat or gentle slope areas (<5° plains, flat areas (L5)—2 sites; >5° open slopes (L6)—3 sites). The remaining 50.82% of LBA sites were located on the concave landforms: 69 sites in U-shaped valleys (L4); 20 on upland drainages/headwaters (L3); 4 sites on middle slope drainages/shallow valleys (L2); and

91 on deeply incised streams (L1). The most represented landform classes are: deeply incised streams (L1)—25.14%; hill tops and high ridges (L10)—21.27%; local ridges/hills in valleys (L8)—19.88%; and U-shaped valleys (L4)—19.06%. The least represented landform classes are the flat or gentle slope areas, most likely due to the wetlands which occupied the floodplains of the Jijia River and its tributaries (e.g., Bahlui, Miletin, Başeu), areas not suitable for habitation but very important for pastoral activities. The different slope position classification outcomes, presented in the previous subchapter, based on the TPI-based DEV landform classification using small-DEV 300 m and large-DEV 1200 m, reveal that half of the LBA sites located on the ridges (Sp6) are actually located on the local ridges/hills in the valley (L8). These results are also highlighted by the high number of the LBA sites with ashmounds (>50 sites), which occur on the local ridges/hills in valley (L8). Therefore, the second method of landform selection for settlements' placements during the LBA was finding the locations in the vicinity of natural channels, such as stream meanders or steep banks inside the major floodplains, or on small hills.

5. Discussion

5.1. Characterization of Habitation Practices during the Late Bronze Age

In this study, an alternative solution for automated relief analysis based on slope positions and landforms patterns of the location of Noua Culture settlements, in the landscape of the Jijia catchment (NE Romania), is presented [24,25,70–72]. The outcomes related to the relationship between the archaeological site locations and the geomorphological features indicate four specific landforms classes preferred by the prehistoric communities throughout the entire Late Bronze Age period: L10—hilltops, high ridges, L8—local ridges/hills in valleys, L4—U-shaped valleys and L1—deeply incised streams. Therefore, the settlements located on the hilltops and ridges, and the settlements located on the local ridges and hills in valleys, indicate the necessity to provide defense for at least one or two sides of the LBA sites and to gain better views of the valleys. These habitation practices characteristics, also identified at other prehistoric communities that occupied the Jijia River basin earlier [31,32], indicate that this was the first criterion used in selecting permanent site locations based on local topography [25]. However, what is specific to Noua Culture and different than other prehistoric cultures in the region is the high number of settlements located inside of U-shaped valleys of the Jijia River and its major tributaries (Bahlui, Miletin and Sitna rivers). For example, during the Late Bronze Age, more than 24% of the sites were identified in the floodplain areas of Jijia River, compared with Precucuteni–Cucuteni Culture which occupied the study area between cal. 5000–3500 BCE, where only 8% of the settlements were located within floodplain areas [25]. These significant differences in habitation characteristics can be explained, firstly, by climatic conditions (the climate changed during the mid-Holocene—a drier period towards the end of Bronze Age period which caused a probable reduction in flood events [73–77]), and secondly, due to the necessity of LBA communities for wide river plains areas that could support the work practiced, namely, cattle shepherding [24,43–55] (Table 5).

Another particularity of the LBA sites is the presence of the ashmounds (see Figure 3), which were observed in 195 sites out of the 362 sites investigated in this study. In this study, we wanted to find out if there were significant differences regarding landform selection between the two different types of LBA sites (see Figure 3). To investigate this issue, we compared the results of the slope position (Figure 7a,b) and landform classification (Figure 7c,d) between the LBA settlements with ashmounds and the ones without.

Table 5. Differences in habitation practices characteristics between Noua Culture (Late Bronze Age) and Precucuteni–Cucuteni Culture (Chalcolithic) [25] highlighted by the archaeological sites occurring over ten specific landform classes in the Jijia catchment (see Figure 6); the main differences are underlined in bold.

[1] Landform Classes Description	[2] Small-DEV Neighborhood Size	[3] Large-DEV Neighborhood Size	Relative Frequency (%)	
			Noua Culture (Late Bronze Age)	Precucuteni–Cucuteni Culture (Eneolithic)
L10: Hill tops, high ridges	$Z_0 > SD$	$Z_0 > SD$	**21.27**	**45.86**
L9: Middle slope ridges, small hills in plains	$Z_0 > SD$	$0 \leq Z_0 \leq SD$	1.10	2.51
L8: Local ridges/hills in valley	$Z_0 > SD$	$Z_0 < -SD$	**19.89**	**10.05**
L7: Upper slopes	$-D \leq Z_0 \leq SD$	$Z_0 > SD$	5.52	4.92
L6: Open slopes (>5°)	$-SD \leq Z_0 \leq SD$	$0 \leq Z_0 \leq SD$	0.55	0.73
L5: Plains, flat areas (<5°)	$-SD \leq Z_0 \leq SD$	$-SD \leq Z_0 < 0$	0.83	0.73
L4: U-shaped valleys	$-SD \leq Z_0 \leq SD$	$Z_0 < -SD$	**19.06**	**7.84**
L3: Upland drainage, headwaters	$Z_0 < -SD$	$Z_0 > SD$	5.52	2.62
L2: Middle slope drainage, shallow valley	$Z_0 < -SD$	$0 \leq Z_0 \leq SD$	1.10	0.63
L1: Deeply incised streams	$Z_0 < -SD$	$-SD \leq Z_0 < 0$	25.14	24.09

[1] Landform classes' descriptions after [70,71] for specific landscape characteristics of the study area; [2] R = 300 m (small radius value around z_0); [3] R = 1000 m (large radius value around z_0).

According to the slope position classification results, there are not significant differences between the two categories, even if the flat areas (Sp3) and lower slopes (Sp2) classes are relatively more representative for LBA sites without visible ashmounds (Table 6). According to the landform classification based on combination of small-DEV and large-DEV, several differences can be observed in U-shaped valleys (L4) and local ridges/hills in valley (L8) classes (Table 7). There are 15% more LBA sites with ashmounds located on the local ridges/hills in valleys (L8) than LBA sites without these structures; and there are 9% more sites without ashmounds located in U-shaped valleys (L4) landforms then LBA sites with ashmounds. The most plausible explanation for these outcomes is the ashmound's structure and how they appeared in the LBA settlements as a result of cattle shepherding activities [76]. Therefore, the reason why there are more LBA sites with ashmounds located on the local ridges/hills in valley (L8) than in U-shaped valleys (L4) is the fact that the wide and open valleys are good for grazing but unsuitable for habitation due to weather exposure and floods events [25].

Table 6. Descriptive statistics of LBA sites (see Figure 7a) vs. LBA sites with ashmounds (see Figure 7b) occurring over six slope position classes in the Jijia catchment.

[1] Slope Position Classes Description	[2] DEV Threshold	Landform Code	Number of LBA Sites	Number of LBA Sites with Ashmounds
Sp6: Ridge	TPI > 1 SD	Sp6	60 (35.93%)	93 (47.69%)
Sp5: Upper slope	0.5 SD < TPI ≤ 1 SD	Sp5	10 (5.99%)	12 (6.15%)
Sp4: Middle slope	−0.5 SD < TPI ≤ 0.5 SD	Sp4	9 (5.39%)	13 (6.67%)
Sp3: Flat area	−0.5 SD < TPI ≤ 0.5 SD	Sp3	**20 (11.98%)**	**7 (3.59%)**
Sp2: Lower slope	−1 SD < TPI ≤ −0.5 SD	Sp2	**15 (8.98%)**	**8 (4.1%)**
Sp1: Valley	TPI ≤ −1 SD	Sp1	53 (31.74%)	62 (31.79%)

[1] Slope position classes adapted after [70,71] for specific landscape characteristics of the study area; [2] R = 300 m (radius value around z_0).

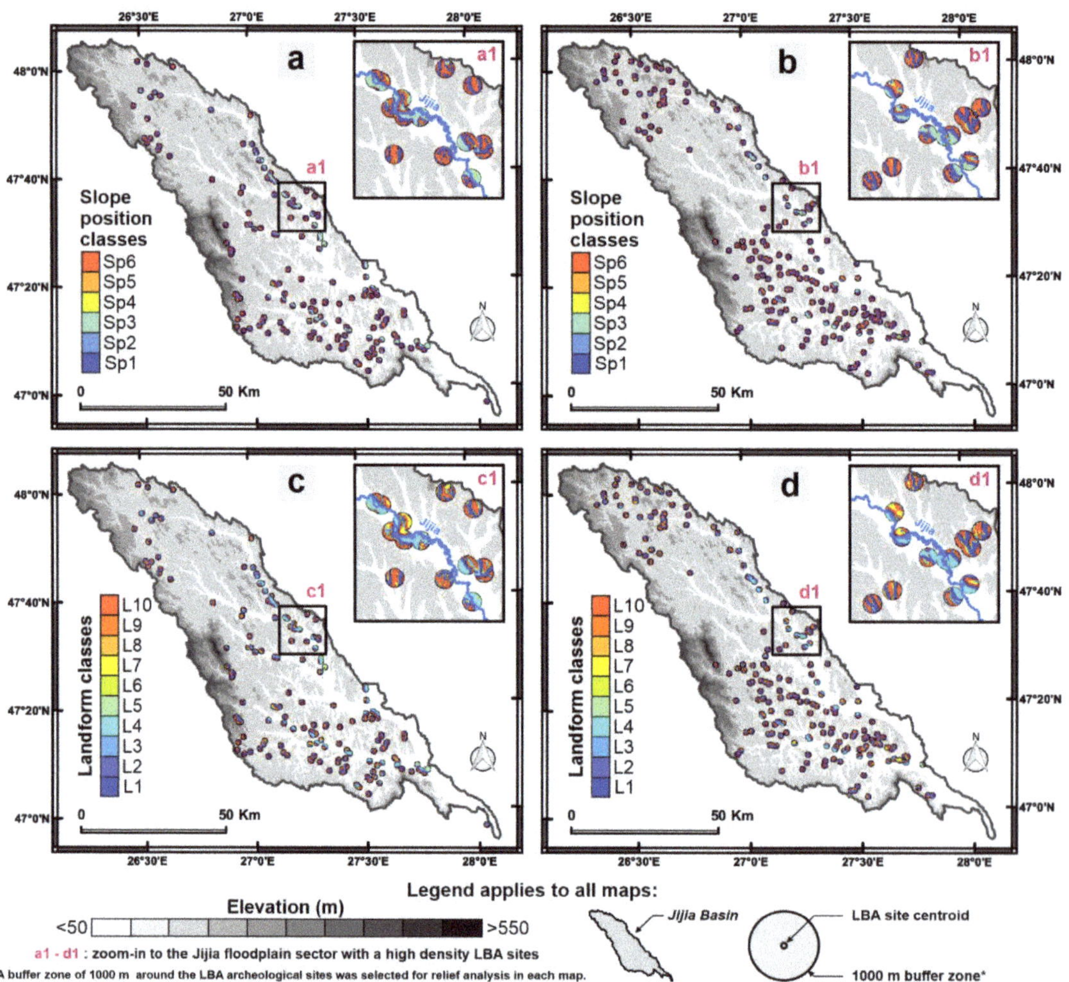

Figure 7. Automated relief analysis of LBA sites in the Jijia catchment: slope position classification based on TPI-derived DEV with six morphological classes (see Table 1) of the (**a**) LBA archaeological sites without ashmounds (167 sites) and (**b**) LBA sites with ashmounds (195 sites); landform classification based on TPI-derived DEV with ten landform types (see Table 2) of the (**c**) LBA archaeological sites without ashmounds (167 sites) and (**d**) LBA sites with ashmounds (195 sites).

Although, usually, the individuals that inhabited the Jijia catchment during the end of the Bronze age are considered to have been mere shepherds, concerned only with performing the mentioned activity, the totality of the preferences manifested by them in selecting the locations for their settlements reveals different behavior. Thus, the appetite for proximity to water sources finds its explanation in the good visibility and control gained over the river found in its vicinity, especially since rivers do not represent only a source of water supply but also a means of communication/transportation. Additionally, this characteristic allows good artefactual mobility, a fact proven by the presence of a large number of LBA sites in the immediate proximity of Jijia River, the most important watercourse in the entire workspace. Bearing in mind the risks of flood events, the LBA human groups chose to be located on the highest landforms present in the Jijia Valley, and the explanation for this lies in the very high visibility they gained, thereby controlling the entire middle sector of the river.

Table 7. Descriptive statistics of LBA sites (see Figure 7c) vs. LBA sites with ashmounds (see Figure 7d) occurring over ten landform classes in the Jijia catchment.

[1] Landform Classes Description	[2] Small-DEV Neighborhood Size	[3] Large-DEV Neighborhood Size	Number of LBA Sites	Number of LBA Sites with Ashmounds
L10: Hill tops, high ridges	$Z_0 > SD$	$Z_0 > SD$	41 (24.55%)	40 (20.51%)
L9: Middle slope ridges, small hills in plains	$Z_0 > SD$	$0 \leq Z_0 \leq SD$	2 (1.2%)	3 (1.54%)
L8: Local ridges/hills in valley	$Z_0 > SD$	$Z_0 < -SD$	**17 (10.18%)**	**50 (25.65%)**
L7: Upper slopes	$-SD \leq Z_0 \leq SD$	$Z_0 > SD$	6 (3.59%)	6 (3.08%)
L6: Open slopes (>5°)	$-SD \leq Z_0 \leq SD$	$0 \leq Z_0 \leq SD$	3 (1.8%)	2 (1.03%)
L5: Plains, flat areas (<5°)	$-SD \leq Z_0 \leq SD$	$-SD \leq Z_0 < 0$	5 (2.99%)	4 (2.05%)
L4: U-shaped valleys	$-SD \leq Z_0 \leq SD$	$Z_0 < -SD$	**40 (23.95%)**	**28 (14.36%)**
L3: Upland drainage, headwaters	$Z_0 < -SD$	$Z_0 > SD$	8 (4.79%)	6 (3.08%)
L2: Middle slope drainage, shallow valley	$Z_0 < -SD$	$0 \leq Z_0 \leq SD$	1 (0.6%)	2 (1.03%)
L1: Deeply incised streams	$Z_0 < -SD$	$-SD \leq Z_0 < 0$	44 (26.35%)	54 (27.69%)

[1] Landform classes description after [70,71] for specific landscape characteristics of the study area; [2] R = 300 m (small radius value around z_0); [3] R = 1000 m (large radius value around z_0).

Last but not least, we have to take into consideration some of the archaeological excavations performed on settlements with no ashmounds visible on the surface, located in areas that were not subject to agricultural processes [44,58]. In some cases, the invasive research conducted at these sites has highlighted the presence of the ash-like soil layer at a depth of 20–30 cm. The presence of the LBA cultural layer so close to the topsoil could represent evidence of the level of destruction [78,79] and an appeal to the necessity of preservation and conservation of the archaeological remains belonging to the end of the Bronze Age.

5.2. Automated Relief Analysis: Applicability and Limitations for Archaeological Studies

In this study, an alternative GIS-based solution for analyzing the spatial patterns and geomorphological characteristics of the settlements belonging to Late Bronze Age (LBA, chronological framework: cal. 1500/1450-1100 BC) was presented. To this end, we used an adapted version of Topographic Position Index (TPI) methodology [70,71], abbreviated DEV after [72], which consists of: application of the standard deviation of TPI for the mean elevation (DEV) around each analyzed LBA site (1000 m buffer zone); classification of the archaeological sites' locations using six slope position classes (first method) and ten morphological classes by combining the parameters from two small-DEV and large-DEV neighborhood sizes (second method). The outcomes produced new and valuable information regarding the 362 archaeological sites belonging to the Noua Culture, which flourished during the Late Bronze Age (cal. 1500/1450-1100 BCE) in the heterogenous landscape of the Jijia catchment (NE Romania). However, even if the automated relief analysis using TPI-based DEV methodology could not replace geomorphological expert opinions, the results bring new insights regarding remote sensing applied in cultural heritage assessments and archaeological predictive modelling.

According to [72], the TPI, or adapted DEV, is an example of a topographic metric that became institutionalized by its integration into widely used tools, such as the ESRI products [72]. This resulted not only in the popularity of this metric parameter, but also in the uncritical application by users without solid geomorphological backgrounds [72], such as archeologists [25]. In this context, there are many examples of when the GIS-based landform classification was easily integrated into various geo-archaeological studies due to its large applicability and low-cost performance [24,25]. Overall, most studies have focused on these automated modeling techniques, especially when the investigation of large archaeological databases was needed for various purposes (e.g., cultural heritage management, habitation practices characterization during different cultural periods, impact assessments of natural and anthropogenic hazards on the archaeological sites and so on) [24–27]; but there are also examples of when the TPI algorithm was applied in small-scale situations (intra-site) when the digital elevation models with high resolution (e.g., LiDAR-derived DEM's) were available [2,3,25]. Therefore, one of the technical limitations of the GIS-based methodology proposed in this study is the low resolution of DEM.

6. Conclusions

The automated relief investigation using the LiDAR-derived DEM and TPI-based DEV landform classification has produced new and valuable information regarding the 362 archaeological sites (195 sites with ashmounds) belonging to the Noua Culture, which flourished during the Late Bronze Age (cal. 1500/1450-1100 BCE) in the heterogenous landscape of the Jijia catchment (NE Romania). Therefore, the main habitation practices characteristics derived from relief analysis of the LBA sites' locations are:

- According to the slope position classification (first method) resulting from R = 300 m of TPI-based DEV and LiDAR-derived slope combination, over 38% of LBA settlements were placed on the concave landforms (Sp1—valleys and Sp2—lower slope); 7.45% of LBA sites were located on the Sp3—flat areas ($\leq 5°$); and over 54% of LBA settlements were placed on the convex landforms (Sp4—middle slope, Sp5—upper slope and Sp6—ridges, tops of the hills).
- According to the TPI-based DEV landform classification using small-DEV 300 m and large-DEV 1200 m combined neighborhood sizes, over 47% of LBA sites were located on convex landforms (L10—hilltops, high ridges, L9—middle slope ridges, small hills in the plains, L8—local ridges/hills in valleys and L7—upper slopes), 1.38% of the LBA settlements were located on the flat or gentle slope areas (L5—plains/flat areas and L6—open slopes) and 50.82% of LBA sites were located on the concave landforms (L4—U-shaped valleys, L3—upland drainages/headwaters, L2—middle slope drainages/shallow valleys and L1—deeply incised streams).
- The very high-density of NC sites located just on the four specific landforms, of which two are convex landforms (L10—hilltops, high ridges and L8—local ridges/hills in valleys) and two are concave landforms (L4—U-shaped valleys and L1—deeply incised streams), indicates habitation practices based on agro-pastoral activities manly induced by the suitability of the local topography. Additionally, the landform patterns highlight a specific eco-cultural niche just for the LBA communities, different from those of the other prehistoric cultures (e.g., Precucuteni-Cucuteni Culture) which flourished in the same workspace.

From the perspective of the GIS-based methodology applied in this study, the automated relief analysis using TPI-based DEV combined with LiDAR-derived DEM and other geomorphological variables (e.g., terrain slope) can be integrated very easily into various geo-archaeological studies (e.g., paleo-environmental reconstructions, eco-cultural niche modelling and archaeological predictive modelling) due to its great applicability. Furthermore, there are many improvements that the GIS-based techniques used in this work bring to the conventional geo-archaeological surveys: the fast and low-cost performance of relief analysis at both small and large scales, the ability to divide the landscape into ten landform classes and replicate the analysis for various archaeological contexts and also the ability to describe prehistoric human behavior based on certain geographical datasets.

Supplementary Materials: The following are available online at https://www.mdpi.com/article/10.3390/rs14102466/s1. Table S1: Noua Culture sites' distribution in the Jijia River basin used in this work for landform classification of settlements' locations and habitation practices characterization for the Late Bronze Age.

Author Contributions: Conceptualization, A.M.-P. and C.B.; methodology, A.M.-P.; software, A.M.-P., C.B. and C.C.S.; validation, A.M.-P. and C.B.; formal analysis, A.M.-P. and C.B.; investigation, A.M.-P. and C.B..; resources, A.M.-P., C.B. and C.C.S.; data curation, A.M.-P. and C.B.; writing—original draft preparation, A.M.-P. and C.B.; writing—review and editing, A.M.-P.; visualization, A.M.-P., C.B. and C.C.S.; supervision, A.M.-P. and C.B.; project administration, A.M.-P. and C.B.; funding acquisition, A.M.-P., C.B. and C.C.S. All authors have read and agreed to the published version of the manuscript.

Funding: This research received no external funding.

Acknowledgments: LiDAR-derived DEM data were provided by NARW-PBWA based on the institutional collaboration protocol between PBWA and UAIC. All information was processed in the Geoarchaeology Laboratory of Institute of Interdisciplinary Research, Science Department (Arheoinvest Centre), University "Alexandru Ioan Cuza" of Iaşi (UAIC).

Conflicts of Interest: The authors declare no conflict of interest. The founding sponsors had no role in the design of the study; in the collection, analyses, or interpretation of data; in the writing of the manuscript, and in the decision to publish the results.

Abbreviations

The following abbreviations are used in this manuscript:

GIS	Geographic Information System
TPI	Topographic Position Index
SD	Standard Deviation
DEV	Standard Deviation of Topographic Position Index
LiDAR	Light Detection and Ranging
DEM	Digital Elevation Model
IDW	Inverse Distance Weighting
R	Radius
$\bar{z}$	Average elevation pixels for various candidate radii
z_0	Central pixel elevation for various candidate radii
BA	Bronze Age
LBA	Late Bronze Age
NC	NOUA Culture
NARW-PBWA	Romanian Waters–Prut-Bîrlad Water Administration
PBRB	Prut–Bîrlad River Basin
UAV	Unmanned Aerial Vehicle
Sp1	Slope position class: Valley
Sp2	Slope position class: Lower slope
Sp3	Slope position class: Flat area
Sp4	Slope position class: Middle slope
Sp5	Slope position class: Upper slope
Sp6	Slope position class: Ridge
L1	Landform class: Deeply incised streams
L2	Landform class: Middle slope drainage, shallow valley
L3	Landform class: Upland drainage, headwaters
L4	Landform class: U-shaped valleys
L5	Landform class: Plains, flat areas (<5°)
L6	Landform class: Open slopes (>5°)
L7	Landform class: Upper slopes
L8	Landform class: Local ridges/hills in valley
L9	Landform class: Middle slope ridges, small hills in plains
L10	Landform class: Hill tops, high ridges

References

1. Butzer, K.W. Challenges for a cross-disciplinary geoarchaeology: The intersection between environmental history and geomorphology. *Geomorphology* **2008**, *101*, 402–411. [CrossRef]
2. Verhagen, P.; Drăguţ, L. Object-based landform delineation and classification from DEMs for archaeological predictive mapping. *J. Archaeol. Sci.* **2012**, *39*, 698–703. [CrossRef]
3. Nicu, I.C.; Mihu-Pintilie, A.; Williamson, J. GIS-Based and Statistical Approaches in Archaeological Predictive Modelling (NE Romania). *Sustainability* **2019**, *11*, 5969. [CrossRef]
4. Biscione, M.; Danese, M.; Masini, N. A framework for cultural heritage management and research: The Cancellara case study. *J. Maps* **2018**, *14*, 576–582. [CrossRef]
5. Gioia, D.; Bavusi, M.; Di Leo, P.; Giammatteo, T.; Schiattarella, T. Geoarchaeology and geomorphology of the Metaponto area, Ionian coastal belt, Italy. *J. Maps* **2020**, *16*, 117–125. [CrossRef]
6. Piloyan, A.; Konečný, M. Semi-Automated Classification of Landform Elements in Armenia Based on SRTM DEM using K-Means Unsupervised Classification. *Quaest. Geogr.* **2017**, *36*, 93–103. [CrossRef]
7. Simensen, T.; Halvorsen, R.; Erikstad, L. Methods for landscape characterization and mapping: A systematic review. *Land Use Policy* **2018**, *75*, 557–569. [CrossRef]
8. Deumlich, D.; Schmidt, R.; Sommer, M. A multiscale soil-landform relationship in the glacial-drift area based on digital terrain analysis and soil attributes. *J. Plant. Nutr. Soil Sci.* **2010**, *173*, 843–851. [CrossRef]
9. Illés, G.; Kovács, G.; Heil, B. Comparing and evaluating digital soil mapping methods in a Hungarian forest reserve. *Can. J. Soil Sci.* **2011**, *91*, 615–626. [CrossRef]
10. Mora-Vallejo, A.; Claessens, L.; Stoorvogel, J.; Heuvelink, G.B.M. Small scale digital soil mapping in southeastern Kenya. *Catena* **2008**, *76*, 44–53. [CrossRef]
11. Pracilio, G.; Smettem, K.; Bennett, D.; Harper, R.; Adams, M. Site assessment of a woody crop where a shallow hardpan soil layer constrained plant growth. *Plant Soil* **2006**, *288*, 113–125. [CrossRef]
12. Dębniak, K.; Mège, D.; Gurgurewicz, J. Geomorphology of Ius Chasma, Valles Marineris, Mars. *J. Maps* **2017**, *13*, 260–269. [CrossRef]
13. Drăguţ, L.; Blaschke, T. Automated classification of landform elements using object-based image analysis. *Geomorphology* **2006**, *81*, 330–344. [CrossRef]
14. McGarigal, K.; Tagil, S.; Cushman, S. Surface metrics: An alternative to patch metrics for the quantification of landscape structure. *Landsc. Ecol.* **2009**, *24*, 433–450. [CrossRef]
15. Tagil, S.; Jenness, J. GIS-based automated landform classification and topographic, landcover and geologic attributes of landforms around the Yazoren Polje, Turkey. *J. App. Sci.* **2008**, *8*, 910–921. [CrossRef]
16. Wilson, M.F.J.; O'Connell, B.; Brown, C.; Guinan, J.C.; Grehan, A.J. Multiscale terrain analysis of multibeam bathymetry data for habitat mapping on the continental slope. *Mar. Geod.* **2007**, *30*, 3–35. [CrossRef]
17. Wright, D.J.; Heyman, W.D. Introduction to the special issue: Marine and coastal GIS for geomorphology, habitat mapping, and marine reserves. *Mar. Geod.* **2008**, *31*, 223–230. [CrossRef]
18. Zieger, S.; Stieglitz, T.; Kininmonth, S. Mapping reef features from multibeam sonar data using multiscale morphometric analysis. *Mar. Geod.* **2009**, *264*, 209–217. [CrossRef]
19. Lesschen, J.P.; Kok, K.; Verburg, P.H.; Cammeraat, L.H. Identification of vulnerable areas for gully erosion under different scenarios of land abandonment in southeast Spain. *Catena* **2007**, *71*, 110–121. [CrossRef]
20. Liu, H.; Bu, R.; Liu, J.; Leng, W.; Hu, Y.; Yang, L.; Liu, H. Predicting the wetland distributions under climate warming in the Great Xing'an Mountains, northeastern China. *Ecol. Res.* **2011**, *26*, 605–613. [CrossRef]
21. Bunn, A.; Hughes, M.; Salzer, M. Topographically modified tree-ring chronologies as a potential means to improve paleoclimate inference. *Clim. Change* **2011**, *105*, 627–634. [CrossRef]
22. Guitet, S.; Cornu, J.-F.; Brunaux, O.; Betbeder, J.; Carozza, J.-M.; Richard-Hansen, C. Landform and landscape mapping, French Guiana (South America). *J. Maps* **2013**, *9*, 325–335. [CrossRef]
23. Fei, S.; Schibig, J.; Vance, M. Spatial habitat modeling of American chestnut at Mammoth Cave National Park. *For. Ecol. Manag.* **2007**, *252*, 201–207. [CrossRef]
24. Argyriou, A.V.; Teeuw, R.M.; Sarris, A. GIS-based landform classification of Bronze Age archaeological sites on Crete Island. *PLoS ONE* **2017**, *12*, e0170727. [CrossRef] [PubMed]
25. Mihu-Pintilie, A.; Nicu, I.C. GIS-based Landform Classification of Eneolithic Archaeological Sites in the Plateau-plain Transition Zone (NE Romania): Habitation Practices vs. Flood Hazard Perception. *Remote Sens.* **2019**, *11*, 915. [CrossRef]
26. De Reu, J.; Bourgeois, J.; De Smedt, P.; Zwertvaegher, A.; Antrop, M.; Bats, M.; De Maeyer, P.; Finke, P.; Van Meirvenne, M.; Verniers, J.; et al. Measuring the relative topographic position of archaeological sites in the landscape, a case study on the Bronze Age barrows in northwest Belgium. *J. Archaeol. Sci.* **2011**, *38*, 3435–3446. [CrossRef]
27. Noviello, M.; Cafarelli, B.; Calculli, C.; Sarris, A.; Mairota, P. Investigating the distribution of archaeological sites: Multiparametric vs probability models and potentials for remote sensing data. *Appl. Geogr.* **2018**, *95*, 34–44. [CrossRef]
28. Schmaltz, E.; Märker, M.; Rosner, H.-J.; Kandel, A.W. The integration of landscape processes in archaeological site prediction in the Mugello basin (Tuscany/Italy). In *Proceedings of the 42nd Annual Conference on Computer Applications and Quantitative Methods in Archaeology CAA*; Giligny, F., Djindjian, F., Costa, L., Moscati, P., Robert, S., Eds.; Paris 1 Panthéon-Sorbonne University: Paris, France, 2014; pp. 451–458.

29. Verhagen, P.; Whitley, T.G. Integrating Archaeological Theory and Predictive Modeling: A Live Report from the Scene. *J. Archaeol. Method Theory* **2012**, *19*, 49–100. [CrossRef]

30. Gioia, D.; Bavusi, M.; Di Leo, P.; Giammatteo, T.; Schiattarella, M. A geoarchaeological study of the Metaponto coastal belt, southern Italy, based on geomorphological mapping and GIS-supported classification of landforms. *Geogr. Fis. E Din. Quat.* **2016**, *39*, 137–148. [CrossRef]

31. Asăndulesei, A. Inside a Cucuteni Settlement: Remote Sensing Techniques for Documenting an Unexplored Eneolithic Site from Northeastern Romania. *Remote Sens.* **2017**, *9*, 41. [CrossRef]

32. Brigan, R.; Weller, O. Neo-Eneolithic settlement pattern and salt exploitation in Romanian Moldavia. *J. Archaeol. Sci. Rep.* **2018**, *17*, 68–78. [CrossRef]

33. Dietrich, L. *Die Mittlere und Späte Bronzezeit und die Ältere Eisenzeit in Südostsiebenbürgen Aufgrund der Siedlung von Rotbav;* European Verlag Dr. Rudolf Habelt GMBH: Bonn, Germany, 2014.

34. Huţanu, E.; Mihu-Pintilie, A.; Urzica, A.; Paveluc, L.E.; Stoleriu, C.C.; Grozavu, A. Using 1D HEC-RAS Modeling and LiDAR Data to Improve Flood Hazard Maps Accuracy: A Case Study from Jijia Floodplain (NE Romania). *Water* **2020**, *12*, 1624. [CrossRef]

35. Stoleriu, C.C.; Urzica, A.; Mihu-Pintilie, A. Improving flood risk map accuracy using high-density LiDAR data and the HEC-RAS river analysis system: A case study from north-eastern Romania. *J. Flood Risk Manag.* **2020**, *13*, e12572. [CrossRef]

36. Bacauanu, V. *Câmpia Moldovei. Studiu Geomorfologic;* Editura Academiei Romane: Bucuresti, Romania, 1968; pp. 1–222.

37. Niculiţă, M.; Mărgărint, M.C.; Santangelo, M. Archaeological evidence for Holocene landslide activity in the eastern Carpathian lowland. *Quat. Int.* **2016**, *415*, 175–189. [CrossRef]

38. Haase, D.; Fink, J.; Haase, G.; Ruske, R.; Pécsi, M.; Richter, H.; Altermann, M.; Jäger, K.-D. Loess in Europe—its spatial distribution based on a European loess map, scale 1:250,000. *Quat. Sci. Rev.* **2007**, *26*, 1301–1312. [CrossRef]

39. Romanescu, G.; Cîmpianu, C.I.; Mihu-Pintilie, A.; Stoleriu, C.C. Historic flood events in NE Romania (post-1990). *J. Maps* **2017**, *13*, 787–798. [CrossRef]

40. Mărgărint, M.C.; Niculiţă, M. Landslide Type and Pattern in Moldavian Plateau, NE Romania. In *Landform Dynamics and Evolution in Romania;* Radoane, M., Vespremeanu-Stroe, A., Eds.; Springer Geography, Springer: Cham, Switzerland, 2017; pp. 271–304. [CrossRef]

41. Nicu, I.C. Is overgrazing really influencing soil erosion? *Water* **2018**, *10*, 1077. [CrossRef]

42. Nicu, I.C. Natural risk assessment and mitigation of cultural heritage sites in North-eastern Romania (Valea Oii river basin). *Area* **2019**, *51*, 142–154. [CrossRef]

43. Florescu, A.C. Contribuţii la cunoaşterea culturii Noua. *Arheol. Mold.* **1964**, *II–III*, 143–216.

44. Dascălu, L. *Bronzul Mijlociu şi Târziu în Câmpia Moldovei;* Editura Trinitas: Iaşi, Romania, 2007.

45. Sava, E. *Aşezări din Perioada târzie a Epocii Bronzului în Spaţiul Pruto-Nistrean (Noua-Sabatinovka);* Biblioteca Tyragetia: Chişinău, Moldova, 2014; Volume XXVI.

46. Zaretskyi, I.A. Za-metka o drevnostyakh Khar'kovskoy gub. Bogodukhovskogo uyezda, slobody Likha-chevki. *Khar'kovskiy Sb.* **1888**, *2*, 229.

47. Gorodčov, V.A. Dnev-nik arkheologicheskikh issledovaniy v Zen'kovskom uyezde Poltavskoy guber- nii v 1906 g. Issledovaniye Bel'skogo gorodishcha. *Tr. 14 Arkheologi-Cheskogo S"Yezda V Chernigove* **1911**, *3*, 93–161.

48. Petrescu-Dîmboviţa, M. Contribuţii la problema sfîrşitului epocii bronzului şi începutului epocii fierului în Moldova. *Stud. Şi Cercet. De Istor. Veche Şi Arheol.* **1953**, *IV*, 443–486.

49. Motzoi-Chicideanu, I. Cenuşar. In *Enciclopedia Arheologiei şi Istoriei Vechi a României, I;* Preda, C., Ed.; Editura Enciclopedică: Bucureşti, Romania, 1994.

50. Gerškovič, J.P. Fenomen zol'nikov belogrodovskogo tipa. *Ross. Arheol.* **2004**, *4*, 104–113.

51. Comşa, E. L'évolution des types d'habitation du territoire de la Roumanie (depuis l'énéolithique jusqu'à la fin de l' âge du bronze). In *Dritter Internationaler Thrakologischer Kongress zu Ehren W. Tomascheks, 2–6 Juni 1980 Wien;* Peschew, A., Popov, D., Jordanov, K., Fol, A., Eds.; Swjat: Sofia, Bulgaria, 1984; pp. 121–137.

52. Berezanskaja, S.S. Komarówskaja kul'tura. *Arheol. Ukrajn'skoj RSR* **1985**, *I*, 428–437.

53. Arnăut, T. *Spaţii sacre şi Practici Funerare din Mileniul I a.Chr. în Arealul Carpato-Balcanic;* Casa Editorial-Poligrafică Bons Offices: Chişinău, Moldova, 2014.

54. Pieniążek, M. Kultische Landschaften in der Steppe. Zu den Anfängen sakraler Architektur im Nordpontikum. *Prähistorische Z.* **2011**, *86*, 8–30. [CrossRef]

55. Sava, E.; Kaiser, E. *Poselenie s „Zolnicami" u Acela Odaia-Miciurin, Respublica Moldova (Arheologhicesne i Estestvennonaucinie issledovaniia) / Die Siedlung mit „Aschenhügeln" Beim Dorf Odaia-Miciurin, Republik Moldova (Archäologische und Naturwissenschaftliche Untersuchungen);* Biblioteca Tyragetia: Chişinău, Moldova, 2011; Volume XIX.

56. SMIS-CSNR 17945 (Water Administration Prut—Bîrlad, Romania) Works for Reducing the Flood Risk in Prut—Bîrlad Basin. Available online: http://www.romair.ro (accessed on 28 March 2022).

57. Diaconu, V. Cultura Noua în Regiunea Vestică a Moldovei. Ph.D. Thesis, Institute of Archaeology, Romanian Academy, Iaşi, Romania, 2014.

58. Florescu, A.C. Repertoriul culturii Noua-Coslogeni din România. In *Cultură şi Civilizaţie la Dunărea de Jos;* Museum of Dunarea de Jos: Călăraşi, Romania, 1991; Volume IX.

59. Chirica, V.; Tanasachi, M. *Repertoriul Arheologic al Judeţului Iaşi, I;* Institutul de Istorie şi Arheologie 'A. D. Xenopol': Iaşi, Romania, 1984.

60. Chirica, V.; Tanasachi, M. *Repertoriul Arheologic al Județului Iași, II*; Institutul de Istorie și Arheologie 'A. D. Xenopol': Iași, Romania, 1985.
61. Păunescu, A.; Șadurschi, P.; Chirica, V. *Repertoriul Arheologic al Județului Botoșani*; CIMEC: București, Romania, 1976.
62. Șovan, O.L. *Repertoriul Arheologic al Județului Botoșani*, 2nd ed.; CIMEC: Botoșani, Romania, 2016.
63. Themistocleous, K. The Use of UAVs for Cultural Heritage and Archaeology. In *Remote Sensing for Archaeology and Cultural Landscapes*; Chini, M., Ehlers, M., Lakshmi, V., Mueller, N., Refice, A., Rocca, F., Skidmore, A., Vadrevu, K., Eds.; Springer: Cham, Switzerland, 2020; pp. 241–269. [CrossRef]
64. Muzirafuti, A.; Cascio, M.; Lanza, S. UAV Photogrammetry-based Mapping the Pocket Beach of Isola Bella, Taormina (Northeastern Sicily). In Proceedings of the 2021 IEEE International Workshop on Metrology for the Sea (MetroSea 2021), Reggio Calabria, Italy, 4–6 October 2021. [CrossRef]
65. Manfreda, S.; McCabe, M.F.; Miller, P.E.; Lucas, R.; Madrigal, V.P.; Mallinis, G.; Dor, E.B.; Helman, D.; Estes, L.; Ciraolo, G.; et al. On the Use of Unmanned Aerial Systems for Environmental Monitoring. *Remote Sens.* **2018**, *10*, 641. [CrossRef]
66. Zimmerman, D.; Pavlik, C.; Ruggles, A.; Armstrong, M.P. An experimental comparison of ordinary and universal Kriging and Inverse Distance Weighting. *Math. Geol.* **1999**, *31*, 375–390. [CrossRef]
67. Lu, G.Y.; Wong, D.W. An adaptive inverse-distance weighting spatial interpolation technique. *Comput. Geosci.* **2008**, *34*, 1044–1055. [CrossRef]
68. Zhou, T.; Popescu, S.; Malambo, L.; Zhao, K.; Krause, K. From LiDARWaveforms to Hyper Point Clouds: A Novel Data Product to Characterize Vegetation Structure. *Remote Sens.* **2018**, *10*, 1949. [CrossRef]
69. Doneus, M. Openness as visualization technique for interpretative mapping of airborne LiDAR derived digital terrain models. *Remote Sens.* **2013**, *5*, 6427–6442. [CrossRef]
70. Weiss, A.D. Topographic Position and Landforms Analysis. In Proceedings of the ESRI User Conference, San Diego, CA, USA, 9–13 July 2001; pp. 227–245. Available online: http://www.jennessent.com/downloads/TPIposter-TNC_18x22.pdf (accessed on 28 March 2022).
71. Jenness, J. Topographic Position Index (tpi_jen.avx) Extension for ArcView 3.x, v. 1.2. Jenness Enterprises. Available online: http://www.jennessent.com/arcview/tpi.htm (accessed on 28 March 2022).
72. De Reu, J.; Bourgeois, J.; Bats, M.; Zwertvaegher, A.; Gelorini, V.; De Smedt, P.; Chu, W.; Antrop, M.; De Maeyer, P.; Finke, P.; et al. Application of the topographic position index to heterogeneous landscapes. *Geomorphology* **2013**, *186*, 39–49. [CrossRef]
73. Watanabe, S.; Hajima, T.; Sudo, K.; Nagashima, T.; Takemura, T.; Okajima, H.; Nozawa, T.; Kawase, H.; Abe, M.; Yokohata, M.; et al. MIROC-ESM 2010: Model description and basic results of CMIP5-20c3m experiments. *Geosci. Model Dev.* **2011**, *4*, 845–872. [CrossRef]
74. Hijmans, R.J.; Cameron, S.E.; Parra, J.L.; Jones, P.G.; Jarvis, A. Very High Resolution Interpolated Climate Surfaces for Global Land Areas. *Int. J. Climatol.* **2005**, *25*, 1965–1978. [CrossRef]
75. Ivanova, M.; De Cupere, B.; Ethier, J.; Marinova, E. Pioneer farming in Southeast Europe during the early sixth millennium BC: Climate-related adaptations in the exploitation of plants and animals. *PLoS ONE* **2018**, *13*, e0197225. [CrossRef]
76. Fick, S.E.; Hijmans, R.J. WorldClim 2: New 1km spatial resolution climate surfaces for global land areas. *Int. J. Climatol.* **2017**, *37*, 4302–4315. [CrossRef]
77. Gent, P.R.; Danabasoglu, G.; Donner, L.J.; Holland, M.M.; Hunke, E.C.; Jayne, S.R.; Lawrence, D.M.; Neale, R.B.; Rasch, F.J.; Vertenstein, M.; et al. The Community Climate System Model Version 4. *J. Climate* **2011**, *24*, 4973–4991. [CrossRef]
78. Serjeantson, D. Intensification of animal husbandry in the Late Bronze Age? The contribution of sheep and pigs. In *The Earlier Iron Age in Britain and the Near Continent*; Haselgrove, C., Pope, R., Eds.; Oxbow Books: Oxford, UK, 2017; pp. 80–93. [CrossRef]
79. Brașoveanu, C. Perioada târzie a Epocii Bronzului în bazinul Jijiei (România). Habitat și materialitate. Ph.D. Thesis, Alexandru Ioan Cuza University of Iași, Iași, Romania, 2021.

Article

Evaluating the Territorial Impact of Built-Up Area Expansion in the Surroundings of Bucharest (Romania) through a Multilevel Approach Based on Landsat Satellite Imagery

Ilinca-Valentina Stoica [1,2], Daniela Zamfir [1,2,*] and Marina Vîrghileanu [1,3]

1   Faculty of Geography, University of Bucharest, 010041 Bucharest, Romania;
    valentina.stoica@unibuc.ro (I.-V.S.); marina.virghileanu@geo.unibuc.ro (M.V.)
2   The Interdisciplinary Center for Advanced Research on Territorial Dynamics (CICADIT),
    University of Bucharest, 030018 Bucharest, Romania
3   Romanian Young Academy, University of Bucharest, 050663 Bucharest, Romania
*   Correspondence: daniela.zamfir@unibuc.ro

Abstract: Assessing the relentless expansion of built-up areas is one of the most important tasks for achieving sustainable planning and supporting decision-making on the regional and local level. In this context, techniques based on remote sensing can play a crucial role in monitoring the fast rhythm of urban growth, allowing the regular appraisal of territorial dynamics. The main aim of the study is to evaluate, in a multi-scalar perspective, the built-up area expansion and the spatio–temporal changes in Ilfov County, which overlaps the surroundings of Bucharest, capital of Romania. Our research focuses on processing multi-date Landsat satellite imagery from three selected time references (2000, 2008, 2018) through the supervised classification process. Further on, the types of built-up area dynamics are explored using LDTtool, a landscape metrics instrument. The results reveal massive territorial restructuring in the 18 years, as the new built-up developments occupy a larger area than the settlements' surface in 2000. The rhythm of the transformations also changed over time, denoting a significant acceleration after 2008, when 75% of the new development occurred. At the regional level, the spatial pattern has become more and more complex, in a patchwork of spatial arrangements characterized by the proliferation of low density areas interspersed with clusters of high density developments and undeveloped land. At the local level, a comparative assessment of the administrative territorial units' pathway was conducted based on the annual growth of built-up areas, highlighting the most attractive places and the main territorial directions of development. In terms of the specific dynamics of built-up areas, the main change patterns are "F—NP increment by gain", followed by "G—Aggregation by gain", both comprising around 80% of the total number of cells. The first type was prevalent in the first period (2000–2008), while the second is identified only after 2008, when it became the most represented, followed in the hierarchy by the previously dominant category. The spatial pattern differentiations were further explored in three complementary case studies investigated in correlation with socioeconomic data, revealing a heterogeneous landscape.

Keywords: built-up growth; remote sensing; post-socialist context; territorial planning; Ilfov County

Citation: Stoica, I.-V.; Zamfir, D.; Vîrghileanu, M. Evaluating the Territorial Impact of Built-Up Area Expansion in the Surroundings of Bucharest (Romania) through a Multilevel Approach Based on Landsat Satellite Imagery. *Remote Sens.* **2021**, *13*, 3969. https://doi.org/10.3390/rs13193969

Academic Editor: Deodato Tapete

Received: 14 August 2021
Accepted: 27 September 2021
Published: 3 October 2021

1. Introduction

The increasing demand for land reclaimed for the incessant process of built-up area expansion represents a major concern around the world [1,2], as it leads to irreversible changes to, and negative impacts on, the environment. In this regard, urban expansion is considered a threat to sustainable development goals [3,4], as previous research reported detrimental consequences related to soil sealing [5], air and water pollution [6,7], biodiversity loss [8,9], land use conflicts [10–13].

However, land is a finite natural resource [14–16], which is why better planning that will control the spread of built-up areas is critical [4,17]. For this reason, studies addressing the territorial impact of urban expansion, its patterns and its driving forces have become essential for researchers, planners and policy makers.

Lately, the topic of urban growth has been the focus of a huge amount of literature [18] in the context of various fields and geographical regions. While its pace differs from one continent to another or even from region to region, the process of urban growth is one of the most visible in the landscape today [19,20]. Although this phenomenon is global, the biggest territorial transformations due to the augmentation of built-up areas are found in Asia [21]. In this respect, there are numerous studies that deal with single metropolitan regions [22–29], cross-sectional comparisons between two or more cities [30–33], as well as studies that focus on much larger areas [34–36]. In the African continent, overall, the tempo of urban development is considered low [21], but is anticipated to increase significantly [8]. The number of studies is more limited [30], and tackle some of the largest urban regions, such as Kinshasa [37], Cairo [38], Nairobi [39], Lagos [40], Lusaka [41] and Pretoria [42]. Meanwhile, other research make comparisons between different places across the continent [43–47]. All these studies emphasize the fast pace of built-up area expansion that occurs in an unplanned and unregulated manner.

On the other hand, Europe has registered a generally moderate rhythm of urban growth over the last few decades [21], but, nevertheless, many regions exhibit a significant increase in built-up areas, which is expected to continue [4]. The phenomenon has been consistently observed even in areas that have experienced population decline [48]. Previous studies have concluded that the different spatial structure and patterns of urban growth are related with historical and political backgrounds, but also with socioeconomic and cultural features [49,50]. In this regard, Central and Eastern European (CEE) countries had a distinct urban development pathway compared with the rest of the continent [49,51–54]. This macro-region is considered one of the most impacted in the world, in the last 30 years, by complex transformations (political, socioeconomic, cultural and institutional) as a result of the communist regime dismantling and the subsequent transition from centralized planning to free market capitalism [55–59]. A relentless process of spatial reconfiguration was registered and the compact shape of the socialist city and the once well-defined boundaries [60–62] faded gradually, along with the dispersion of built-up areas in fragmented, less regular development in the surrounding regions [63,64]. In fact, recent studies revealed that the CEE city regions are generally characterized by a more dispersed pattern of development than Western Europe [52,56] in view of higher rates of urban growth [65]. The main challenges are the inefficient policies and loose regulatory framework [66,67] that foster an uncontrolled development of built-up areas in the vicinity of large cities [56,68].

Still, beyond this general trajectory, regional variations related to the management of territorial transformations within CEE countries have been observed, variations that were deeply connected to national place-specific factors [52,68]. This situation is also rooted in the European Union's approach, which assigns the land use management and spatial planning responsibilities to each member state [16,17]. In this respect, robust analysis at regional level, which will provide in-depth knowledge regarding the spatial patterns of growth and their determinants, is essential [69]. Such investigations will be extremely useful, if not downright necessary, for further comparative analysis.

Previous studies have outlined the pressing need for the quantitative assessment of the urban expansion territorial impact in CEE countries [70], as the majority of research uses qualitative approaches to describe the political and socioeconomic context, as well as the factors that influence built-up area growth and its overall characteristics [56,62,63,71–73]. Even though qualitative research brings valuable findings, quantitative appraisal is essential in the support of better governance, which will ultimately lead to sustainable planning. In this regard, it has been emphasized that the spatial impact of urban expansion around large cities, as well as its socioeconomic and environmental effects, have been insufficiently investigated [54].

Bucharest, the capital of Romania, is considered the largest post-socialist city [74,75] in the European Union and its neighboring Ilfov County the most dynamic, predominantly urban, region in terms of the growth rate of inhabitants for the period 2004–2014 [76]. The first decade after the fall of communism was marked by modest territorial transformations [57] followed, after 2000, by an ongoing process of unplanned built-up area expansion into the predominantly agricultural landscape of Ilfov County. Given the rapid pace of territorial change, more information about the resultant complex spatial patterns is needed to substantiate further decision-making on the local and regional levels.

Lately, various studies have pointed out Landsat satellite images as valuable resources for investigating the territorial impact of built-up area expansion, as they allow a comprehensive, integrated regional overview [77–83]. One of the strengths of using geospatial information is that it enables longitudinal analysis, in view of the long temporal coverage based on consistent archives with data [80]. They can play a critical role in depicting various spatio–temporal patterns [84]; evaluating the impact of enforced policies, urban plans and strategies [85]; and sustaining effective measures to improve territorial management and land use [4,49,79,86,87].

Of the few studies that applied quantitative measurements in CEE countries, several have utilized the Landsat dataset to explore urban growth from different perspectives in different contexts: the structural changes in Berlin and its surroundings [88], urban change detection in the Ljubljana region [89], the built-up area development in the surroundings of Prague [90]. Another paper addressed the dynamics of urban expansion and its spatial patterns in a cross-national comparison of several post-socialist cities [70]. The aforementioned studies revealed the general pattern of development, while the current research goes beyond this perspective, using a more complex approach and seeking deeper insights into the expansion of the built-up area through a multi-scalar territorial vision.

Regarding Bucharest and its surroundings, previous studies used Landsat to outline the overall trends of built-up area expansion until 2010, focused on the capital and the first ring of administrative territorial units (ATUs) located nearby [75], or on its fringes—20–25 km distance from the city center [91]. A more recent study treated Bucharest and Ilfov County as a laboratory region for comparing Corine Land Cover and Landsat datasets with each other and the ground truth, thus revealing their strengths and limitations in assessing the built-up area's changes [64]. The current study is distinguished by a more comprehensive framework: a gradual evaluation of the expansion of built-up areas on a regional level followed by comparative assessment of the local ATU's trajectory. The potential for using the results as a basis for qualitative analysis is explored in correlation with ancillary, socioeconomic data. Moreover, the research provides up-to-date information by investigating the most dynamic period (after 2000) in terms of land cover changes.

In any case, in-depth knowledge about the built-up area expansion process around Bucharest is limited compared with other post-socialist capitals [57,61]. One of the reasons for this may be that the impact of Bucharest's urban expansion became visible in the landscape much later—after the turn of the millennium—whereas the deconcentration of urban functions took place more rapidly in some of its CEE counterparts (such us Prague and Budapest).

Considering all of the above, the goal of this study is to contribute to filling this knowledge gap by analyzing the territorial transformations induced by the built-up area expansion that reconfigured the Bucharest surroundings after 2000, through a multi-scalar approach.

The paper follows several stages: (a) mapping built-up area coverage for three years: 2000, 2008, 2018; (b) investigating the intensity of built-up area expansion and associated spatio–temporal changes for two distinct periods: 2000–2008 and 2008–2018 on the macroterritorial (regional perspective) and medium level; (c) exploring the distribution patterns of the built-up area.

The results of the study reveal an overview of the territorial impact of built-up area expansion, providing a valuable scientific approach and data support which can be used for better planning, sustainable land use management and the design of efficient development plans.

2. Materials and Methods

2.1. Study Area

The study focuses on Ilfov County, located in the southern part of Romania, overlapping the surrounding area of Bucharest (Figure 1). From an administrative point of view, Ilfov County is divided into eight urban settlements and 32 communes, covering a total area of 1583.3 km^2 [92]. The registered population is around 430,000 inhabitants (Romanian National Institute of Statistics—NIS) but there are estimates, even from local authorities, that the actual number of people is much higher, since there is no obligation to register changes of legal residence [57]. However, between 2000 and 2018, the number of people more than doubled and this trend is expected to continue, stimulated by the proximity to Bucharest [64].

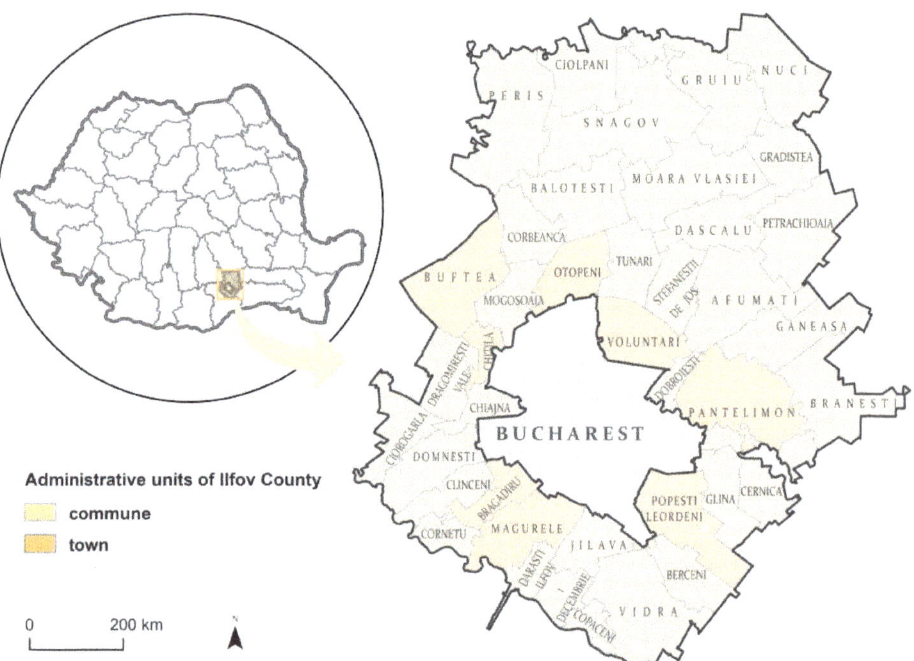

Figure 1. Location map of the study area.

Traditionally, Ilfov County was known for its agricultural profile, as around 70% of the total surface was composed of agricultural land in the 1990s (NIS). The settlements featured a high degree of compactness—a result of highly regulated communist central planning [75]. The area registered numerous spatial transformations, mainly after 2000, under the auspices of a free market economy, as new built-up areas (residential and non-residential) emerged in the landscape. Furthermore, Ilfov, together with Bucharest, is considered the richest region of the country [50,93], attracting newcomers and investors which leads to a continuous process of territorial reorganisation. This process is not uniform, as there is spatial differentiation between the local entities in terms of attractiveness, natural resources, level of economic development, etc. One of the main challenges remains the uncontrolled growth of built-up areas in a fragmented spatial pattern, with no integrated vision or strategy for governing urban expansion at the regional level.

2.2. Methodology

The methodology includes several phases to capture the spatio–temporal changes of the built-up area through a multi-scalar perspective. Key time points (2000, 2008, 2018) were selected for mapping the extent of the built-up area to allow a stepwise evaluation of the recorded transformations.

The year 2000 was chosen as a reference point because the urban expansion spilled over significantly into the surrounding territory of Bucharest only after the turn of the 21st century, significantly transforming the area's previous spatial pattern.

The year 2008 was selected as a benchmark in light of the global economic crisis, which hit Romania in the second half of the year. Previous evidence reported a subsequent sharp decline in construction in CEE countries [66,90,94]. Thus, the long term territorial impact of the crisis was investigated a decade after its occurrence, in 2018.

Our analysis is based on Landsat satellite imagery at 30 m spatial resolution, as it constitutes the longest record of observation data with free availability [95]. In this regard, up to three cloud free scenes from the summer–autumn seasons were selected from different Landsat sensors for each reference moment: Landsat 7 Enhanced Thematic Mapper (ETM+) for 2000, Landsat 5 Thematic Mapper (TM) for 2008 and Landsat 8 Operational Land Imager (OLI) for 2018. All the images, downloaded at Level-2 processing in terms of surface reflectance from the USGS archives, helped create multi-date stacks of Landsat imagery for each temporal moment. The multi-date approach was preferred because of the demonstrated higher performance of the classification algorithms, as opposed to single-date imagery [96], especially in areas with confusion between some built-up features and bare lands because of the spectral similarities, as is the case in our study area.

The approach follows several steps, as presented in Figure 2.

First, the training sample datasets for the land cover features were created for each reference moment. This was performed by visual interpretation of the input satellite images in different band combinations together with independent data sources, including orthophotos and field observations, with special focus on built-up features.

Second, a thematic supervised classification method was applied on each multi-date stacked dataset, resulting in a thematic classification of land cover features, grouped later in two main classes: built-up area and non-built-up area. Different methods of supervised classification were used and tested on multispectral data in the last 30 years, in order to find the most efficient classifier for data extraction from remote sensing images. However, the effectiveness of the classification algorithms strongly depends on various factors, including the image properties and resolutions, the complexity of land cover/use classes, and the training datasets. Our purpose is to accurately isolate the built-up features from other land cover/use classes for a large area with heterogeneous landscape. The most challenging task was to reduce the confusions caused by the spectral similarities between built-up features and barren lands, less arable vegetation plots, and landfills. In this context, several supervised classification learning methods were tested and compared, including the parametric algorithm Maximum Likelihood and the non-parameter algorithms Support Vector Machine and Neural Network. The Neural Network classifier returned better results for built-up area discrimination in our study area, with fewer confusions and a higher precision of the features. As several other studies previously mentioned, the Neural Network algorithm has the ability to solve large scale complex problems, such as pattern recognition, nonlinear modelling, association and control [97–99]. Contrary to parametric classifiers, which are highly dependent upon statistical distribution [100], non-parametric classifiers, including Neural Network, can handle large amounts of noisy data from dynamic and nonlinear systems [101], and identify the relationship from given patterns [100]. This is why the Neural Network classifier was preferred for built-up feature extractions in different study areas [102,103].

Third, a double validation of the binary classifications was performed, by comparison with the input images and independent datasets (such as Google Earth imagery, orthophotos and field observation). The objective of the qualitative approach of visual comparison between resulted classification and independent data was to identify and correct the mis-classifications generated by the spectral similarities between built-up features and other land cover classes, especially bare lands or landfills.

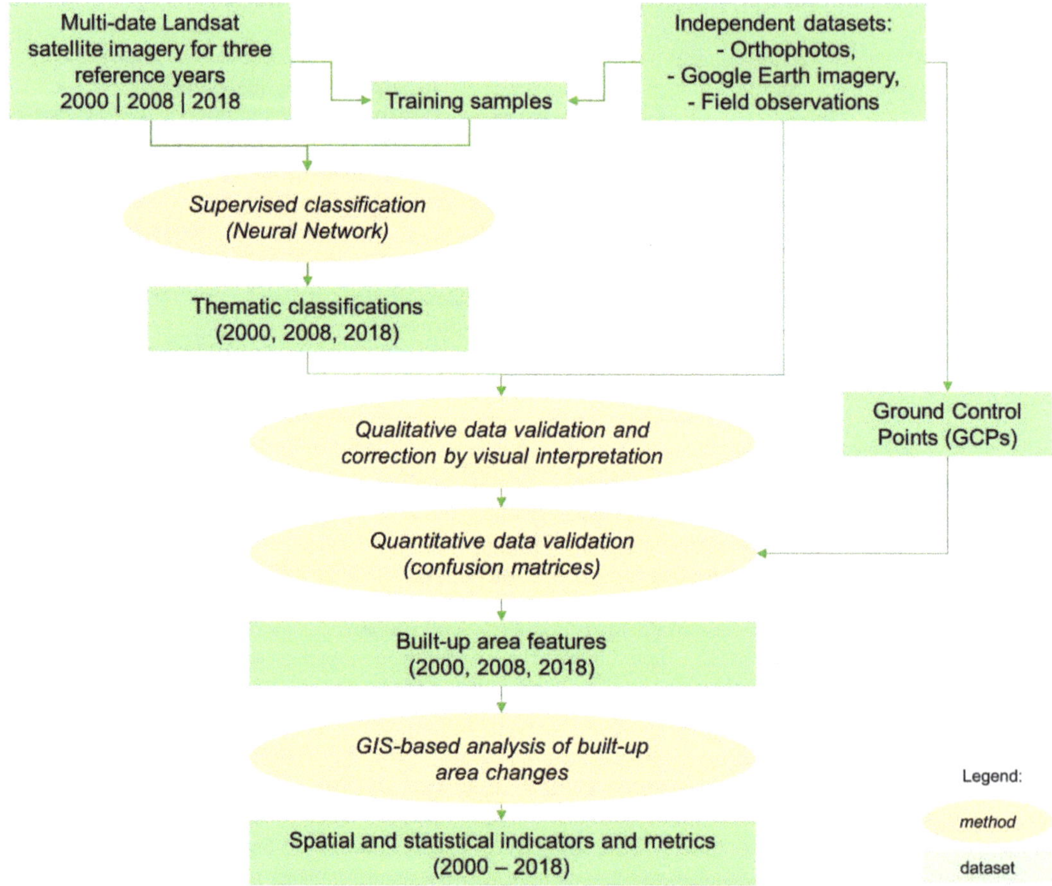

Figure 2. Analysis workflow.

The quantitative method of confusion matrices [104,105] was based on a number of 300 ground control points (GCPs) randomly distributed over the study area, with binary attributes (built-up and non-built-up), created for each reference moment through visual interpretation of input imagery and independent datasets. The number of the GCPs was determined through a statistically solid sampling scheme [104], accounting for various parameters, such as the level of probability assigned to the estimation, the estimated percentage of successes, the percentage of errors and the level of error allowed. In the case of classified images, where the variable is categorical, the preferred approach is a bimodal distribution of probability [104]. The results consisted in statistical indicators of accuracies, such as overall accuracy (%) and a kappa coefficient.

Finally, geographical information systems (GIS) techniques were used for built-up area change detection analysis and interpretation, in a spatial and statistical manner, at the regional and local administrative levels, as well as for data mapping.

In order to quantify the magnitude of urban expansion and the main territorial directions of development at the ATU level, the annual growth (AG) of built-up areas was calculated for two periods: 2000–2008 and 2008–2018. This indicator can be defined as the measurement of the newly developed built-up areas as the annual average over a given period and is expressed as follows:

$$AG = (At2 - At1)/n \ (ha/year),$$

where At2 and At1 represent the size of the built-up area at the final and initial years, and n refers to the time span between the two dates. The values were ranked in four categories to reveal the spatial differentiation (under 10 ha/year; 10–20 ha/year; 20–30 ha/year; over 30 ha/year) and to allow further comparisons.

The type of built-up area dynamics between 2000 and 2018 was evaluated with the help of LDTtool, a landscape metrics instrument that focuses on accounting for both composition and configuration changes of the binary land cover classes [106]. This tool is based on the combination of the characteristics of two spatial metrics (area and number of patches—NP) that can increase, decrease or remain the same between both reference times, resulting in a set of landscape dynamics types. The analysis was run using the built-up area polygons from all three moments of reference (2000, 2008 and 2018) on an artificial sampling grid. The chosen grid cell size was 250 m, a value we consider large enough (comparing to building dimensions) to ensure a proper spatial resolution and, at the same time, to minimize patch intersection with the spatial grid which could lead to an increased number of patches.

From the group of ATUs recording the highest annual growth of built-up area (over 30 ha/year) in the period 2008–2018, three complementary case studies denoting various spatial patterns of development were chosen, in connection with socioeconomic data. The development patterns erected in the post-socialist context were examined using high resolution Google Earth Images.

3. Results

3.1. Spatial and Temporal Features of Built-Up Area Expansion

The spatial configuration of the built-up area in the three key temporal moments (2000, 2008, 2018) is represented in Figure 3.

The accuracy assessment of the satellite based spatial data was performed by both qualitative and quantitative approaches. First, the resulted datasets were visually compared to independent geospatial layers as orthophotos and Google Earth imagery. Second, confusion matrices were used to explain the evaluation in a quantitative formula, based on a set of 300 GCPs generated with a random distribution over the study area.

Table 1 provides a synthetic view over the statistical validation of the built-up datasets extracted from Landsat satellite images with the GCPs.

Table 1. Main accuracies of the built-up datasets extracted from Landsat imagery.

	2000	2008	2018
	Landsat 7	**Landsat 5**	**Landsat 8**
Overall accuracy (%)	95.6667	96.0000	95.6667
Kappa coefficient	0.9356	0.9344	0.9409

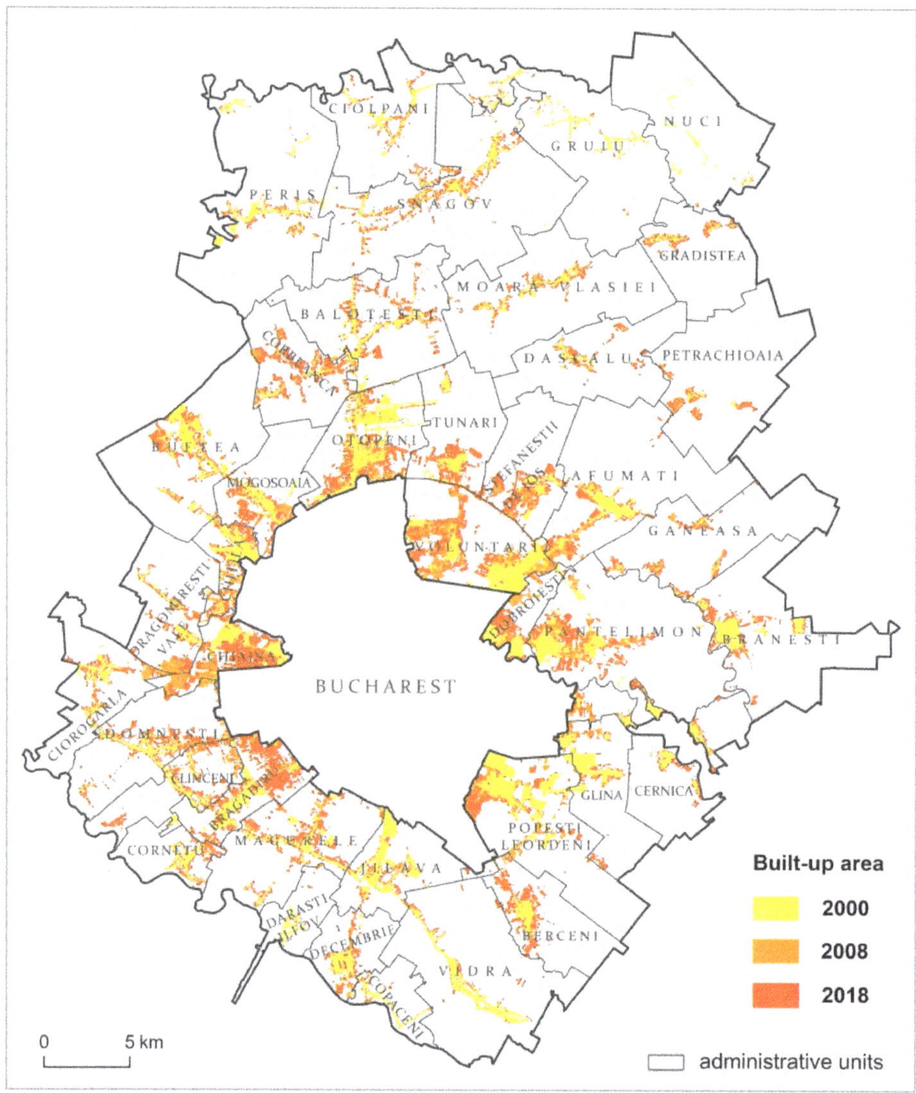

Figure 3. Built-up area coverage in Ilfov County at different temporal moments.

The values of the global accuracies and of the calculated Kappa coefficients (over 95% and over 0.9, respectively) are encouraging; they illustrate the results' performance.

The result map highlights the process of territorial restructuring with a significant increase in built-up areas on previously undeveloped land, mainly through the conversion of greenfield sites.

To better understand the changes, the analysis was further centered on two phases: 2000–2008; 2008–2018. In 2000 the built-up area covered around 9822.3 ha, after which it increased by 31%, reaching 12,891.4 ha in 2008 (Table 2). Analysis of the spatial configuration revealed that the spread of newly built-up areas followed certain directions of development, more obviously along major communication routes and on the edges of previously existent settlements. Additionally, some significant nuclei of development in terms of land consumption emerged far away, highlighting a fragmented spatial pattern. Usually, these clusters were comprised of residential complexes, but non-residential expansion has also been observed.

Table 2. The built-up area expansion and the annual growth increase (2000–2018).

	2000	2008	2018
Built up area (ha)	9822.3	12,891.4	21,948.3

	2000–2008	2008–2018
Built up area expansion (ha)	3069.1	9057.0
Annual growth (ha/years)	383.6	905.7

Overall, a higher concentration of newly built-up areas is noted in several places located near Bucharest. Still, generally, some settlements in the southern, south-western and even western side proved to be less appealing.

Between 2008 and 2018, the amplitude of built-up area expansion heightened, as the newly developed constructions increased by 70%, meaning 9057 ha. In 2018, the built-up area encompassed 21,948.3 ha. The map highlights large area construction increases and a general tendency for more spread development into the territory. The most noticeable changes occurred nearby the compact edges of settlements or next to previously post-socialist nuclei of development. Additionally, an infilling process can be observed in several places, as well as an attractiveness of the open spaces close to the boundaries of Bucharest. Still, the emergence of new fragmented clusters can be observed almost all over the county.

As far as the general spatial pattern is concerned, the preferred territorial directions seem to have preserved the pre-existing radial layout. Nevertheless, compared to the previous period, it is obvious that the overall configuration is moving towards a more circular development in the immediate vicinity of Bucharest.

The territorial extent of the newly built-up area between 2000 and 2018 encompasses 12,126 ha, revealing a massive spatial reconfiguration as the new developments are occupying a larger area than the settlements' surface in 2000. Further analysis of the map shows that this process is more than visible in many places, such as Chiajna, Bragadiru, Corbeanca, etc. In fact, for 65% of ATUs, the surface of the new constructions exceeded the initial size of settlements at the turn of the millennium. Overall, the territorial reconversion did not follow the same pace, considering that between 2000–2008 only 3069.1 ha were built, while in the next decade the impact was much higher—9057 ha were converted into new buildings. In fact, an enormous 75% of the entire development took place in the last decade. Analysis of the annual growth of built-up area confirmed the substantial acceleration of urban decentralisation in the second interval, 136% higher than the values recorded between 2000 and 2008.

The initially prevalent compact form of the settlements with clearly established boundaries, a legacy of the communist policy, has faded in the areas experiencing the most dynamic development, as the open spaces between them have been swallowed up by new constructions. As a result, in several places, a relative continuum of built-up areas, including clusters of isolated buildings, has become visible in the landscape. In addition, low density developments proliferated in a scattered way all over the territory. This situation is also supported by the conversion of a large part of the agricultural surface to land available for construction, boosting the building sector.

From a morphological point of view, the settlements' structures became increasingly complex: in addition to the main core, more and more new secondary built-up areas have been identified. It is important to investigate these features to evaluate the financial resources needed to ensure adequate infrastructure and basic services. Meanwhile, other settlements maintain their previously linear structure, along the roads, even though new built-up areas emerged. This pattern is distributed mostly in the northern part of the county, but also in the south and south-west.

3.2. Comparison of Built-Up Area Expansion in a Medium-Scale Perspective

Extending the analysis to the local level is relevant for exploring the spatial differentiations and for identifying the places with the most significant development of new constructions. This type of analysis can be useful for planners and policy makers as the basis for further assessment of the local land use management, urban plans and strategies. For an accurate evaluation of the territorial dynamics, the annual growth of built-up areas was calculated for two periods (2000–2008; 2008–2018).

In the first interval, between 2000 and 2008, the majority of ATUs (65%) presented a limited spread of built-up areas, under 10 ha each (Table 3). Higher values were registered in several places in the first and second ring of ATUs near Bucharest, mainly in the northern, north-eastern, and north-western side. Only two attractive places are located in the south-eastern (Popeşti-Leordeni) and south-western part (Bragadiru), both granted with an urban status (Figure 4). This denotes a prevalence of development in certain areas. New developments, comprising over 20 ha, were registered in places with a good accessibility to Bucharest, in settlements located in the first ring of ATUs. The town of Voluntari, located in the north-eastern part of Romania's capital, recorded the highest increase for this period (about 50 ha/year).

Table 3. Annual growth of built-up areas (%).

	2000–2008	2008–2018
under 10 ha	65	17.5
10–20 ha	22.5	27.5
20–30 ha	10	27.5
over 30 ha	2.5	27.5

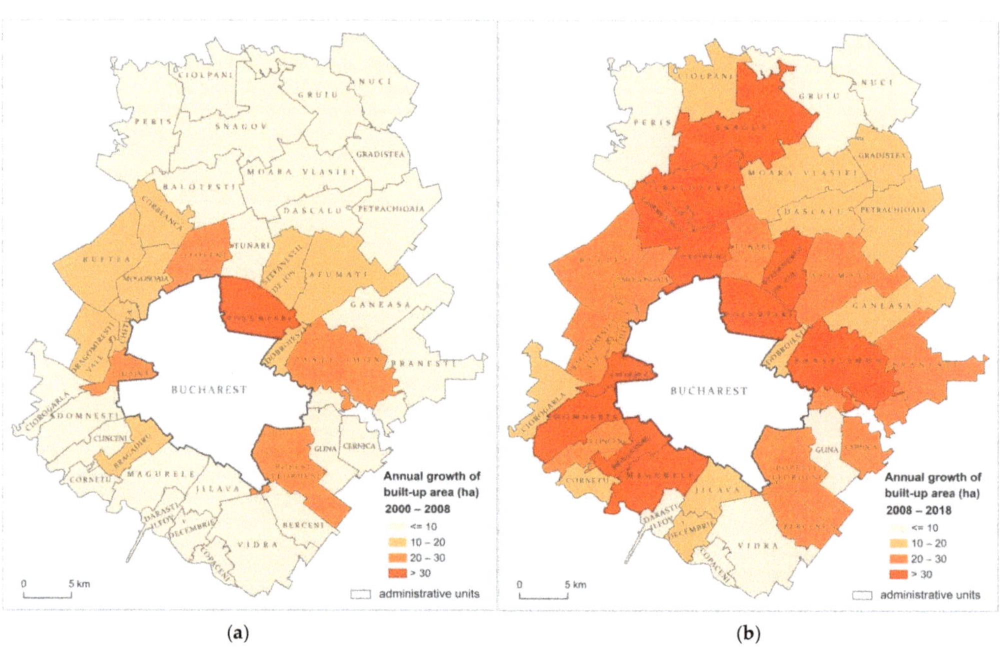

(a) (b)

Figure 4. Annual growth of built-up area at the administrative unit level, 2000–2008 (**a**) and 2008–2018 (**b**).

In the second interval (2008–2018) the majority of ATUs in Ilfov County saw an accelerated expansion of built-up areas. In this regard, just 17.5% of the ATUs were less appealing, located in the north or south extremities, and in the south-east—Glina, known for its environmental issues. In the first place in terms of attractiveness is a town located in the southwestern side (Bragadiru) with about 59 ha/year land consumption, followed by Voluntari (about 55 ha/year) and Chiajna (about 52 ha/year). Analysis of the map reveals that the most attractive places are located near Bucharest or along an axis of territorial development which extends to the northern part of the county, following one of the main communication routes to the capital.

Thus, comparing the maps enables the conclusion that, in the second period, significant real estate investments expanded to other parts of the county as well. For example, some settlements in the south and south-west side which did not use to be appealing now attract investors and Bucharest residents.

Confronting the annual growth of built-up areas for the two intervals, only a few ATUs have maintained a limited extent of transformations over the whole period—below 10 ha—although, even in these cases, a slight increase has been recorded. Meanwhile, the others maintain the same, relatively high, pace of transformations (Popești-Leordeni) or one substantially amplified (e.g., Bragadiru went from about 14 ha/year to about 59 ha/year). This process gradually leads to a trend of densification in certain areas, even though divergent spatial patterns can be noticed in many places, through a mixture of low density developments with high-rise complexes, sometimes inside the same settlement.

3.3. Identifying the Types of Built-Up Area Dynamics

The types of built-up area dynamics, for the period 2000–2018, as revealed by LDTtool, is presented in Figure 5. The map denotes the intensive spatio–temporal transformation patterns of built-up area expansion, which is further interpreted according to the characterizations of dynamics types established by Machado et al. in 2020 [106].

Overall, the main change patterns are represented by "F—NP increment by gain", including 3435 cells, followed by "G—Aggregation by gain" with 2835 cells (Figure 6), both visible almost all over the county. These two forms of dynamics comprise around 80% of the total number of cells. The first type expresses the development of multiple new patches of built-up areas not contiguous to the existing ones, more frequent in some of the ATUs located in the first or second ring near Bucharest, such as: Chiajna, Bragadiru, Ștefăneștii de Jos, Corbeanca. The second most identified type of dynamics—"G—Aggregation by gain"—denotes situations in which new nuclei of built-up areas or the augmentation of previous clusters generate the fusion of patches.

At the same time, Ilfov County registered some losses, framed especially in the "H—NP decrement by loss" class, identified in 658 cells, more obvious in the south-eastern and northern parts. This category can reflect a variety of situations, presumably disused former construction sites or disassembled ruins of economic units from the communist period which have yet not been reused. Another example is temporary agricultural constructions that have been dismantled.

In the general hierarchy, the following type of built-up area dynamics is "D—gain", which occurred in 493 cells, dispersed among several ATUs, expressing an increase in the built-up area with no influence on the NP.

A comparison of the number of cells between 2000–2008 and 2008–2018 denotes some significant differences concerning the dynamics of built-up area types. One of the most notable changes is related to the "G—Aggregation by gain" type, identified only in the second interval, when the largest number of cells—3172—was registered. In the first period, the "F—NP increment by gain" category was predominant, presumably boosting the fusion of patches after 2008. Indeed, this type reached the second highest number of cells in the last period. Meanwhile, in the first interval, the category "A—No change" was identified in 1090 cells, leading to the assumption that no modifications were registered. Subsequently, after 2008, only 193 cells were found, expressing a significant increase in the construction sector. The categories "B—Fragmentation per se" and "C—Aggregation per se" reflect pure geometric variations, without amount changes [106,107].

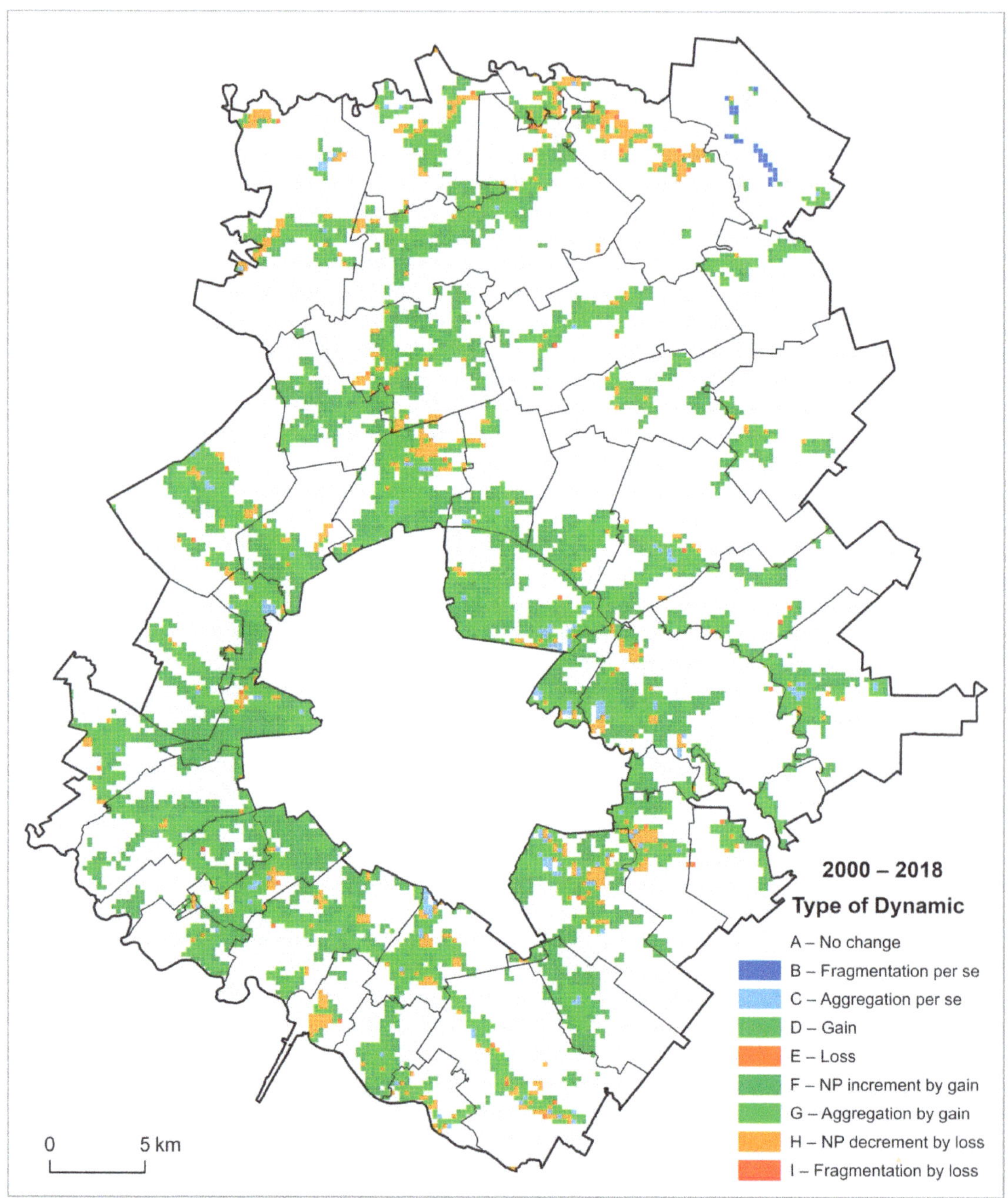

Figure 5. Built-up area types of dynamics (2000–2018).

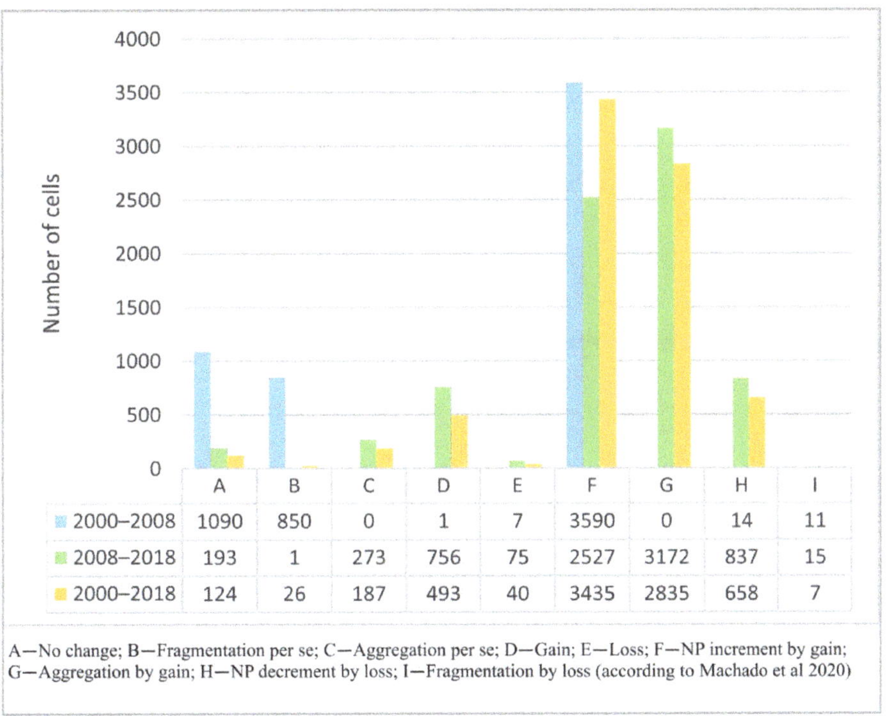

A—No change; B—Fragmentation per se; C—Aggregation per se; D—Gain; E—Loss; F—NP increment by gain;
G—Aggregation by gain; H—NP decrement by loss; I—Fragmentation by loss (according to Machado et al 2020)

Figure 6. Number of cells assigned to each built-up area dynamic type.

3.4. Various Development Patterns of Built-Up Area at the Local Level

Our study is enriched with complementary case studies, in order to explore the spatial pattern differentiations and to better understand the utility of Landsat datasets for sustainable planning and for substantiating further qualitative analysis. The case studies illustrate different administrative statuses and geographical locations: an urban settlement—Bragadiru, and a commune—Chiajna, nearby Bucharest; a commune in the second ring of ATUs—Corbeanca, further away from the capital. In all three ATUs (Bragadiru, Chiajna and Corbeanca) the human pressure has been unprecedented and the spatial pattern was completely redesigned as the new construction exceeds the built-up area outlined in 2000.

Bragadiru, a town granted with urban status in 2005, is located in the south-west part of Ilfov County. Traditionally, this settlement was well known for industrial and agricultural functions (around 91% agricultural land in 2000) (Figure 7).

As a legacy of the communist policy of the territorial reorganization of rural areas, Bragadiru's central part includes several blocks of flats. Single-family houses and nonresidential constructions were also diffused into the territory. Subsequently, during the 1990s, limited transformations were reported. Meanwhile, a large share of agricultural plots was rezoned for urban use, with residential developments and other types of constructions for secondary and tertiary activities. In the first interval (2000–2008), the annual growth of built-up area was reduced (about 14 ha/year), mainly along the main roads towards Bucharest or at the edge of the existent settlement. Isolated nuclei of new constructions emerged in a fragmented pattern in a few places.

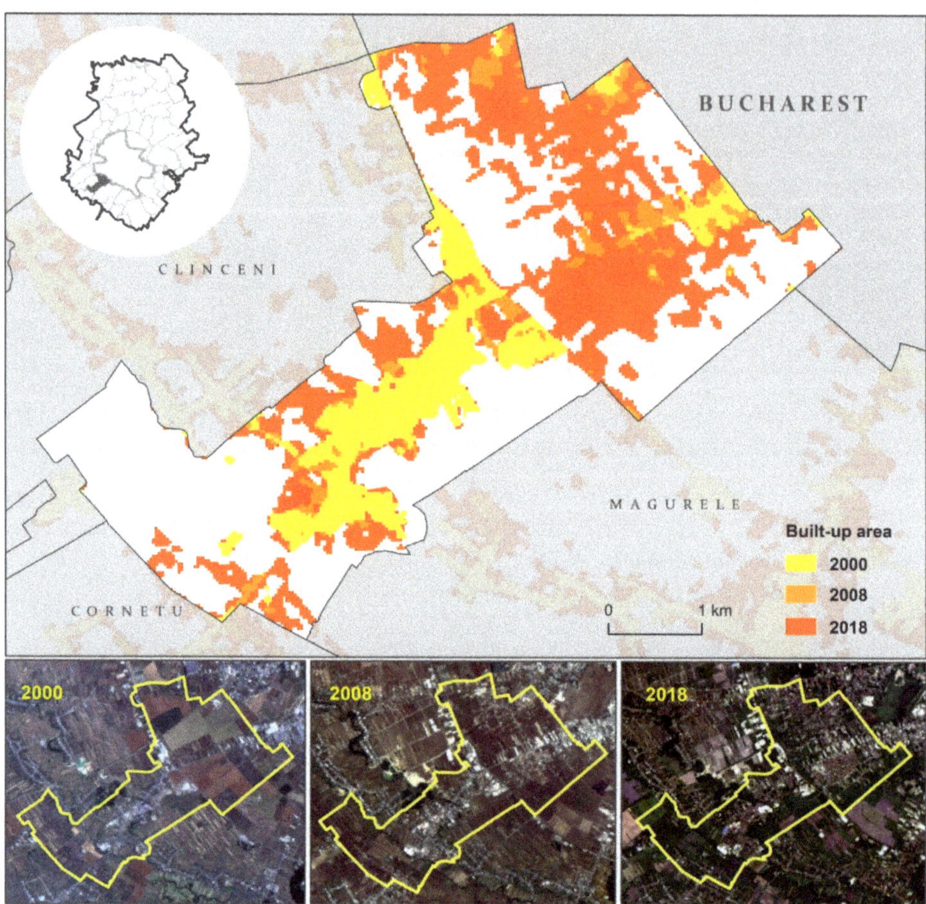

Figure 7. Changes in the built-up area configuration in Bragadiru, on the south-western edge of Bucharest (Source: Landsat imagery archives, earthexplorer.usgs.gov, accessed 24 March 2021).

In the second period (2008–2018), Bragadiru saw a fast rhythm of development, presenting the highest annual growth of built-up land in Ilfov County (about 59 ha/year). The highest development was registered on greenfield sites located between Bucharest's limits and the beltway. In this regard, this communication route fragments the spatial pattern, as it is widely known for frequent traffic jams. In addition, a more limited surface occupied by new constructions is noticed in the expansion of Bragadiru's old core, on the western side, possibly related with its proximity to a leisure area. Overall, a surface of 699 ha of land was consumed for urban growth between 2000 and 2018. The spatial pattern is complex, comprised of an overall mixture of residential and non-residential areas (commercial, industrial, logistics, etc.). The latter are usually concentrated in several places, sometimes at the edge of built-up areas, but interspersed among residential neighborhoods.

In terms of residential expansion, clusters of single-family houses or multi-family residential complexes stand out in several places, but a more common feature is the disordered intermingling of the two. Some of the newly emerged residential clusters are the result of large scale projects developed almost entirely after 2008, including several blocks of flats.

Another case study explores the spatial dynamics of three villages—Chiajna, Dudu and Roșu—in the Chiajna commune, on the western edge of Bucharest (Figure 8).

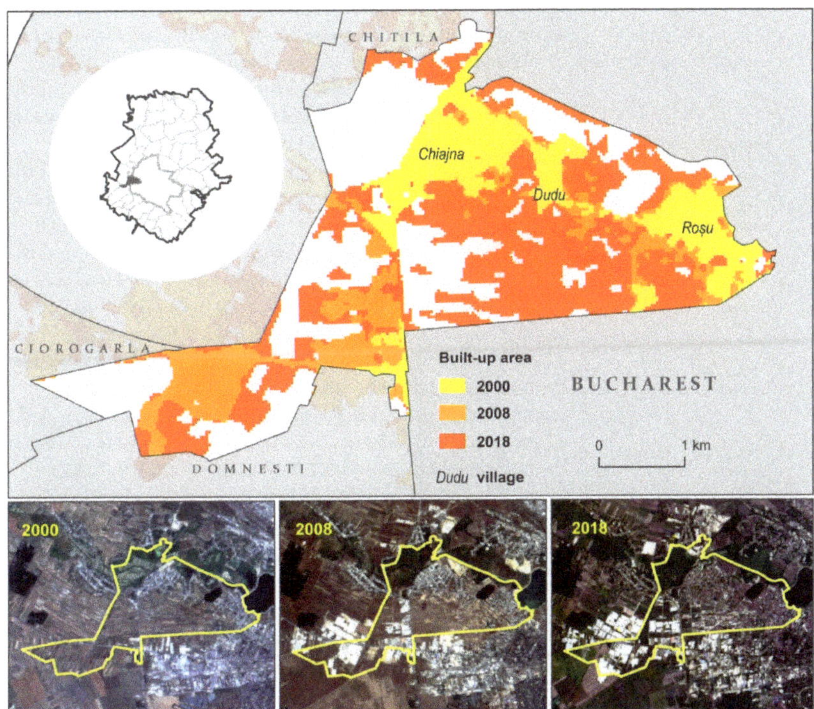

Figure 8. Changes in the built-up area configuration in Chiajna commune, on the western edge of Bucharest (Source: Landsat imagery archives, earthexplorer.usgs.gov, accessed 24 March 2021).

In 2000 the largest part of the territory was composed of agricultural land (67%), typical villages and some small scale post-socialist developments. The latter included a few commercial and logistics spaces at the junction of the Bucharest beltway and the A1 highway. Subsequently, this commune has become one of the most attractive places for the Bucharest population and investors, transforming it into an area of intense economic activity and residential development. Thus, between 2000–2008, the annual growth of built-up area was the second highest registered in Ilfov County (about 30 ha/year).

Over the next period, this trend intensified, reaching around 52 ha/year between 2008–2018, placing it third in the region. The total surface converted into built-up area between 2000 and 2018 was around 750 ha. From a functional point of view, two major areas of development can be delineated: the largest one, in the eastern part, predominantly recorded a spread of residential buildings; meanwhile, in the western side non-residential activities were concentrated.

A previous analysis even noted that some of the largest logistic spaces and clusters of logistics parks in Ilfov County are located in the western side of Chiajna [108]. These developments were established along the Bucharest beltway and the A1 highway. In terms of the types of land consumption, between 2000 and 2008, it was mostly non-residential areas that expanded. New houses were developed mainly in an unplanned, fragmented way at the edges of the pre-existent villages, but also expanding largely into the nearby open space. After 2008, even though the deconcentration of non-residential activities continued, residential buildings made up the lion's share of built-up area expansion, spilling over into greenfield sites. Large scale projects and multi-family residential complexes are definitive characteristics of this area in recent years.

The last case study addresses the territorial reconfiguration of the Corbeanca commune, in the second ring of ATUs, north of Bucharest, and composed of four villages:

Corbeanca, Ostratu, Petrești and Tamași. In terms of accessibility, these villages are located around 20 km from the capital center. In 2000, rural settlements were characterized by a linear structure surrounded by agricultural plots (84% of the entire land surface), forests and bodies of water. Distinguished by attractive natural settings, this area was among the earliest to be developed for residential expansion, with some new single-family houses built in the first post-socialist decade. However, in terms of land consumption, in the period 2000–2008, Corbeanca presented a modest spread of built-up areas (about 11 ha/year), compared with 40 ha/year between 2008–2018, making it the sixth fastest expanding place in the region. The total surface transformed into built-up area between 2000 and 2018 encompasses 492 ha. Regarding the growth patterns, in the first interval the spread of constructions generally went beyond the edges of the old villages, in a more scattered way into greenfield sites (Figure 9). Some of these new developments were placed on isolated patches far away from the old built-up areas. After 2008 there was a new wave of construction, creating a significant nucleus of new built-up area in the open space between the four villages—a greater diffusion of new buildings into the territory. This led to the creation of a more articulated spatial pattern, comprised of almost a continuum of built space, fragmented by patches of forest, bodies of water and agricultural land. Gradually, the pressure on land near forest intensified, much like it did for waterfront plots. A rarer trend is the development of new houses interspersed between previously built dwellings. The predominant functional use of the land is residential constructions, followed by developments in the service sector. Generally, the landscape is dominated by single-family homes mixed with residential complexes with detached houses built as part of several large scale projects. Still, a few blocks of flats may be observed in some areas.

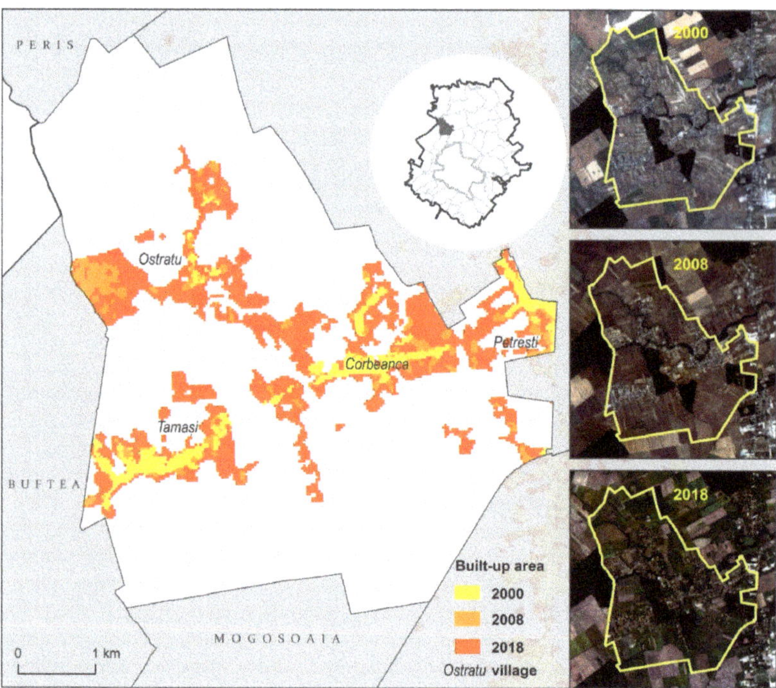

Figure 9. Changes in the built-up area configuration in Corbeanca commune, north-west of Bucharest (Source: Landsat imagery archives, earthexplorer.usgs.gov, accessed 24 March 2021).

4. Discussion

The result maps illustrate the gradual expansion of the built-up area in the surroundings of Bucharest, which allows an overview of the structural changes that have occurred after the turn of the millennium. Previous research noted that, generally in CEE countries, the uncontrolled expansion of built-up areas was embraced as a positive measure, regardless of the long lasting detrimental effects [5,109]. Thus, such quantitative approaches based on Landsat datasets enable the understanding of the urban development pathway, the resulting spatial outcome and related challenges, which can then be used to improve policy making that might sustain better planning.

Our results reinforce previous findings that indicated substantial territorial transformations in the Bucharest city region in the post-communist period, under the increasing pressure of real estate developers, investors and the growing population [64,75,91,110]. The large amount of greenfield land and its lower costs, as compared with the capital, are considered the main attractive factors for establishing new developments [111]. Moreover, the preferences for the open space adjacent to large cities is common in CEE countries [5,53,63], fostered by the interest of local authorities to increase their budget revenues [68].

Deepening our analysis, we have revealed spatial and temporal variations of processes related to urban growth, both at regional and local level, shedding light on the trajectory of the largest post-socialist city in the European Union. The employed up-to-date information brings a new, integrated perspective to the last decade's territorial changes. Considering all this, the resulted knowledge can serve as a starting point for further qualitative approaches.

The noteworthy differentiations recorded between the two analyzed periods are anchored in the socioeconomic context. The fall of communism was followed by economic restructuring, a profound downturn and the late implementation of the reforms necessary for the transition from a centralized economy to a free market. As a result, the economic recovery was late [50], stimulated also by the mandatory reforms required for joining the European Union (2007). In this context, between 2000 and 2008, the development of newly built-up areas has augmented progressively—a total increase of 31%, according to our findings.

Initially the urban expansion process and the development of large scale projects were thought to be delayed in this region compared to its CEE counterparts [91], but now it seems that those times are long gone. In fact, after 2008, the annual growth of built-up development significantly accelerated, to a concerning degree, and it shows no signs of slowing down; on the contrary, there are hints of an upward trend. The human pressure has become greater and greater as the number of players in the real estate market multiplied. The 2008 economic crisis was expected to trigger a significant decrease in built-up area expansion, as reported in other CEE countries [63]. Moreover, the deep economic recession detrimentally impacted the Romanian construction sector, as noted in previous research [112,113]. Still, a decade later, an assessment of the territorial impact of built-up area expansion reveals that the share of land overtaken for urban growth is significantly higher than in the previous period, encompassing 75% of the entire development. One of the main reasons for this unexpected trajectory was the governmental decisions made in 2009 to encourage the construction sector, still in effect today, which boosted the residential sector.

This situation confirms previous findings that national policies play an essential role in guiding built-up area expansion [63,83]. In this context, the legislation in the field of spatial and urban planning is considered inefficient [114,115], even though a framework law was adopted in 2001. This normative act registered several structural changes, especially beginning with 2008. Still, these legal provisions failed in ensuring effective regulations for limiting urban expansion.

The analysis of the development trends and urban growth intensity at the local level outlines that, prior to 2008, the most dynamic territorial transformations (over 20 ha/year) took place in 12.5% of the ATUs located in the first ring near Bucharest. In the last decade, however, a similar intensity of development was registered in 55% of the ATUs dispersed across Ilfov County. The newly emerging clusters were more territorially concentrated in the first period, but, after 2008, amid the increasing magnitude of urban expansion, other parts of the county which initially experienced modest development also became very appealing. These results confirm several intraregional differences, broadly suggested by other studies [75], as well as the significant rise of land consumption.

In addition, LDTtool provides valuable findings. Through quantitative measurements, it identifies the specific dynamics of built-up area and the overall fragmented spatial way of development hinted in other studies [57,75]. Still, some interesting differentiations were revealed, as the prevailing change pattern was "F—NP increment by gain" before 2008, but afterwards the "G—Aggregation by gain" type of dynamics became the most represented, followed by the previously dominant category. Thus, we notice an obvious shift in the last decade toward a spatial homogenization in several places, as the fusion of patches fosters this process [106]. Still, overall, the results demonstrate the maintenance of the general development pathway, with new built-up areas springing up further away from the existing ones, but which can, afterward, support a process of densification. These findings demonstrate the area's unsustainable urban expansion, given that one of the countermeasures indicated on the European level is to clearly designate building zones to sustain compact growth [4].

Exploring the three complementary case studies, we have captured several similarities and differences between the patterns of development at the local level. In Bragadiru and Chiajna the overall pattern is much more complex than in Corbeanca, with residential development and substantial areas occupied by non-residential activities (commercial, industrial, logistics, etc.). In Bragadiru, concentrations of non-residential activities are interspersed among residential neighbourhoods in several places, while in Chiajna they are mostly clustered on the western side, at the junction between the Bucharest beltway and the A1 highway. In Chiajna, the communication routes seem to have stimulated new activities; meanwhile, in Bragadiru, the beltway has fragmented the spatial pattern, as the largest development took place between the beltway and Bucharest's city limits. In Corbeanca, the majority of newly emerged areas are residential, with the prevalence of single-family detached houses. In addition, in Chiajna and Bragadiru, clusters of multi-family residential complexes emerged in several places, developed almost entirely after 2008. These three case studies illustrate various development patterns, even inside the same settlement boundaries—a result of an incoherent planning policy, as suggested by previous studies [110,116,117].

Concerning the planning process, while the elaboration of a master plan for Ilfov County has been publicly discussed since 2016, it has yet to be completed. However, such a master plan will only provide a guiding character to the territorial impact of socioeconomic development. Clearly enforceable regulations for land use management and built-up area development are established only at the local level entities through specific plans. Thus, the local authorities at the ATU level decide land development independently, being in permanent competition with each other to attract more investments. The lack of coordination between these entities, coupled with the absence of a joint development plan at the regional level, has led to uncontrolled development, labelled even as chaotic [118]. This situation is likely to persist, as there are no plans for cooperation in the foreseeable future. In this sense, it seems that the space is structured more according to the private interests of developers and newcomers, than for the public's benefit. Previous research reported these practices as common in other CEE countries as well [109], even though in several places progress has been made toward more sustainable planning [5,66].

Uncontrolled urban expansion is perceived as a negative way of development [4,94], in and of itself. It is stimulated by the fast pace of transformations, illustrating the lack of proper planning, characterized by ad-hoc decisions [5,67,68] without coordination between different projects [111], which all lead, more often than not, to a patchwork of spatial arrangements. One of the most visible results is the lack of clear functional zoning that fosters a heterogeneous mixture of residential (single-family houses or multi-family residential complexes) and non-residential areas (commercial, logistical, high-rise offices, industrial, etc.), sometimes tightly intertwined on the same parcels of land. Another serious negative dysfunctionality is the emergence of residential development nuclei near landfills. In this sense, previous studies have drawn attention to inappropriate solid waste management in the surroundings of Bucharest [119].

Other detrimental consequences are related to the increasing human pressure over attractive natural amenities, considering that new clusters have appeared near or even inside forests and along bodies of water [64], some of which are protected areas. This type of aggression is a continuous process, leading to the progressive degradation of natural resources, as well as limiting recreational areas for Bucharest residents. Another common dysfunctionality in many places is the lack of, or undersized, basic infrastructure and services, which further on may lead to water and soil pollution. Additionally, frequent traffic congestion was reported [118] which seems to lead to a higher attractiveness of the more accessible places located near Bucharest.

In terms of land consumption, the urban growth has led to the fragmentation of greenfield land and proliferation of low density areas interspersed with clusters of high-density developments and undeveloped land. The latter is not usually used for agricultural production [111], but is more often acquired for speculative purposes, and abandoned until profitable opportunities arise for investors.

Previous studies have argued that, in Europe, discontinuous spatial development represents an intermediary phase toward a more contiguous pattern [52,120]. It is true that, in several places in Ilfov County lately, as revealed by our analysis, one of the tendencies has been to build on the open spaces between previously dispersed buildings. Still, this potential evolution does not erase the long term detrimental impacts of uncontrolled development in terms of the degradation of natural resources, pollution, inefficient land use, etc., all of which have negative repercussions on the inhabitants' quality of life.

These findings reinforce the conclusions of previous studies that call for urgent action [4,17,121] and for up-to-date information that will allow the design of appropriate policies and strategies for a more sustainable future. Local urban plans are updated once every few years, usually after more than a decade, and they focus on the territorial development trends at the time of their elaboration. Thus, complementary instruments, such as the Landsat dataset, have become essential for the constant monitoring of the fast paced dynamics of built-up areas, as revealed in previous researches [22,27,31,34,44,122–126]. In this regard, different studies highlighted the importance of analyzing temporal and spatial characteristics [22,126], the main directions of urban expansion [25,123] and the spatial pattern of land changes [31,34,44,124]. All these studies focused on different places and are in the same line of thought, illustrating the fast rhythm of urban growth, as we have identified in the case of Bucharest. In addition, our results confirm previous findings that point out that investigating built-up area dynamics based on remote sensing can serve as a basis for decision-making [26,29,35,122,127]. For getting deeper insights about territorial transformations, LDTtool has proven suitable for detecting the types of local dynamics, as suggested in previous research [106,107]. Still, there are some limitations related to the modifiable areal unit problem (MAUP) which refers "to the fact that the areal units can be set arbitrarily and are modifiable" [106].

5. Conclusions

The study explores the built-up area expansion process and the resulting spatio–temporal changes between 2000 and 2018 in Ilfov County, which overlaps the surroundings of Bucharest, capital of Romania. The research is approached from a multi-scalar perspective on a regional and local level, through the comparative assessment of the ATUs' pathway, based on Landsat satellite data at a medium spatial resolution of 30 m. Additionally, the types of built-up area dynamics were identified by using the LDT tool.

The results illustrate a massive spatial reconfiguration that took place under the relatively recent conditions of free market capitalism, as the built-up area has more than doubled in the 18 years. The pace of territorial transformations has also changed, with a significant acceleration after 2008, when 75% of the entire development was registered. Regarding the overall growth pattern between 2000 and 2008, prevalent directions of development with a higher expansion concentrated in several places closer to Bucharest. After 2008, a more widespread development trend was noticed in the territory, even though some areas remain more appealing than others. The annual growth of built-up areas at the local level reflect that the greatest territorial impact (over 20 ha/year) was recorded in 12.5% of the ATUs in the first period, but in the last interval a similar increase was registered in 55% of the entities. In some cases, tremendous changes were recorded: from a limited transformation in the first period to substantial development after 2008.

Concerning the specific dynamics of built-up area for the entire period, the number of cells classified as dynamics related to gain is predominant ("F-NP increment by gain", "G-Aggregation by gain", "D—gain"), encompassing around 87% of the total.

The overall spatial configuration became more and more complex and heterogeneous, in a mosaic of spatial arrangements, with low density developments interspersed with high density clusters and undeveloped areas. The analysis of the three complementary case studies illustrates various development patterns, even inside the same settlement boundaries. The results reveal that the largest part of the newly built-up area emerged through fragmentation of greenfield land.

Our research enriches the knowledge on CEE countries by bringing the largest post-socialist city in the European Union into the spotlight, and thus contributing to filling the recognized gap for more quantitative assessment of the built-up area's territorial impact. At the same time, the results draw attention to the fast pace of transformations in this macro-region, and to the crucial role of Landsat dataset and other related tools in capturing spatio–temporal changes, especially in the context of lacking relevant socioeconomic data. Compared with other studies that focused on Bucharest's surroundings, or even beyond, the approach implies a more comprehensive framework, displaying the multilevel territorial dynamics for two distinct periods.

The present study highlights the importance of deeper insights into the trajectory of different CEE city regions to understand the fast rhythm of changes and monitor the spatial outcome. These findings provide a valuable scientific approach that can be exploited by decision-makers to evaluate the territorial impact of implemented plans and strategies, as well as to inform future development decisions for better planning.

In addition, the approach can be used to investigate the trajectories of other places with similar development patterns. Moreover, it provides a basis for further comparative analysis in the wider post-socialist framework.

Future research should link these findings with other types of data from a more comprehensive perspective, to understand the local impact and the causal mechanisms that triggered the territorial reconfiguration.

Author Contributions: All authors have equally contributed to the article. In more detail, the contributions for the different sections are: conceptualization, I.-V.S.; methodology, formal analysis and investigation, I.-V.S., M.V., D.Z; writing—original draft preparation, I.-V.S., M.V.; writing—review and editing, I.-V.S., M.V., D.Z. All authors have read and agreed to the published version of the manuscript.

Funding: The article was partially funded by the University of Bucharest.

Institutional Review Board Statement: Not applicable.

Informed Consent Statement: Not applicable.

Conflicts of Interest: The authors declare no conflict of interest. The funders had no role in the design of the study; in the collection, analyses, or interpretation of data; in the writing of the manuscript, or in the decision to publish the results.

References

1. Angel, S.; Parent, J.; Civco, D.L.; Blei, A.; Potere, D. The Dimensions of Global Urban Expansion: Estimates and Projections for All Countries, 2000–2050. *Prog. Plan.* **2011**, *75*, 53–107. [CrossRef]
2. Liu, X.; Pei, F.; Wen, Y.; Li, X.; Wang, S.; Wu, C.; Cai, Y.; Wu, J.; Chen, J.; Feng, K.; et al. Global Urban Expansion Offsets Climate-Driven Increases in Terrestrial Net Primary Productivity. *Nat. Commun.* **2019**, *10*, 5558. [CrossRef]
3. European Environment Agency. *Urban Sprawl in Europe: The Ignored Challenge*; EEA Report; European Environment Agency; Office for Official Publications of the European Communities: Copenhagen, Denmark; Luxembourg, 2006.
4. European Environment Agency; Swiss Federal Office for the Environment (FOEN). *Urban Sprawl in Europe: Joint EEA-FOEN Report*; Publications Office: Luxembourg, 2016.
5. Pichler-Milanović, N. Confronting Suburbanization in Ljubljana: From "Urbanization of the Countryside" to Urban Sprawl. In *Confronting Suburbanization*; Stanilov, K., Sýkora, L., Eds.; John Wiley & Sons, Ltd.: Chichester, UK, 2014; pp. 65–96. [CrossRef]
6. Larkin, A.; van Donkelaar, A.; Geddes, J.A.; Martin, R.V.; Hystad, P. Relationships between Changes in Urban Characteristics and Air Quality in East Asia from 2000 to 2010. *Environ. Sci. Technol.* **2016**, *50*, 9142–9149. [CrossRef]
7. Zaharia, L.; Ioana-Toroimac, G.; Cocoş, O.; Ghiţă, F.A.; Mailat, E. Urbanization Effects on the River Systems in the Bucharest City Region (Romania). *Ecosyst. Health Sustain.* **2016**, *2*, e01247. [CrossRef]
8. Seto, K.C.; Guneralp, B.; Hutyra, L.R. Global Forecasts of Urban Expansion to 2030 and Direct Impacts on Biodiversity and Carbon Pools. *Proc. Natl. Acad. Sci. USA* **2012**, *109*, 16083–16088. [CrossRef]
9. McDonald, R.I.; Güneralp, B.; Huang, C.-W.; Seto, K.C.; You, M. Conservation Priorities to Protect Vertebrate Endemics from Global Urban Expansion. *Biol. Conserv.* **2018**, *224*, 290–299. [CrossRef]
10. Matei, E.; Manea, G.; Cocoş, O.; Vijulie, I.; Tîrlă, L.; Bogan, E.; Tiscovschi, A. Sustainable Development as a Solution for Agriculture and Human Settlements Competition in Ilfov County. In Proceedings of the 14th International Multidisciplinary Scientific Geoconference SGEM 2014, Sofia, Bulgaria, 19–25 June 2014; Volume 3, pp. 415–422. [CrossRef]
11. Nae, M.; Dumitrache, L.; Suditu, B.; Matei, E. Housing Activism Initiatives and Land-Use Conflicts: Pathways for Participatory Planning and Urban Sustainable Development in Bucharest City, Romania. *Sustainability* **2019**, *11*, 6211. [CrossRef]
12. Ianoş, I.; Merciu, F.C.; Merciu, G.; Zamfir, D.; Stoica, I.-V.; Vlăsceanu, G. Unclear Perspectives for a Specific Intra-Urban Space: Văcăreşti Lake Area (Bucharest City). *Carpathian J. Earth Environ. Sci.* **2014**, *9*, 215–224.
13. Merciu, F.C.; Sîrodoev, I.; Merciu, G.; Zamfir, D.; Schvab, A.; Stoica, I.V.; Paraschiv, M.; Saghin, I.; Cercleux, A.L.; Văidianu, N.; et al. The "Văcăreşti Lake" Protected Area, a Neverending Debatable Issue? *Carpathian J. Earth Environ. Sci.* **2017**, *12*, 463–472.
14. Louwagie, G.; European Environment Agency. Land Recycling in Europe: Approaches to Measuring Extent and Impacts. *EEA Rep.* **2016**, *31*, 51.
15. Dallhammer, E.; Schuh, B.; Caldeira, I. *Urban Impact Assessment Report. Implementation of the 2030 Agenda. The Influence of SDG 11.3 on Urban Development through Spatial Planning*; European Committee of the Region: Luxembourg, 2018.
16. European Environment Agency. Land Take in Europe. Available online: https://www.eea.europa.eu/data-and-maps/indicators/land-take-3 (accessed on 6 April 2021).
17. Meiner, A.; Pedroli, G.B.M.; European Environment Agency. *Landscapes in Transition: An Account of 25 Years of Land Cover Change in Europe*; EEA Publications Office: Luxembourg, 2017.
18. Xie, H.; Zhang, Y.; Duan, K. Evolutionary Overview of Urban Expansion Based on Bibliometric Analysis in Web of Science from 1990 to 2019. *Habitat Int.* **2020**, *95*, 102100. [CrossRef]
19. Wang, L.; Brenden, T.; Seelbach, P.; Cooper, A.; Allan, D.; Clarck, R.; Wiley, M. Landscape Based Identification of Human Disturbance Gradients and Reference Conditions for Michigan Streams. *Environ. Monit. Assess.* **2008**, *144*, 483–484. [CrossRef]
20. Zhang, Z.; Liu, F.; Zhao, X.; Wang, X.; Shi, L.; Xu, J.; Yu, S.; Wen, Q.; Zuo, L.; Yi, L.; et al. Urban Expansion in China Based on Remote Sensing Technology: A Review. *Chin. Geogr. Sci.* **2018**, *28*, 727–743. [CrossRef]
21. He, C.; Liu, Z.; Gou, S.; Zhang, Q.; Zhang, J.; Xu, L. Detecting Global Urban Expansion over the Last Three Decades Using a Fully Convolutional Network. *Environ. Res. Lett.* **2019**, *14*, 034008. [CrossRef]
22. Xiao, J.; Shen, Y.; Ge, J.; Tateishi, R.; Tang, C.; Liang, Y.; Huang, Z. Evaluating Urban Expansion and Land Use Change in Shijiazhuang, China, by Using GIS and Remote Sensing. *Landsc. Urban Plan.* **2006**, *75*, 69–80. [CrossRef]
23. Li, H.; Wei, Y.; Huang, Z. Urban Land Expansion and Spatial Dynamics in Globalizing Shanghai. *Sustainability* **2014**, *6*, 8856–8875. [CrossRef]
24. Luo, T.; Tan, R.; Kong, X.; Zhou, J. Analysis of the Driving Forces of Urban Expansion Based on a Modified Logistic Regression Model: A Case Study of Wuhan City, Central China. *Sustainability* **2019**, *11*, 2207. [CrossRef]
25. Zeng, C.; Zhang, M.; Cui, J.; He, S. Monitoring and Modeling Urban Expansion—A Spatially Explicit and Multi-Scale Perspective. *Cities* **2015**, *43*, 92–103. [CrossRef]
26. Kumar, A.; Pandey, A.C.; Hoda, N.; Jeyaseelan, A.T. Evaluating the Long-Term Urban Expansion of Ranchi Urban Agglomeration, India Using Geospatial Technology. *J. Indian Soc. Remote Sens.* **2011**, *39*, 213–224. [CrossRef]
27. Kantakumar, L.N.; Kumar, S.; Schneider, K. Spatiotemporal Urban Expansion in Pune Metropolis, India Using Remote Sensing. *Habitat Int.* **2016**, *51*, 11–22. [CrossRef]
28. Rustiadi, E.; Pravitasari, A.E.; Setiawan, Y.; Mulya, S.P.; Pribadi, D.O.; Tsutsumida, N. Impact of Continuous Jakarta Megacity Urban Expansion on the Formation of the Jakarta-Bandung Conurbation over the Rice Farm Regions. *Cities* **2021**, *111*, 103000. [CrossRef]

29. Dhanarak, K.; Angadi, D.P. Urban Expansion Quantification from Remote Sensing Data for Sustainable Land-Use Planning in Mangaluru, India. *Remote Sens. Appl. Soc. Environ.* **2021**, *23*, 100602. [CrossRef]
30. Xu, G.; Jiao, L.; Liu, J.; Shi, Z.; Zeng, C.; Liu, Y. Understanding Urban Expansion Combining Macro Patterns and Micro Dynamics in Three Southeast Asian Megacities. *Sci. Total Environ.* **2019**, *660*, 375–383. [CrossRef]
31. Wu, W.; Zhao, S.; Zhu, C.; Jiang, J. A Comparative Study of Urban Expansion in Beijing, Tianjin and Shijiazhuang over the Past Three Decades. *Landsc. Urban Plan.* **2015**, *134*, 93–106. [CrossRef]
32. Zhang, Q.; Su, S. Determinants of Urban Expansion and Their Relative Importance: A Comparative Analysis of 30 Major Metropolitans in China. *Habitat Int.* **2016**, *58*, 89–107. [CrossRef]
33. Zhang, S.; Fang, C.; Kuang, W.; Sun, F. Comparison of Changes in Urban Land Use/Cover and Efficiency of Megaregions in China from 1980 to 2015. *Remote Sens.* **2019**, *11*, 1834. [CrossRef]
34. Yang, C.; Li, Q.; Zhao, T.; Liu, H.; Gao, W.; Shi, T.; Guan, M.; Wu, G. Detecting Spatiotemporal Features and Rationalities of Urban Expansions within the Guangdong–Hong Kong–Macau Greater Bay Area of China from 1987 to 2017 Using Time-Series Landsat Images and Socioeconomic Data. *Remote Sens.* **2019**, *11*, 2215. [CrossRef]
35. Rimal, B.; Zhang, L.; Stork, N.; Sloan, S.; Rijal, S. Urban Expansion Occurred at the Expense of Agricultural Lands in the Tarai Region of Nepal from 1989 to 2016. *Sustainability* **2018**, *10*, 1341. [CrossRef]
36. Shi, L.; Liu, F.; Zhang, Z.; Zhao, X.; Liu, B.; Xu, J.; Wen, Q.; Yi, L.; Hu, S. Spatial Differences of Coastal Urban Expansion in China from 1970s to 2013. *Chin. Geogr. Sci.* **2015**, *25*, 389–403. [CrossRef]
37. De Boeck, F. Urban Expansion, the Politics of Land, and Occupation as Infrastructure in Kinshasa. *Land Use Policy* **2020**, *93*, 103880. [CrossRef]
38. Hereher, M.E. Analysis of Urban Growth at Cairo, Egypt Using Remote Sensing and GIS. *Nat. Sci.* **2012**, *4*, 355–361. [CrossRef]
39. Mundia, C.N.; Aniya, M. Analysis of Land Use/Cover Changes and Urban Expansion of Nairobi City Using Remote Sensing and GIS. *Int. J. Remote Sens.* **2005**, *26*, 2831–2849. [CrossRef]
40. Braimoh, A.K.; Onishi, T. Spatial Determinants of Urban Land Use Change in Lagos, Nigeria. *Land Use Policy* **2007**, *24*, 502–515. [CrossRef]
41. Simwanda, M.; Murayama, Y. Spatiotemporal Patterns of Urban Land Use Change in the Rapidly Growing City of Lusaka, Zambia: Implications for Sustainable Urban Development. *Sustain. Cities Soc.* **2018**, *39*, 262–274. [CrossRef]
42. Magidi, J.; Ahmed, F. Assessing Urban Sprawl Using Remote Sensing and Landscape Metrics: A Case Study of City of Tshwane, South Africa (1984–2015). *Egypt. J. Remote Sens. Space Sci.* **2019**, *22*, 335–346. [CrossRef]
43. Terfa, B.K.; Chen, N.; Liu, D.; Zhang, X.; Niyogi, D. Urban Expansion in Ethiopia from 1987 to 2017: Characteristics, Spatial Patterns, and Driving Forces. *Sustainability* **2019**, *11*, 2973. [CrossRef]
44. Hou, H.; Estoque, R.C.; Murayama, Y. Spatiotemporal Analysis of Urban Growth in Three African Capital Cities: A Grid-Cell-Based Analysis Using Remote Sensing Data. *J. Afr. Earth Sci.* **2016**, *123*, 381–391. [CrossRef]
45. Linard, C.; Tatem, A.J.; Gilbert, M. Modelling Spatial Patterns of Urban Growth in Africa. *Appl. Geogr.* **2013**, *44*, 23–32. [CrossRef]
46. Xu, G.; Dong, T.; Cobbinah, P.B.; Jiao, L.; Sumari, N.S.; Chai, B.; Liu, Y. Urban Expansion and Form Changes across African Cities with a Global Outlook: Spatiotemporal Analysis of Urban Land Densities. *J. Clean. Prod.* **2019**, *224*, 802–810. [CrossRef]
47. Forget, Y.; Shimoni, M.; Gilbert, M.; Linard, C. Mapping 20 Years of Urban Expansion in 45 Urban Areas of Sub-Saharan Africa. *Remote Sens.* **2021**, *13*, 525. [CrossRef]
48. Haase, D.; Kabisch, N.; Haase, A. Endless Urban Growth? On the Mismatch of Population, Household and Urban Land Area Growth and Its Effects on the Urban Debate. *PLoS ONE* **2013**, *8*, e66531. [CrossRef]
49. Salvati, L.; Zambon, I.; Chelli, F.M.; Serra, P. Do Spatial Patterns of Urbanization and Land Consumption Reflect Different Socioeconomic Contexts in Europe? *Sci. Total Environ.* **2018**, *625*, 722–730. [CrossRef] [PubMed]
50. Ianoş, I.; Petrişor, A.-I.; Zamfir, D.; Cercleux, A.L.; Stoica, I.V.; Tălângă, C. In Search of a Relevant Index Measuring Territorial Disparities in a Transition Country. Romania as a Case Study. *ERDE—J. Geogr. Soc. Berl.* **2013**, *144*, 69–81. [CrossRef]
51. Hirt, S. Suburbanizing Sofia: Characteristics of Post-Socialist Peri-Urban Change. *Urban. Geogr.* **2007**, *28*, 755–780. [CrossRef]
52. Taubenböck, H.; Gerten, C.; Rusche, K.; Siedentop, S.; Wurm, M. Patterns of Eastern European Urbanisation in the Mirror of Western Trends—Convergent, Unique or Hybrid? *Environ. Plan. B Urban. Anal. City Sci.* **2019**, *46*, 1206–1225. [CrossRef]
53. Stanilov, K.; Hirt, S. Sprawling Sofia: Postsocialist Suburban Growth in the Bulgarian Capital. In *Confronting Suburbanization*; Stanilov, K., Sýkora, L., Eds.; John Wiley & Sons, Ltd.: Chichester, UK, 2014; pp. 163–191. [CrossRef]
54. Kovács, Z.; Farkas, Z.J.; Egedy, T.; Kondor, A.C.; Szabó, B.; Lennert, J.; Baka, D.; Kohán, B. Urban Sprawl and Land Conversion in Post-Socialist Cities: The Case of Metropolitan Budapest. *Cities* **2019**, *92*, 71–81. [CrossRef]
55. Matei, E.; Dumitrache, L.; Nae, M.; Vijulie, I.; Onetiu, N. Evaluating Sustainability of Urban Development of the Small Towns in Romania. In Proceedings of the SGEM Conference, Albena, Bulgaria, 20–25 June 2011; Volume 3, pp. 1065–1072. [CrossRef]
56. Hirt, S.; Stanilov, K. Revisiting Urban Planning in the Transitional Countries. Regional Study Prepared for Planning Sustainable Cities Global Report on Human Settlements 2009. Available online: http://www.unhabitat.org/grhs/2009 (accessed on 22 April 2021).
57. Dumitrache, L.; Zamfir, D.; Nae, M.; Simion, G.; Stoica, V. The Urban Nexus: Contradictions and Dilemmas of (Post) Communist (Sub) Urbanization in Romania. *Hum. Geogr.—J. Stud. Res. Hum. Geogr.* **2016**, *10*, 38–50. [CrossRef]

58. Ursu, A.; Burtila, R.; Minea, V.; Marius, A.; Ichim, P. Urban Public Transportation System Changes, in Post Communist Period in Iasi Municipality. In Proceedings of the SGEM 2015 Conference, Albena, Bulgaria, 16–24 June 2015; Volume 2, pp. 615–622. [CrossRef]
59. Stoica, I.-V.; Tulla, A.F.; Zamfir, D.; Petrișor, A.-I. Exploring the Urban Strength of Small Towns in Romania. *Soc. Indic. Res.* **2020**, *152*, 843–875. [CrossRef]
60. Stanilov, K. Urban Planning and the Challenges of Post-Socialist Transformation. In *The Post-Socialist City: Urban Form and Space Transformations in Central and Eastern Europe After Socialism*; GeoJournal Library; Springer: Dordrecht, The Netherlands, 2007.
61. Sýkora, L.; Bouzarovski, S. Multiple Transformations: Conceptualising the Post-Communist Urban Transition. *Urban. Stud.* **2011**, *49*, 43–60. [CrossRef]
62. Slaev, A.D.; Nedović-Budić, Z.; Krunić, N.; Petrić, J.; Daskalova, D. Suburbanization and Sprawl in Post-Socialist Belgrade and Sofia. *Eur. Plan. Stud.* **2018**, *26*, 1389–1412. [CrossRef]
63. Stanilov, K.; Sýkora, L. Postsocialist Suburbanization Patterns and Dynamics: A Comparative Perspective. In *Confronting Suburbanization*; Stanilov, K., Sýkora, L., Eds.; John Wiley & Sons, Ltd.: Chichester, UK, 2014; pp. 256–295. [CrossRef]
64. Stoica, I.-V.; Vîrghileanu, M.; Zamfir, D.; Mihai, B.-A.; Săvulescu, I. Comparative Assessment of the Built-Up Area Expansion Based on Corine Land Cover and Landsat Datasets: A Case Study of a Post-Socialist City. *Remote Sens.* **2020**, *12*, 2137. [CrossRef]
65. Hirt, S. Whatever Happened to the (Post) Socialist City? *Cities* **2013**, *32*, S29–S38. [CrossRef]
66. Kovács, Z.; Tosics, I. Urban Sprawl on the Danube: The Impacts of Suburbanization in Budapest. In *Confronting Suburbanization*; Stanilov, K., Sýkora, L., Eds.; John Wiley & Sons, Ltd.: Chichester, UK, 2014; pp. 33–64. [CrossRef]
67. Hirt, S. Planning during Post-Socialism. In *International Encyclopedia of Social and Behavioral Sciences*, 2nd ed.; Elsevier: London, UK, 2015; Volume 18, pp. 187–192.
68. Stanilov, K.; Sýkora, L. Managing Suburbanization in Postsocialist Europe. In *Confronting Suburbanization*; Stanilov, K., Sýkora, L., Eds.; John Wiley & Sons, Ltd.: Chichester, UK, 2014; pp. 296–320. [CrossRef]
69. Ferenčuhová, S. Explicit Definitions and Implicit Assumptions about Post-Socialist Cities in Academic Writings: Explicit Definitions and Implicit Assumptions. *Geogr. Compass* **2016**, *10*, 514–524. [CrossRef]
70. Poghosyan, A. Quantifying Urban Growth in 10 Post-Soviet Cities Using Landsat Data and Machine Learning. *Int. J. Remote Sens.* **2018**, *39*, 8688–8702. [CrossRef]
71. Leetmaa, K.; Tammaru, T.; Anniste, K. From Priority-Led to Market-Led Suburbanisation In A Post-Communist Metropolis. *Tijdschr. Voor Econ. En Soc. Geogr.* **2009**, *100*, 436–453. [CrossRef]
72. Krisjane, Z.; Berzins, M. Post-Socialist Urban Trends: New Patterns and Motivations for Migration in the Suburban Areas of Rīga, Latvia. *Urban. Stud.* **2012**, *49*, 289–306. [CrossRef]
73. Spórna, T.; Krzysztofik, R. 'Inner' Suburbanisation—Background of the Phenomenon in a Polycentric, Post-Socialist and Post-Industrial Region. Example from the Katowice Conurbation, Poland. *Cities* **2020**, *104*, 102789. [CrossRef]
74. Gentile, M.; Marcińczak, S. Housing Inequalities in Bucharest: Shallow Changes in Hesitant Transition. *GeoJournal* **2014**, *79*, 449–465. [CrossRef]
75. Ianoş, I.; Sîrodoev, I.; Pascariu, G.; Henebry, G. Divergent Patterns of Built-up Urban Space Growth Following Post-Socialist Changes. *Urban. Stud.* **2016**, *53*, 3172–3188. [CrossRef]
76. *Urban Europe: Statistics on Cities, Towns and Suburbs*, 2016 ed.; Koceva, M.M.; Brandmüller, T.; Lupu, I.; Önnerfors, Å.; Corselli-Nordblad, L.; Coyette, C.; Johansson, A.; Strandell, H.; Wolff, P.; Europäische Kommission (Eds.) Statistical books/Eurostat; Publications Office of the European Union: Luxembourg, 2016.
77. Schneider, A.; Woodcock, C.E. Compact, Dispersed, Fragmented, Extensive? A Comparison of Urban Growth in Twenty-Five Global Cities Using Remotely Sensed Data, Pattern Metrics and Census Information. *Urban. Stud.* **2008**, *45*, 659–692. [CrossRef]
78. Netzband, M.; Jürgens, C. Urban and Suburban Areas as a Research Topic for Remote Sensing. In *Remote Sensing of Urban and Suburban Areas*; Rashed, T., Jürgens, C., Eds.; Remote Sensing and Digital Image Processing; Springer: Dordrecht, The Netherlands, 2010; Volume 10, pp. 1–9. [CrossRef]
79. Alqurashi, A.; Kumar, L.; Sinha, P. Urban Land Cover Change Modelling Using Time-Series Satellite Images: A Case Study of Urban Growth in Five Cities of Saudi Arabia. *Remote Sens.* **2016**, *8*, 838. [CrossRef]
80. Firozjaei; Sedighi; Kiavarz; Qureshi; Haase; Alavipanah. Automated Built-Up Extraction Index: A New Technique for Mapping Surface Built-Up Areas Using LANDSAT 8 OLI Imagery. *Remote Sens.* **2019**, *11*, 1966. [CrossRef]
81. Firozjaei, M.K.; Sedighi, A.; Kiavarz, M.; Qureshi, S.; Haase, D.; Alavipanah, S.K. Monitoring of Urban Sprawl and Densification Processes in Western Germany in the Light of SDG Indicator 11.3.1 Based on an Automated Retrospective Classification Approach. *Remote Sens.* **2021**, *13*, 1694. [CrossRef]
82. Dolean, B.-E.; Bilașco, Ș.; Petrea, D.; Moldovan, C.; Vescan, I.; Roșca, S.; Fodorean, I. Evaluation of the Built-Up Area Dynamics in the First Ring of Cluj-Napoca Metropolitan Area, Romania by Semi-Automatic GIS Analysis of Landsat Satellite Images. *Appl. Sci.* **2020**, *10*, 7722. [CrossRef]
83. Cheng, C.; Yang, X.; Cai, H. Analysis of Spatial and Temporal Changes and Expansion Patterns in Mainland Chinese Urban Land between 1995 and 2015. *Remote Sens.* **2021**, *13*, 2090. [CrossRef]
84. Khanal, N.; Uddin, K.; Matin, M.; Tenneson, K. Automatic Detection of Spatiotemporal Urban Expansion Patterns by Fusing OSM and Landsat Data in Kathmandu. *Remote Sens.* **2019**, *11*, 2296. [CrossRef]
85. Maktav, D.; Erbek, F.S.; Jürgens, C. Remote Sensing of Urban Areas. *Int. J. Remote Sens.* **2005**, *26*, 655–659. [CrossRef]

86. Leinenkugel, P.; Deck, R.; Huth, J.; Ottinger, M.; Mack, B. The Potential of Open Geodata for Automated Large-Scale Land Use and Land Cover Classification. *Remote Sens.* **2019**, *11*, 2249. [CrossRef]
87. Ranagalage, M.; Estoque, R.; Handayani, H.; Zhang, X.; Morimoto, T.; Tadono, T.; Murayama, Y. Relation between Urban Volume and Land Surface Temperature: A Comparative Study of Planned and Traditional Cities in Japan. *Sustainability* **2018**, *10*, 2366. [CrossRef]
88. Ellen, D.; Patrick, H. Assessing Post-Socialist Urban Change with Landsat Data; Case Study Berlin, Germany. In *2007 Urban Remote Sensing Joint Event*; IEEE: Paris, France, 2007; pp. 1–4. [CrossRef]
89. Kanjir, U.; Veljanovski, T.; Ostir, K. Change Detection of Urban Areas—the Ljubljana, Slovenia Case Study. In *2011 Joint Urban Remote Sensing Event*; IEEE: Munich, Germany, 2011; pp. 425–428. [CrossRef]
90. Franke, D. Development of Suburbanization in the Hinterland of Prague Monitored by Remote Sensing. In Proceedings of the SGEM 2015 Conference, Albena, Bulgaria, 16–24 June 2015; Volume 1, pp. 1027–1034. [CrossRef]
91. Mihai, B.; Nistor, C.; Simion, G. Post-Socialist Urban Growth of Bucharest, Romania—a Change Detection Analysis on Landsat Imagery (1984–2010). *Acta Geogr. Slov.* **2015**, *55*, 223–234. [CrossRef]
92. Institutul Național de Statistică (National Institute of Statistics). *Anuarul Statistic al României—Serii de Timp (CD-ROM)*; Bucharest, Romania, 2020.
93. Pîrvu, R.; Bădîrcea, R.; Manta, A.; Lupăncescu, M. The Effects of the Cohesion Policy on the Sustainable Development of the Development Regions in Romania. *Sustainability* **2018**, *10*, 2577. [CrossRef]
94. Sýkora, L.; Mulíček, O. Prague: Urban Growth and Regional Sprawl. In *Confronting Suburbanization*; Stanilov, K., Sýkora, L., Eds.; John Wiley & Sons, Ltd.: Chichester, UK, 2014; pp. 133–162. [CrossRef]
95. Hansen, M.C.; Loveland, T.R. A Review of Large Area Monitoring of Land Cover Change Using Landsat Data. *Remote Sens. Environ.* **2012**, *122*, 66–74. [CrossRef]
96. Rujoiu-Mare, M.-R.; Olariu, B.; Mihai, B.-A.; Nistor, C.; Săvulescu, I. Land Cover Classification in Romanian Carpathians and Subcarpathians Using Multi-Date Sentinel-2 Remote Sensing Imagery. *Eur. J. Remote Sens.* **2017**, *50*, 496–508. [CrossRef]
97. Yoshida, T.; Omatu, S. Neural Network Approach to Land Cover Mapping. *IEEE Trans. Geosci. Remote Sens.* **1994**, *32*, 1103–1109. [CrossRef]
98. Dreiseitl, S.; Ohno-Machado, L. Logistic Regression and Artificial Neural Network Classification Models: A Methodology Review. *J. Biomed. Inform.* **2002**, *35*, 352–359. [CrossRef]
99. Ndehedehe, C.; Ekpa, A.; Simeon, O.; Nse, O. Understanding the Neural Network Technique for Classification of Remote Sensing Data Sets. *NY Sci. J.* **2013**, *6*, 26–33.
100. Debojit, B.J.H.; Arora Manoj, K.; Balasubramanian, R. Study and Implementation of a Non-Linear Support Vector Machine Classifier. *Int. J. Earth Sci. Eng.* **2011**, *4*, 985–988.
101. Ndehedehe, C.E.; Oludiji, S.M.; Asuquo, I. Supervised Learning Methods in the Mapping of Built up Areas from Landsat-Based Satellite Imagery in Part of Uyo Metropolis. *NY Sci. J.* **2013**, *6*, 45–52.
102. Mukherjee, A.; Kumar, A.A.; Ramachandran, P. Development of New Index-Based Methodology for Extraction of Built-Up Area from Landsat7 Imagery: Comparison of Performance With SVM, ANN, and Existing Indices. *IEEE Trans. Geosci. Remote Sens.* **2021**, *59*, 1592–1603. [CrossRef]
103. Zhang, T.; Tang, H. Evaluating the Generalization Ability of Convolutional Neural Networks for Built-up Area Extraction in Different Cities of China. *Optoelectron. Lett.* **2020**, *16*, 52–58. [CrossRef]
104. Chuvieco, E. *Fundamentals of Satellite Remote Sensing: An Environmental Approach*, 2nd ed.; CRC Press: Boca Raton, FL, USA, 2016. [CrossRef]
105. Lewis, H.G.; Brown, M. A Generalized Confusion Matrix for Assessing Area Estimates from Remotely Sensed Data. *Int. J. Remote Sens.* **2001**, *22*, 3223–3235. [CrossRef]
106. Machado, R.; Bayot, R.; Godinho, S.; Pirnat, J.; Santos, P.; de Sousa-Neves, N. LDTtool: A Toolbox to Assess Landscape Dynamics. *Environ. Model. Softw.* **2020**, *133*, 104847. [CrossRef]
107. Machado, R.; Godinho, S.; Pirnat, J.; Neves, N.; Santos, P. Assessment of Landscape Composition and Configuration via Spatial Metrics Combination: Conceptual Framework Proposal and Method Improvement. *Landsc. Res.* **2018**, *43*, 652–664. [CrossRef]
108. Strategia de Dezvoltare a Județului Ilfov 2020–2030, Sinteza. Consiliul Județean Ilfov, The World Bank. March 2020.
109. Sýkora, L.; Stanilov, K. The Challenge of Postsocialist Suburbanization. In *Confronting Suburbanization*; John Wiley & Sons, Ltd.: Chichester, UK, 2014; pp. 1–32. [CrossRef]
110. Gheorghe, K.; Ines, G. Urban Growth in the Bucharest Metropolitan Area: Spatial and Temporal Assessment Using Logistic Regression. *J. Urban. Plan. Dev.* **2018**, *144*, 05017013. [CrossRef]
111. Pătroescu, M.; Vânău, G.; Niță, M.R.; Iojă, C.; Iojă, A. Land Use Change in the Bucharest Metropolitan Area and Its Impacts on the Quality of the Environment in Residential Developments. *Forum Geogr.* **2011**, *10*, 177–186. [CrossRef]
112. Flanders Investment&Trade. *Construction-Real Estate Sector in Romania*; Flanders Investment&Trade: Bucharest, Romania, 2018.
113. Negescu, O. Efectele Crizei Financiare Și Ale Evoluției Pieței Imobiliare Asupra Sectorului Construcțiilor Din România. In *Analiză și Evaluare Economico-Financiară. Provocări pe Piața Imobiliară*; Academia de Studii Economice București: Bucharest, Romania, 2013; pp. 28–42.
114. Tosa, C.; Mitrea, A.; Sato, H.; Miwa, T.; Morikawa, T. Economic Growth and Urban Metamorphosis: A Quarter Century of Transformations within the Metropolitan Area of Bucharest. *J. Transp. Land Use* **2018**, *11*, 273–295. [CrossRef]

115. Pascariu, G. Overview of Romanian Planning Evolution. In *Proceedings of AESOP 26th Annual Congress*; METU: Ankara, Turkey, 2012.
116. Ianoş, I.; Sorensen, A.; Merciu, C. Incoherence of Urban Planning Policy in Bucharest: Its Potential for Land Use Conflict. *Land Use Policy* **2017**, *60*, 101–112. [CrossRef]
117. Grigorescu, I.; Kucsicsa, G.; Popovici, E.-A.; Mitrică, B.; Mocanu, I.; Dumitraşcu, M. Modelling Land Use/Cover Change to Assess Future Urban Sprawl in Romania. *Geocarto Int.* **2021**, *36*, 721–739. [CrossRef]
118. Stoica, I.V.; Tălângă, C.; Braghină, C.; Zamfir, D. Ways of Managing the Urban-Rural Interface. Case Study: Bucharest. *Analele Univ. Din. Oradea—Ser. Geogr.* **2011**, *21*, 313–322.
119. Ianos, I.; Zamfir, D.; Stoica, V.; Cercleux, L.; Schvab, A.; Pascariu, G. Municipal Solid Waste Management For Sustainable Development Of Bucharest Metropolitan Area. *Environ. Eng. Manag. J.* **2012**, *11*, 359–369. [CrossRef]
120. PeiSer, R. Decomposing Urban Sprawl. *Town Plan. Rev.* **2001**, *72*, 275–298. [CrossRef]
121. Rusu, A.; Ursu, A.; Stoleriu, C.C.; Groza, O.; Niacşu, L.; Sfîcă, L.; Minea, I.; Stoleriu, O.M. Structural Changes in the Romanian Economy Reflected through Corine Land Cover Datasets. *Remote Sens.* **2020**, *12*, 1323. [CrossRef]
122. Jat, M.K.; Garg, P.K.; Khare, D. Monitoring and Modelling of Urban Sprawl Using Remote Sensing and GIS Techniques. *Int. J. Appl. Earth Obs. Geoinf.* **2008**, *10*, 26–43. [CrossRef]
123. Dewan, A.M.; Yamaguchi, Y. Land Use and Land Cover Change in Greater Dhaka, Bangladesh: Using Remote Sensing to Promote Sustainable Urbanization. *Appl. Geogr.* **2009**, *29*, 390–401. [CrossRef]
124. Estoque, R.C.; Murayama, Y. Intensity and Spatial Pattern of Urban Land Changes in the Megacities of Southeast Asia. *Land Use Policy* **2015**, *48*, 213–222. [CrossRef]
125. Zeng, C.; Liu, Y.; Liu, Y.; Qiu, L. Urban Sprawl and Related Problems: Bibliometric Analysis and Refined Analysis from 1991 to 2011. *Chin. Geogr. Sci.* **2014**, *24*, 245–257. [CrossRef]
126. Luo, J.; Xing, X.; Wu, Y.; Zhang, W.; Chen, R.S. Spatio-Temporal Analysis on Built-up Land Expansion and Population Growth in the Yangtze River Delta Region, China: From a Coordination Perspective. *Appl. Geogr.* **2018**, *96*, 98–108. [CrossRef]
127. Krishna, A.P.; Mitra, S.K. *Geoinformatics Based Environmental Quality Assessment of Physical Parameters of Urbanization: Case Study of Ranchi City, India*; Map Asia: Kuala Lumpur, Malaysia, 2007.

remote sensing

Article

Rapid Glacier Shrinkage and Glacial Lake Expansion of a China-Nepal Transboundary Catchment in the Central Himalayas, between 1964 and 2020

Yan Zhong [1,2,3], Qiao Liu [1,3,*], Liladhar Sapkota [1,2], Yunyi Luo [1,2], Han Wang [1,2], Haijun Liao [1,2] and Yanhong Wu [1,3]

[1] Institute of Mountain Hazards and Environment, Chinese Academy of Sciences, Chengdu 610041, China; zhongyan19@mails.ucas.ac.cn (Y.Z.); liladhar.767517@cdg.tu.edu.np (L.S.); luoyunyi20@mails.ucas.ac.cn (Y.L.); wanghan202@mails.ucas.ac.cn (H.W.); liaohaijun@imde.ac.cn (H.L.); yhwu@imde.ac.cn (Y.W.)

[2] College of Resources and Environment, University of Chinese Academy of Sciences, Beijing 100049, China

[3] Branch of Sustainable Mountain Development, Kathmandu Center for Research and Education, Chinese Academy of Sciences-Tribhuvan University, Chengdu 610041, China

* Correspondence: liuqiao@imde.ac.cn; Tel.: +86-135-4788-7225

Abstract: Climate warming and concomitant glacier recession in the High Mountain Asia (HMA) have led to widespread development and expansion of glacial lakes, which reserved the freshwater resource, but also may increase risks of glacial lake outburst floods (GLOFs) or debris floods. Using 46 moderate- and high-resolution satellite images, including declassified Keyhole and Landsat missions between 1964 and 2020, we provide a comprehensive area mapping of glaciers and glacial lakes in the Tama Koshi (Rongxer) basin, a highly glacierized China-Nepal transnational catchment in the central Himalayas with high potential risks of glacier-related hazards. Results show that the 329.2 ± 1.9 km^2 total area of 271 glaciers in the region has decreased by 26.2 ± 3.2 km^2 in the past 56 years. During 2000–2016, remarkable ice mass loss caused the mean glacier surface elevation to decrease with a rate of -0.63 m a^{-1}, and the mean glacier surface velocity slowed by ~25% between 1999 and 2015. The total area of glacial lakes increased by 9.2 ± 0.4 km^2 (~180%) from 5.1 ± 0.1 km^2 in 1964 to 14.4 ± 0.3 km^2 in 2020, while ice-contacted proglacial lakes have a much higher expansion rate (~204%). Large-scale glacial lakes are developed preferentially and experienced rapid expansion on the east side of the basin, suggesting that in addition to climate warming, the glacial geomorphological characters (aspect and slope) are also key controlling factors of the lake growing process. We hypothesize that lake expansion will continue in some cases until critical local topography (i.e., steepening icefall) is reached, but the lake number may not necessarily increase. Further monitoring should be focused on eight rapidly expanding proglacial lakes due to their high potential risks of failure and relatively high lake volumes.

Keywords: climate change; glacier change; glacial lake change; GLOFs; central Himalaya; Tama Koshi

Citation: Zhong, Y.; Liu, Q.; Sapkota, L.; Luo, Y.; Wang, H.; Liao, H.; Wu, Y. Rapid Glacier Shrinkage and Glacial Lake Expansion of a China-Nepal Transboundary Catchment in the Central Himalayas, between 1964 and 2020. *Remote Sens.* **2021**, *13*, 3614. https://doi.org/10.3390/rs13183614

Academic Editors: Cristian Constantin Stoleriu and Adrian Ursu

Received: 4 August 2021
Accepted: 8 September 2021
Published: 10 September 2021

Publisher's Note: MDPI stays neutral with regard to jurisdictional claims in published maps and institutional affiliations.

1. Introduction

Glaciers play essential roles in the global cryosphere and Earth's water cycle [1] by buffering freshwater supply [2], modifying mountain landforms [3], and regulating fluvial sediment flux [4]. Glacier and glacier-related processes are highly vulnerable to climate change [5,6]. The Tibet Plateau and its surroundings, namely the High Mountain Asia (HMA), host the largest volume of glaciers outside of the polar regions [7,8] and is referred to as the Asian water tower [9,10]. Among all, approximately 20,431 (covering 19,679 km^2) glaciers and 4950 (covering 455.3 km^2) glacial lakes are developed along with the Himalayas [11–14]. Nearly 800 million people in the Indus, Ganges and Yarlung Tsangpo river (i.e., Brahmaputra River) basins are benefited from the Himalayan glacial meltwaters [15,16].

With a warming global climate, Himalayan glaciers have shrunk severely [16–18], accelerating the development and evolution of glacial lakes in this region [19]. The existence of glacial lakes could cause a lag effect of glacial runoff that affects downstream river discharges, hydrological processes, channel erosions, and landscape evolutions [20]. They could also induce glacier-related hazards, e.g., glacial lake outburst floods [19], which is one of the main climate hazards caused by warming in high altitude regions [21]. Himalayan glacial lakes are mainly located between 4000 and 5700 m a.s.l., and expanded by approximately 14.1% from 1990 to 2015 [11]. The widespread expansion of the Himalayan glacial lakes has been the focus of attention by both scientific researchers and social communities [19,22–30].

Owing to frequent geo-hazards, regional development delays, poor road conditions, and low access to glaciers and glacial lakes, field investigations are difficult to conduct in the HMA. However, with the recent rapid development of remote sensing technology and cloud-based geospatial data distribution, efficient and reliable mapping methods have been widely used in cryosphere science [31,32]. To better assess and mitigate the risks of GLOFs hazard and their impact, it is crucial to investigate the evolution of the glaciers and glacial lakes and predict their future development trends under climate change [33]. A higher temporal resolution (e.g., annual) of glacial lake mapping and the expansion process of different types of glacial lakes are still limited in previous studies [34]. In this study, 46 mid- and high-resolution remote sensing images based on Landsat and Keyhole series satellite were used to provides a complete mapping (1964–2020) of glacial lakes and glaciers in the Tama Koshi (Rongxer) basin, a less reported China-Nepal transnational catchment in the central Himalayas. We aim to (1) analyze the dynamics of different types of glaciers and glacial lakes and (2) discuss the glacier-lake interactions, possible forcing mechanisms of glacial lakes expansion, their future evolution, and GLOF risk implications.

2. Study Area

The Tama Koshi basin (a sub-basin of the Koshi River Basin, 28°N, 86°E, Figure 1) is a long, narrow, and deep valley extending northeast-southwest, located on the south side of the central Himalayas. The total basin area is 4117 km^2 (about 1/3 is in China), and the elevation ranges from less than 500 m a.s.l. at Beni Ghat, where the Tama Koshi joins the Sunkoshi, to above 8000 m a.s.l. adjacent to Mt. Cho Oyu (8188 m a.s.l.). Due to the enormous elevation difference (more than 6800 m), the ecological environment in the basin is unique, complex, and diversified. Data from the European Centre for Medium-Range Weather Forecasts (ECMWF, see Section 3.4) show that the mean annual temperature of the entire basin was 5.42 °C, and the mean annual precipitation is 1380 mm, and both showed increasing trends in the past 40 years, by about 1 °C and 300 mm, respectively (Figure 1B).

Glaciers and glacial lakes are widely distributed in the upper basin. According to the Randolph Glacier Inventory (RGI) 6.0 [13], a total of 279 glaciers are developed above 4600 m a.s.l., covering a total area of 334.8 km^2 (8% of the entire basin). According to the latest inventory by Wang et al. [12], glacial lakes in the basin have a total number of 229 and an area of 14.6 km^2 in 2018, which can be classified as five major types: cirque lake (11%), end moraine-dammed lake (54%), lateral moraine-dammed lake (5%), moraine thaw lake (1%) and supraglacial lake (16%). As reported by ICIMOD in 2020 [21], 42 lakes in the Koshi basin have been identified as potentially dangerous glacial lakes (PDGL), of which 8 are located in the Tama Koshi basin. A study by Rounce et al. [35] also noted 11 very high-risk lakes in Nepal, one of them is located in this basin. Glacier and glacial lake changes in the Tama Koshi basin have been reported previously, but most with a low spatial-temporal resolution [34,36,37]. One of the most reported glacial lakes in upper Tama Koshi is the Tsho Rolpa (Rolwaling Valley), which has been evaluated as the most dangerous glacial lake in Nepal [38]. The Tsho Rolpa has been monitored, measured, and modeled since the 1990s with comprehensive details about its expansion mechanisms [39] and failure risks [40]. Various risk prevention measurements, e.g., siphon system, automatic early warning system, and artificial channel excavation, have been carried out at the Tsho Rolpa [41].

Several studies have predicted the possible response of glaciers and glacial lakes under the climate change in the Tama Koshi basin, provided the management experience and scheme of dangerous glacial lakes and discussed the potential risks of GLOFs [37,40,42,43].

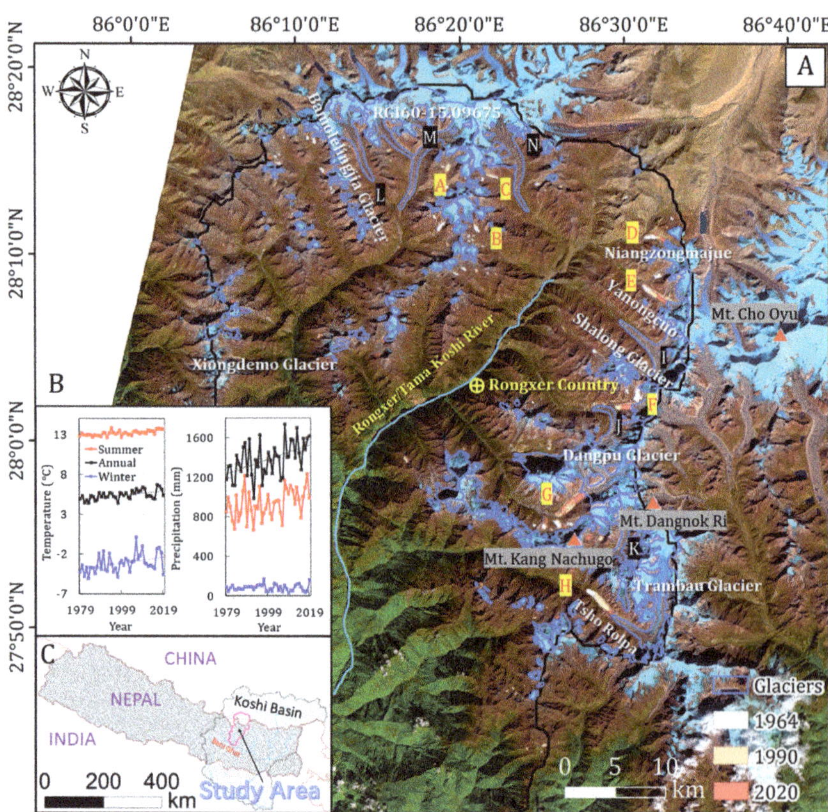

Figure 1. (**A**) Distribution of glaciers and glacial lakes in the Tama Koshi basin with the background image of Landsat 8 (24 October 2018), letters indicate detailly investigated glaciers and glacial lakes. The inset (**B**) shows the mean annual (red), summer (blue), and winter (black) air temperature (**left**) and precipitation (**right**) extracted from ERA5. The inset (**C**) indicates the location of the study area.

3. Data and Methods

3.1. Data Set Used

The remote sensing images used in this study were obtained from the United States Geological Survey (USGS, https://earthexplorer.usgs.gov/; accessed on 26 October 2020), Geospatial Data Cloud (http://www.gscloud.cn/; accessed on 30 October 2020) and PlanetLabs (https://www.planet.com/; accessed on 21 October 2020) between 1964 and 2020. A total of 6 KH-4A and KH-9 Keyhole (KH) images (2.7–9 m), 36 Landsat (TM/ETM+/OLI) surface reflectance images (30 m), and 4 PlanetScope (PL) Ortho Tile imagery (3.125 m) were used to manually delineated glaciers in 1964, 1980, 1990, 2000, 2010, 2018, and 2020 and glacial lakes in 1964, 1980, and automatically delineated glacial lakes in 1987–2020, respectively. Images were mainly acquired in post-monsoon and near the end of ablation seasons (August to December) with less cloud cover and seasonal snow cover. The orthorectified PL products were used as base images for the registration of KH images.

Other data sets we have used include: (1) RGI 6.0 [44], (2) Glacial Lake Inventory of High-mountain Asia in 1990 and 2018 [12], (3) ERA5 monthly aggregates from 1979 to

2019 (latest climate reanalysis produced by ECMWF/Copernicus Climate Change Service (ECCCS)), (4) the 30 m Shuttle Radar Topography Mission Digital Elevation Model (SRTM DEM), (5) 2007 ALOS PALSAR DEM (12.5 m), (6) 1975–2000, 2000–2016 glacier surface elevation change for the HMA derived by differencing KH-9 DEM (generated from optical stereo) and ASTER DEMs [18], (7) 1999–2003, 2013–2015 glacier surface flow velocity for the HMA derived by optical satellite images via feature-tracking algorithm [45], (8) Land-Scan: global population distribution data for 2000 and 2018 [46,47], and (9) 1:72,000 scale topographic map (base map from Esri). The glacier and glacial lake data sets were used to extract relevant attributes (for example, aspect and ID) and provide a reference for shaping delineations, and the DEM and glacier thickness change data were used to extract terrain parameters and glacier surface elevation changes. The attribute information and purpose of each data set are detailed in Supplementary Materials, Table S1.

3.2. Glacier and Glacial Lake Outline Delineation

The KH series images were characterized by high resolution, single band (panchromatic), and lack of spatial coordinates and projection information [48,49], which needs to be preprocessed before mapping. Four high-resolution orthorectified PL images (as the base image) from September to October 2020 were used for the preliminary spatial registration of KH images (as the warp image) using the ENVI Image-to-Image Registration Tool. A total of 103 (for KH-4A) and 72 (for KH-9) ground control points were selected in stable terrains such as ridgelines, river intersections, and moraine ridgelines. According to the ground control point pairs in the two images, a functional relationship is established, and the geometric correction of the distorted image is completed using coordinate transformation.

To monitor the detailed annual change of the rapidly expanding glacial lakes (Figure 1A, A–H), we automatically delineated the glacial lakes (Figure S1) in 1987–2020 (September–November) based on the Google Earth Engine (GEE) remote sensing cloud processing platform [50] Python Application Programming Interface (API) [51]. The Normalized Difference Water Index (*NDVI*; Equation (1)) and Normalized Difference Snow Index (*NDSI*; Equation (2)) were combined and applied to multi-temporal Landsat (TM/ETM+/OLI) images to reduce the errors caused by sediments and seasonally frozen lake water. Then, it is complemented with manual editing to extract glacial lake boundaries. The combination of this object-oriented image processing method and expert knowledge has shown suitable performance in glacial lake delineation in the Himalayas [52].

$$NDWI = \frac{green - NIR}{green + NIR} \tag{1}$$

$$NDSI = \frac{green - SWIR1}{green + SWIR1} \tag{2}$$

where *green*, *NIR*, and *SWIR*1 are the green, near-infrared, and shortwave-infrared 1 bands, respectively. By using the Python API interactive interface, the classification thresholds can be adjusted more accurately than the Javascript API interface based on the on-screen inspection of the original images. All Landsat and PlanetScope images were atmospheric and topographic corrected products. Additionally, an error of ±0.5 pixels was estimated to calculate the area uncertainties (i.e., multiply linear error and perimeter, e.g., the linear error of Landsat images are 15 m, and of KH-4A images are ~1.35 m) [33,53]. For supraglacial lakes, only the base-level lakes with perennial water storage were retained, while the perched lakes with large seasonal variations were removed [54].

3.3. Glacier Surface Dynamics Analysis

Eight lake-terminating glaciers (Figure 1A, A–H) and six land-terminating glaciers (Figure 1A, I–N) were selected to compare their dynamics and relationships with glacio-geomorphology or lake-ice interactions. Based on the 2000 SRTM, the two periods of glacier surface elevation in 1975 and 2016 were calculated by combining the 1975–2000 and

2000–2016 glacier surface elevation change. Surface gradient was calculated based on the 2007 ALOS PALSAR DEM using the QGIS slope analysis module, by which we extracted values between the terminal moraine to the mid-point position of the glacier tongue to better explore the slope of the downstream tongue. Subsequently, equidistant points at a 30 m spacing were generated along the entire glacier center flow line to identify and visualize the elevation, flow velocity, and slope. For the elevation and velocity analysis of the ice/lake area, we extended the line to a further 300–800 m in front of the moraine dam.

3.4. Meteorological Data Extractions

Since there is no meteorological observation in the Tama Koshi basin, we used the ERA5 monthly aggregates data to analyze the impact of meteorological factors. The ERA5 data set, with a spatial resolution of 0.25 radian, is the fifth generation ECMWF atmospheric reanalysis of the global climate [55]. Its grided data set is now incorporated in the GEE data catalog, and we extracted the temperature and precipitation data inside the basin domain between 1979 and 2019 pixel by pixel.

4. Results

4.1. Glacier Distribution and Changes

4.1.1. Distribution of Glaciers

Our glacier inventories show (Figure 2A) a total of 271 glaciers with a total area of 329.2 ± 1.9 km^2 in 2020 (Table 1). The largest one is Glacier M (42.6 ± 0.1 km^2), which is a debris-covered glacier facing south and located in the upper Lapche, one of the western tributaries of the Tama Koshi (Figure 1, Randolph Glacier Inventory ID: RGI60-15.09675). The mean elevation of all glaciers ranges from 4760 to 6270 m a.s.l. (Figure 2B) and most (96%) glacier areas are distributed between 5200 and 5800 m a.s.l. The lowest glacier tongue, which is close to the Rongxer country (ID: RGI60-15.09760), extends to an altitude of 4475 m a.s.l. Due to the northeast (high altitude) to the southwest (low altitude) orientation of the basin, southern oriented glaciers have the maximum area coverage (84.9 ± 0.4 km^2, ~26%, Figure 2C). Northwest-oriented glaciers have the second largest area of 68.2 ± 0.3 km^2 (~21%), whereas the glaciers with the east orientation are the smallest (10.3 ± 0.1 km^2, 3.1%).

Table 1. Glacier inventories between 1964 and 2020 of the Tama Koshi basin.

Study Period (Sensor)	Glacier Number	Total Area (km^2)	Area Change (% a^{-1})	West Side of Basin		East Side of Basin	
				Number	Area (km^2)	Number	Area (km^2)
1964 (KH-4A)	282	355.3 ± 1.8	–	141	142.0 ± 0.8	141	213.4 ± 1.0
1980 (KH-9)	283	350.4 ± 5.9	-0.09	143	141.2 ± 2.6	140	209.3 ± 3.4
1990 (Landsat TM)	279	337.5 ± 19.0	-0.37	139	136.2 ± 8.1	140	201.3 ± 10.9
2000 (Landsat TM)	279	335.9 ± 19.0	-0.05	139	135.5 ± 8.1	140	200.5 ± 10.9
2010 (Landsat ETM+)	279	335.0 ± 18.9	-0.03	139	135.2 ± 8.1	140	199.8 ± 10.8
2018 (Landsat OLI)	271	331.2 ± 18.5	-0.14	133	133.9 ± 7.9	138	197.4 ± 10.6
2020 (PlanetScope)	271	329.2 ± 1.9	-0.31	133	133.8 ± 0.8	138	195.4 ± 1.1

Glacier distributions are slightly different between the basin's east and west sides, divided by the Rongxer/Tama River corridor (Figure 1 and Table 1). On the west side of the corridor, the three largest debris-covered glaciers (L, M, and N in Figure 1) occupied ~63% of the total glacier area (133.7 km^2) compared to the other 131 relatively smaller glaciers (50.0 km^2). In contrast, mean glacier areas in the east (1.4 km^2 for the 138 glaciers) are larger than the west (1.0 km^2), with seven glaciers are >5 km^2 and five are >10 km^2. Glaciers on the east side distribution with a wide altitude range (mainly between 5200 and 5800 m a.s.l.), whereas the west side glacier areas are concentrated between 5600 and 5700 m a.s.l. All five lake-terminating glaciers are located on the east side, but the supraglacial lakes are more frequently found on the lower part of the three largest debris-covered glaciers.

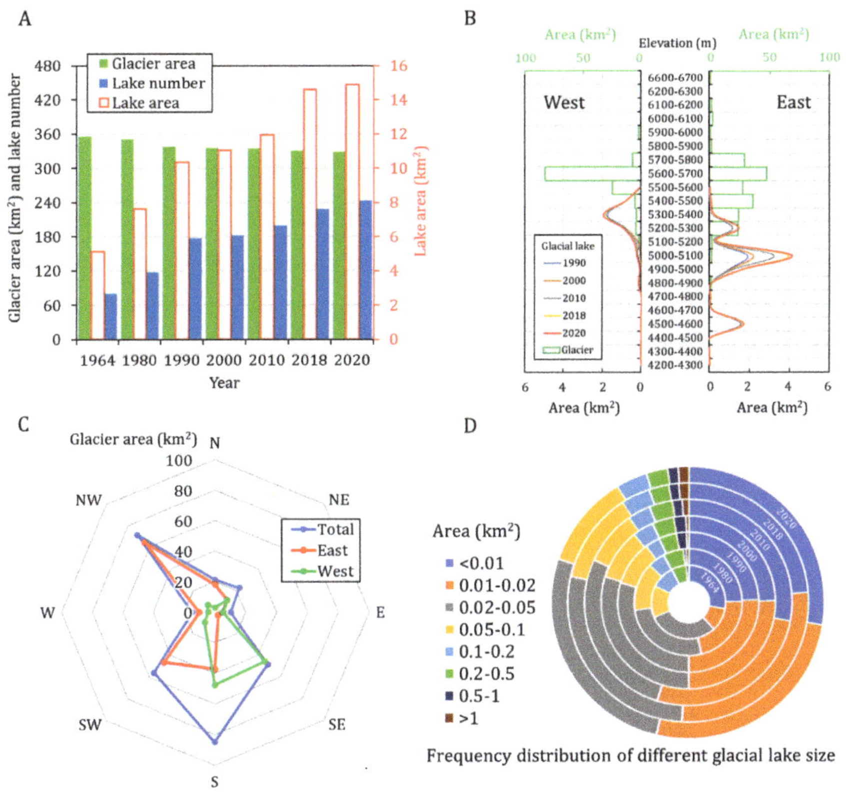

Figure 2. Overview of glaciers and glacial lakes. (**A**) Number and area changes of glaciers and glacial lakes from 1964 to 2020. (**B**) Elevation distribution of glaciers (in 2020) and glacial lakes (between 1990 and 2020) on the west and east side of the basin. (**C**) Glacier distribution with aspect in the total (blue), east (red), and west (green) of the basin. (**D**) Frequency of glacial lakes in various size classes.

4.1.2. Glacier Changes

From 1964 to 2020, the total area of glaciers in the Tama Koshi basin decreased by 26.2 ± 3.2 km^2 (0.13% a^{-1}) in 56 years (Tables 1 and 2), with a mean glacier terminus retreat rate of 3.2 m a^{-1}. Glacier area loss rates in the periods of 1980–1990 and 2018–2020 were 0.37% a^{-1} and 0.31% a^{-1}, respectively, which were much higher than the rest study period (0.03–0.14% a^{-1}, Table 1). Eight small glaciers covering an area of 0.03–0.29 km^2 had disappeared between 1980 and 1990. The mean retreat rate of lake-terminating glaciers (37.7 m a^{-1}) is about 14 times higher than those land-terminating ones (2.6 m a^{-1}). Glacier G, which terminates in lake G, experienced the longest terminus retreat (68 m a^{-1}) between 1964 and 2020. On the east side of the basin, glacier area loss rate (0.15% a^{-1}) and retreat rate (4.3 m a^{-1}) are generally higher than those on the west (0.10% a^{-1} and 2.08 m a^{-1}, respectively, Table 2).

Table 2. Glacier changes between 1964 and 2020 for different glacier types and locations.

	Study Period	Total	Land-Terminating	Lake-Terminating	West Side of Basin	East Side of Basin
Area change rate (% a^{-1})	1964–2020	0.13	0.13	0.14	0.10	0.15
Retreat rate (m a^{-1})	1964–2020	3.2	2.6	37.7	2.1	4.3
Elevation change rate (m a^{-1})	1975–2000	−0.32	−0.30	−0.40	−0.26	−0.33
	2000–2016	−0.63	−0.62	−0.85	−0.69	−0.62
Mean Velocity (m a^{-1})	1999–2003	5.3	4.8	6.9	5.4	5.2
	2013–2015	4.0	3.4	6.1	4.1	3.8

Note: From 1975 to 2000, glacier-wide mean elevation changes were extracted using the glacier boundary in 1980 and from 2000 to 2016 using the boundary in 2000. From 1999 to 2003, mean velocities were extracted using glacier boundaries in 2000 and from 2013 to 2015 using glacier boundaries in 2010.

Meanwhile, glaciers also showed a trend of thinning and slowdown (Table 2). The mean change rates of glacier surface elevation during 1975–2000 and 2000–2016 were −0.32 m a^{-1} and −0.63 m a^{-1}, respectively, indicating an accelerated thinning of the ice. Glacier elevation change rates on the western side (from −0.26 m a^{-1} in the period 1975–2000 to -0.69 m a^{-1} in the period 2000–2016) shows more remarkable than the east (−0.33 m a^{-1} in the period 1975–2000 to 0.62 m a^{-1} in the period 2000–2016) (Table 2). The most downwasted glacier on the west side of the basin is the Glacier L, one of the largest land-terminating glaciers (debris-covered, with a length of 12.7 km) in the upper Tama Koshi. Its mean elevation change rate increased from −0.45 m a^{-1} in 1975–2000 to −0.97 m a^{-1} in 2000–2016 (Figure 3), higher than that of Glacier E, I, and G on the east side (−0.78 m a^{-1}, −0.60 m a^{-1} and −0.89 m a^{-1} in 2000–2016, respectively). Nevertheless, the mean elevation change rates of lake-terminating glaciers (−0.40 m a^{-1} in the period 1975–2000 and −0.85 m a^{-1} in the period 2000–2016) in the basin were higher than that of land-terminating glaciers (−0.30 m a^{-1} in the period 1975–2000 and −0.62 m a^{-1} in the period 2000–2016) during the two study periods.

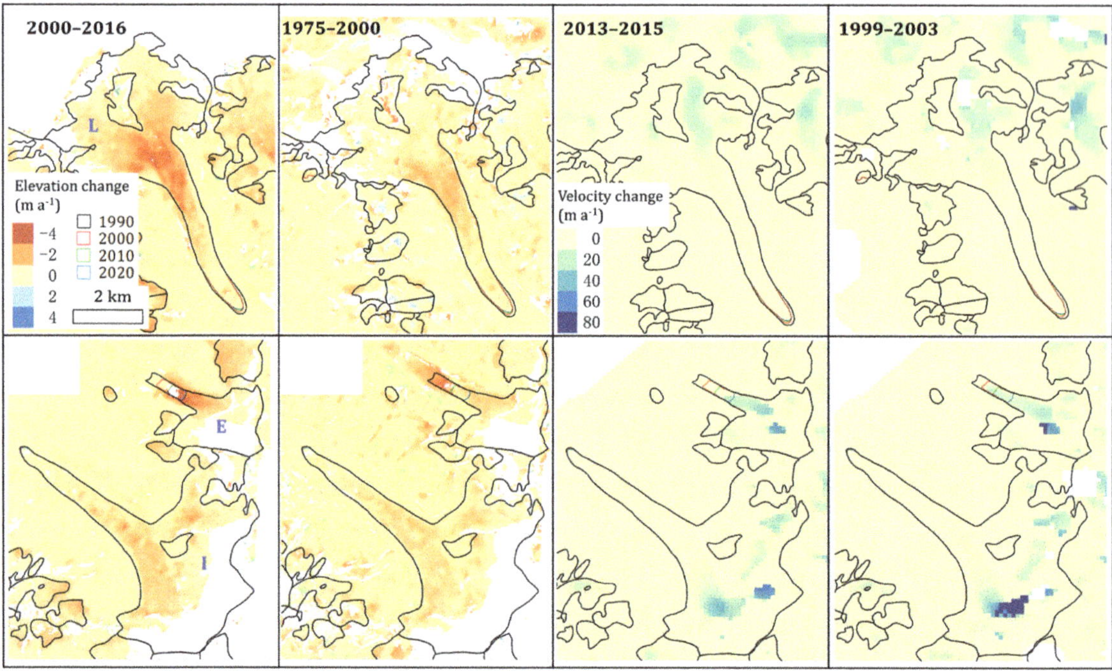

Figure 3. Comparisons of elevation change rates and velocities of three typical land- and lake-terminating glaciers (Glacier E, I, and L) on the east and west sides of the basin.

The mean glacier surface velocity of the Tama Koshi basin decreased by ~25% from 5.3 m a^{-1} in 1999–2003 to 4.0 m a^{-1} in 2013–2015. Overall, mean glacier surface velocities on the west side (5.4 m a^{-1} in the period 1999–2003 and 4.1 m a^{-1} in the period 2013–2015) were slightly higher than on the east side (5.2 m a^{-1} in the period 1999–2003 and 3.8 m a^{-1} in the period 2013–2015), while both have decelerated during the two study periods. Lake-terminating glaciers (with a velocity of 6.9 m a^{-1} in the period 1999–2003 and 6.1 m a^{-1} in the period 2013–2015) moved faster than land-terminating glaciers (4.8 m a^{-1} in the period 1999–2003 and 3.4 m a^{-1} in the period 2013–2015). Even though the surface velocity declined by 11%, it was much slower than that of land-terminating glaciers (−29%).

4.2. Glacial Lake Distribution and Changes

4.2.1. Distribution of Glacial Lakes

The statistical results of glacial lake inventories between 1964 and 2020 are presented in Table 3. A total of 196 glacial lakes (more than 50% are moraine-dammed proglacial lakes) are developed in the basin, with a total area of 14.4 ± 0.3 km^2 and a mean area of 0.07 km^2 in 2020. These values are slightly lower than the glacial lake inventory in 2018 (see Section 2) published by Wang [12] but much higher than Wu's study in 2005 (lake number: 28; lake area: 7.87 km^2) [34]. There are 107 glacial lakes (~55% of the total number of glacial lakes) with an area of >0.02 km^2 (Figure 2D), which occupy 13.3 km^2 (~93%) of the total lake area. Most glacial lakes (14.2 ± 0.3 km^2, ~99%) are distributed between 4500 and 5500 m a.s.l. (Figure 2B) and most are located on the east side of the corridor. The total area of the glacial lake on the east side (10.0 ± 0.1 km^2, ~69%) is much larger than that of the west (4.4 ± 0.1 km^2). The altitudes of the glacial lakes on the west side (5100–5500 m a.s.l.) are generally higher than the east (4500–5300 m a.s.l.). Nevertheless, there are a total of 119 glacial lakes on the western side, accounting for ~61% of the total number of our mapped glacial lakes. However, all of them are less than 0.3 km^2 in area.

Table 3. Glacial lake inventories between 1964 and 2020 and changes in meteorological conditions in the Tama Koshi basin.

Study Period (Sensor)	Lake Number	Total Area (km^2)	Area Change (% a^{-1})	West Side of Basin		East Side of Basin		Temperature Change (°C a^{-1})	Precipitation Change (mm a^{-1})
				Number	Area (km^2)	Number	Area (km^2)		
1964 (KH-4A)	78	5.1 ± 0.1	-	51	2.0 ± 0.1	27	3.1 ± 0.1	-	-
1980 (KH-9)	115	7.6 ± 0.5	3.0	76	3.0 ± 0.6	39	4.5 ± 0.2	-	-
1990 (Landsat TM)	171	10.3 ± 2.2	3.6	108	3.8 ± 1.1	63	6.5 ± 1.1	-	-
2000 (Landsat TM)	172	10.7 ± 2.2	0.47	109	4.0 ± 1.1	63	6.9 ± 1.1	0.033	0.48
2010 (Landsat ETM+)	175	11.7 ± 2.3	0.85	112	3.8 ± 1.1	63	7.8 ± 1.1	0.038	6.4
2018 (Landsat OLI)	198	14.2 ± 2.7	2.8	118	4.4 ± 1.3	80	9.8 ± 1.4	0.0007	14.6
2020 (PlanetScope)	196	14.4 ± 0.3	0.49	119	4.4 ± 0.1	77	10.0 ± 0.1	-	-
Total									
Lake growth (dA%)	151	-		180	133	123	185	217	-	-
Lake growth rate (% a^{-1})	2.7	-		3.2	2.4	2.2	3.3	3.9	-	-

Note: Temperature and precipitation changes were obtained by calculating the difference between ten-year averages.

4.2.2. Glacial Lake Changes

From 1964 to 2020, the total number and area of glacial lakes in the Tama Koshi basin increased considerably (by ~151% in number and ~180% or 3.2% a^{-1} in area, Figure 2A). Lake expansion between 1980 and 1990 reached the highest rate of 3.5% a^{-1} (Table 3). Statistically, a total of 12 glacial lakes disappeared during 2000–2020, 36 glacial lakes were newly formed, 10 glacial lakes shrank (7 in the west), and 26 glacial lakes expanded. The glacial lakes that develop between 5000 and 5100 m a.s.l. have experienced the highest expansion rate, with a rate of 4.0% a^{-1}.

In 1964, there were only 27 glacial lakes on the eastern side of the basin, with a total area of 3.1 ± 0.1 km^2 (Table 3). This number in 2020 has increased to 77 by a rate of 3.3% a^{-1} (185%), which is higher than the 2.4% a^{-1} (133%) on the western side (from 51 in 1964 to 119 in 2020, Table 3). Their overall expansion rate (3.9% a^{-1}) was much higher than the west side (2.2% a^{-1}). According to Table 4, moraine-dammed ice-contacted glacial lakes show a much higher expansion rate (~204%) than non-moraine-dammed ice-contacted lakes (~175%). According to our multi-decade inventories (Table 4), the total number of moraine-dammed ice-contacted lakes was peaked around 1980 at 17, an increase from 10 in 1964. After 1980, the number of moraine-dammed ice-contacted glacial lakes gradually decreased to five in 2020; the other 12 were disconnected from the retreating mother glacier. The annual expansion rates during periods after 1980 (0.4% a^{-1}–2.2% a^{-1}) were also lower than the period before1980 (5.7% a^{-1}, 1964–1980). In contrast, for the non-ice-contacted lakes, their numbers showed a continuous increase from 68 in 1964 to 191 in 2020, while their mean annual area expansion rate (3.1% a^{-1}) is close to those moraine-dammed ice-contacted lakes (3.7% a^{-1}).

Table 4. The expansion differences between ice-contact lake and non-ice-contact lake in the Tama Koshi basin.

Study Period (Sensor)	Moraine-Dammed Ice-Contact Lake			Non-Moraine-Dammed Ice-Contact Lake		
	Number	Area (km^2)	Expansion Rate (% a^{-1})	Number	Area (km^2)	Expansion Rate (% a^{-1})
1964 (KH-4A)	10	2.1 ± 0.1	-	68	3.0 ± 0.1	-
1980 (KH-9)	17	3.8 ± 0.2	5.7	98	3.7 ± 0.3	1.5
1990 (Landsat TM)	11	4.0 ± 0.4	0.40	160	6.3 ± 1.8	6.9
2000 (Landsat TM)	12	4.9 ± 0.6	2.2	160	5.9 ± 1.7	−0.65
2010 (Landsat ETM+)	10	5.3 ± 0.5	0.83	165	6.4 ± 1.9	0.85
2018 (Landsat OLI)	5	5.8 ± 0.5	1.3	193	8.4 ± 2.4	4.0
2020 (PlanetScope)	5	6.1 ± 0.1	2.2	191	8.3 ± 0.3	−0.71
Total						
Lake growth (dA%)	−50	-	204	181	-	175
Lake growth rate (% a^{-1})	−0.9	-	3.7	3.2	-	3.1

Between 1964 and 2020, the lake size distribution has also varied with time (Figure 2D). Glacial lakes with an area of <0.01 km^2 and between 0.05 and 0.2 km^2 showed the most unstable, while lake sizes between 0.01 and 0.05 km^2 gradually increased. For those glacial lakes with larger sizes, e.g., lakes with a size of 0.2–0.5 km^2 were relatively stable, and lake size of >1 km^2 showing more expansive during 1964–1990 following by relatively stable after 1990.

In this study, we found that two types of glacial lakes have experienced rapid expansion: moraine-dammed ice-contact lake and supraglacial lake (Figures 4 and 5, and Table 4). Based on the results of potentially dangerous glacial lakes evaluation from ICIMOD [21], the actual changes of the glacial lakes, and the past GLOF events (see Section 5.4), we identified eight (seven (B-H) of them are same with ICIMOD's report) rapidly expanding moraine-dammed ice-contact lakes (Lake A–H, Figures 1 and 6) as PDGLs. Lake G is a moraine-dammed lake formed around in 2009 by the expanding and coalescing with several supraglacial ponds (most of which are base-level lakes) in the 1980s. The total lake area over the entire glacier basin has increased by 9.0% a^{-1} from 0.08 km^2 in 1980 to 1.92 km^2 in 2020. Lakes A–C, located on the west side of the basin, expanded rapidly before 1990 but then gradually stabilized in their size. Lake A has shrunk significantly between 2018 and 2020, with a rate of ~−36% (or 0.08 km^2 a^{-1}). It was confirmed that the lake burst in 2018 according to the comparison of satellite images (more details are given in Section 5.4 and Figure S2). The expansion rate of Lake D (Niangzongmajue) was maintained at 1.3% a^{-1} before 2014 and then remained relatively stable, with no evident expansion in the following six years. Lake E (Yalongcuo) has experienced steady growth (1.9% a^{-1}) over the past 56 years. The expansion rate of Lake F increased rapidly after 2004, from 3.8% a^{-1} between 1990 and 2004 to 5.8% a^{-1} between 2004 and 2020, with an overall area growth of 52.9%. Lake H (Tsho Rolpa) expanded clearly before 1990 at a rate of ~3.5% a^{-1}, and then its area remained virtually unchanged after the manual intervention (drainage to lower down the lake level) in the 1990s. However, after 2012, despite continued intervention, it still showed a slight expansion trend with a rate of 0.8% a^{-1}. The above results are summarized in Table 5.

Table 5. Rates of changes and other related information for several lake-terminating glaciers (Glacier A-H) and land-terminating glaciers (Glacier I-N).

ID	Name (Lake or Glacier ID)	(1964–2020) Rate of Change in Lake Area (% a^{-1})	(1999–2015) Rate of Change in Glacier Velocity (m a^{-1})	(1975–2016) Change in Glacier Elevation (m a^{-1})	(2007) Average Slope (°)
A	RGI60-15.09690	−1.9	2.1	−0.55	5.5
B	RGI60-15.09714	0.33	4.8	−0.55	5.8
C	RGI60-15.09721	0.84	9.5	−0.49	5.2
D	Niangzongmajue	1.1	25.4	−0.40	5.5
E	Yalongcuo	1.9	10.4	−0.80	3.1
F	Dangpu Lake	4.2	2.8	−1.2	7.7
G	RGI60-15.09771	9.0	7.4	−1.1	5.5
H	Tsho Rolpa	0.60	9.7	−0.71	3.0
I	Shalong Glacier	0.04	5.1	−0.64	7.9
J	Dangpu Glacier	-	3.7	−0.47	10.5
K	RGI60-15.03428	-	2.5	−0.71	7.9
L	Bamolelingjia Glacier	-	5.7	−0.86	8.2
M	RGI60-15.09675	-	8.9	−0.64	7.8
N	RGI60-15.09729	-	14.7	−0.26	8.1

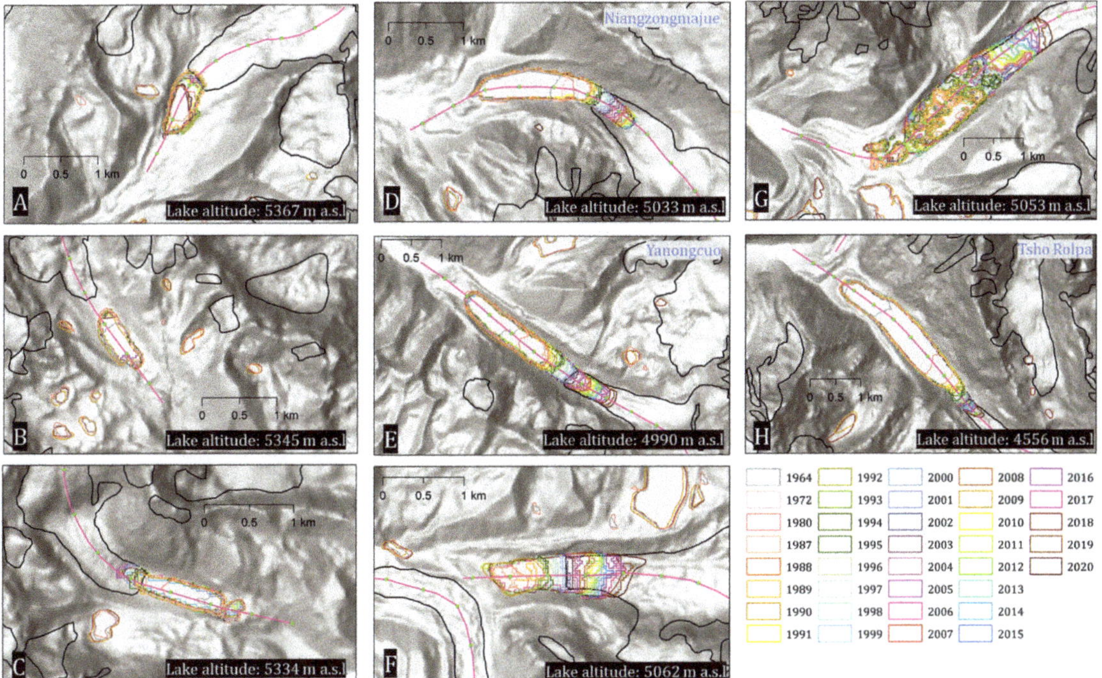

Figure 4. Changes of eight (Lake **A–H**) rapidly expanding proglacial lakes in the past 56 years overlayed on the Esri topographic map.

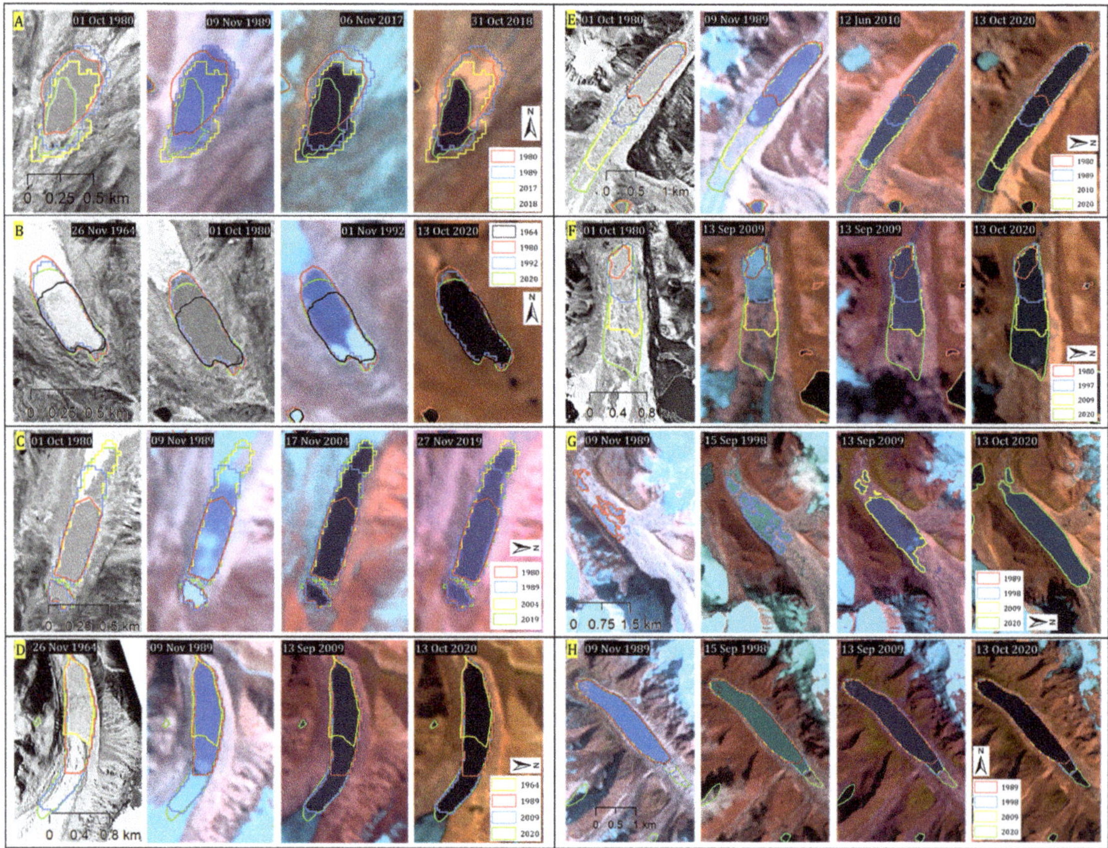

Figure 5. The expansion process of eight (Lake **A–H**) rapidly expanding proglacial lakes with the background of KH and Landsat images.

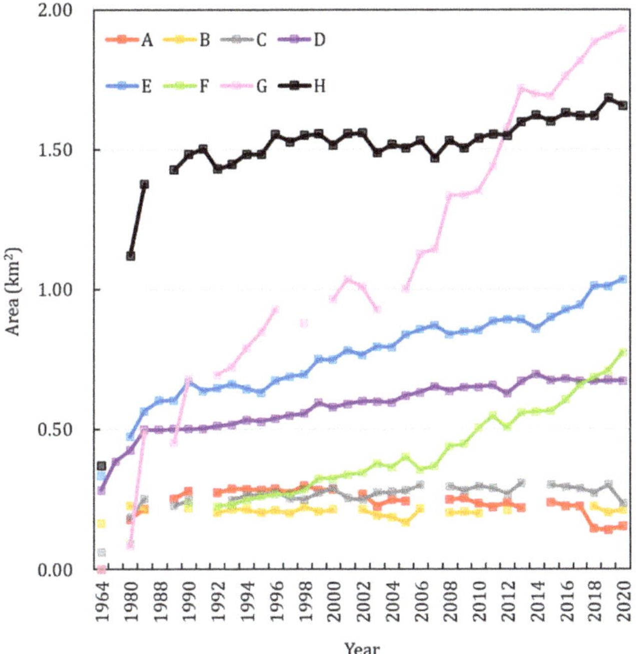

Figure 6. Area changes of eight (Lake **A–H**) rapidly expanding proglacial lakes.

5. Discussion

5.1. Glacier and Glacial Lake Variations and Climate Changes

As most HMA lakes are marginally influenced by direct human activities, glacial lake variations are closely related to climate change. Air temperature increase is considered to be the main driving force for glacier retreat and wasting [56]. Previous studies have shown significant warming in the Himalayas. From 1979 to 2019, the temperature rising rate in the Tama Koshi basin was measured at 0.01 °C a^{-1}, and for winter and summer is 0.002 °C a^{-1} and 0.02 °C a^{-1}, respectively, according to the analysis of ERA5. The precipitation changes during the past decades in the entire Himalayas were not uniform [57]. For the Tama Koshi basin, the basin-scale annual precipitation between 1979 and 2019 showed an increasing trend with a rate of 11.0 mm a^{-1}, and during the winter and summer seasons are 1.8 and 3.4 mm a^{-1}, respectively. This trend was similar to the meteorological observations at a nearby Tingri Station [11]. Although both temperature and humidity will affect evapotranspiration of the lake, the increase in temperature and precipitation has elevated freezing levels in the post-monsoon and early winter periods [58] and may boost the water supply of glacial lakes due to increased glacier ablation and runoff, which cause a negative effect on evapotranspiration [59], and eventually cause the lake level to rise (for example Lake Tsho Rolpa [60]) or induce a GLOF (please see Lake A in Section 5.4).

5.2. Glacier-Lake Interactions

Compared to the land-terminating glaciers, the mean change rates of area, length, elevation, and velocity of lake-terminating glaciers are remarkably higher (Table 2). In the Tama Koshi basin, length changes of most land-terminating glaciers (mostly are debris-covered) were not obvious during the study period (e.g., Glacier I, J, K, on the eastern side and Glacier L, M, N on the western side). The decrease in ice thickness, on the other hand, was their main result of mass loss. This is contrasted with those rapid retreats of lake-terminating glaciers (Figure 3). For example, for those rapidly expanding moraine-dammed lakes (Figure 5), they are all currently connecting with glaciers or have been in contact with ice. Such significant differences in glacial retreating rates were widely observed in the HMA regions [25,61,62] since the lake-calving is suggested as one of the important mass loss processes of the lake-terminating glaciers [63,64]. When those proglacial lakes finished their rapid expansion stage and gradually separated from the mother glacier, their expansion rates become gradually decreased (e.g., Glacier A, B, C). However, the retreat rates of the ice tongue are still faster than those initial land-terminating glaciers.

In addition, glaciers terminating lakes also showed notable changes in their thicknesses and flow velocities. The surface elevations of the glaciers that are connecting with Lakes A, B, E, F, G, and H have shown notably higher lowering rates than those land-terminating debris-covered glaciers (Tables 2 and 5, e.g., I, J, K). The largest elevation change occurred on the lake-terminating glacier connected with Lake F, reaching -1.2 m a^{-1} between 1975 and 2016 (Table 5). The remarkable high lowering rate (-0.64 m a^{-1}, Table 5) occurred at the largest debris-covered land-terminating glacier (M) is likely caused by the high-density distribution of supraglacial lakes (most of them are perched lakes) and ice cliffs (Figure S3) on its lower tongue, as these features acting as ablation hot spots on debris-covered glaciers [65–67] are unstable and seasonal. Although both lake- and land-terminating glaciers have shown a slowdown of the velocities since 1999, the decelerating rates of lake-terminating glaciers are much lower than the land-terminating glaciers (Table 2), indicating the modulation effect of proglacial calving processes on the maintenance of high ice flow rate of the lake-terminating glaciers [61].

5.3. Evolution and Future Development of Glacial Lakes Controlled by Topography

In general, the topography where the glacial lake develops (the pink rectangle highlighted area in Figure 7) generally is relatively flat. The mean slope of these lake-terminating glaciers is around 5°, while the mother glacier of Lake E has the lowest slope (~3°), and the mother glacier of Lake F has the highest surface gradient but still no more than 8° (Table 5). When the valley between receding glacier fronts and terminal moraines provides a sufficient and achievable area for glacial lake development, e.g., Lake D–H, the lake could continually expand. However, when the glacier gradually retreats to the region with higher elevation and steeper slope (Table 5), the expansion rate of the glacial lake gradually slows down, e.g., lake A–C has been fully loaded (or have reached their maximum upper limit). For Lake F and G, although they have high-gradient ice tongues, they are surviving thanks to their moraine dams supported by adjacent glacier-moraine complexes (Figure 4). From the longitudinal profile analysis (Figure 7), the height of the moraine dam in front of Lake F is the highest (50 m), which offers a strong bearing capacity for lake growth. On the other hand, despite the evident mass loss (with the mean glacier surface elevation change rate of -0.6 m a^{-1}), the land-terminating glaciers (I–N) are still difficult to store water due to the lack of stable dams (Figure 5) or gentle slopes (with the mean slope of 8.4°; Table 5). Supraglacial lakes on these debris-covered glaciers generally are perched lakes thus could only be filled seasonally [54], or englacial or subglacial drainage system in the glacier maybe exist that could drain the lakes periodically [68]. The number and area of supraglacial ponds on Glacier M and N are increasing slightly and may merge into a large supraglacial lake or moraine-dammed lake in the future once those perched lakes are located in the lowest part of the glacier change to base-level lakes as the glacier downwasting. Additionally, some cirque lakes may have formed after small cirque glaciers seriously downwasted or disappeared. We therefore argue that topography is not only an essential factor affecting the expansion of glacial lakes, but also a major condition for the development of glacial lakes. Moreover, the altitude and orientation of the proglacial lake may also affect the thermal condition and water balance of the glacial lake by influencing the absorption of solar radiation and ablation processes of adjacent ice.

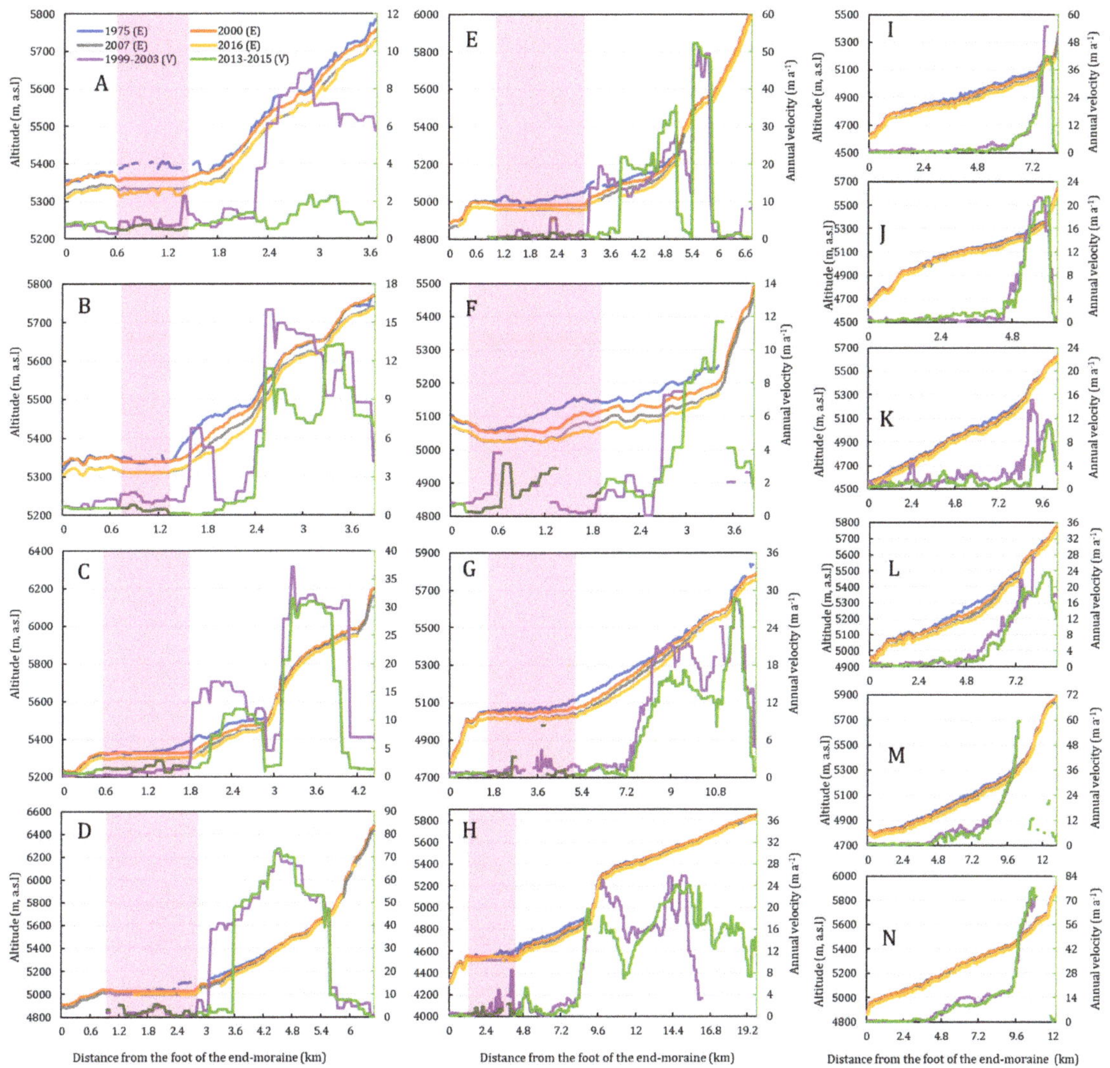

Figure 7. Changes in elevation and velocity of the lake-terminating glaciers (**A–H**) and land-terminating glaciers (**I–N**). "E" is elevation, "V" is velocity, the pink rectangle highlighted is the area where the glacial lake develops.

5.4. Implications for the Risk of GLOFs

Over 50 GLOF events have been documented in the Himalayas due to glaciers recession [69], 3 of which are located in the Tama Koshi basin (Figure 8, Table 6). The first one is entitled Chubung GLOF (86°27′38″E, 27°52′37″N), located in the Rolwaling Valley in Nepal, in front of Rolpa Tsho (true northwest). This GLOF was triggered by an ice avalanche that fell into the lake on 22 July 1991 and caused the lake to completely drain [38]. The second GLOF event (86°26′49″E, 27°55′46″N), which is named Upper Langbu Tsho, is located in the upper Dagazhuoma River, Rongxer country, Tibet, China, near Lake G (true right). The outburst of this lake from September to November 1992 was possibly triggered by an ice avalanche [69]). The third GLOF event was identified in this study by a continuous satellite tracking of Lake A (86°19′16″E, 27°14′43″N, Figure S2), which is located in the upper Labujikongzangbu River. Remote sensing monitoring showed that Lake A was covered by thick clouds for 13 days (28 July–9 August 2018) before its failure. Meteorological results also showed that the summer precipitation in 2018 was remarkably higher than in other years (Figure 1B, with a total of 1234 mm between June and August). We speculate that the high level of precipitation triggered this GLOF.

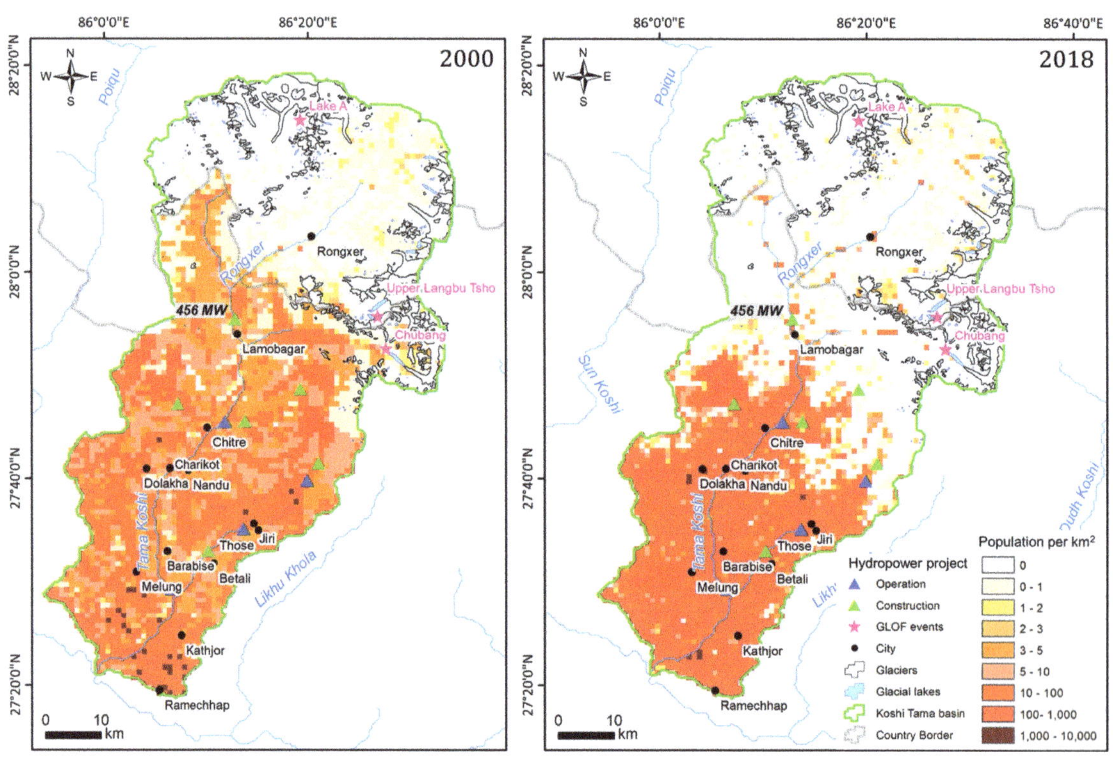

Figure 8. GLOF events, hydropower projects, and population tendency (left 2000 and right 2018) in the basin.

Table 6. Historical GLOF events of the basin.

Name	Date of Outburst	Location	(Possible) Trigger	Source
Chubung	12 July 1991	Rolwaling Valley, Nepal	Ice avalanche	(Reynolds 1999)
Upper Langbu Tsho	22 September 1992~ 17 November 1992	Dagazhuoma River, China	Ice avalanche	(Nie et al., 2018)
Lake A	30 July 2018~9 August 2018	Labujikongzangbu River, China	Extreme precipitation	This study

With the increase in human activities (the population grew from 34 per km² in 2000 to 59 per km² in 2018, Figure 8) and socioeconomic development in both Nepal and Chinese parts of the Tama Koshi basin, it is demanded to construct hydropower facilities to reduce the dependence on biofuels and fossil fuels imports. At present, approximately 10 hydropower projects (the largest hydropower project has an installed capacity of 456 MW, which is also the largest hydropower project of Nepal) in the lower reaches of Tama Koshi permits have been issued, of which 6 are under construction (Figure 8). However, eight rapidly expanding glacial lakes are developed in the basin, many of which are close to the roads, village, and other infrastructures, the straight-line distance between them and the nearest hydropower station is about 15 km. Future climate warming will amplify the consequence and increase the risk of GLOFs hazard [14,40,42,70]. Meanwhile, with the improved living standard and increased public awareness of disaster prevention and mitigation (increased awareness of GLOF hazards) in the Tama Koshi basin, numerous residents have migrated from rural areas (the upper valley) to urban areas, resulting in a significant increase in population density in downstream cities (Figure 8). This avoids some small-scale GLOFs but increases the effect when large-scale GLOFs occur. For example, a recent study [60] has shown that the Tsho Rolpa lake elevation has increased in recent years. If the lake area gradually increases, as we expect (see Section 5.3), the water pressure on the dam body will be greatly increased. Once a large-scale outburst occurs, it will bring disastrous effects to the highly populated areas in Lamobagar.

Remote sensing is a vital tool in quickly monitor the glacial dynamics and hazard risk assessment [71]. While this study has obtained detailed information on glacial lakes area changes, basic data on glacial lakes (e.g., lake basin topography, type of dam construction material, the internal structure of moraine dams, etc.) are still lacking and field investigations are needed. To date, the lake basin parameters and dam characteristics were measured only at Tsho Rolpa in Nepal [39]. Detail investigations and research are relatively rarely conducted in the Chinese Tibet part, where some glacial lakes with rapid expansion and their GLOFs could lead to transnational impacts on downstream Nepal. For those lakes with high risks (e.g., Niangzongmajue and Yanongcuo in China), hazard assessment is crucial, and, if necessary, artificial drainage measurements could be adopted to reduce their GLOF risks. For example, Tsho Rolpa in Nepal has been automatically drained by the construction of an open channel and siphon since 1995, and its outburst risk has been basically controlled, and even keep stable during the Gorkha earthquake in 2015 [72].

6. Conclusions and Remarks

Using 42 moderate- and high-resolution remote sensing images between 1964 and 2020, we completed a comprehensive investigation of glaciers and glacial lakes of the entire Tama Koshi basin, central Himalayas. We mapped a total of 271 glaciers and 196 glacial lakes with coverage of 329.2 ± 1.9 km^2 and 14.4 ± 0.3 km^2, respectively, in 2020 across the basin. During the past 56 years, the total area of glaciers in the basin has shrunk by 26.2 ± 3.2 km^2 (0.13% a^{-1}) and thinned by ~20 m on average, with their mean velocity has gradually slowed down from 5.3 m a^{-1} in 1999–2003 to 4.0 m a^{-1} in 2013–2015. The total area of glacial lakes increased by 9.2 ± 0.4 km^2 (~180%), with ice-contacted proglacial lakes showing a much higher expansion rate (~204%) than the others. Moreover, large-scale glacial lakes are developed preferentially and experienced rapid expansion on the east side of the basin. For land-terminating debris-covered glaciers, the decrease in ice thickness is the main cause of its mass loss. However, for lake-terminating glaciers, rapid calving retreat of the terminus was the dominant process. Preliminary analysis shows that the increase in temperature and precipitation caused the glacier shrinkage and glacial lake expansion in the Tama Koshi basin. Meanwhile, topography also acted as a vital factor that not only affects the expansion of glacial lakes but also controls the development of glacial lakes. We hypothesize that lake expansion will continue in some cases until the critical local topography of steepening icefall is reached, but the number of lakes may not necessarily increase. Many residents (~58 per km^2 in 2018) and hydropower stations (~10 in 2020) are sited in the lower reaches of the basin. The impact of the hazard will be amplified if a transnational GLOF occurs. To better assess and reduce the GLOF risks, further detailed monitoring should be focused on the eight rapidly expanding proglacial lakes in the upper reaches in Chinese Tibet Rongxer, e.g., obtaining the basic information (such as lake bathymetry, lake area/level fluctuation, dam characteristics and downstream high-resolution terrain) of these lakes via field and in situ investigations.

Supplementary Materials: The following are available online at https://www.mdpi.com/article/10.3390/rs13183614/s1, Figure S1: An example of extracting the rapid expansion glacial lakes on Landsat images using Google Earth Engine Python API, Figure S2: Identified Lake A GLOF event by

PlanetScope imagery, Figure S3: Supraglacial lakes and ice cliffs on Glacier M base on PlanetScope image, Table S1: Data used for this study.

Author Contributions: Conceptualization, Q.L.; Data curation, Y.Z.; Formal analysis, Y.Z.; Funding acquisition, Q.L. and Y.W.; Investigation, Y.Z.; Methodology, Y.Z.; Project administration, Q.L. and Y.W.; Visualization, Y.Z.; Writing—original draft, Y.Z.; Writing—review and editing, Y.Z., Q.L., L.S., Y.L., H.W., H.L. and Y.W. All authors have read and agreed to the published version of the manuscript.

Funding: This research was funded by the "Chinese Academy of Sciences Overseas Institutions Platform Project, grant number 131C11KYSB20200033" and the "IAEA Interregional Technical Cooperation Project, grant number INT/5/156"; Our research was also supported by the "National Natural Science Foundation of China Project, grant number 41871069" and the "Sichuan Science and Technology Programs, grant number 2021JDJQ0009 and 2020JDJQ0002".

Data Availability Statement: The data sets of the glacier and glacial lake in the Tama Koshi basin and the Python code for this study can be found and downloaded from https://doi.org/10.5281/zenodo.5496959.

Acknowledgments: The authors gratefully acknowledge the PlanetLab for the provision of PlanetScope imagery and U.S. Geological Survey for Landsat and Keyhole images, and the authors would also like to thank developers in the GEE community and Qiusheng Wu from the University of Tennessee for sharing very helpful sample codes.

Conflicts of Interest: The authors declare no conflict of interest.

References

1. Siegert, M.J. Role of Glaciers and Ice Sheets in Climate and the Global Water Cycle. *Encycl. Hydrol. Sci.* **2006**, *4*, 164. [CrossRef]
2. Kääb, A.; Berthier, E.; Nuth, C.; Gardelle, J.; Arnaud, Y. Contrasting patterns of early twenty-first-century glacier mass change in the Himalayas. *Nature* **2012**, *488*, 495–498. [CrossRef] [PubMed]
3. Carton, A.; Baroni, C. The Adamello-Presanella and Brenta Massifs, Central Alps: Contrasting High-Mountain Landscapes and Landforms. In *Landscapes and Landforms of Italy*; Soldati, M., Marchetti, M., Eds.; Springer International Publishing: Cham, Switzerland, 2017; pp. 101–112.
4. Jansson, P.; Rosqvist, G.; Schneider, T. Glacier fluctuations, suspended sediment flux and glacio-lacustrine sediments. *Geogr. Ann. Ser. A Phys. Geogr.* **2005**, *87*, 37–50. [CrossRef]
5. Huss, M.; Bookhagen, B.; Huggel, C.; Jacobsen, D.; Bradley, R.S.; Clague, J.J.; Vuille, M.; Buytaert, W.; Cayan, D.R.; Greenwood, G.; et al. Toward mountains without permanent snow and ice. *Earth's Future* **2017**, *5*, 418–435. [CrossRef]
6. Nie, Y.; Zhang, Y.; Ding, M.; Liu, L.; Wang, Z. Lake change and its implication in the vicinity of Mt. Qomolangma (Everest), central high Himalayas, 1970–2009. *Environ. Earth Sci.* **2012**, *68*, 251–265. [CrossRef]
7. Gardner, A.S.; Moholdt, G.; Cogley, J.G.; Wouters, B.; Arendt, A.A.; Wahr, J.; Berthier, E.; Hock, R.; Pfeffer, W.T.; Kaser, G.; et al. A Reconciled Estimate of Glacier Contributions to Sea Level Rise: 2003 to 2009. *Science* **2013**, *340*, 852–857. [CrossRef] [PubMed]
8. Wu, G.; Yao, T.; Wang, W.; Zhao, H.; Yang, W.; Zhang, G.; Li, S.; Yu, W.; Lei, Y.; Hu, W. Glacial hazards on Tibetan Plateau and surrounding alpines. *Bull. Chin. Acad. Sci.* **2019**, *34*, 1285–1292. [CrossRef]
9. Immerzeel, W.W.; Beek, L.P.H.v.; Bierkens, M.F.P. Climate change will affect the Asian water towers. *Science* **2010**, *328*, 1382–1385. [CrossRef] [PubMed]
10. Immerzeel, W.W.; Lutz, A.F.; Andrade, M.; Bahl, A.; Biemans, H.; Bolch, T.; Hyde, S.; Brumby, S.; Davies, B.J.; Elmore, A.C.; et al. Importance and vulnerability of the world's water towers. *Nature* **2020**, *577*, 364–369. [CrossRef]
11. Nie, Y.; Sheng, Y.; Liu, Q.; Liu, L.; Liu, S.; Zhang, Y.; Song, C. A regional-scale assessment of Himalayan glacial lake changes using satellite observations from 1990 to 2015. *Remote Sens. Environ.* **2017**, *189*, 1–13. [CrossRef]
12. Wang, X.; Guo, X.; Yang, C.; Liu, Q.; Wei, J.; Zhang, Y.; Liu, S.; Zhang, Y.; Jiang, Z.; Tang, Z. Glacial lake inventory of high-mountain Asia in 1990 and 2018 derived from Landsat images. *Earth Syst. Sci. Data* **2020**, *12*, 2169–2182. [CrossRef]
13. Pfeffer, W.T.; Arendt, A.A.; Bliss, A.; Bolch, T.; Cogley, J.G.; Gardner, A.S.; Hagen, J.-O.; Hock, R.; Kaser, G.; Kienholz, C.; et al. The Randolph Glacier Inventory: A globally complete inventory of glaciers. *J. Glaciol.* **2017**, *60*, 537–552. [CrossRef]
14. Nie, Y.; Pritchard, H.D.; Liu, Q.; Hennig, T.; Wang, W.; Wang, X.; Liu, S.; Nepal, S.; Samyn, D.; Hewitt, K.; et al. Glacial change and hydrological implications in the Himalaya and Karakoram. *Nat. Rev. Earth Environ.* **2021**, *2*, 91–106. [CrossRef]
15. Yao, T.; Yu, W.; Wu, G.; Xu, B.; Yang, W.; Zhao, H.; Wang, W.; Li, S.; Wang, N.; Li, Z.; et al. Glacier anomalies and relevant disaster risks on the Tibetan Plateau and surroundings. *Chin. Sci. Bull.* **2019**, *64*, 2770–2782.
16. Bolch, T.; Kulkarni, A.; Kääb, A.; Huggel, C.; Paul, F.; Cogley, J.G.; Frey, H.; Kargel, J.S.; Fujita, K.; Scheel, M.; et al. The State and Fate of Himalayan Glaciers. *Science* **2012**, *336*, 310. [CrossRef] [PubMed]
17. Brun, F.; Berthier, E.; Wagnon, P.; Kääb, A.; Treichler, D. A spatially resolved estimate of High Mountain Asia glacier mass balances from 2000 to 2016. *Nat. Geosci.* **2017**, *10*, 668–673. [CrossRef]

18. Maurer, J.M.; Schaefer, J.M.; Rupper, S.; Corley, A. Acceleration of ice loss across the Himalayas over the past 40 years. *Sci. Adv.* **2019**, *5*, eaav7266. [CrossRef] [PubMed]
19. Westoby, M.J.; Glasser, N.F.; Brasington, J.; Hambrey, M.J.; Quincey, D.J.; Reynolds, J.M. Modelling outburst floods from moraine-dammed glacial lakes. *Earth-Sci. Rev.* **2014**, *134*, 137–159. [CrossRef]
20. Harrison, S.; Kargel, J.S.; Huggel, C.; Reynolds, J.; Shugar, D.H.; Betts, R.A.; Emmer, A.; Glasser, N.; Haritashya, U.K.; Klimeš, J.; et al. Climate change and the global pattern of moraine-dammed glacial lake outburst floods. *Cryosphere* **2018**, *12*, 1195–1209. [CrossRef]
21. Bajracharya, S.; Maharjan, S.B.; Shrestha, F.; Sherpa, T.C. *Inventroy of Glacial Lakes and Identification of Potentially Dangerous Glacial Lakes in the Koshi, Gandaki, and Karnali River Basins of Nepal, the Tibet Autonomous Region of China, and India*; Research Report; ICIMOD, UNDP: Kathmandu, Nepal, 2020.
22. Cook, K.L.; Andermann, C.; Gimbert, F.; Adhikari, B.R.; Hovius, N. Glacial lake outburst floods as drivers of fluvial erosion in the Himalaya. *Science* **2018**, *362*, 53. [CrossRef]
23. Wang, W.; Yao, T.; Yang, X. Variations of glacial lakes and glaciers in the Boshula mountain range, southeast Tibet, from the 1970s to 2009. *Ann. Glaciol.* **2017**, *52*, 9–17. [CrossRef]
24. Yao, X.; Liu, S.; Sun, M.; Zhang, X. Study on the glacial lake outburst flood events in Tibet since the 20th Century. *J. Nat. Resour.* **2014**, *29*, 1377–1390.
25. Liu, Q.; Guo, W.; Nie, Y.; Liu, S.; Xu, J. Recent glacier and glacial lake changes and their interactions in the Bugyai Kangri, southeast Tibet. *Ann. Glaciol.* **2016**, *57*, 61–69. [CrossRef]
26. Allen, S.K.; Zhang, G.; Wang, W.; Yao, T.; Bolch, T. Potentially dangerous glacial lakes across the Tibetan Plateau revealed using a large-scale automated assessment approach. *Sci. Bull.* **2019**, *64*, 435–445. [CrossRef]
27. Liu, J.; Zhang, J.; Gao, B.; Li, Y.; Mengyu, L.; Wujin, D.; Zhou, L. An overview of glacial lake outburst flood in Tibet, China. *J. Glaciol. Geocryol.* **2019**, *41*, 1335–1347.
28. Fan, J.; An, C.; Zhang, X.; Li, X.; Tan, J. Hazard assessment of glacial lake outburst floods in Southeast Tibet based on RS and GIS technologies. *Int. J. Remote Sens.* **2019**, *40*, 4955–4979. [CrossRef]
29. Cui, P.; Dang, C.; Cheng, Z.; Scott, K.M. Debris Flows Resulting from Glacial-Lake Outburst Floods in Tibet, China. *Phys. Geogr.* **2010**, *31*, 508–527. [CrossRef]
30. Cheng, Z.L.; Liu, J.J.; Liu, J.K. Debris flow induced by glacial lake break in southeast Tibet. *WIT Trans. Eng. Sci.* **2010**, *67*, 101–111. [CrossRef]
31. Avian, M.; Bauer, C.; Schlögl, M.; Widhalm, B.; Gutjahr, K.-H.; Paster, M.; Hauer, C.; Frießenbichler, M.; Neureiter, A.; Weyss, G.; et al. The Status of Earth Observation Techniques in Monitoring High Mountain Environments at the Example of Pasterze Glacier, Austria: Data, Methods, Accuracies, Processes, and Scales. *Remote Sens.* **2020**, *12*, 1251. [CrossRef]
32. Chen, F.; Zhang, M.; Tian, B.; Li, Z. Extraction of Glacial Lake Outlines in Tibet Plateau Using Landsat 8 Imagery and Google Earth Engine. *IEEE J. Sel. Top. Appl. Earth Obs. Remote Sens.* **2017**, *10*, 4002–4009. [CrossRef]
33. Haritashya, U.K.; Kargel, J.S.; Shugar, D.H.; Leonard, G.J.; Strattman, K.; Watson, C.S.; Shean, D.; Harrison, S.; Mandli, K.T.; Regmi, D. Evolution and Controls of Large Glacial Lakes in the Nepal Himalaya. *Remote Sens.* **2018**, *10*, 798. [CrossRef]
34. Wu, S.; Yao, Z.; Huang, H.; Liu, Z.; Liu, G. Responses of glaciers and glacial lakes to climate variation between 1975 and 2005 in the Rongxer basin of Tibet, China and Nepal. *Reg. Environ. Chang.* **2012**, *12*, 887–898. [CrossRef]
35. Rounce, D.; Watson, C.; McKinney, D. Identification of Hazard and Risk for Glacial Lakes in the Nepal Himalaya Using Satellite Imagery from 2000–2015. *Remote Sens.* **2017**, *9*, 654. [CrossRef]
36. Shilpakar, R.; Shakya, N.M.; Hiratsuka, A. Impact of Climate Change on Snowmelt Runoff: A Case Study of Tamakoshi Basin in Nepal. In Proceedings of the International Symposium on Society for Social Management Systems, SSMS, Kochi, Japan, 5–7 March 2009. SMS09-124.
37. Khadka, D.; Babel, M.S.; Shrestha, S.; Tripathi, N.K. Climate change impact on glacier and snow melt and runoff in Tamakoshi basin in the Hindu Kush Himalayan (HKH) region. *J. Hydrol.* **2014**, *511*, 49–60. [CrossRef]
38. Reynolds, J.M. Glacial hazard assessment at Tsho Rolpa, Rolwaling, Central Nepal. *Q. J. Eng. Geol. Hydrogeol.* **1999**, *32*, 209. [CrossRef]
39. Sakai, A.; Chikita, K.; Yamada, T. Expansion of a moraine-dammed glacial lake, Tsho Rolpa, in Rolwaling Himal, Nepal Himalaya. *Limnol. Oceanogr.* **2000**, *45*, 1401–1408. [CrossRef]
40. Shrestha, B.B.; Nakagawa, H.; Kawaike, K.; Zhang, H. Glacial and Sediment Hazards in the Rolwaling Valley, Nepal. *Int. J. Eros. Control Eng.* **2012**, *5*, 123–133. [CrossRef]
41. Dahal, K.R.; Hagelman, R. People's risk perception of glacial lake outburst flooding: A case of Tsho Rolpa Lake, Nepal. *Environ. Hazards* **2011**, *10*, 154–170. [CrossRef]
42. Shrestha, S.; Bajracharya, A.R.; Babel, M.S. Assessment of risks due to climate change for the Upper Tamakoshi Hydropower Project in Nepal. *Clim. Risk Manag.* **2016**, *14*, 27–41. [CrossRef]
43. Khadka, M.; Kayastha, R.B.; Kayastha, R. Future projection of cryospheric and hydrologic regimes in Koshi River basin, Central Himalaya, using coupled glacier dynamics and glacio-hydrological models. *J. Glaciol.* **2020**, *66*, 831–845. [CrossRef]
44. Randolph Glacier Inventory Consortium. *Randolph Glacier Inventory (RGI)—A Dataset of Global Glacier Outlines: Version 6.0*; Technical Report; Global Land Ice Measurements from Space; Digital Media: Boulder, CO, USA, 2017. [CrossRef]

45. Dehecq, A.; Gourmelen, N.; Gardner, A.S.; Brun, F.; Goldberg, D.; Nienow, P.W.; Berthier, E.; Vincent, C.; Wagnon, P.; Trouvé, E. Twenty-first century glacier slowdown driven by mass loss in High Mountain Asia. *Nat. Geosci.* **2019**, *12*, 22–27. [CrossRef]

46. Rose, A.N.; McKee, J.J.; Urban, M.L.; Bright, E.A.; Sims, K.M. *LandScan 2018*; Oak Ridge National Laboratory: Oak Ridge, TN, USA, 2019.

47. Bright, E.A.; Coleman, P.R. *LandScan 2000*; Oak Ridge National Laboratory: Oak Ridge, TN, USA, 2001.

48. Surazakov, A.; Aizen, V. Positional accuracy evaluation of declassified hexagon KH-9 mapping camera imagery. *Photogramm. Eng. Remote Sens.* **2010**, *76*, 603–608. [CrossRef]

49. Goerlich, F.; Bolch, T.; Mukherjee, K.; Pieczonka, T. Glacier mass loss during the 1960's and 1970's in the Ak-Shirak range (Kyrgyzstan) from multiple stereoscopic Corona and Hexagon imagery. *Remote Sens.* **2017**, *9*, 275. [CrossRef]

50. Gorelick, N.; Hancher, M.; Dixon, M.; Ilyushchenko, S.; Thau, D.; Moore, R. Google Earth Engine: Planetary-scale geospatial analysis for everyone. *Remote Sens. Environ.* **2017**, *202*, 18–27. [CrossRef]

51. Wu, Q. geemap: A Python package for interactive mapping with Google Earth Engine. *J. Open Source Softw.* **2020**, *5*, 2305. [CrossRef]

52. Nie, Y.; Liu, Q.; Liu, S. Glacial lake expansion in the central Himalayas by Landsat images, 1990–2010. *PLoS ONE* **2013**, *8*, e83973. [CrossRef]

53. Salerno, F.; Thakuri, S.; D'Agata, C.; Smiraglia, C.; Manfredi, E.C.; Viviano, G.; Tartari, G. Glacial lake distribution in the Mount Everest region: Uncertainty of measurement and conditions of formation. *Glob. Planet. Chang.* **2012**, *92*, 30–39. [CrossRef]

54. Mertes, J.R.; Thompson, S.S.; Booth, A.D.; Gulley, J.D.; Benn, D.I. A conceptual model of supra-glacial lake formation on debris-covered glaciers based on GPR facies analysis. *Earth Surf. Process. Landf.* **2017**, *42*, 903–914. [CrossRef]

55. Hersbach, H.; Bell, B.; Berrisford, P.; Biavati, G.; Horányi, A.; Muñoz Sabater, J.; Nicolas, J.; Peubey, C.; Radu, R.; Rozum, I.; et al. ERA5 Monthly Averaged Data on Single Levels from 1979 to Present. Copernicus Climate Change Service (C3S) Climate Data Store (CDS). 2019. Available online: https://cds.climate.copernicus.eu/cdsapp#!/dataset/10.24381/cds.f17050d7?tab=overview (accessed on 26 October 2020).

56. Yao, T.; Thompson, L.; Yang, W.; Yu, W.; Gao, Y.; Guo, X.; Yang, X.; Duan, K.; Zhao, H.; Xu, B.; et al. Different glacier status with atmospheric circulations in Tibetan Plateau and surroundings. *Nat. Clim. Chang.* **2012**, *2*, 663–667. [CrossRef]

57. Maussion, F.; Scherer, D.; Mölg, T.; Collier, E.; Curio, J.; Finkelnburg, R. Precipitation Seasonality and Variability over the Tibetan Plateau as Resolved by the High Asia Reanalysis. *J. Clim.* **2014**, *27*, 1910–1927. [CrossRef]

58. Pelto, M.; Panday, P.; Matthews, T.; Maurer, J.; Perry, L.B. Observations of Winter Ablation on Glaciers in the Mount Everest Region in 2020–2021. *Remote Sens.* **2021**, *13*, 2692. [CrossRef]

59. Song, C.; Huang, B.; Richards, K.; Ke, L.; Hien Phan, V. Accelerated lake expansion on the Tibetan Plateau in the 2000s: Induced by glacial melting or other processes? *Water Resour. Res.* **2014**, *50*, 3170–3186. [CrossRef]

60. Thakuri, S.; Chauhan, R.; Baskota, P. Glacial Hazards and Avalanches in High Mountains of Nepal Himalaya. *J. Tour. Himalaya Adventures* **2020**, *2*, 87–104.

61. Liu, Q.; Mayer, C.; Wang, X.; Nie, Y.; Wu, K.; Wei, J.; Liu, S. Interannual flow dynamics driven by frontal retreat of a lake-terminating glacier in the Chinese Central Himalaya. *Earth Planet. Sci. Lett.* **2020**, *546*, 116450. [CrossRef]

62. Basnett, S.; Kulkarni, A.V.; Bolch, T. The influence of debris cover and glacial lakes on the recession of glaciers in Sikkim Himalaya, India. *J. Glaciol.* **2013**, *59*, 1035–1046. [CrossRef]

63. King, O.; Dehecq, A.; Quincey, D.; Carrivick, J. Contrasting geometric and dynamic evolution of lake and land-terminating glaciers in the central Himalaya. *Glob. Planet. Chang.* **2018**, *167*, 46–60. [CrossRef]

64. Watson, C.S.; Kargel, J.S.; Shugar, D.H.; Haritashya, U.K.; Schiassi, E.; Furfaro, R. Mass Loss from Calving in Himalayan Proglacial Lakes. *Front. Earth Sci.* **2020**, *7*, 342. [CrossRef]

65. Miles, E.S.; Willis, I.; Buri, P.; Steiner, J.F.; Arnold, N.S.; Pellicciotti, F. Surface Pond Energy Absorption Across Four Himalayan Glaciers Accounts for 1/8 of Total Catchment Ice Loss. *Geophys. Res. Lett.* **2018**, *45*, 10464–10473. [CrossRef]

66. Steiner, J.F.; Buri, P.; Miles, E.S.; Ragettli, S.; Pellicciotti, F. Supraglacial ice cliffs and ponds on debris-covered glaciers: Spatio-temporal distribution and characteristics. *J. Glaciol.* **2019**, *65*, 617–632. [CrossRef]

67. Buri, P.; Miles, E.S.; Steiner, J.F.; Ragettli, S.; Pellicciotti, F. Supraglacial ice cliffs can substantially increase the mass loss of debris-covered glaciers. *Geophys. Res. Lett.* **2021**, *48*, e2020GL092150. [CrossRef]

68. Bolch, T.; Buchroithner, M.F.; Peters, J.; Baessler, M.; Bajracharya, S. Identification of glacier motion and potentially dangerous glacial lakes in the Mt. Everest region/Nepal using spaceborne imagery. *Nat. Hazards Earth Syst. Sci.* **2008**, *8*, 1329–1340. [CrossRef]

69. Nie, Y.; Liu, Q.; Wang, J.; Zhang, Y.; Sheng, Y.; Liu, S. An inventory of historical glacial lake outburst floods in the Himalayas based on remote sensing observations and geomorphological analysis. *Geomorphology* **2018**, *308*, 91–106. [CrossRef]

70. Terink, W.; Immerzeel, W.W.; Lutz, A.F.; Droogers, P.; Khanal, S.; Nepal, S.; Shrestha, A.B. Hydrological and Climate Change Assessment for Hydropower Development in the Tamakoshi River Basin, Nepal. 2017. Available online: https://www.futurewater.nl/wp-content/uploads/2017/05/final_report_WT_v2.pdf (accessed on 30 October 2020).

71. Quincey, D.J.; Lucas, R.M.; Richardson, S.D.; Glasser, N.F.; Hambrey, M.J.; Reynolds, J.M. Optical remote sensing techniques in high-mountain environments: Application to glacial hazards. *Prog. Phys. Geogr. Earth Environ.* **2005**, *29*, 475–505. [CrossRef]
72. Kargel, J.S.; Leonard, G.J.; Shugar, D.H.; Haritashya, U.K.; Bevington, A.; Fielding, E.J.; Fujita, K.; Geertsema, M.; Miles, E.S.; Steiner, J.; et al. Geomorphic and geologic controls of geohazards induced by Nepal's 2015 Gorkha earthquake. *Science* **2016**, *351*, 140–150. [CrossRef] [PubMed]

Article

The Relative Contributions of Climate and Grazing on the Dynamics of Grassland NPP and PUE on the Qinghai-Tibet Plateau

Huilin Yu, Qiannan Ding, Baoping Meng, Yanyan Lv, Chang Liu, Xinyu Zhang, Yi Sun, Meng Li * and Shuhua Yi

School of Geographic Sciences, Nantong University, Nantong 226007, China;
2017320004@stmail.ntu.edu.cn (H.Y.); 1822011046@stmail.ntu.edu.cn (Q.D.); mengbp09@lzu.edu.cn (B.M.);
lvyy18@ntu.edu.cn (Y.L.); 1921110165@stmail.ntu.edu.cn (C.L.); 1822011018@stmail.ntu.edu.cn (X.Z.);
sunyi@ntu.edu.cn (Y.S.); yis@ntu.edu.cn (S.Y.)
* Correspondence: limeng@ntu.edu.cn

Abstract: Net primary productivity (NPP) and precipitation-use efficiency (PUE) are crucial indicators in understanding the responses of vegetation to global change. However, the relative contributions of climate change and human interference to the dynamics of NPP and PUE remain unclear. During the past few decades, the impacts of climate change and human activities on alpine grasslands on the Qinghai-Tibet Plateau (QTP) have been intensifying. The aims of the study were to investigate the spatiotemporal patterns of grassland NPP and PUE on the QTP during 2000–2017 and quantify how much of the variance in NPP and PUE can be attributed to the climatic factors (precipitation and temperature) and grazing intensity. The results showed that: (1) grassland NPP significantly increased with a rate of 0.6 g C m^{-2} year^{-1} over the past 18 years, mainly induced by the increased temperature and the enhanced precipitation. The temperature was the dominant factor for NPP interannual variation in mid-eastern QTP, and precipitation restrained vegetation growth most in the southwest and northeast. (2) The PUE was higher on the eastern and western parts of the plateau, but lower at the center. Regarding grassland types, the PUE of alpine steppe (0.19 g C m^{-2} mm^{-1}) was significantly lower than those of alpine meadow (0.31 g C m^{-2} mm^{-1}) and desert steppe (0.32 g C m^{-2} mm^{-1}). (3) Precipitation was significantly and negatively correlated with PUE and contributed the most to the temporal variation of grassland PUE on the QTP (52.7%). (4) Furthermore, we found that the grazing activities had the lowest contributions to both NPP and PUE interannual variation, compared to temperature and precipitation. Thus, it is suggested that climate variability rather than grazing activities dominated vegetation changes on the QTP.

Keywords: alpine grassland; carbon and water cycles; climate change; human activities; Tibetan Plateau

Citation: Yu, H.; Ding, Q.; Meng, B.; Lv, Y.; Liu, C.; Zhang, X.; Sun, Y.; Li, M.; Yi, S. The Relative Contributions of Climate and Grazing on the Dynamics of Grassland NPP and PUE on the Qinghai-Tibet Plateau. *Remote Sens.* 2021, 13, 3424. https://doi.org/10.3390/rs13173424

Academic Editors: Adrian Ursu and Cristian Constantin Stoleriu

Received: 5 August 2021
Accepted: 26 August 2021
Published: 28 August 2021

1. Introduction

Due to climate change and human activities, terrestrial ecosystems have undergone unprecedented changes, including the changes of carbon sources and sinks [1], land cover and land use [2–4], and loss of biodiversity [5]. The human–climate–vegetation interaction has become one of the major environmental paradigms [6,7]. Particularly in arid and semi-arid regions, anthropogenic activities and climate change can easily cause ecosystem degradation [8]. The grassland ecosystem is one of the most widely distributed vegetation types, accounting for about one-fifth of the world's surface area [9]. It plays a vital role in maintaining biogeochemical cycles, protecting biodiversity, and supporting animal husbandry and food production [10]. Due to global warming and overgrazing, nearly 50% of global grasslands have degraded [11]. Grassland degradation is threatening many habitats and even affecting ecological security. Although the dynamics of grasslands have been explored by intensive studies, the studies on relative contributions of climatic and anthropogenic drivers in influencing grassland variations are limited.

Net primary productivity (NPP) represents the amount of carbon fixed by plants in a certain period of time and characterizes the energy production and conversion status [12,13]. Precipitation-use efficiency (PUE) is the ratio of NPP to precipitation, and reflects the relationship between energy production and water consumption [14]. NPP and PUE are essential indicators for the relationship between terrestrial water and carbon cycles [15]. NPP is positively correlated with precipitation in almost all terrestrial ecosystems [16], particularly in grassland ecosystems [17]. Accelerated warming can also promote NPP by prolonging the growing season [18,19]. PUE is generally considered to decrease with spatial precipitation [15]. It has also been suggested that PUE decreases from arid biomes to humid biomes, changing with the level of precipitation [20]. Meanwhile, human activities can profoundly affect vegetation changes and even cause detrimental or catastrophic damage to plant growth [21]. For example, overgrazing could lead to decreased vegetation productivity and trigger severe grassland degradation [22,23]. Therefore, understanding how NPP and PUE respond to climate and grazing is essential for mitigating environmental damages.

The Qinghai-Tibet Plateau (QTP) is the highest Plateau in the world, with an average elevation of higher than 4000 m. Alpine grasslands cover most areas of the QTP and feature fragile environmental conditions [24,25]. The air temperature has increased about twice as fast as the global average during the past three decades [26], and the livestock number almost tripled in the 1990s from the 1950s [27]. The dramatic climate change and intensive grazing activities have caused half of the grasslands to degrade [28]. Recently, studies on how climate change and human activities influence the changes in NPP or PUE on the Tibetan Plateau are emerging. For example, Chen et al. [29] found that human activities dominated the NPP variations after 2001, and Lehnert et al. [30] suggested that climate variability was the primary driver for NPP changes. In terms of PUE, Yang et al. [31] stated that spatial PUE in Tibetan grasslands exhibited a unimodal pattern across broad precipitation gradients. However, Zhao et al. [32] argued that the spatial PUE had a U-shaped relationship with precipitation. These differences may be caused by the different climate gradients and human activity intensity levels [32]. Therefore, it is important to identify the spatiotemporal changes of NPP and PUE across the QTP and quantify the relative contributions of the climate factors and human activities to their dynamics.

The study aimed to answer: (1) How did grassland NPP and PUE vary from 2000 to 2017? (2) How much can the variances in NPP and PUE be attributed to the corresponding climatic factors and grazing intensity, respectively? To answer these questions, firstly, we simulated grassland NPP and PUE across the QTP from 2000 to 2017 and explored the changes in climatic factors (precipitation and temperature) and in grazing intensity over time. Secondly, we investigated the relationships of NPP and PUE with temperature, precipitation, and grazing intensity. Finally, we quantified the relative importance of climate and grazing to influencing interannual NPP and PUE variations.

2. Materials and Methods

2.1. Study Area

The QTP is located in southwestern China and accounts for more than 26.8% of the total land area of China (Figure 1a). Known as the "roof of the world" and the "Asian water tower", this plateau is a crucial security barrier for China and even Asia [33]. The QTP lies in the alpine zone and has a mean annual temperature below 0 °C. The distribution of precipitation is uneven. The mean annual precipitation is greater than 1000 mm in the southeastern plateau and less than 100 mm in the northwestern part [34]. These unique climatic conditions have developed various alpine ecosystems on the QTP. The grasslands are mainly composed of humid alpine meadow, semi-arid alpine steppe, and arid desert steppe (Figure 1b). The grassland areas on the QTP cover 1.3×10^8 hectares, accounting for about 32.5% of the grassland area in China [35]. Herders have existed on rgw QTP for tens of thousands of years, and livestock husbandry is their primary source of income [26].

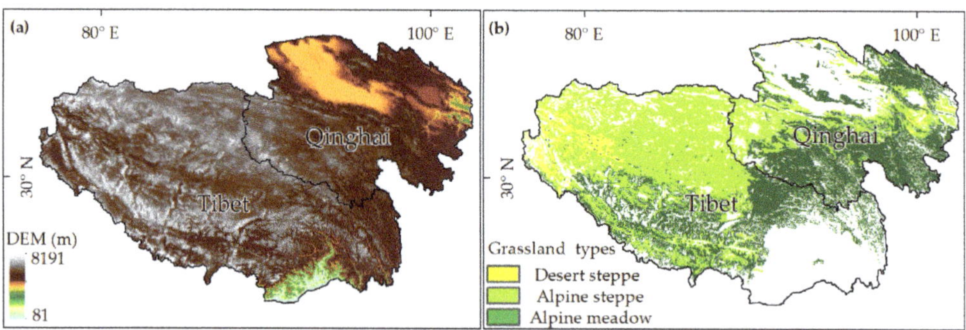

Figure 1. (**a**) The altitude of the study region and (**b**) the spatial pattern of the grassland types. DEM indicates digital elevation model.

2.2. Data Collection and Process Method

The monthly MOD13A3C6 normalized difference vegetation index (NDVI) data were obtained from the NASA LP DAAC (Land Processes Distributed Active Archive Center) website (https://lpdaac.usgs.gov/get_data/data_pool, accessed on 1 May 2020). This dataset were calculated by the maximum value composite method at a 1000 m resolution and calibrated for geometrical, atmospheric, and radiation influences. The MODIS Re-projection Tool (MRT) was used to define geographical coordinates as WGS1984 and the projection coordinate as Albers conical equal area projection [36]. The NDVI data were handled with the TIMESAT software further to eliminate the effects of clouds and atmospheric contamination. This is a software package for processing and assessing the time-series data of vegetation dynamics, and curve fits and runs phenological/seasonality metrics [37,38]. Based on an asymmetric Gaussian function-fitting, we adopted the Savitzky–Golay (SG) smoothing model to eliminate the influence of random noise, using data filtering and reconstruction [39].

The meteorological data, including daily temperature, precipitation, and sunshine duration, were downloaded from the National Meteorological Information Center of China Meteorological Administration (http://geodata.cn, accessed on 1 May 2020). The radiation was calculated based on geographical position and sunshine duration [40]. Firstly, the daily climatic data were integrated into monthly data. Then, ANUSPLIN4.3 was used to interpolate monthly precipitation, temperature, and radiation into the raster with a spatial resolution of 1000 m [41].

Socio-economic data with livestock numbers from 2000 to 2017 were obtained from statistical yearbooks for the Tibet Autonomous Region and Qinghai province. We collected livestock inventory and the number of livestock slaughtered for each county on the QTP. Different types of livestock animals were converted to standard sheep units (SHU) based on a method in which one sheep or one goat is equal to one SHU, and one large livestock animal (yak, donkey, and horse) is equal to four SHUs [42]. Grassland distribution was determined by the China Vegetation Atlas with a scale of 1:1,000,000, which was derived from the Resource and Environment Data Cloud Platform (http://www.resdc.cn/data.aspx?DATAID=122, accessed on 1 May 2020).

2.3. NPP Simulation and Validation

Along with the development of remote sensing techniques and geographic information systems, some models based on light use efficiency (LUE) and remote sensing data were well developed to study NPP and estimate carbon dynamics at global or regional scales. Monteith et al. [43] and Potter et al. [44] found that vegetation productivity is correlated with the amount of photosynthetically active radiation absorbed or intercepted by green foliage and then built CASA (Carnegie–Ames–Stanford approach) model. The CASA runs on a monthly time interval to model the seasonal carbon cycle. In recent decades, this model has been modified by many researchers [45–47]. In CASA, NPP on grid cell x in month t is determined by absorbed photosynthetically active radiation (APAR) and the actual LUE (ε). The formulas are as follows:

$$NPP(x, t) = APAR\ (x, t) \times \varepsilon\ (x, t) \tag{1}$$

$$APAR(x, t) = SOL(x, t) \times FPAR(x, t) \times 0.5 \tag{2}$$

where x is spatial location; t is time; SOL is total solar radiation MJ/m^2 at grid cell x in month t; 0.5 represents the ratio between photosynthetically active solar radiation and total solar radiation. FPAR is the fraction of photosynthetically active radiation absorbed by vegetation, which can be quantified by NDVI and vegetable types.

The light energy conversion (ε) is influenced by water and temperature. The calculation formula is:

$$\varepsilon(x, t) = T_{\varepsilon 1}(x, t) \times T_{\varepsilon 2}(x, t) \times W_\varepsilon(x, t) \times \varepsilon^* \tag{3}$$

$T_{\varepsilon 1}(x, t)$ and $T_{\varepsilon 2}(x, t)$ represent temperature stress. $T_{\varepsilon 1}(x, t)$ reflects the inherent biochemical limitations on photosynthesis in plants at very low and very high temperatures. $T_{\varepsilon 2}(x, t)$ represents the change of the light energy conversion rate when the temperature changes from the optimum temperature to a high or low temperature. $W_\varepsilon(x, t)$ represents the effect of available water conditions on the light energy conversion. $W_\varepsilon(x, t)$ gradually increases with the amount of water available in the environment [45,48]. ε^* indicates the maximum possible LUE under ideal conditions. At the global scale, the value of ε^* is 0.389 g C·MJ^{-1} [49]. However, ε^* showed significant differences among vegetation types [50]. For alpine grasslands on the QTP, ε^* can be fixed at 0.56 g C·MJ^{-1} according to the finding by Zhang et al. [51].

$$T_{\varepsilon 1}(x, t) = 0.8 + 0.02 \times T_{opt}(x, t) - 0.0005 \times T_{opt}(x, t) \tag{4}$$

$$T_{\varepsilon 2}(x, t) = 1.1814/\{1 + exp[0.2 \times (T_{opt}(x, t) - 10 - T(x, t))]\}/\{1 + exp[0.3 \times (-T_{opt}(x, t) - 10 + T(x, t))]\} \tag{5}$$

$$W_\varepsilon(x, t) = 0.5 + 0.5 \times EET(x, t)\ /PET(x, t) \tag{6}$$

where $T_{opt}(x, t)$ is the monthly average temperature when the NDVI reaches the highest value. If the temperature on grid cell x in month t is below $-10\ °C$, $T_{\varepsilon 1}$ is equal to 0. PET refers to potential evapotranspiration (mm), and EET refers to estimated evapotranspiration (mm).

The performance of CASA model has been validated by observed NPP data in the previous study. We found that the simulated NPP matched well with the measurement [34,52].

2.4. PUE Calculation

PUE has been calculated in two common ways. First, according to PUE's definition, it was calculated directly by the ratio of NPP and precipitation [53]. Second, PUE was estimated based on the slope of the NPP–precipitation relationship [20]. We accepted the first method using the ratio of grassland NPP and corresponding precipitation (PPT) because this method has been widely used on the QTP [15,31]. The formula is as follows:

$$PUE = \frac{NPP}{PPT} \tag{7}$$

where PUE is precipitation use efficiency (g C m^{-2} mm^{-1}), NPP is net primary production (g C m^{-2}), and PPT is precipitation in one year (mm).

2.5. Grazing Intensity Calculation

Grazing intensity (GI) is usually calculated as the ratio of livestock number to natural grassland, and the formula is as follows:

$$GI = \frac{C_n + C_h}{A} \tag{8}$$

where C_n is the livestock inventory in a given year, and C_h is the number of livestock slaughtered in a given year. A represents the area of available natural grassland (ha) in each county.

2.6. Data Analysis

We took alpine grassland as the research object. Before further analysis, we eliminated the non-grassland pixels and null-value pixels due to unavailable livestock data in the center of QTP. These data are indicated by white color in the figures. We used the ordinary least squares (OLS) linear regression method to calculate the temporal trend in grassland NPP and PUE during 2000–2017 for each pixel. The spatial analysis was calculated by *raster package* (https://cran.r-project.org/web/packages/raster/index.html, accessed on 1 October 2020). The differences in NPP and PUE among different grassland types were analyzed by the analysis of variance (ANOVA) using *agricolae* package (https://cran. r-project.org/web/packages/agricolae/index.html, accessed on 1 October 2020). The Spearman correlation test was used to calculate the relationships of NPP and PUE with climate and grazing.

Recently, the correlation metrics [54], principal component analyses [27], and generalized linear model (GLM) [30] were used to discriminate the relative contributions of climate change and human activities to vegetation changes. Among them, the GLM is simple and widely recommended [30]. In this study, the relative contributions of grazing and climate change to the temporal dynamics of NPP and PUE in each pixel were decomposed from the GLM. In the GLM model, we selected NPP and PUE as response variablesl and mean annual temperature (MAT), mean annual precipitation (MAP), and grazing intensity (GI) as predictor variables. The percentage of NPP or PUE variance explained by each predictor variable in the GLM was interpreted as the contribution of each variable to NPP or PUE variation. All statistical analyses were performed in the environment R4.02 (https://www.r-project.org/) and mapped in ArcGIS 10.2 (ESRI, Inc. 380 New York Street Redlands, CA 92373 USA).

3. Results

3.1. Spatiotemporal Patterns of Climate Factors and Grazing Intensity

From 2000 to 2017, the MAT in the QTP increased significantly with a rate of 0.05 °C/year ($p < 0.01$) (Figure 2a), particularly in western Tibet and southwestern Qinghai Province (Figure 2b). The MAP also showed a slight but non-significant increasing trend at a rate of 1.8 mm/year ($p > 0.05$) (Figure 2c). However, the spatial pattern of MAP trend demonstrated heterogeneity across the plateau. MAP significantly increased in the northeast of the QTP, and non-significantly decreased in the south of the Plateau (Figure 2d). The mean GI of the QTP showed unimodal dynamics from 2000 to 2017. The highest GI was generated in 2007 (0.72 SSU/ha), and the lowest value occurred in 2016 (0.64 SHU/ha) (Figure 2e). Evident spatial variability of the GI trend was observed across the QTP. The GI decreased in the southwestern region but increased in the northeastern part (Figure 2f).

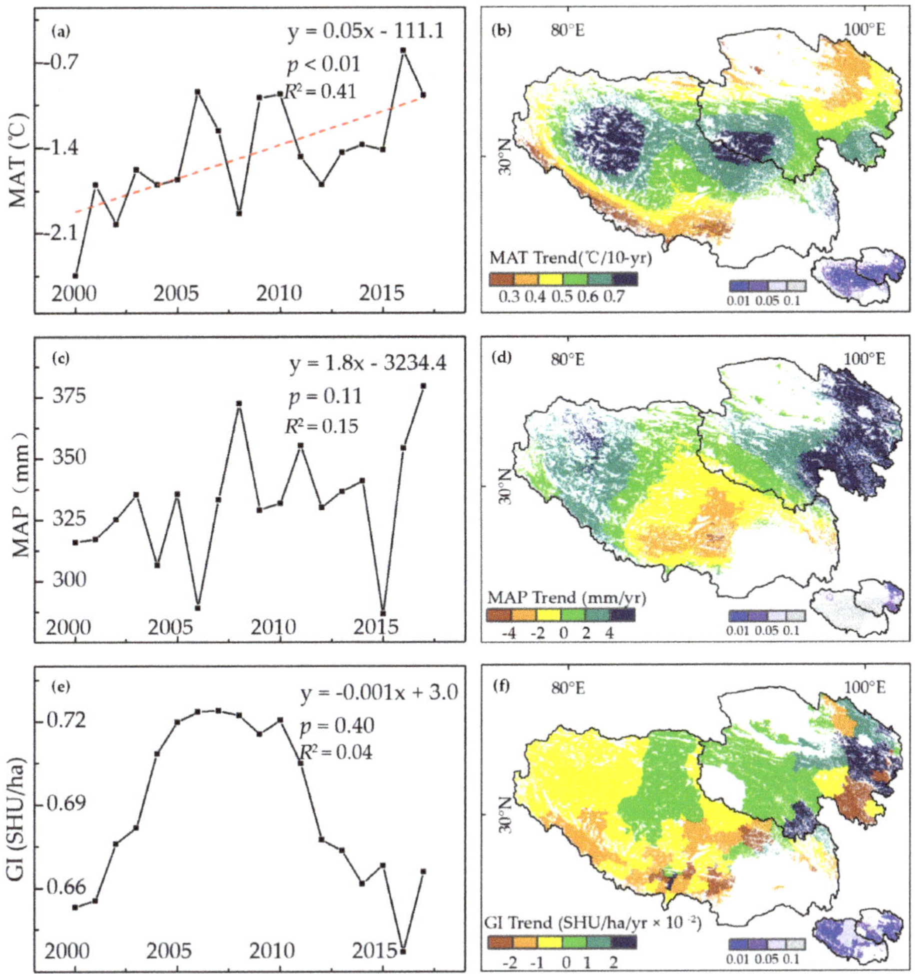

Figure 2. The spatiotemporal patterns of MAT (**a,b**), MAP (**c,d**), and GI (**e,f**) on the QTP from 2000 to 2017. MAT, MAP, and GI represent mean annual temperature, mean annual precipitation, and grazing intensity, respectively. Small maps at the bottom represent the ranges of significance levels.

3.2. Spatiotemporal Patterns of NPP

The overall mean annual grassland NPP during 2000–2017 on the QTP was 88.5 g C m^{-2}, and the highest and lowest NPP were found in 2010 (98.5 g C m^{-2}) and 2001 (80.7 g C m^{-2}), respectively. The spatial distribution of grassland NPP presented an increasing pattern from northwest to southeast of the QTP (Figure 3a). Alpine meadow on the southeastern plateau had the largest NPP (133.6 g C m^{-2}) due to the favorable hydrothermal conditions, followed by the alpine steppe (50.7 g C m^{-2}) and desert steppe (45.1 g C m^{-2}) (Table 1).

From 2000 to 2017, grassland NPP on the QTP increased significantly at a rate of 0.6 g C m^{-2} year^{-1} (Figure 3a). Statistically, NPP significantly increased with 31.34% of grasslands, mainly in the north of the plateau. Grasslands with a significant decreasing trend in NPP only accounted for 0.68% of the total area. In different grassland types, 45.3% of desert steppe showed a significant increased trend in NPP, almost twice than that of alpine meadow and 1.5 times of alpine steppe (Table 1).

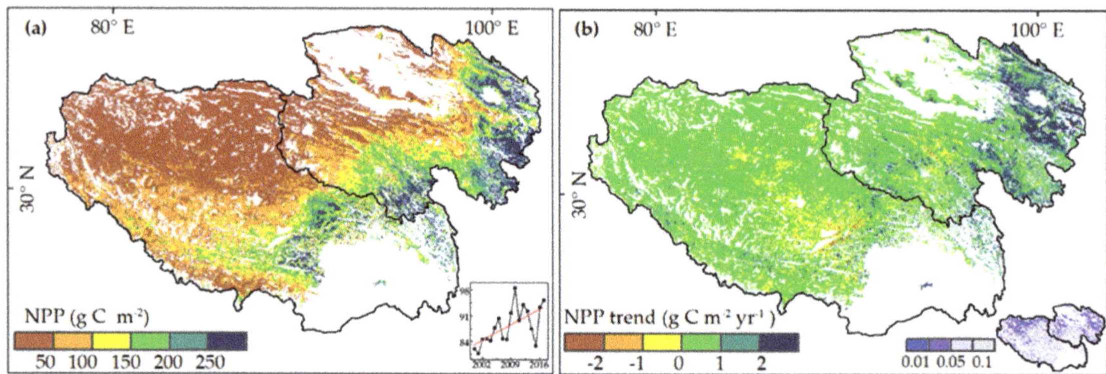

Figure 3. The spatiotemporal patterns of the grassland NPP on the QTP. (**a**) The mean of grassland NPP. The small map at the bottom shows the grassland NPP dynamics from 2000 to 2017. (**b**) The distribution of the trend in grassland NPP. The small map at the bottom shows the range of significance levels.

Table 1. Percentages of grasslands with different variation trends in NPP across the whole plateau and within different grassland types from 2000 to 2017. AM, AS, and DS indicate alpine meadow, alpine steppe, and desert steppe, respectively. Different letters (a, b, c) denote significant differences between them ($p < 0.05$).

	Mean of NPP (g C m⁻²)	Trend of NPP (%)			
		Significant Increasing	Insignificant Increasing	Significant Decreasing	Insignificant Decreasing
AM	133.6 ± 81.5 a	24.63	60.59	0.60	14.17
AS	50.7 ± 50.76 b	36.43	51.57	0.79	11.21
DS	45.1 ± 46.2 c	45.36	50.03	0.21	4.40
total	88.5 ± 78.4	31.34	55.65	0.68	12.33

3.3. Spatiotemporal Patterns of PUE

The grassland PUE showed an evident spatial pattern across the QTP (Figure 4a). The highest PUE were observed in the eastern and western plateau, with values higher than 0.5 g C m⁻² mm⁻¹, and the central plateau had the lowest PUE (<0.1 g C m⁻² mm⁻¹). From 2000 to 2017, the annual mean PUE was about 0.25 g C m⁻² mm⁻¹; the lowest value was 0.21 g C m⁻² mm⁻¹ in 2008 and the highest value was 0.28 g C m⁻² mm⁻¹ in 2009 (Table 2). At the grassland type level, the PUE of alpine steppe (0.19 g C m⁻² mm⁻¹) was significantly lower than those of alpine meadow (0.31 g C m⁻² mm⁻¹) and desert steppe (0.32 g C m⁻² mm⁻¹) (Table 2).

There was no significant change in the PUE during 2000–2017 (Figure 4b). PUE significantly increased and decreased only in 2.96% and 1.59% of grasslands. Within different grassland types, 3.36% of alpine steppe showed a significant increasing trend, and 2.50% of alpine meadow showed a significant decreasing trend in PUE (Table 2).

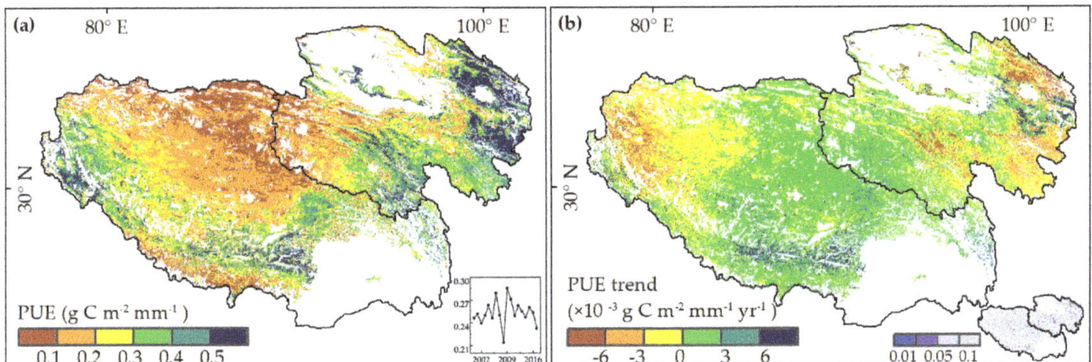

Figure 4. The spatiotemporal patterns of the PUE on the QTP. (**a**) The mean of PUE. The small map at the bottom is the dynamics of the PUE from 2000 to 2017. (**b**) The distribution of the trend in PUE. The small map at the bottom is the range of significance levels.

Table 2. Percentages of grasslands with different variation trends in PUE across the whole plateau and within different grassland types from 2000 to 2017. AM, AS, and DS indicate alpine meadow, alpine steppe, and desert steppe, respectively. Different letters (a, b) denote significant differences between them ($p < 0.05$).

	Mean of PUE ($g\ C\ m^{-2}\ mm^{-1}$)	Trend of PUE (%)			
		Significant Increasing	Insignificant Increasing	Significant Decreasing	Insignificant Decreasing
AM	0.31 ± 0.15 a	2.55	55.33	2.50	39.62
AS	0.19 ± 0.14 b	3.36	53.35	0.86	42.43
DS	0.32 ± 0.15 a	2.32	25.80	0.38	71.51
total	0.25 ± 0.17	2.96	53.31	1.59	42.15

3.4. Relationships of NPP and PUE with Climatic Factors and Grazing Intensity

Figure 5 displays the correlation coefficients of MAT, MAP, and GI with grassland NPP and PUE during 2000–2017. NPP was positively correlated with MAT ($R_{NPP\_MAT} = 0.33$) and MAP ($R_{NPP\_MAP} = 0.32$) but negatively correlated with GI ($R_{NPP\_GI} = -0.06$). Statistical analysis showed that 36.1% and 34.6% of grasslands showed positive and significant ($p < 0.05$) R_{NPP_MAT} and R_{NPP_MAP}, respectively.

The PUE had a weak correlation with MAT ($R_{PUE\_MAT} = 0.2$) and GI ($R_{PUE\_GI} = 0.09$) but a strong correlation with MAP ($R_{PUE\_MAP} = -0.71$). From the space, R_{PUE_MAT} was significant at levels below 0.05 in the southeastern QTP, and R_{PUE_MAT} was significant in almost the entire plateau. The grasslands with significant R_{PUE_MAT} and R_{PUE_GI} only occupied 15.6% and 4.6% of grasslands. However, PUE in over 88% of grasslands had a negative and significant correlation with MAP.

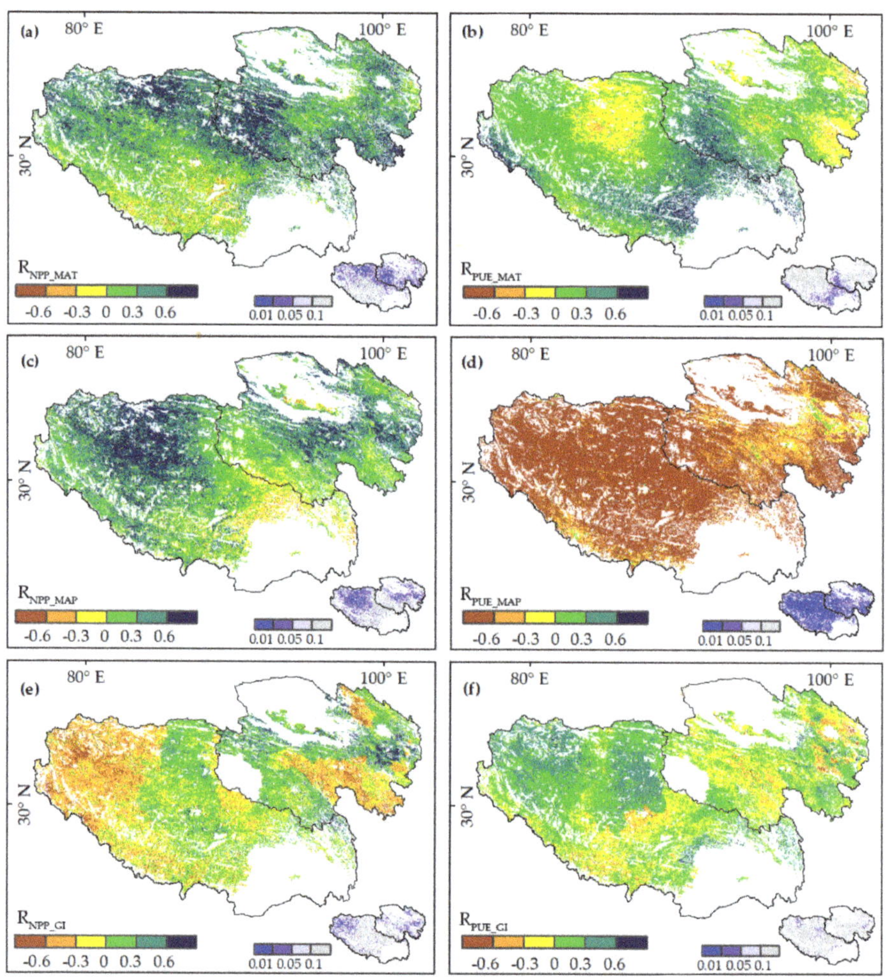

Figure 5. The correlation coefficients of MAT, MAP, and GI with grassland NPP (**a,c,e**) and PUE (**b,d,f**) during 2000–2017. Small maps at the bottom are the ranges of significance levels.

3.5. Relative Contributions of Climatic Factors and Grazing Intensity

On the entire QTP, MAT, MAP, and GI explained 17.5%, 14.8%, and 5.5% of NPP variance, respectively. MAT explained more NPP variance than that of MAP in alpine meadow. However, in alpine steppe and desert steppe, MAP explained most of the NPP variance (Table 3). From a spatial perspective, the C_{MAT_NPP} was highest in the northern plateau and lowest in the southwest (Figure 6a). The C_{MAP_NPP} in the southwestern and northeastern parts of the plateau were higher than in other regions (Figure 6b). The C_{GI_NPP} was generally lower than 10% in most areas of the QTP (Figure 6c). Overall, climatic factors dominated the interannual variation of NPP on the QTP (Figure 6d). Statistically, the percentage of MAT-dominated grasslands was the highest (48.9%), followed by MAP-dominated (40%), and the lowest was GI-dominated (11.1%).

In terms of the contributions of MAT, MAP, and GI to PUE interannual variation (C_{MAT_PUE}, C_{MAP_PUE}, C_{GI_PUE}), we found that the C_{MAP_PUE} (52.7%) was significantly higher than C_{MAT_PUE} (9.6%) and C_{GI_PUE} (3.1%) (Figure 7a–c, Table 4). Overall, the percentage of MAP-dominated grasslands was 87.9%, and MAT-dominated and GI-dominated areas were 8.8% and 3.3%, respectively (Figure 7d).

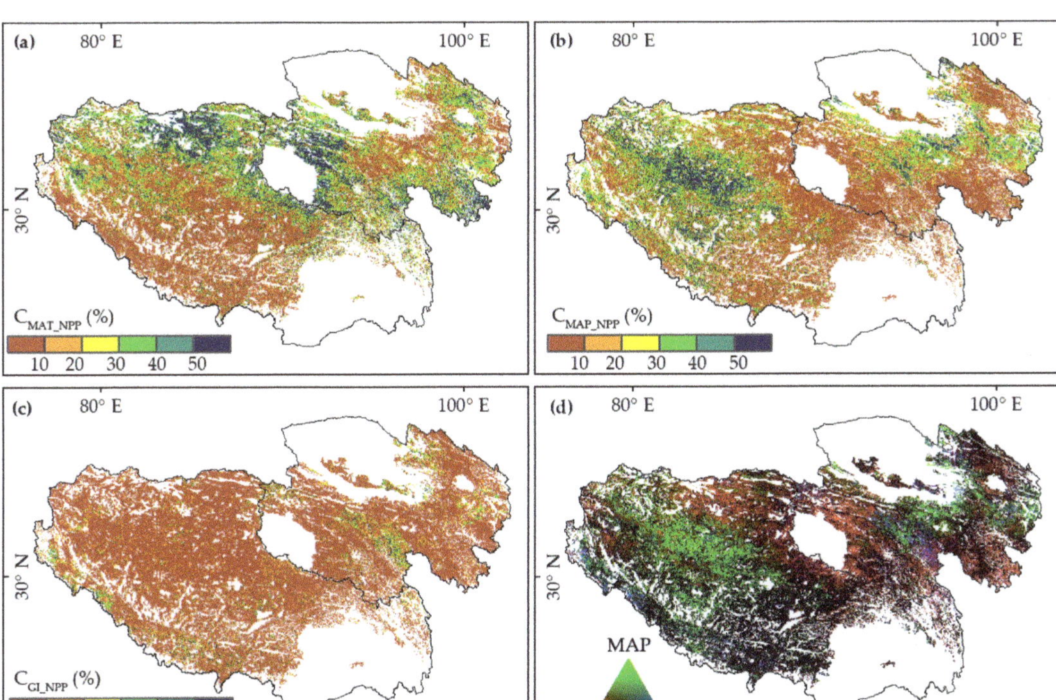

Figure 6. Relative contributions of (**a**) MAT, (**b**) MAP, and (**c**) GI to the NPP interannual variation estimated by generalized linear models (GLMs). (**d**) Spatial pattern of the dominant variables, which was displayed by RGB composite (MAT: red; MAP: green; and GI: blue).

Table 3. Means ($\pm$SE) of the relative contributions (%) of mean annual temperature (C_{MAT_NPP}), mean annual precipitation (C_{MAP_NPP}), and grazing intensity (C_{GI_NPP}) to NPP variation estimated by generalized linear model (GLM) across the entire QTP and within alpine meadow (AM), alpine steppe (AS), and desert steppe (DS).

	QTP	AM	AS	DS
C_{MAT_NPP}	17.5 ± 0.01	16.7 ± 0.02	18.3 ± 0.02	16.3 ± 0.07
C_{MAP_NPP}	14.8 ± 0.01	10.3 ± 0.02	18.5 ± 0.02	20.8 ± 0.07
C_{GI_NPP}	5.5 ± 0.01	5.7 ± 0.01	5.1 ± 0.01	6.8 ± 0.04

Table 4. Means ($\pm$SE) of the relative contributions (%) of mean annual temperature (C_{MAT_PUE}), mean annual precipitation (C_{MAP_PUE}), and grazing intensity (C_{GI_PUE}) to PUE variation estimated by the generalized linear model (GLM) across the entire QTP and within alpine meadow (AM), alpine steppe (AS), and desert steppe (DS).

	QTP	AM	AS	DS
C_{MAT_PUE}	9.6 ± 0.01	13.3 ± 0.02	6.6 ± 0.01	9.6 ± 0.05
C_{MAP_PUE}	52.7 ± 0.02	45.1 ± 0.03	59.8 ± 0.03	56.4 ± 0.11
C_{GI_PUE}	3.1 ± 0.01	3.5 ± 0.01	3.0 ± 0.01	2.7 ± 0.03

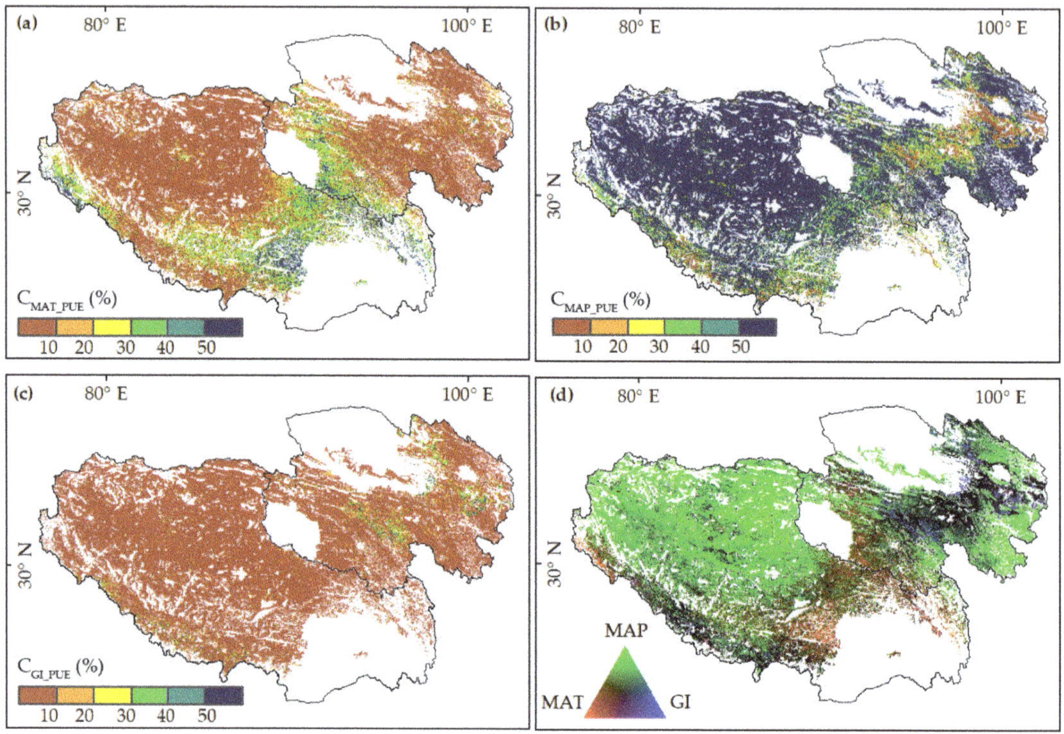

Figure 7. Relative contributions of (**a**) MAT, (**b**) MAP, and (**c**) GI to the PUE interannual variation. (**d**) The spatial pattern of the dominant variables, which was displayed by RGB composite (MAT: red; MAP: green; and GI: blue).

4. Discussion

4.1. Warmer and Wetter Climate Promotes NPP Increases on the QTP

From 2000 to 2017, grassland NPP on the QTP significantly increased at a rate of $0.6\,\mathrm{g\,C\,m^{-2}\,year^{-1}}$, with 31.3% and 0.68% of the total grasslands showing significant increasing and decreasing trend, respectively. Spatially, grasslands with increased NPP were mainly distributed in the north of the QTP, and grasslands with decreased NPP were concentrated in the central plateau. The increasing trend in grassland productivity after 2000 had been widely documented by previous studies. For example, Chen et al. [29] found that the mean grassland NPP of the Tibetan Plateau slightly increased from 2000 to 2011. Liu et al. [55] also suggested that the grassland aboveground biomass increased in 70% of the grasslands on the QTP during 2000–2012, mainly concentrated in Qinghai Province. The increases in NPP can be attributed to warmer and wetter climate conditions and decreased grazing intensity. First, the MAT and MAP on the QTP increased with a rate of 0.05 °C/year and 1.8 mm/year from 2000 to 2017, respectively. The warming temperature increases the photosynthesis rate before the vegetation reaches its optimum temperature [56,57]. As water-limited grasslands are greatly sensitive to precipitation, the increased precipitation could stimulate NPP by providing a sufficient water supply, which has been demonstrated by in situ monitoring, satellite monitoring, and manipulative experiments on the QTP [58,59]. Second, the ecological protection projects on the QTP also positively affect grassland productivity [60]. Some conservation efforts, such as fencing grasslands and reducing livestock numbers, have gradually recovered degraded grasslands and promoted grassland productivity [29]. Especially after 2009, the grazing intensity showed a significant declining trend, causing a noticeable reduction of human-consumed NPP [61].

Given the spatial differentiation of climate change and grazing intensity, the dominant factor for the NPP interannual variation was spatially heterogeneous. On the one hand, climate factors overwhelmed grazing to control grassland NPP dynamics in most of the QTP's grassland areas. A similar finding was made by Lehnert et al. [30], who suggested that although grazing activities could exacerbate the negative effects of the global change, climate factors were still dominant for NPP changes on the QTP. Li et al. [34] also found that the impact of climate change on grasslands NPP was more significant than that of grazing activities. This was possibly caused by the lower intensity of human activity on the QTP than on the rest of the world, due to its vast and sparsely populated environment [4,62]. Since 2000, to protect the ecological environment of this plateau, the Chinese government had launched various policies, such as limiting grazing and fencing degraded grasslands, which have further alleviated the human disturbances on natural grasslands [52]. On the other hand, temperature-dominated and precipitation-dominated grasslands showed a clear spatial pattern. The temperature was the limiting factor in mid-eastern QTP. Precipitation dominated the grassland changes in the southwestern and northeastern parts of the plateau. This spatial pattern was similar to the findings in Li et al. [63] and Huang et al. [64], showing that the mid-eastern grasslands were more sensitive to temperature, and the southwestern and northeastern grasslands were more sensitive to precipitation. The spatially varied limiting factors were closely correlated with the complex meteorological and biotic conditions of the QTP [65]. In arid and semi-arid regions such as the alpine steppe in northern Tibet, vegetation can quickly respond to precipitation. However, in humid areas such as alpine meadow in southern Qinghai, the temperature became prominent due to adequate moisture [63,66].

4.2. The Spatiotemporal Variation of PUE

The averaged PUE of the alpine grasslands, over 2000–2017, was 0.25 g C m^{-2} mm^{-1}, which is within the range of global grasslands (0.05–1.81 g m^{-2} mm^{-1}) reported by Le Houérou H.N et al. [67] and within the range of China's grasslands (0.13–0.64 g m^{-2} mm^{-1}) reported by Hu et al. [15]. Further, it was found in this study that the QTP's PUE was grassland type-dependent. Alpine steppe's PUE (0.19 g C m^{-2} mm^{-1}) was significantly lower than those of alpine meadow (0.31 g C m^{-2} mm^{-1}) and desert steppe (0.32 g C m^{-2} mm^{-1}). This may be caused by vegetation composition and environmental conditions. For example, the high species richness can respond to high precipitation and use the available water completely [31,68], leading to a larger PUE in the alpine meadow. The soil texture is another factor that affects the PUE of alpine grassland by altering the soil water retention. The high silt content and low sand content in alpine meadows generally cause a high water-holding capacity, which is conducive to shaping a high PUE [31,69]. However, it should be noted that desert steppe with a low species richness and silt content also had a high PUE. The potential explanation is that plants in desert steppe usually can also absorb deep water to adapt to an extremely dry climate [70]. Plants in arid regions can resist water stress through some trade-offs between functional traits, such as changing the thickness and size of leaves [32,71].

From 2000 to 2017, the grassland's PUE on the QTP showed fluctuations with non-significant interannual variations. Precipitation was significantly and negatively correlated with PUE and contributed to most variations of PUE. These are in line with the findings of previous studies in different regions. For example, Liu and Huang [72] proposed that the PUE of alpine grasslands on the QTP showed a significantly negative correlation with precipitation. Bai et al. [53] suggested that temperate grasslands' PUE in the Inner Mongolian steppe region decreased with increased annual precipitation. Huxman et al. [20] reported that grasslands in North and South America had maximal PUE in the driest years. The underlying mechanisms of PUE interannual variations were involved with the interactions among precipitation, nutrient, and biotic factors [53]. First, long-term saturated soil moisture is likely to reduce root and soil microbial activity by limiting oxygen supply and then curbing vegetation growth [73]. Second, although increased precipitation alleviates the available water limitations, nitrogen availability limitations may be strengthened to constrain the response of production to increased precipitation, resulting in a lower PUE [53,74]. Finally, the species in arid and semi-arid regions generally have relatively low growth rates to promote their drought-enduring ability, which enables limiting the response of vegetation to extreme drought events [75]. Therefore, the PUE of alpine grasslands in the south of QTP showed an increasing trend despite the MAP decreased across this region.

4.3. The Effects of GI on Vegetation Growth

Compared with MAT and MAP, the GI had the lowest contributions to NPP and PUE variation. However, 11.1% and 3.3% of the grasslands' NPP and PUE were dominated by GI. For grasslands on the QTP, grazing is one of the most typical human disturbances, significantly affecting native species diversity and ecosystem stability [76]. From 2000 to 2017, the GI on this plateau presented "converse-U" changes. The highest GI was generated in 2007 (0.72 SSU/ha) and then declined significantly due to the strict livestock reduction policy. In this period, NPP showed a negative correlation with GI, suggesting that this decrease in GI had a positive effect on vegetation growth. This information about the relationship between grazing activities and vegetation productivity is beneficial to policymakers and herders to develop specific policies and sustainable management strategies. However, the relationship between grazing activities and vegetation growth was complicated due to its complex and diverse processes, such as forage selection and trampling. Numerous studies viewed grazing as detrimental to plant growth [77]. However, the moderate interference hypothesis suggested that moderate grazing intensity may lead to compensatory growth and then positively affect grassland productivity [78,79]. Undoubtedly, overgrazing could, directly and indirectly, destroy the structure and function of grassland ecosystems. Thus, for grasslands where grazing dominates vegetation growth, it is necessary to investigate how grazing intensity affects grassland changes and speculate on the possible underlying mechanisms based on grazing-manipulated experiments.

4.4. Limitations and Uncertainties

This study aimed to reveal the relative contributions of climate factors (precipitation and temperature) and grazing to interannual NPP and PUE variations on the QTP. It should be noted that the effects of land cover changes on the dynamics of NPP and PUE were not considered, although they have modified or amplified the biogeochemical cycling of grassland ecosystems [26]. In the future, to deepen our understanding of the mechanism underlying the climate–vegetation–human relationship, the roles of human-induced land use/cover change in influencing the variations of NPP or PUE should be taken into consideration. Besides, the simulated NPP had some shortcomings; for example, the maximum LUE was treated as a constant for the entire grassland area instead of treating the alpine meadow, alpine steppe, and desert steppe separately. We thus emphasize the need for complementary studies at the local scale with field surveys to calculate the maximum LUE for each grassland type on the QTP. In addition, the CASA model was run using temperature, precipitation, and solar radiation, resulting in the autocorrelations of simulated NPP with climate factors. To quantify the influences of the autocorrelation on the results of GLMs, we further used growing-season NDVI (GSNDVI) as a surrogate of NPP to construct GLMs. We found that the relative contributions of temperature, precipitation, and grazing to GSNDVI were similar to those to NPP (Figure S1), suggesting that the autocorrelations between NPP and climate factors had no significant effects on the conclusions.

5. Conclusions

This study investigated the spatial and temporal patterns of grassland NPP and PUE on the QTP from 2000 to 2017. It was found that the NPP decreased from the southeast to the northwest on this plateau. However, the PUE was higher at the eastern and western ends of the plateau, but was lower in the center. Over time, NPP showed a significant increasing trend, and PUE showed a non-significant increasing trend. The relative contributions of climatic factors (temperature and precipitation) and grazing intensity to the temporal variations of NPP and PUE were quantified by GLMs. Although numerous studies claimed that overgrazing impaired the structure and function of alpine grasslands, our research found climatic factors played a dominant role in vegetation changes on the QTP.

Supplementary Materials: The following are available online at https://www.mdpi.com/article/10.3390/rs13173424/s1, Figure S1: Relative contributions of (a) MAT, (b) MAP, and (c) GI to the growing-season NDVI (GSNDVI) interannual variation estimated by generalized linear models (GLMs). (d) Spatial pattern of the dominant variables, which was displayed by RGB composite (MAT: red; MAP: green; and GI: blue).

Author Contributions: M.L. conceptualized this study. M.L. and H.Y. collected and analyzed the data. H.Y. led the writing. Q.D., B.M., Y.L., C.L., X.Z., Y.S. and S.Y. interpreted the results and revised the text. All authors have read and agreed to the published version of the manuscript.

Funding: The study was supported by the National Key R&D Program of China (2017YFA0604801) and the National Nature Science Foundation of China (31901393).

Institutional Review Board Statement: Not applicable.

Informed Consent Statement: Not applicable.

Data Availability Statement: The monthly MOD13A3C6 NDVI data were obtained from the NASA LP DAAC (Land Processes Distributed Active Archive Center) website (https://lpdaac.usgs.gov/get_data/data_pool, accessed on 1 May 2020). The meteorological data were downloaded from the National Meteorological Information Center of China Meteorological Administration (http://geodata.cn, accessed on 1 May 2020). Grassland distribution was determined by the China Vegetation Atlas with a scale of 1:1,000,000, which was derived from the Resource and Environment Data Cloud Platform (http://www.resdc.cn/data.aspx?DATAID=122, accessed on 1 May 2020).

Acknowledgments: We are very grateful to the Three-River-Source National Park Administration for their support of our field survey. We sincerely appreciate the editor and three anonymous reviewers for their constructive comments and suggestions. We acknowledge the support of all co-authors for their constructive and helpful comments and organization of this study.

Conflicts of Interest: The authors declare that the research was conducted in the absence of any commercial or financial relationships that could be construed as a potential conflict of interest.

References

1. Radu, D.D.; Duval, T.P. Precipitation Frequency Alters Peatland Ecosystem Structure and CO_2 Exchange: Contrasting Effects on Moss, Sedge, and Shrub Communities. *Glob. Chang. Biol.* **2018**, *24*, 2051–2065. [CrossRef]
2. DeFries, R. Past and Future Sensitivity of Primary Production to Human Modification of the Landscape. *Geophys. Res. Lett.* **2002**, *29*, 36-1–36-4. [CrossRef]
3. Vitousek, P.M.; Mooney, H.A.; Lubchenco, J.; Melillo, J.M. Human Domination of Earth's Ecosystems. *Science* **1997**, *277*, 494–499. [CrossRef]
4. Venter, O.; Sanderson, E.W.; Magrach, A.; Allan, J.R.; Beher, J.; Jones, K.R.; Possingham, H.P.; Laurance, W.F.; Wood, P.; Fekete, B.M. Sixteen Years of Change in the Global Terrestrial Human Footprint and Implications for Biodiversity Conservation. *Nat. Commun.* **2016**, *7*, 12558. [CrossRef] [PubMed]
5. Seiferling, I.; Proulx, R.; Wirth, C. Disentangling the Environmental-Heterogeneity Species-Diversity Relationship Along a Gradient of Human Footprint. *Ecology* **2014**, *95*, 2084–2095. [CrossRef] [PubMed]
6. Piao, S.; Wang, X.; Park, T.; Chen, C.; Lian, X.; He, Y.; Bjerke, J.W.; Chen, A.; Ciais, P.; Tømmervik, H.; et al. Characteristics, Drivers and Feedbacks of Global Greening. *Nat. Rev. Earth Environ.* **2020**, *1*, 14–27. [CrossRef]
7. Gil, M.A.; Baskett, M.L.; Munch, S.B.; Hein, A.M. Fast Behavioral Feedbacks Make Ecosystems Sensitive to Pace and Not Just Magnitude of Anthropogenic Environmental Change. *Proc. Natl. Acad. Sci. USA* **2020**, *117*, 25580–25589. [CrossRef]

8. Chen, T.; Bao, A.M.; Jiapaer, G.; Guo, H.; Zheng, G.X.; Jiang, L.L.; Chang, C.; Tuerhanjiang, L. Disentangling the Relative Impacts of Climate Change and Human Activities on Arid and Semi-Arid Grasslands in Central Asia during 1982–2015. *Sci. Total Environ.* **2019**, *653*, 1311–1325. [CrossRef]
9. Curlock, J.M.O.S.; Hall, D.O. The Global Carbon Sink: A Grassland Perspective. *Glob. Chang. Biol.* **2010**, *4*, 229–233. [CrossRef]
10. Xie, G.D.; Zhang, Y.L.; Lu, C.X. Study on Valuation of Rangeland Ecosystem Services of China. *J. Natl. Resour.* **2001**, *16*, 47–53.
11. Gang, C.; Zhou, W.; Chen, Y.; Wang, Z.; Sun, Z.; Li, J.; Qi, J.; Odeh, I. Quantitative Assessment of the Contributions of Climate Change and Human Activities on Global Grassland Degradation. *Environ. Earth Sci.* **2014**, *72*, 4273–4282. [CrossRef]
12. Gao, Y.; Zhou, X.; Wang, Q.; Wang, C.; Zhan, Z.; Chen, L.; Yan, J.; Qu, R. Vegetation Net Primary Productivity and Its Response to Climate Change During 2001–2008 in the Tibetan Plateau. *Sci. Total Environ.* **2013**, *444*, 356–362. [CrossRef]
13. Nemani, R.R.; Keeling, C.D.; Hashimoto, H.; Jolly, W.M.; Piper, S.C.; Tucker, C.J.; Myneni, R.B.; Running, S.W. Climate-Driven Increases in Global Terrestrial Net Primary Production from 1982 to 1999. *Science* **2003**, *300*, 1560–1563. [CrossRef] [PubMed]
14. Ling, L.; Xin, L.; Huang, C.L.; Veroustraete, F. Analysis of the Spatio-Temporal Characteristics of Water Use Efficiency of Vegetation in West China. *J. Glaciol. Geocryol.* **2007**, *29*, 777–784.
15. Hu, Z.; Yu, G.; Fan, J.; Zhong, H.; Wang, S.; Li, S. Precipitation-Use Efficiency Along a 4500-km Grassland Transect. *Glob. Ecol. Biogeogr.* **2010**, *19*, 842–851.
16. Siepielski, A.M.; Morrissey, M.B.; Buoro, M.; Carlson, S.M.; Caruso, C.M.; Clegg, S.M.; Coulson, T.; DiBattista, J.; Gotanda, K.M.; Francis, C.D.; et al. Precipitation Drives Global Variation in Natural Selection. *Science* **2017**, *355*, 959–962. [CrossRef]
17. Knapp, A.K.; Fay, P.A.; Blair, J.M.; Collins, S.L.; Smith, M.D.; Carlisle, J.D.; Harper, C.W.; Danner, B.T.; Lett, M.S.; McCarron, J.K. Rainfall Variability, Carbon Cycling, and Plant Species Diversity in a Mesic Grassland. *Science* **2002**, *298*, 2202–2205. [CrossRef] [PubMed]
18. Wang, S.; Zhang, B.; Yang, Q.; Chen, G.; Yang, B.; Lu, L.; Shen, M.; Peng, Y. Responses of Net Primary Productivity to Phenological Dynamics in the Tibetan Plateau, China. *Agric. For. Meteorol.* **2017**, *232*, 235–246. [CrossRef]
19. Piao, S.; Friedlingstein, P.; Ciais, P.; Viovy, N.; Demarty, J. Growing Season Extension and Its Impact on Terrestrial Carbon Cycle in the Northern Hemisphere Over the Past 2 Decades. *Glob. Biogeochem. Cycles* **2007**, *21*, 116–123. [CrossRef]
20. Huxman, T.E.; Smith, M.D.; Fay, P.A.; Knapp, A.K.; Shaw, M.R.; Loik, M.E.; Smith, S.D.; Tissue, D.T.; Zak, J.C.; Weltzin, J.F.; et al. Convergence Across Biomes to a Common Rain-Use Efficiency. *Nature* **2004**, *429*, 651–654. [CrossRef]
21. Rockström, J.; Steffen, W.; Noone, K.; Persson, Å.; Chapin, F.S.; Lambin, E.F.; Lenton, T.M.; Scheffer, M.; Folke, C.; Schellnhuber, H.J.; et al. Safe Operating Space for Humanity. *Nature* **2009**, *461*, 472–475. [CrossRef] [PubMed]
22. Niu, Y.; Zhu, H.; Yang, S.; Ma, S.; Zhou, J.; Chu, B.; Hua, R.; Hua, L. Overgrazing Leads to Soil Cracking That Later Triggers the Severe Degradation of Alpine Meadows on the Tibetan Plateau. *Land Degrad. Dev.* **2019**, *30*, 1243–1257. [CrossRef]
23. Han, Q.; Luo, G.; Li, C.; Xu, W. Modeling the Grazing Effect on Dry Grassland Carbon Cycling with Biome-BGC Model. *Ecol. Complex.* **2014**, *17*, 149–157. [CrossRef]
24. Wang, C.S.; Meng, F.D.; Li, X.E.; Jiang, L.L.; Wang, S.P. Responses of Alpine Grassland Ecosystem on Tibetan Plateau to Climate Change: A Mini Review. *Chin. J. Ecol.* **2013**, *32*, 1587–1595.
25. Yao, T.; Wu, F.; Ding, L.; Sun, J.; Zhu, L.; Piao, S.; Deng, T.; Ni, X.; Zheng, H.; Ouyang, H. Multispherical Interactions and Their Effects on the Tibetan Plateau's Earth System: A Review of the Recent Researches. *Natl. Sci. Rev.* **2015**, *2*, 468–488. [CrossRef]
26. Chen, H.; Zhu, Q.; Peng, C.; Wu, N.; Wang, Y.; Fang, X.; Gao, Y.; Zhu, D.; Yang, G.; Tian, J.; et al. The Impacts of Climate Change and Human Activities on Biogeochemical Cycles on the Qinghai-Tibetan Plateau. *Glob. Chang. Biol.* **2013**, *19*, 2940–2955. [CrossRef]
27. Fan, J.; Yong, X.U.; Wang, C.S.; Niu, Y.F.; Chen, D.; Sun, W. The Effects of Human Activities on the Ecological Environment of Tibet Over the Past Half Century. *Chin. Sci. Bull.* **2015**, *60*, 3057–3066. [CrossRef]
28. Dong, S.K.; Li, J.P.; Li, X.Y.; Wen, L.; Zhu, L.; Li, Y.Y.; Ma, Y.S.; Shi, J.J.; Dong, Q.M.; Wang, Y.L. Application of Design Theory for Restoring the "Black Beach" Degraded Rangeland at the Headwater Areas of the Qinghai-Tibetan Plateau. *Afr. J. Agric. Res.* **2010**, *5*, 3542–3552.
29. Chen, B.; Zhang, X.; Tao, J.; Wu, J.; Wang, J.; Shi, P.; Zhang, Y.; Yu, C. The Impact of Climate Change and Anthropogenic Activities on Alpine Grassland Over the Qinghai-Tibet Plateau. *Agric. For. Meteorol.* **2014**, *189*, 11–18. [CrossRef]
30. Lehnert, L.W.; Wesche, K.; Trachte, K.; Reudenbach, C.; Bendix, J. Climate Variability Rather Than Overstocking Causes Recent Large Scale Cover Changes of Tibetan Pastures. *Sci. Rep.* **2016**, *6*, 24367. [CrossRef]
31. Yang, Y.H.; Fang, J.Y.; Fay, P.A.; Bell, J.E.; Ji, C.J. Rain Use Efficiency Across a Precipitation Gradient on the Tibetan Plateau. *Geophys. Res. Lett.* **2010**, *37*. [CrossRef]
32. Zhao, G.; Liu, M.; Shi, P.; Zong, N.; Wang, J.; Wu, J.; Zhang, X. Spatial–Temporal Variation of ANPP and Rain-Use Efficiency Along a Precipitation Gradient on Changtang Plateau, Tibet. *Remote Sens.* **2019**, *11*, 325. [CrossRef]
33. Sun, H.; Zheng, D.; Yao, T.; Zhang, Y. Protection and Construction of the National Ecological Security Shelter Zone on Tibetan Plateau. *Acta Geogr. Sin.* **2012**, *67*, 3–12.
34. Li, M.; Wu, J.S.; Feng, Y.F.; Niu, B.; He, Y.T.; Zhang, X.Z. Climate Variability Rather Than Livestock Grazing Dominates Changes in Alpine Grassland Productivity Across Tibet. *Front. Ecol. Evol.* **2021**, *9*. [CrossRef]
35. Long, R.J.; Apori, S.O.; Castro, F.B.; Orskov, E.R. Feed Value of Native Forages of the Tibetan Plateau of China. *Anim. Feed Sci. Tech.* **1999**, *80*, 101–113. [CrossRef]
36. Mirjanka, L.; Brankica, M. Albers Conical Equal-Area Projection. *Geod. List* **1996**, *2*, 161–170.

37. Jonsson, P.; Eklundh, L. TIMESAT-A Program for Analyzing Time-Series of Satellite Sensor Data. *Comput. Geosci.-UK* **2004**, *30*, 833–845. [CrossRef]
38. Eklundh, L.; Jönsson, P. TIMESAT: A Software Package for Time-Series Processing and Assessment of Vegetation Dynamics. In *Remote Sensing Time Series*; Springer: Cham, Switzerland, 2015.
39. Luo, J.; Ying, K.; Bai, J. Savitzky–Golay Smoothing and Differentiation Filter for Even Number Data. *Signal. Process.* **2005**, *85*, 1429–1434. [CrossRef]
40. Allan, R.; Pereira, L.; Smith, M. *Crop. Evapotranspiration-Guidelines for Computing Crop Water Requirements*; FAO: Rome, Italy, 1998.
41. Hutchinson, M. *Anusplin Version 4.3. Centre for Resource and Environmental Studies*; Australian National University: Canberra, Australia, 2004.
42. Cao, Y.; Wu, J.; Zhang, X.; Niu, B.; He, Y. Comparison of Methods for Evaluating the Forage-Livestock Balance of Alpine Grasslands on the Northern Tibetan Plateau. *J. Resour. Ecol.* **2020**, *11*, 272–282.
43. Monteith, J.L.; Moss, C.J.; Cooke, G.W.; Pirie, N.W.; Bell, G.D.H. Climate and the Efficiency of Crop Production in Britain. *Philos. Trans. R. Soc. London B Biol. Sci.* **1977**, *281*, 277–294.
44. Potter, C.S.; Randerson, J.T.; Field, C.B.; Matson, P.A.; Vitousek, P.M.; Mooney, H.A.; Klooster, S.A. Terrestrial Ecosystem Production: A Process Model Based on Global Satellite and Surface Data. *Glob. Biogeochem. Cycles* **1993**, *7*, 811–841. [CrossRef]
45. Potter, C. Predicting Climate Change Effects on Vegetation, Soil Thermal Dynamics, and Carbon Cycling in Ecosystems of Interior Alaska. *Ecol. Model.* **2004**, *175*, 1–24. [CrossRef]
46. Piao, S.; Fang, J.; He, J. Variations in Vegetation Net Primary Production in the Qinghai-Xizang Plateau, China, from 1982 to 1999. *Clim. Chang.* **2006**, *74*, 253–267. [CrossRef]
47. Yuan, J.; Niu, Z.; Wang, C. Vegetation NPP Distribution Based on MODIS Data and CASA Model-A Case Study of Northern Hebei Province. *Chin. Geogr. Sci.* **2006**, *16*, 334–341. [CrossRef]
48. Field, C.B.; Randerson, J.T.; Malmström, C.M. Global Net Primary Production: Combining Ecology and Remote Sensing. *Remote Sens. Environ.* **1995**, *51*, 74–88. [CrossRef]
49. Field, C.B.; Behrenfeld, M.J.; Randerson, J.T.; Falkowski, P. Primary Production of the Biosphere: Integrating Terrestrial and Oceanic Components. *Science* **1998**, *281*, 237–240. [CrossRef] [PubMed]
50. Wang, H.; Jia, G.; Fu, C.; Feng, J.; Zhao, T.; Ma, Z. Deriving Maximal Light Use Efficiency From Coordinated Flux Measurements and Satellite Data for Regional Gross Primary Pro-Duction Modeling. *Remote Sens. Environ.* **2010**, *114*, 2248–2258. [CrossRef]
51. Zhang, Y.; Qi, W.; Zhou, C.; Ding, M.; Liu, L.; Gao, J.; Bai, W.; Wang, Z.; Zheng, D. Spatial and Temporal Variability in the Net Primary Production of Alpine Grassland on the Tibetan Plateau Since 1982. *J. Geogr. Sci.* **2014**, *24*, 269–287. [CrossRef]
52. Li, M.; Zhang, X.; Wu, J.; Ding, Q.; He, Y. Declining Human Activity Intensity on Alpine Grasslands of the Tibetan Plateau. *J. Environ. Manag.* **2021**, *296*, 113198. [CrossRef]
53. Bai, Y.; Wu, J.; Xing, Q.; Pan, Q.; Huang, J.; Yang, D.; Han, X. Primary Production and Rain Use Efficiency Across a Precipitation Gradient on the Mongolia Plateau. *Ecology* **2008**, *89*, 2140–2153. [CrossRef]
54. Zhang, L.X.; Fan, J.W.; Shao, Q.Q.; Tang, F.P.; Zhang, H.Y.; Yu-Zhe, L.I. Changes in Grassland Yield and Grazing Pressure in the Three Rivers Headwater Region Before and After the Implementation of the Eco-Restoration Project. *Acta Pratacult. Sin.* **2014**, *23*, 116–123.
55. Liu, S.; Cheng, F.; Dong, S.; Zhao, H.; Hou, X.; Wu, X. Spatiotemporal Dynamics of Grassland Aboveground Biomass on the Qinghai-Tibet Plateau Based on Validated MODIS NDVI. *Sci. Rep.* **2017**, *7*, 4182. [CrossRef]
56. Zhao, J.; Hartmann, H.; Trumbore, S.; Ziegler, W.; Zhang, Y. High Temperature Causes Negative Whole-Plant Carbon Balance under Mild Drought. *New Phytol.* **2013**, *200*, 330–339. [CrossRef]
57. Hudson, J.; Henry, G.; Cornwell, W.K. Taller and Larger: Shifts in Arctic Tundra Leaf Traits After 16 Years of Experimental Warming. *Glob. Chang. Biol.* **2011**, *17*, 1013–1021. [CrossRef]
58. Qun, G.; Zhongmin, H.; Shenggong, L.; Guirui, Y.; Xiaomin, S.; Leiming, Z.; Songlin, M.; Xianjin, Z.; Yanfen, W.; Yingnian, L.; et al. Contrasting Responses of Gross Primary Productivity to Precipitation Events in a Water-Limited and a Temperature-Limited Grassland Ecosystem. *Agric. For. Meteorol.* **2015**, *214*, 169–177.
59. Fu, G.; Shen, Z.X.; Zhang, X.Z. Increased Precipitation Has Stronger Effects on Plant Production of an Alpine Meadow Than Does Experimental Warming in the Northern Tibetan Plateau. *Agric. For. Meteorol.* **2018**, *249*, 11–21. [CrossRef]
60. Zhang, Q.; Li, M.A.; Zhang, Z.; Wenhua, X.U.; Zhou, B.; Song, M.; Qiao, A.; Wang, F.; She, Y.; Yang, X. Ecological Restoration of Degraded Grassland in Qinghai-Tibet Alpine Region: Degradation Status, Restoration Measures, Effects and Prospects. *Acta Ecol. Sin.* **2019**, *39*, 7441–7451.
61. Zhang, Y.; Pan, Y.; Zhang, X.; Wu, J.; Yu, C.; Li, M.; Wu, J. Patterns and Dynamics of the Human Appropriation of Net Primary Production and Its Components in Tibet. *J. Environ. Manag.* **2018**, *210*, 280–289. [CrossRef]
62. Li, S.; Wu, J.; Jian, G.; Li, S. Human Footprint in Tibet: Assessing the Spatial Layout and Effectiveness of Nature Reserves. *Sci. Total Environ.* **2018**, *621*, 18–29. [CrossRef]
63. Li, L.; Zhang, Y.; Liu, L.; Wu, J.; Wang, Z.; Li, S.; Zhang, H.; Zu, J.; Ding, M.; Paudel, B. Spatiotemporal Patterns of Vegetation Greenness Change and Associated Climatic and Anthropogenic Drivers on the Tibetan Plateau during 2000–2015. *Remote Sens.* **2018**, *10*, 1525. [CrossRef]
64. Huang, K.; Zhang, Y.; Zhu, J.; Liu, Y.; Zu, J.; Zhang, J. The Influences of Climate Change and Human Activities on Vegetation Dynamics in the Qinghai-Tibet Plateau. *Remote Sens.* **2016**, *8*, 876. [CrossRef]

65. Molnar, P.; Boos, W.; Battisti, D. Orographic Controls on Climate and Paleoclimate of Asia: Thermal and Mechanical Roles for the Tibetan Plateau. *Annu. Rev. Earth Planet. Sci.* **2010**, *38*, 77–102. [CrossRef]
66. Xu, X.; Chen, H.; Levy, J.K. Spatiotemporal Vegetation Cover Variations in the Qinghai-Tibet Plateau under Global Climate Change. *Chin. Sci. Bull.* **2008**, *53*, 915–922. [CrossRef]
67. Le Houérou, H.N.; Bingham, R.L.; Skerbek, W. Relationship Between the Variability of Primary Production and the Variability of Annual Precipitation in World Arid Lands. *J. Arid Environ.* **1988**, *15*, 1–18. [CrossRef]
68. Paruelo, J.M.; Lauenroth, W.K.; Burke, I.C.; Sala, O.E. Grassland Precipitation-Use Efficiency Varies Across a Resource Gradient. *Ecosystems* **1999**, *2*, 64–68. [CrossRef]
69. Epstein, H.E.; Lauenroth, W.K.; Burke, I.C. Effects of Temperature and Soil Texture on ANPP in the U.S. Great Plants. *Ecology* **1997**, *78*, 2628–2631. [CrossRef]
70. Jobbágy, E.; Sala, O. Controls of Grass and Shrub Aboveground Production in the Patagonian Steppe. *Ecol. Appl.* **2000**, *10*, 541–549. [CrossRef]
71. Wei, H.; Wu, B.; Yang, W.; Luo, T. Low Rainfall-Induced Shift in Leaf Trait Relationship within Species Along a Semi-Arid Sandy Land Transect in Northern China. *Plant Biol.* **2011**, *13*, 85–92. [CrossRef]
72. Liu, Z.; Huang, M. Assessing Spatio-Temporal Variations of Precipitation-Use Efficiency Over Tibetan Grasslands Using MODIS and In-Situ Obser-Vations. *Front. Earth Sci.* **2016**, *10*, 784–793. [CrossRef]
73. Shi, S.B. The Photosynthesis of Plant Community in Kobresia humilis Meadow. *Chin. J. Plant Ecol.* **1996**, *20*, 225–234.
74. Knapp, A.K.; Smith, M.D. Variation Among Biomes in Temporal Dynamics of Aboveground Primary Production. *Science* **2001**, *291*, 481–484. [CrossRef]
75. Ogaya, R.; Peuelas, J. Comparative Field Study of Quercus ilex and Phillyrea latifolia: Photosynthetic Response to Experimental Drought Conditions. *Environ. Exp. Bot.* **2003**, *50*, 137–148. [CrossRef]
76. Wu, J.; Zhang, X.; Shen, Z.; Shi, P.; Xu, X.; Li, X. Grazing-Exclusion Effects on Aboveground Biomass and Water-Use Efficiency of Alpine Grasslands on the Northern Tibetan Plateau. *Rangeland Ecol. Manag.* **2013**, *66*, 454–461. [CrossRef]
77. Mikola, J.; Setälä, H.; Virkajärvi, P.; Saarijärvi, K.; Ilmarinen, K.; Voigt, W.; Vestberg, M. Defoliation and Patchy Nutrient Return Drive Grazing Effects on Plant and Soil Properties in a Dairy Cow Pasture. *Ecol. Monogr.* **2009**, *79*, 221–244. [CrossRef]
78. Mazancourt, C.D.; Loreau, M.; Abbadie, L. Grazing Optimization and Nutrient Cycling: When Do Herbivores Enhance Plant Production? *Ecology* **1998**, *79*, 2242–2252. [CrossRef]
79. Luo, G.; Han, Q.; Zhou, D.; Li, L.; Xi, C.; Yan, L.; Hu, Y.; Li, B.L. Moderate Grazing Can Promote Aboveground Primary Production of Grassland under Water Stress. *Ecol. Complex.* **2012**, *11*, 126–136. [CrossRef]

remote sensing

Article

Variation Tendency of Coastline under Natural and Anthropogenic Disturbance around the Abandoned Yellow River Delta in 1984–2019

Zhipeng Sun and Xiaojing Niu *

State Key Laboratory of Hydroscience and Engineering, Department of Hydraulic Engineering,
Tsinghua University, Beijing 100084, China; szp20@mails.tsinghua.edu.cn
* Correspondence: nxj@tsinghua.edu.cn; Tel.: +86-1381-073-5396

Abstract: The coast around the Abandoned Yellow River Delta underwent significant changes under anthropogenic disturbance. This study aims to reveal the variation of the coastline, tidal flat area, and intertidal zone slope before, during, and after extensive reclamation during the period of 1984–2019 using satellite remote sensing images. In order to eliminate the influence of the varying water level, a new coastline correction algorithm had been proposed under the condition of insufficient accurate slope and water level data. The influence of seawalls on slope estimation were considered in it. The spatiotemporal evolution of coast had been analyzed and confirmed to be reasonable by comparing with the observed data. The results show that the coast can be roughly divided into a north erosion part and a south deposition part. Affected by reclamation, their tidal flat area in 2019 is reduced to only 43 and 27% of original area in 1984, respectively, which results in a continuous decrease in the tidal flat width. The adjustment of the tidal flat profile makes the slopes steeper in the erosion part, while the slopes in the deposition part remain stable. The reclamation has stimulated a cumulative effect as the disappearance of the intertidal zone, which may lead to the destruction of biological habitats.

Keywords: Abandoned Yellow River Delta; remote sensing images; spatiotemporal changes; correction algorithm; coastline; tidal flat

Citation: Sun, Z.; Niu, X. Variation Tendency of Coastline under Natural and Anthropogenic Disturbance around the Abandoned Yellow River Delta in 1984–2019. *Remote Sens.* **2021**, *13*, 3391. https://doi.org/10.3390/rs13173391

Academic Editors: Cristian Constantin Stoleriu and Adrian Ursu

Received: 17 July 2021
Accepted: 24 August 2021
Published: 26 August 2021

Publisher's Note: MDPI stays neutral with regard to jurisdictional claims in published maps and institutional affiliations.

1. Introduction

Tidal flat is a transition zone between marine and terrestrial ecosystems and contains abundant natural resources. It can not only provide biological habitats, buffer the impact of extreme climates, and maintain the diversity and stability of the ecosystem, but it can also create an important marine economic development zone that integrates port trade, landscape tourism, sea salt production, and other functions; therefore, it has great development potential [1–3]. The development value of tidal flat resources has laid the foundation for the rapid growth of population and social economy in recent decades. However, the rapid industrialization, urbanization, and the extensive anthropogenic disturbances have severely influenced the changes of coastlines and the reduction in tidal flat resources, which bring a series of environmental and ecological problems, such as habitat destruction, soil degradation, biodiversity loss, etc. [4–7]. Therefore, study on the spatiotemporal changes of coastline and tidal flat resources is of strategic significance for future coastal cities to keep the balance of the development and protection of tidal flat resources, alleviate the substantial conflicts between economic development and environmental protection, and achieve sustainable development [8–11].

The Yellow River, known as "the cradle of Chinese civilization", has historically been recognized for the high suspended sediment load that inspired its name [12]. Therefore, a large amount of sediment is easily carried by the Yellow River and transported to the estuary, thereby the fine-grained sediment deposits form a delta [13]. Throughout history, the downstream of the Yellow River in the North China Plain has been diverted many times. The Abandoned Yellow River Delta, as one of the most typical modern abandoned deltas in the world, was formed between 1128 and 1855, when the Yellow River flew into the Yellow Sea from Jiangsu Province [14]. After 1855, the Yellow River debouched into the Bohai Sea, thus the Abandoned Yellow River Delta lost its main sediment source and, consequently, began to erode [15]. Since the Ancient Yellow River carried a large amount of sediment and deposited around the Abandoned Yellow River Delta, Jiangsu Province, where the Abandoned Yellow River Delta, is located has about one quarter of the total tidal flat area of China. The abundant tidal flat resources become a unique advantage for the development of Jiangsu Province. Since China's reform and opening up in the late 1970s, extensive anthropogenic reclamation has been conducted and resulted in a rapid reduction in the natural tidal flat area. In the last decade, the local policy has shifted to strict protection of the coastline, e.g., strict control of reclamation. It is of interest to reveal the variation tendency of the coastline, tidal flat area, and intertidal zone slope before, during, and after extensive reclamation, to show the influence of the policy for the utilization and conservation of the coastal zone in China over the past several decades.

Based on satellite remote sensing images to extract and analyze the coastlines, there have been a lot of studies in recent years [16–22]. For example, Luijendijk et al. [22] successfully made a global-scale analysis of the occurrence of sandy beaches and rates of shoreline change therein using the satellite-derived shorelines (SDS). Their method is similar to that of Hagenaars et al. [23], by using the moving average composite images to decrease errors from clouds, waves, sensor corrections, georeferencing, and improve the accuracy of SDS to subpixel precision. The method is of high efficiency to obtain the satellite-derived shorelines. It should be noted that influenced by tides and waves, the instantaneous water edges are always in dynamic change at different temporal and spatial scales [24]. Especially for the coast with a relatively gentle slope or a large tidal range, the instantaneous water edge changes dramatically and the method based on moving average composite images needs more images to eliminate the effect of the varying water level. Castelle et al. [25] pointed out that publicly available satellite derived data are associated with great uncertainty for beaches experiencing a large tidal range (>2 m). In order to better describe the coastline changes in the case of gentle slope and large water level variation, two types of coastline indicators are commonly used [26], one is based on a visually discernible coastal feature such as the high tide line [27], and the other one is datum-based shoreline indicators [28,29]. Although the high tide lines are not easily affected by sea level fluctuations compared with the instantaneous shorelines, Crowell et al. [30] pointed out that the actual high tide lines were not obvious. Therefore, the mean sea level tidal-datum-based coastline is widely adopted by researchers and also considered to evaluate the changes of coastline around the Abandoned Yellow River Delta in this study. Actually, accurate intertidal zone slope and water level data are key to obtain the datum-based coastline.

However, due to the relative lack of measured slope data in most areas of China, in many studies of coastline change, the slope data required for coastline correction is usually calculated by two or three satellite images, and the endpoint rate is used to reflect the coastline change [28]. However, the endpoint rate is easily affected by the error of the shoreline position in the image; therefore, there is a large error between the endpoint rate and the actual coastline change rate. For this reason, in order to improve the reliability of the analysis results, statistical analysis based on a large number of instantaneous water edge positions is used to study the coastline migration rate [31]. However, without eliminating the influence of the varying water level, the statistical analysis is difficult to observe the changes of the coastline in a short period of time. In addition, even if the intertidal zone slope estimation and tidal level correction are carried out, the results still have a large deviation from the actual coastlines without considering the change of the seawall or temporary dike.

In order to better analyze the spatial and temporal changes of the coastline around the Abandoned Yellow River Delta in 1984–2019, this study used Landsat series satellite remote sensing images to extract the instantaneous water edges as the data basis. At the same time, the new coastline correction algorithm was proposed based on the intertidal zone slope estimation and water level correction, and the natural and anthropogenic disturbance were further considered in it. The algorithm can not only effectively achieve the coastline correction, thereby reflecting the spatiotemporal changes of the coastline, but also is generally applicable to areas without accurate data of water level and slope. The study results have strategic significance for revealing the past development problems around the Abandoned Yellow River Delta and achieving future sustainable economic development planning and coastal ecological protection.

2. Study Area and Data Source

2.1. Study Area

The study area is around the abandoned Yellow River delta from the Shaoxiang Estuary to Dongtai Estuary, as shown in Figure 1, which is the middle part of the coast in Jiangsu Province and located to the west of the Yellow Sea. This part of the coast in Jiangsu Province is silt coast, and generally has a quite gentle intertidal zone slope. The cause of the tidal flat is closely related to the Ancient Yellow River, and it is greatly affected by the diversion of the Ancient Yellow River. The variation tendency of the coastline in this area was dramatic under both natural and anthropogenic disturbance. Since China's reform and opening up, the Jiangsu policies have shifted from extensive reclamation to strict protection, and this area underwent an extensive reclamation that has had a huge influence on the changes of the coastline.

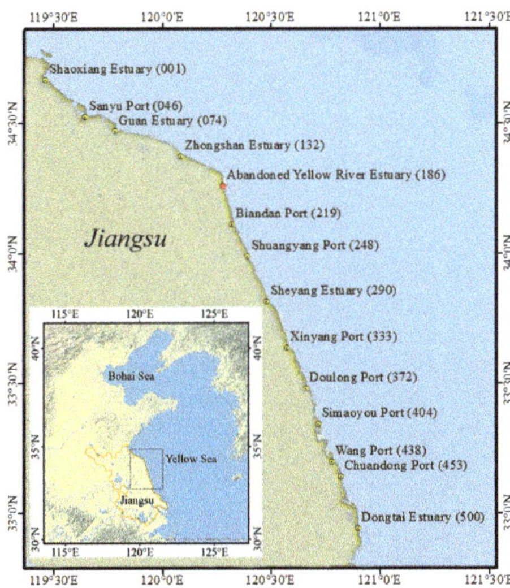

Figure 1. Study area and division of coast transect. The number in brackets is the transect number for main estuaries.

It is hard to obtain long-term observed data on the hydrodynamics of this area. From the literatures and the public ocean tidal model, it can be known that the tidal range is relatively large in the study area and varying from the north part to the south part. The maximum possible tidal range is 5–7 m off the southern coast, while the maximum possible tidal range off the northern coast is about 4 m [32]. The spatial variation of tidal range cannot be ignored in such a large area. The slope in this area is about 1/1000 to

1/10,000. Combining with the gentle slope and large tidal range, it should be noted that the instantaneous water edges in this area are greatly affected by the varying water level.

2.2. Coastal Baseline Selection and Transect Division

The coastal baseline selection and transect division are the basis for obtaining instantaneous water edges and seawall positions, and further analyzing the spatial and temporal changes of the coastline. The coastline baselines are determined based on the principle that they are approximately parallel to the coast direction and placed in the sea side. Based on the coastline baselines, the relative baseline distance from the instantaneous water edges to the coastline baselines can be determined. The positive direction points to the sea side. The relative baseline distance from the land side is negative, while from the sea side it is positive. The changes of the relative baseline distance reflect the coastline changes. The coastline baselines are determined based on the principle that they are approximately parallel to the coast direction. In order to better fit the coast direction, the segmented baselines are adopted. According to the tortuous characteristic of the coast around the Abandoned Yellow River Delta, a total of seven coastline baseline segments are set, as shown in Table 1. In order to reflect the spatiotemporal changes of the coastline, transects with the equal interval of 500 m are arranged along the baseline at the location of each coast transect as much as possible. In total, 500 transects are divided around the Abandoned Yellow River Delta from Shaoxiang Estuary, denoted by number 001, to Dongtai Estuary, denoted by number 500.

Table 1. Coastline baselines basic settings.

Path	Row	Baseline	Transect	Initial Coordinate Position	Final Coordinate Position	Length (m)
		1	001–077	119.501° E, 34.710° N	119.851° E, 34.526° N	38,071
		2	077–125	119.851° E, 34.526° N	120.090° E, 34.437° N	24,082
120	36	3	125–171	120.090° E, 34.437° N	120.307° E, 34.335° N	22,962
		4	171–219	120.307° E, 34.335° N	120.371° E, 34.124° N	24,169
		5	219–233	120.371° E, 34.124° N	120.404° E, 34.065° N	7215
119	37	6	233–363	120.404° E, 34.065° N	120.696° E, 33.532° N	64,781
		7	363–500	120.696° E, 33.532° N	120.973° E, 32.963° N	71,805

2.3. Satellite Remote Sensing Image Data

As an important way to obtain geographic information, remote sensing technology has the advantages of a relatively low data acquisition cost, wide coverage, and high spatial resolution; thus, it has been widely used in shoreline extraction for a long time. Pardo-Pascual et al. [33] believed that the Landsat images' detection accuracy of instantaneous water edges was comparable to high-resolution technology; therefore, the extraction of instantaneous water edges from Landsat series satellite images could relatively accurately describe landform features. In this study, the cloud density and the image clarity were used as the standard for satellite images screening. On this basis, a total of 501 satellite images, including Landsat5 TM(L5), Landsat7 ETM+(L7), and Landsat8 OLI_TIRS(L8), were obtained around the Abandoned Yellow River Delta from 1984 to 2019 through the Google Earth Engine platform. All the selected satellite images were used to extract the instantaneous water edges and seawalls.

Due to the large span of the study area, two satellite images were required to achieve full coverage. A total of 274 satellite images were screened out with path and row number (120, 36), of which the number of satellite images from L5, L7, and L8 were 134, 105, and 35, respectively. A total of 227 satellite images were screened out with path and row number (119, 37), of which the number of satellite images from L5, L7, and L8 were 112, 86, and 29, respectively. The details of all the satellite remote sensing image data are shown in Figure 2.

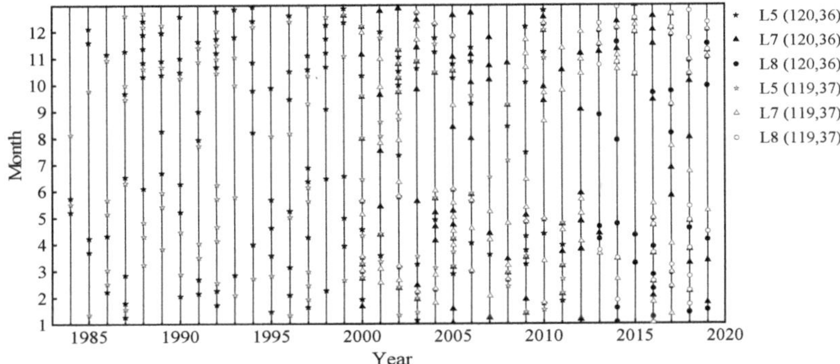

Figure 2. Satellite remote sensing image data.

2.4. Validation Data

In order to verify the rationality and accuracy of the results from the remote sensing analysis, a series of the coastline change rate data from historical observation is used as validation data. From 1980 to 1997, the Jiangsu Tidal Flat Management Bureau and Jiangsu Agricultural Resources Development Bureau conducted 18 years of continuous observations on the changes of the coastline in Jiangsu. They arranged a set of transects perpendicular to the local seawall on the representative coast. Along each transect, a number of cement piles were set from the seawall to the lowest tidal level as far as possible. Every year the elevations of the top of piles were measured, and the height of each pile above the surface of tidal flat was measured every season, as to obtain the profile of tidal flat at each transect. The data on the annual migration rate of average high tide level are available. The data can be found in the report "Research on Comprehensive Development Strategy of Jiangsu Coastal Area. Tidal Flat Volume. Evaluation and Reasonable Development and Utilization of Tidal Flat Resources in Jiangsu Coastal Area" [32].

3. Methodology

The semi-automated shoreline extraction method proposed by Daniels [34] is used to extract instantaneous water edges from satellite images as the basic data, while a visual interpretation is more convenient to extract the positions of seawalls or temporary dikes. Due to the fact the slope is very gentle around the Abandoned Yellow River Delta, the instantaneous water edge is greatly affected by the water level at the moment that the satellite image was taken. Without eliminating the influence of the fluctuating water level, it is nearly impossible to reliably analyze the actual spatiotemporal changes of the coastline using the instantaneous water edges. Thus, a correction to eliminate the influence of the fluctuating water level is first conducted, in order to obtain the datum-based coastline from the instantaneous water edge. There are the following two difficulties: (1) the slope in this area is varying spatiotemporally, and no available data can be used for each transect; (2) although we can obtain the tidal level from some public databases, it is not possible to obtain the accurate local water level at every moment. Moreover, the public tidal data are known to be not very accurate in the nearshore shallow water area, which has been confirmed by comparing them with the limited available measure data. To solve these problems, a new coastline correction algorithm that contains an intertidal zone slope estimation and water level correction is used to obtain the datum-based coastlines.

3.1. Coastline Correction Algorithm Process

Affected by the varying water level, the long-term variation trend of the coastline is concealed under the drastic change of instantaneous water edges. The distance of the instantaneous water edge from the baseline (denoted by C_{IWE}) at transect number s at time t can be divided into two parts, as expressed in Equation (1), including the distance of the mean sea level datum-based coastline (denoted by $C_{MSL}(s,t)$) and a deviation due to the fluctuation of water level (C_{WLF}).

$$C_{IWE}(s,t) = C_{MSL}(s,t) + C_{WLF}(s,t) \tag{1}$$

The object is to determine C_{MSL}, which is also a function of location and time. As C_{IWE} is known after image data extraction, the key is to estimate C_{WLF}. The variation of the water level can be divided into several components, including the long-term variation of the mean sea level, daily variation due to the tide, and the higher frequency variation corresponding to surges and wind waves. Assuming the tidal flat profile at each transect is approximately a slope, basically C_{WLF} can be calculated using Equation (2).

$$C_{WLF}(s,t) = \frac{H_{total}(s,t)}{S(s,t)} = \frac{H_T(s,t) + H_M(t) + H_R(s,t)}{S(s,t)} \tag{2}$$

In which, $H_{total}(s,t)$ is the local water level at transect number s and time t. $S(s,t)$ is the intertidal zone slope. $H_{total}(s,t)$ is further divided into three components. $H_T(s,t)$ is the tidal level, and is varying with time as well as locations, because the tidal range within the study area varies largely. The tidal level can be obtained directly from the public ocean tidal model [35], which includes the main shore-period tidal constituents. $H_M(t)$ is the mean sea level changes, which is treated as a constant within the area and only varying with time. The monthly mean sea level is approximately adopted, which is obtained from the "2019 China Sea Level Bulletin" [36]. $H_R(s,t)$ is called the random fluctuation of the water level, contributed to by surges and waves, as well as errors in the estimation of tidal level.

However, the random fluctuation of the water level and intertidal zone slope are unknown. An estimation for the two variables should be given firstly. Then, the procedure is divided into the following three major steps:

(1)   to estimate the random fluctuation of water level $H_R(s,t)$;
(2)   to obtain the intertidal zone slope at each transaction and time $S(s,t)$;
(3)   to calculate the mean sea level datum-based coastline (MSL coastline) $C_{MSL}(s,t)$.

The algorithm is designed based on the predictor–corrector technique. The detailed procedure of the coastline correction algorithm proposed in this study is shown in Figure 3, each step will be introduced in the following.

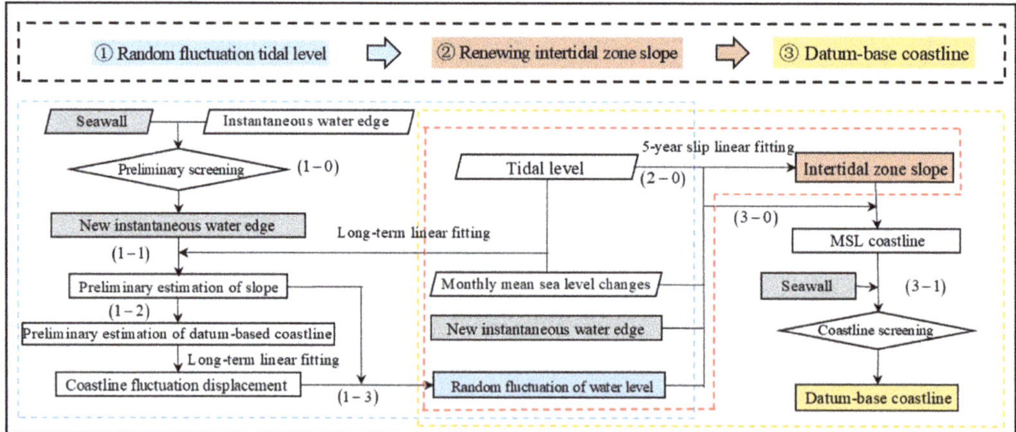

Figure 3. Procedure of the coastline correction algorithm.

3.2. Random Fluctuation of Water Level

The random fluctuation of water level is also reflected in the instantaneous water edge data, this section aims to decouple it from the variation of water edges. It includes several sub-steps, which are shown in the blue box of Figure 3.

Firstly, a preliminary estimation of the time averaged slope at each transect is given in sub-step (1–1). Generally, the instantaneous water edge approaches the coast at high tidal level and recedes to the sea at low tidal level, which provides the basis for the estimation of the intertidal zone slope. The influence of mean sea level H_M and tidal level H_T on the changes of the instantaneous water edge C_{IWE} is the most significant. Although the tidal level obtained from the public ocean tidal model may have phase lag and deviation in tidal range in shallow water area, the major tidal characteristics can be well given. Therefore, the gradient obtained from the linear regression between C_{IWE} and $H_M + H_T$ can be used for the preliminary estimation of intertidal zone slope.

Using linear regression is based on the assumption that the profile of tidal flat is a slope, which can be a reasonable approximation for the tidal flat in a natural environment. However, when the seawalls are located in the intertidal zone, the instantaneous water edges touch the seawalls at middle or high tidal level, and the slope directly estimated using the previous calculation method would introduce huge errors. Therefore, before the estimation of slope, transects at each image are checked and marked whether the positions of the seawall and instantaneous water edge are overlapping. The sub-step is called data screening and numbered as (1–0). If a transect is marked "overlap", it would not be considered in the linear regression for estimating the slope, in order to minimize the influence of anthropogenic disturbance. Considering the resolution of image, the distance between the instantaneous water edge and the seawall less than 60 m is used as the critical value for data screening.

Secondly, a preliminary correction of coastline is obtained considering $H_M + H_T$ as the first approximation of the water level, in order to roughly eliminate the variation of datum-based coastline. The preliminary estimation of datum-based coastline is calculated using Equation (3) in sub-step (1–2).

$$C'_{MSL} = C_{IWE} - \frac{H_T + H_M}{\overline{S_0}} \tag{3}$$

where C'_{MSL} is the preliminary estimation of the datum-based coastline, and $\overline{S_0}$ is the preliminary estimations of intertidal zone slope after the spatial slip average calculation. To weaken the influence of errors in the preliminary estimation of slope, a spatial average among the adjacent transects within a range of 2 km is adopted to obtain $S_0(s)$.

Thirdly, the random fluctuation of water level is estimated as sub-step (1–3). The calculation is based on Equation (2), but because we only have the preliminary slope and preliminary estimation of datum-based coastline, the formula is expressed as follows:

$$H_g = \left(C_{IWE} - \overline{C'_{MSL}}\right)\overline{S_0} - (H_T + H_M) \tag{4}$$

$\overline{C'_{IWE}}$ is the preliminary estimation of datum-based coastline after long-term linear fitting, to weaken the influence of errors induced in previous sub-steps. Considering the major part in the random fluctuation of water level has local spatial consistency, a spatial average among the adjacent transects within a range of 2 km is adopted to obtain the random fluctuation of water level for further analysis. Then, an estimation of the water level at each transect s and time t is given as follows:

$$H_{total}(s, t) = H_T + H_M + \overline{H_g} \tag{5}$$

3.3. Renewing Intertidal Zone Slope and Datum-Based Coastline

The core of the second step is to improve the accuracy of intertidal zone slope, which is shown in the red box in the middle of Figure 3. After the new estimation of water level is obtained, the slope at each transect can be renewed, through the relationship between the total water level and instantaneous water edges.

It should be noted that the slope of each transect changes with time. Theoretically, we can obtain the time-averaged slope between the taken moments of two images, when the instantaneous water edge and corresponding water level are accurate enough. However, it is hard to be absolutely accurate. Considering the errors induced by image resolution and others, a sliding window fitting is adopted to obtain a multi-year-average slope in sub-step (2–0). In this study, the sliding window is set to be 5 years corresponding to China's five-year plan.

Generally, the intertidal zone slopes will not change significantly within a certain spatial range, in order to further improve the accuracy of the intertidal zone slopes of each coastal transect, the 2-kilometer spatial scale is used for the spatial slip average calculation of the intertidal zone slopes.

Then, using the renewing slope and the estimation of water level, the datum-based coastline C_{MSL} can be calculated using Equations (1) and (2) in sub-step (3–0).

Considering the influence of anthropogenic disturbance on the MSL coastlines, it is necessary to compare the MSL coastlines with the seawalls. When the seawalls overlap the MSL coastlines, the seawalls are taken as datum-based coastlines, otherwise, the MSL coastlines are taken as the datum-based coastlines. The coastline screening is established as sub-step (3–1).

In addition, the instantaneous water edges and seawalls always overlap in the satellite images of certain coastal transects in some years, resulting in the lack of instantaneous water edge data for that year at preliminary screening. For this phenomenon, although the coastal transects with missing data are impossible to conduct post-analysis, in terms of probability, it is quite certain that the MSL coastlines overlaps with the seawalls. Therefore, the seawalls can be directly used as the datum-based coastlines in this type of the coastal transects.

3.4. Validity Analysis of Coastline Correction Algorithm

By establishing a relationship between the coastlines and time, it can reflect the temporal changes of coastline position. Figure 4 shows the temporal changes of coastline and instantaneous water edges at the location of some representative coastal transects around the Abandoned Yellow River Delta. Before the correction, the distributions of instantaneous water edge are relatively discrete; therefore, it is difficult to judge the changes of the coastline. However, after processing by the coastline correction algorithm proposed in this study, the changes of coastline can be clearly seen.

It can be seen that the present coastline correction algorithm can perform well in areas with large tidal range variation. Compared with the method based on the moving average composite images [22,23], both have their own advantages. The method based on the moving average composite images is not limited by the number of images used and has a high degree of automation based on the platform of the Google Earth Engine; therefore, it is more efficient to analyze the variation tendency of coastline. This type of method has been extended to the analysis of global coastlines, but the accuracy of the analysis results for specific areas with a high tidal range or gentle slope needs to be further verified. The present algorithm is a little complicated but has good applicability for coastline analysis in areas without accurate data of water level and slope.

In the analysis process of coastline change, the whole time series least-squares fitting is used to analyze the variation tendency of coastline in many studies. Actually, most of the coastal transects around the Abandoned Yellow River Delta often show staged changes, as shown in Figure 5, instead of linear changes, as shown in Figure 4. For this reason, when analyzing the spatiotemporal changes of coastline later, the process of coastline change analysis has considered the different development stages.

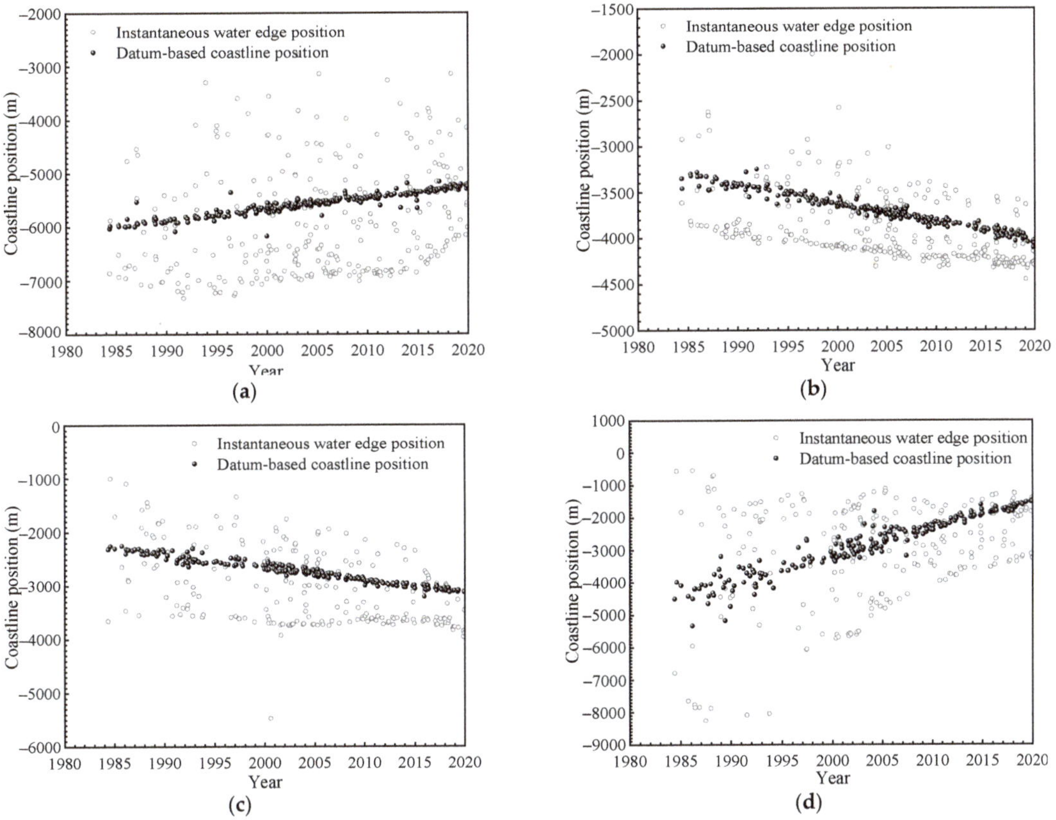

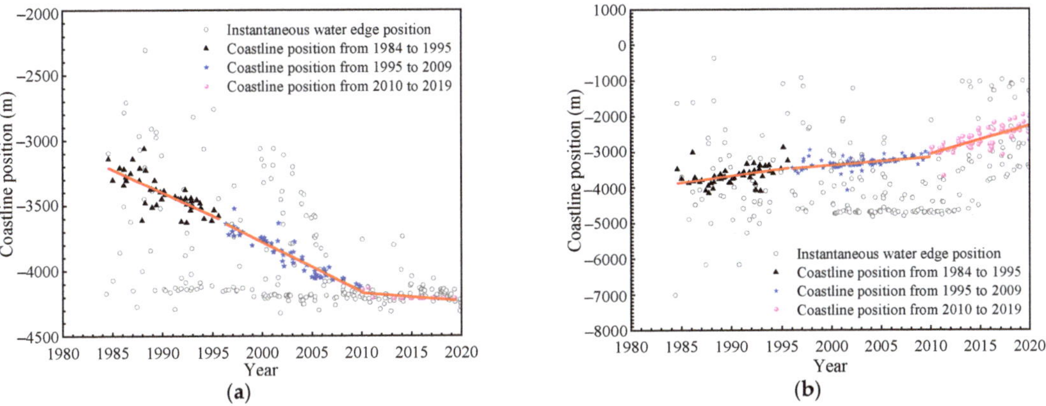

Figure 4. (**a**) The temporal changes of coastline position at the location of transect 81; (**b**) The temporal changes of coastline position at the location of transect 229; (**c**) The temporal changes of coastline position at the location of transect 331; (**d**) The temporal changes of coastline position at the location of transect 431.

Figure 5. (**a**) The temporal changes of coastline position at the location of transect 235; (**b**) The temporal changes of coastline position at the location of transect 292.

4. Results and Discussion

4.1. Phased Anthropogenic Reclamation

In order to obtain the overall variation tendency of a tidal flat resource under natural and anthropogenic disturbance around the Abandoned Yellow River Delta in 1984–2019, the natural tidal flat area, the reclamation area, and the total coastal area are defined and calculated. The natural tidal flat area in this study refers to the area between the seawall and the datum-based coastline, and the area between the baseline and the seawall (or temporary dike) is defined as the reclamation area. The area baseline is defined as the initial seawall in 1984. The total coastal area is the summation of the natural tidal flat area and the reclamation area, which is the area between the area baseline and the datum-based coastline. The changes of the area between the baseline and seawall, the datum-based coastline, over time, reflects the changes of the tidal flat resource under natural and anthropogenic disturbance.

As the policy for the utilization and conservation of the coastal zone in China has been changing over the past several decades, the coast around the abandoned Yellow River underwent a period of extensive reclamation, which can be taken as a typical coast with significant changes under anthropogenic disturbance. In order to reveal the specific changes of anthropogenic reclamation, the reclamation area in each year is calculated by accumulating all the areas at each transect, which are the product of the longitudinal distance between the seawall and the area baseline at each transect and the lateral distance between adjacent transects. The changes of the anthropogenic reclamation area around the Abandoned Yellow River Delta in 1984–2019 are shown in Figure 6. According to the change of the anthropogenic reclamation area in Figure 6, it is easily found that the net increase in the anthropogenic reclamation area is relatively slow before 1995 and after 2010, while relatively fast from 1995 to 2010. After China' reform and opening up, Jiangsu Province formulated a series of coastal tidal flat development strategies to guide people to strengthen the development and utilization of tidal flat resources, in order to promote the development of the coastal economy. Before 1995 or the end of the "Eighth Five-Year Plan" period, the development and utilization of tidal flat resources in Jiangsu entered a period with the main purpose of pursuing economic benefits. From the "Ninth Five-Year Plan" to the "Eleventh Five-Year Plan" period (1996–2010), a series of large-scale tidal flat development projects were initiated, which left an extensive reclamation period in history. After 2010, Jiangsu Province implemented the "Outline of Jiangsu Coastal Flat Reclamation Development and Utilization Plan" and emphasized the sustainable development of tidal flat resources. Therefore, the changes of anthropogenic reclamation around the Abandoned Yellow River Delta are divided into the following three stages: 1984 to 1995 is the stage before extensive reclamation, 1996 to 2009 is the extensive reclamation stage, and 2010 to 2019 is the stage after extensive reclamation.

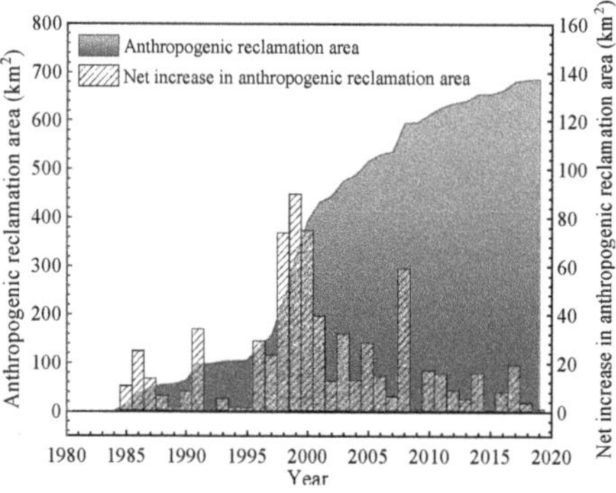

Figure 6. Change of anthropogenic reclamation area around the Abandoned Yellow River Delta.

4.2. Spatial Distribution of Coastline Migration before Extensive Reclamation

As the intensity of anthropogenic reclamation is relatively low in the early stage, anthropogenic disturbance has a relatively weak influence on the coastline migration, and natural evolution is the main cause of the coastal migration. Therefore, the changes of coastline in the early stage are used to show the spatial distribution characteristics of coastline migration under the natural dynamic condition. As shown in Figure 7, the ordinate represents the coastline migration rate, and the abscissa represents the transect number that gradually increases from 1 to 500 from north to south. The dotted line in the middle is the boundary of the satellite images with different path numbers. Although the coastline migration rate varies greatly from one transect to another, there is still some characteristics. It can be seen obviously that the coastline shows erosion tendency at most of transects in the north part (transect number 1–342) and shows deposition tendency in the south part (transect number 343–500). The transitional area is between Xinyang Port (transect number 333) and Doulong Port (transect number 372). It agrees well with the spatial distribution of the deposition and erosion of the coast reported by Yu and Huang [32].

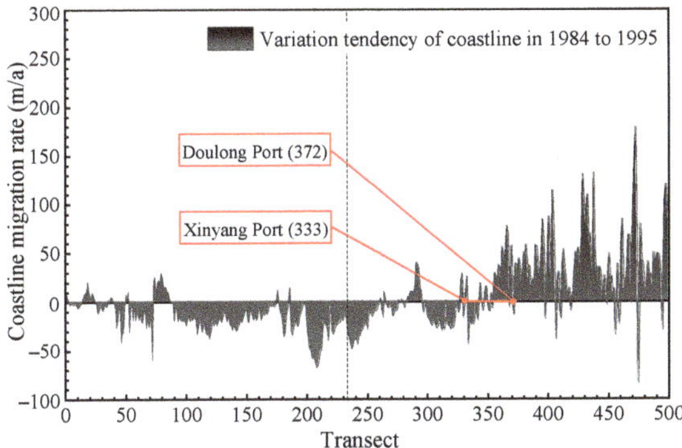

Figure 7. Variation tendency of coastline in 1984 to 1995.

Figure 8 shows a comparison of the coastline migration rate with the observed data collected from Yu and Huang [32]. It can be seen that the two reasonably agree with each other. The overall characteristics of the coastal migration obtained from the remote sensing images are consistent with the field survey data. As the survey data are obtained from the measurement of the tidal flat profile at several years during 1980–1997, some differences may occur due to the calculation method of coastline migration in the field survey and present study.

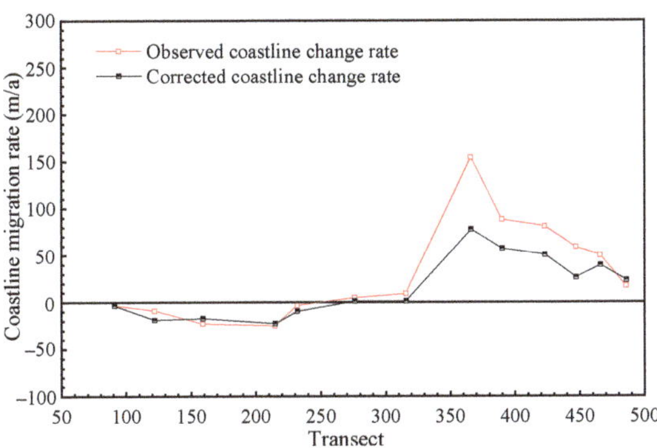

Figure 8. Validation of algorithm results.

Local characteristics of coastline migration are summarized in Table 2. Between two river estuaries, the mean coastline migration rate is computed, and the trend is given. Generally, the north part, from Shaoxiang Estuary to Xinyang Port, is more significantly influenced by the diversion of the Ancient Yellow River in 1855. After the sources of sediment transportation were cut off, this part of the coast has undergone persistent erosion. Under the influence of coastal currents, some of the eroded sediment was transported to the south and deposited. Therefore, the southern part, from the Doulong Port to Dongtai Estuary, shows an overall deposition trend. It would be interesting to investigate the response of the natural tidal flat to the extensive reclamation and to see the difference under erosion conditions and deposition conditions. In the following, the concerned coast is divided into the north erosion part and the south deposition part.

Table 2. Spatial distribution characteristics of each coast.

Area	Starting Coast	Ending Coast	Transect	Changes of Coast
North	Shaoxiang Estuary	Sanyu Port	001–046	Erosion
	Sanyu port	Guan Estuary	046–074	Erosion
	Guan Estuary	Zhongshan Estuary	074–132	Erosion
	Zhongshan Estuary	Abandoned Yellow River Estuary	132–186	Erosion
	Abandoned Yellow River Estuary	Biandan Port	186–219	Erosion
	Biandan Port	Shuangyang Port	219–248	Erosion
	Shuangyang port	Sheyang Estuary	248–290	Erosion
	Sheyang Estuary	Xinyang Port	290–333	Erosion
	Xinyang Port	Doulong Port	333–372	Transition
South	Doulong Port	Simaoyou Port	372–404	Deposition
	Simaoyou Port	Wang Port	404–438	Deposition
	Wang Port	Chuandong Port	438–453	Deposition
	Chuandong port	Dongtai Estuary	453–500	Deposition

4.3. Variation of Tidal Flat Area with Extensive Reclamation

To show the influence of reclamation, the natural tidal flat area in the north part and the south part at each year has been calculated. The variation tendency of the natural tidal flat area around the Abandoned Yellow River Delta is shown in Figure 9. Before the coastline correction, the changes of the instantaneous tidal flat area with time are scattered; therefore, it is almost impossible to observe the variation tendency of the tidal flat area. After the coastline correction, it can be clearly seen that the tidal flat area basically decreased in both the north part and the south part. It is obvious that the continuous reclamation has greatly reduced the natural tidal flat area. For the north part, the area in 2019 is reduced to only 43% of the original area in 1984. While the reduction in the south part was severer, there is only 27% of the natural tidal flat area left. The decrease in the natural tidal flat area each year greatly corresponds with the reclamation, as shown in Figure 10. In Figure 10, the changes of the anthropogenic reclamation area in the north part and south part are given. Corresponding to the three stages of reclamation, the change of the natural tidal flat area also has three stages, both in the north part and the south part. The change trend shows a relatively slow decrease before an extensive reclamation period, and a rapid decrease during the period, and a slow decrease again in the later period. For the negative increase in anthropogenic reclamation in Figure 10, it is mainly due to the destruction of temporary dikes or the coastal restoration, which has led to a slight increase in the tidal flat area in some years in Figure 9a.

When comparing the north part and the south part, the difference in the variation of the natural tidal flat area can be seen. In the north part, except the years affected by the destruction of temporary dikes, all the years shows a decreasing trend, even in the years with little reclamation. However, it can be seen that in the south part, the natural tidal area can recover after reclamation in some years, for example 1992–1995, after the large reclamation in 1991. We can also see a slight increasing trend in recent years after the extensive period. It is obvious that the natural tidal flat in the deposition coast can slowly recover, but in the erosion coast it would disappear forever after being occupied by reclamation.

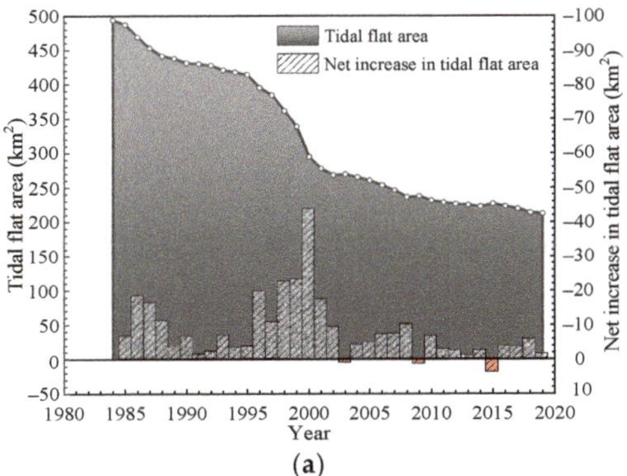
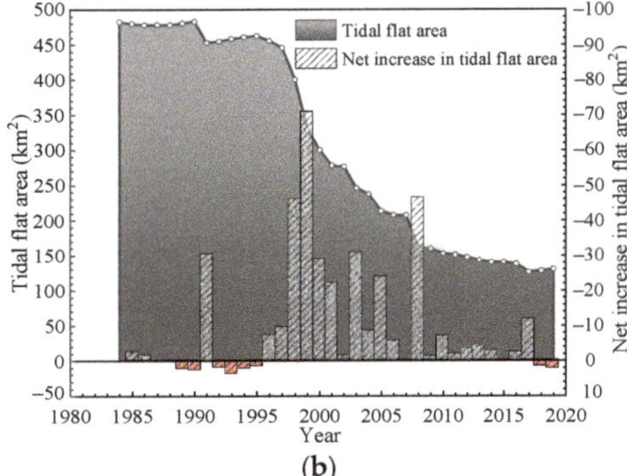

Figure 9. (**a**) The variation tendency of natural tidal flat area in the north erosion part; (**b**) The variation tendency of natural tidal flat area in the south deposition part.

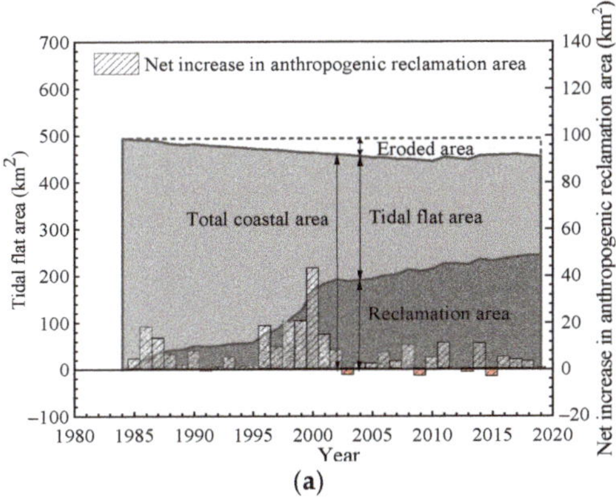
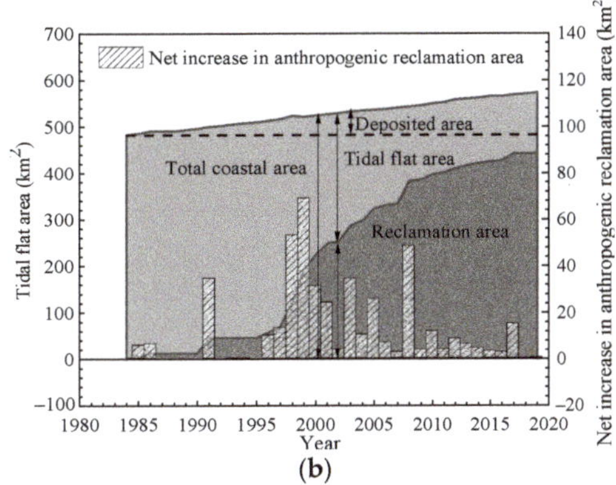

Figure 10. (**a**) The changes of anthropogenic reclamation area in the north erosion part; (**b**) The changes of anthropogenic reclamation area in the south deposition part.

Both anthropogenic reclamation and natural evolution have an influence on the variation of the tidal flat, but it is clear that anthropogenic reclamation plays the most important role. The reduction in the natural tidal flat area corresponds to the narrowing of the tidal flat width. Figure 11 shows the tidal flat width at each transect in 1984, 1995, 2010, and 2019. It should be noted that the tidal flat width defined here is the distance of the seawall and datum-base coastline, which is smaller than the width of the intertidal zone.

Whether before, during, or after extensive reclamation, the average width of the tidal flat in the north erosion part is relatively narrower, while relatively wider in the south deposition part. It can be seen that the reduction ratio in the erosion part is smaller than the deposition part. Certainly, the main reason is that reclamation in the deposition transects is much more than that in the erosion transects. However, it is also because of the adjustment of the tidal flat profile. By the coastline correction algorithm, we also can obtain an estimation of the slope of intertidal zone. In the next section, the characteristics of the intertidal zone slope and its response to the extensive reclamation are discussed.

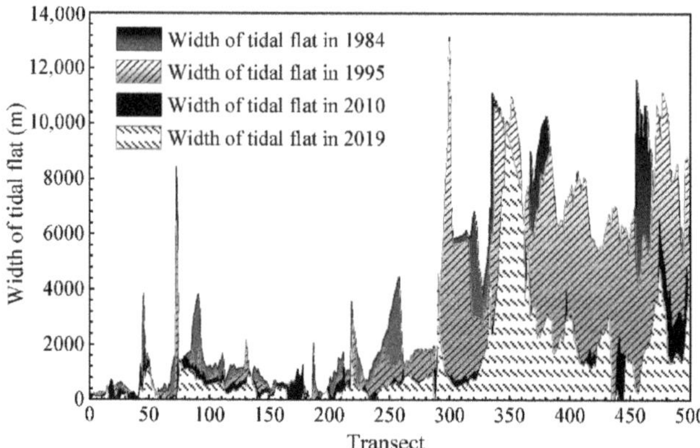

Figure 11. The spatial distribution of tidal flat width with time.

4.4. Spatiotemporal Changes of Intertidal Zone Slope

In order to analyze the spatial distributions of the intertidal zone slope, a temporal mean slope at each transect is given, as shown in Figure 12. Generally, the slope in the study area is very gentle within the range of 5.5–0.3‰. Although the variation with different transects is large, it can still see as a general feature that the northern erosion coast with a narrow tidal flat has relatively steep slopes, while the southern deposition coast with a wide tidal flat has relatively gentle slopes. The intertidal zone slope from north to south gradually becomes gentler. It should be noted that the slope is calculated along each transect, which may not completely perpendicular to the isobaths.

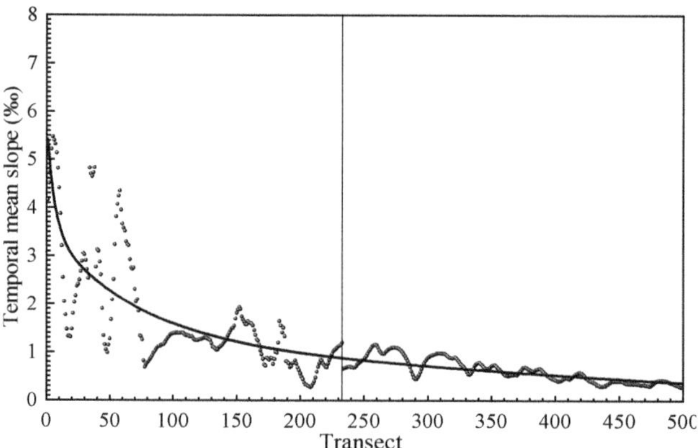

Figure 12. The spatial distributions of intertidal zone slope.

As revealed in the previous section, extensive reclamation has been conducted in this area and the width of the tidal flat continues to shrink with time. During this process, the tidal flat profile has also been continuously evolving under natural dynamic conditions. A representative parameter is the intertidal zone slope. In order to see the variation tendency of the intertidal zone slope over time, the averaged intertidal zone slopes in the north part and the south part are calculated, as shown in Figure 13, in order to show the response of the tidal flat profile to the extensive reclamation.

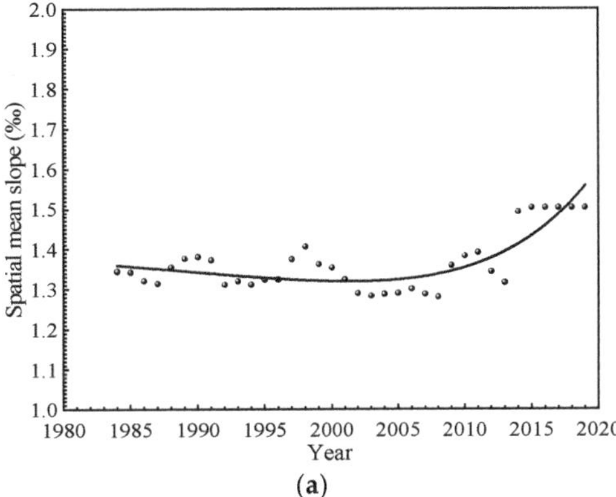 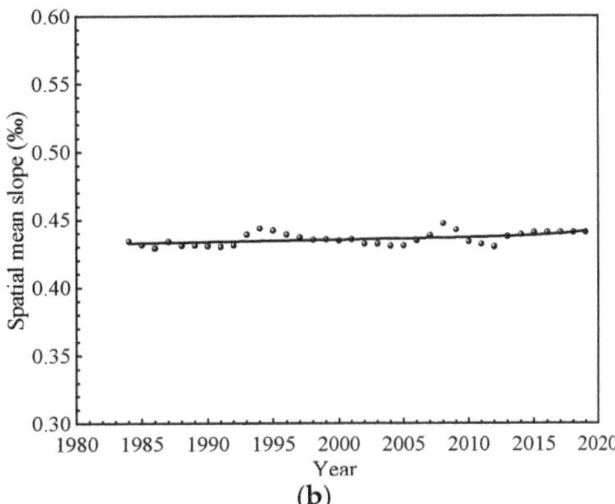

Figure 13. (**a**) The temporal distribution of intertidal zone slope in the north erosion part; (**b**) The temporal distribution of intertidal zone slope in the south deposition part.

The difference between the erosion part and the deposition part can be seen. As shown in Figure 13, intertidal zone slopes gradually become steeper and steeper with time in the north erosion part. A clear steepening trend can be seen after the extensive reclamation period. However, in the south deposition part, the overall slope did not change much, although in both parts, the width of the tidal flat gradually becomes narrower. That is because the original tidal flat width is relatively wide in the south part and natural deposition can alleviate the impact of anthropogenic reclamation on the narrowing of tidal flat width to a certain extent. It is further demonstrated that reclamation in deposition coasts has relatively less influence than that in the erosion coast.

4.5. Influence of Extensive Reclamation Cumulative Effect on the Biological Habitats

Although extensive reclamation has great impact on the coastline changes, when the reclamation intensities are within a certain range, it will not affect the natural coastline changes. However, it should be noticed that continued reclamation has a cumulative effect. As some species need to survive in a supratidal zone, the disappearance of the supratidal zone may lead to the destruction of biological habitats and have a negative impact on the biodiversity of coastal ecosystems.

In order to explore the year when extensive reclamation has a sharp impact on the changes of the coastline around the Abandoned Yellow River Delta, we counted the number of transects whose supratidal zones disappeared in each year, which is defined in this study as when the datum-based coastline overlaps with the seawall position. The percentage of those transects in the north part and the south part is used to further reveal the influence of extensive reclamation on the coastline changes after 1984, as shown in Figure 14.

It can be seen from Figure 14 that the percentage of coast without a supratidal zone shows an overall increasing trend with time in the north part, among which change trend shows a relatively slow increase before an extensive reclamation period, and a rapid increase during the period, and a slow increase again in the later period. However, in the south deposition part, a sudden change occurs in the year 2010. It is always zero before 2010 and then increases rapidly, which means that the extensive reclamation has sharp influence on coastline changes in 2010.

Both anthropogenic and natural factors in the north erosion part have contributed to the continuous reduction in the tidal flat area, among which extensive reclamation has played a stronger leading role. Although the natural factors in the south deposition part have contributed to the increase in the tidal flat area, the extensive reclamation has also led to the decrease in the tidal flat area. Whether in the north erosion part or the south deposition part, extensive reclamation is the main reason for the decrease in the tidal flat area, which is also the main driving force for the cumulative effect.

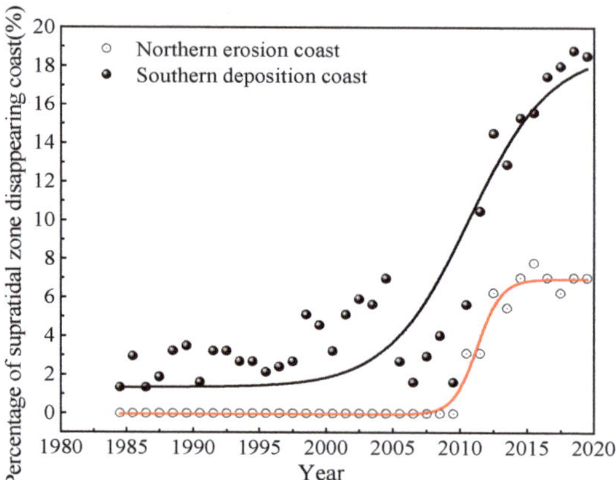

Figure 14. Percentage of coast without supratidal zone around the Abandoned Yellow River Delta.

Due to the fact that the width of the tidal flat in the north erosion part is narrow, reclamation has already stimulated the cumulative effect in the early stage, but the cumulative effect is relatively weak, the influence of reclamation on the coastline changes is relatively small; therefore, the percentage of the supratidal zone disappearing coast is relatively low. During the extensive reclamation period, the cumulative effect surged; extensive reclamation has a great impact on the coastline changes, which leads to a further rapid increase in the percentage of the supratidal zone disappearing coast. In the later period, affected by the marine protection policy, the intensity of reclamation becomes lower, the increase in the cumulative effect becomes slow; therefore, the percentage of the supratidal zone disappearing coast slowly increases again.

Although extensive reclamation has led to the reduction in the tidal flat area on the southern deposition coast, due to the fact that the width of tidal flat in the south deposition part is wide and the natural evolution has continuous deposition, reclamation has not stimulated the cumulative effect before an extensive reclamation period and during the period. With the accumulation of the reclamation area in the south deposition part, the cumulative effect was stimulated in 2010; therefore, a sudden change of the percentage of the supratidal zone disappearing coast occurs in the year 2010. However, affected by the marine protection policy in the later period, the cumulative effect in the south part has not been fully increased before it has been flattened. Therefore, the extensive reclamation has greater influence on the coastline changes in the erosion coast with a narrow tidal flat than the deposition coast with a wide tidal flat.

When the extensive reclamation cumulative effect is stimulated, the datum-based coastline of some coasts overlaps with the seawall, leading to the disappearance of the tidal flat habitat in the supratidal zone. With the further accumulation of extensive reclamation, the cumulative effect is gradually obvious, and more coastal biological habitats are lost. Moreover, the study area is an important protected area for migratory bird migration and habitats in the world, and endangered animals such as red-crowned cranes and elk mainly live here; therefore, the destruction of a large area of biological habitats may have a negative impact on the biodiversity of coastal ecosystems. Therefore, appropriate marine protection measures must be strictly strengthened to control the development of the cumulative effect, in order to guide the sustainable development of tidal flat resources around the Abandoned Yellow River Delta in the future.

5. Conclusions

In order to analyze the variation tendency of the coastline, tidal flat area, and tidal flat slope under natural and anthropogenic disturbance around the Abandoned Yellow River Delta before, during, and after extensive reclamation during the period of 1984–2019, this study used semi-automatic shoreline extraction technology and a visual interpretation correction method to extract the instantaneous water edges and seawalls from the 500 Landsat series satellite remote sensing images. At the same time, a new coastline correction algorithm had been proposed to obtain the datum-based coastline, aimed to eliminate the influence of varying water level and to better show the variation tendency of the coastline. In addition, the influence of seawalls on the coastlines and the slope estimation were taken into consideration in the algorithm. Based on the obtained coastline and seawall position, the variation tendency of the coastline around the Abandoned Yellow River Delta had been analyzed, comprehensively considering the natural evolution and anthropogenic disturbance.

The results show the variation tendency of the very representative tidal flat coast in recent decades. Affected by the historical diversion of the abandoned Yellow River estuary, this section of the coast shows the characteristics of erosion tendency in the north part and deposition tendency in the south part, which provides a good research object for the comparative analysis of the response of the north erosion coast and the south deposition coast. In addition, this section of the coast has a continuous anthropogenic reclamation from 1984 to 2019, of which the reclamation intensity is the most intense in 1995–2010. Regarding 1995–2010 as the extensive reclamation period, satellite remote sensing images show the changes of the tidal flat area before, during, and after the extensive reclamation period, especially the sharp decrease in the tidal flat area during the extensive reclamation stage. Both anthropogenic and natural factors in the north erosion part have contributed to the continuous reduction in the tidal flat area, among which extensive reclamation has played a stronger leading role and the tidal flat area in 2019 was reduced to only 43% of the original area in 1984. Although the natural factors in the south deposition coast have contributed to the increase in the tidal flat area, the extensive reclamation has led to the decrease in the tidal flat area and the tidal flat area in 2019 was reduced to only 27% of the original area in 1984.

Under the anthropogenic disturbance, we can also see the response of natural evolution to anthropogenic reclamation. Due to the fact that the reclamation area in the south deposition part is larger than the north erosion area, the reduction ratio of tidal flat width in the south deposition part is higher. Actually, the spatial distribution of tidal flat width is relatively narrower in the north erosion part and relatively wider in the south deposition part all the time; even if the reduction ratio of tidal flat width in the south deposition part is higher, the tidal flat width is still relatively wider than the north erosion part. Affected by the extensive reclamation, the adjustment of the tidal flat profile makes the south deposition part have relatively gentler slopes than the north erosion part, and the intertidal zone slopes in the erosion part become steeper along with the extensive reclamation period, while the slope in the deposition part remains stable.

The cumulative effect of anthropogenic reclamation is worthy of attention. The disappearance of the supratidal zone will have influence on the loss of biological habitats and the destruction of coastal ecosystems. The results indicate that there is obvious supratidal zone disappearance during or after the extensive reclamation period in this area. Fortunately, affected by the marine protection policy in the later period, the cumulative effect has been flattened. Therefore, the policy around the Abandoned Yellow River Delta is adjusted in time to avoid further degradation of the intertidal zone, which is benefit to the future protection of biodiversity and the sustainable development of coastal ecosystems.

Author Contributions: Conceptualization, Z.S. and X.N.; methodology, Z.S.; validation, Z.S.; resources, Z.S. and X.N.; writing—original draft preparation, Z.S.; writing—review and editing, Z.S. and X.N.; supervision, X.N.; funding acquisition, X.N. Both authors have read and agreed to the published version of the manuscript.

Funding: This research was funded by National Key Research and Development Program of China, grant number 2018YFC0407504.

Data Availability Statement: All the Landsat data are available at no cost here: https://earthexplorer.usgs.gov/ (accessed on 17 July 2021).

Acknowledgments: The authors would like to acknowledge the support by National Key Research and Development Program of China under grant No. 2018YFC0407504. Meanwhile, the authors sincerely thank all anonymous reviewers and the editors for their constructive and excellent reviews that greatly improve the article quality.

Conflicts of Interest: The authors declare no conflict of interest.

References

1. Galbraith, H.; Jones, R.; Park, R.; Park, R.; Clough, J.; Herrod-Julius, S.; Harrington, B.; Page, G. Global Climate Change and Sea Level Rise: Potential Losses of Intertidal Habitat for Shorebirds. *Waterbirds* **2002**, *25*, 173–183. [CrossRef]
2. Liu, C.L.; Zhang, Z.K.; Zhang, X.Q.; Ren, H.; Liu, B.L.; Bao, Z.L.; Zhang, X.X. Coastal development history and tidal flat ecosystem conservation along the coast of Jiangsu Province, China. *J. Coast. Conserv.* **2018**, *23*, 857–867. [CrossRef]
3. Zhang, T.H.; Niu, X.J. Analysis on the utilization and carrying capacity of coastal tidal flat in bays around the Bohai Sea. *Ocean Coast. Manag.* **2020**, *203*, 105449. [CrossRef]
4. Imhoff, M.L.; Bounoua, L.; DeFries, R.; Lawrence, W.T.; Stutzer, D.; Tucker, C.J.; Ricketts, T. The consequences of urban land transformation on net primary productivity in the United States. *Remote Sens. Environ.* **2004**, *89*, 434–443. [CrossRef]
5. Johnson, B.G.; Zuleta, G.A. Land-use land-cover change and ecosystem loss in the Espinal ecoregion, Argentina. *Agric. Ecosyst. Environ.* **2013**, *181*, 31–40. [CrossRef]
6. Zhang, Y.; Wang, T.W.; Cai, C.F.; Li, C.G.; Liu, Y.J.; Bao, Y.Z.; Guan, W.H. Landscape pattern and transition under natural and anthropogenic disturbance in an arid region of northwestern China. *Int. J. Appl. Earth Obs. Geoinf.* **2016**, *44*, 1–10. [CrossRef]
7. Zhou, Y.K.; Ning, L.X.; Bai, X.L. Spatial and temporal changes of human disturbances and their effects on landscape patterns in the Jiangsu coastal zone, China. *Ecol. Indic.* **2018**, *93*, 111–122. [CrossRef]
8. Zhao, S.D.; Zhang, Y.M. Ecosystems and human well-being: The achievements, Contributions and Prospects of the Millennium Ecosystem Assessment. *Adv. Earth Sci.* **2006**, *21*, 895–902.
9. Liu, J.Y.; Zhang, Z.X.; Xu, X.L.; Kuang, W.H.; Zhou, W.C.; Zhang, S.W.; Li, R.D.; Yan, C.Z.; Yu, D.S.; Wu, S.X.; et al. Spatial patterns and driving forces of land use change in China during the early 21st century. *J. Geogr. Sci.* **2010**, *20*, 483–494. [CrossRef]
10. Madeira, C.; Mendonca, V.; Leal, M.C.; Flores, A.A.V.; Cabral, H.N.; Diniz, M.S.; Vinagre, C. Environmental health assessment of warming coastal ecosystems in the tropics—Application of integrative physiological indices. *Sci. Total Environ.* **2018**, *643*, 28–39. [CrossRef]
11. Sowman, M.; Raemaekers, S. Socio-ecological vulnerability assessment in coastal communities in the BCLME region. *J. Mar. Syst.* **2018**, *188*, 160–171. [CrossRef]
12. An, S.Q.; Li, H.B.; Guan, B.H.; Zhou, C.F.; Wang, Z.S.; Deng, Z.F.; Zhi, Y.B.; Liu, Y.H.; Xu, C.; Fang, S.B.; et al. China's natural wetlands: Past problems, current status, and future challenges. *Ambio* **2007**, *36*, 335–342. [CrossRef]
13. Wang, H.J.; Yang, Z.S.; Saito, Y.; Liu, J.P.; Sun, X.X.; Wang, Y. Stepwise decreases of the Huanghe (Yellow River) sediment load (1950–2005): Impacts of climate change and human activities. *Glob. Planet. Chang.* **2007**, *57*, 331–354. [CrossRef]
14. Liu, T.; Shi, X.F.; Li, C.X.; Yang, G. The reverse sediment transport trend between abandoned Huanghe River (Yellow River) Delta and radial sand ridges along Jiangsu coastline of China-an evidence from grain size analysis. *Acta Oceanol. Sin.* **2012**, *31*, 83–91. [CrossRef]
15. Liu, Z.X.; Xia, D.X. *Tidal Sands in the China Seas*; China Ocean Press: Beijing, China, 2004. (In Chinese)
16. Zhu, Q.T.; Li, P.; Li, Z.H.; Pu, S.X.; Wu, X.; Bi, N.S.; Wang, H.J. Spatiotemporal Changes of Coastline over the Yellow River Delta in the Previous 40 Years with Optical and SAR Remote Sensing. *Remote Sens.* **2021**, *13*, 1940. [CrossRef]
17. Wang, X.G.; Yan, F.Q.; Su, F.Z. Changes in coastline and coastal reclamation in the three most developed areas of China, 1980–2018. *Ocean Coast. Manag.* **2021**, *204*, 105542. [CrossRef]
18. Cao, W.T.; Zhou, Y.Y.; Li, R.; Li, X.C. Mapping changes in coastlines and tidal flats in developing islands using the full time series of Landsat images. *Remote Sens. Environ.* **2020**, *239*, 111665. [CrossRef]
19. Li, W.Y.; Gong, P. Continuous monitoring of coastline dynamics in western Florida with a 30-year time series of Landsat imagery. *Remote Sens. Environ.* **2016**, *179*, 196–209. [CrossRef]
20. Ford, M. Shoreline changes interpreted from multi-temporal aerial photographs and high resolution satellite images: Wotje Atoll, Marshall Islands. *Remote Sens. Environ.* **2013**, *135*, 130–140. [CrossRef]
21. Konlechner, T.M.; Kennedy, D.M.; O'Grady, J.J.; Leach, C.; Ranasinghe, R.; Carvalho, R.C.; Luijendijk, A.P.; McInnes, K.L.; Ierodiaconou, D. Mapping spatial variability in shoreline change hotspots from satellite data; a case study in southeast Australia. *Estuar. Coast. Shelf Sci.* **2020**, *246*, 107018. [CrossRef]
22. Luijendijk, A.; Hagenaars, G.; Ranasinghe, R.; Baart, F.; Donchyts, G.; Aarninkhof, S. The State of the World's Beaches. *Sci. Rep.* **2018**, *8*, 6641. [CrossRef]

23. Hagenaars, G.; de Vries, S.; Luijendijk, A.P.; de Boer, W.P.; Reniers, A.J.H.M. On the accuracy of automated shoreline detection derived from satellite imagery: A case study of the sand motor mega-scale nourishment. *Coast. Eng.* **2018**, *133*, 113–125. [CrossRef]
24. Eman, G.; Jehan, M.; Douglas, G.; Joanne, H.; Mostafa, A. Nile Delta exhibited a spatial reversal in the rates of shoreline retreat on the Rosetta promontory comparing pre- and post-beach protection. *Geomorphology* **2015**, *228*, 1–14.
25. Castelle, B.; Masselink, G.; Scott, T.; Stokes, C.; Konstantinou, A.; Marieu, V.; Bujan, S. Satellite-derived shoreline detection at a high-energy meso-macrotidal beach. *Geomorphology* **2021**, *383*, 107707. [CrossRef]
26. Boak, E.H.; Turner, I.L. Shoreline definition and detection: A review. *J. Coast. Res.* **2005**, *21*, 688–703. [CrossRef]
27. Chu, Z.X.; Sun, X.G.; Zhai, S.K.; Xu, K.H. Changing pattern of accretion/erosion of the modem Yellow River (Huanghe) subaerial delta, China: Based on remote sensing images. *Mar. Geol.* **2006**, *227*, 13–30. [CrossRef]
28. Chen, W.W.; Chang, H.K. Estimation of shoreline position and change from satellite images considering tidal variation [J]. Estuarine. *Coast. Shelf Sci.* **2009**, *84*, 54–60. [CrossRef]
29. Liu, Y.X.; Huang, H.J.; Qiu, Z.F.; Fan, J.Y. Detecting coastline change from satellite images based on beach slope estimation in a tidal flat. *Int. J. Appl. Earth Obs. Geoinf.* **2013**, *23*, 165–176. [CrossRef]
30. Crowell, M.; Leatherman, S.P.; Buckley, M.K. Historical shoreline change: Error analysis and mapping accuracy. *J. Coast. Res.* **1991**, *7*, 839–852.
31. Maiti, S.; Bhattacharya, A.K. Shoreline change analysis and its application to prediction: A remote sensing and statistics based approach. *Mar. Geol.* **2009**, *257*, 11–23. [CrossRef]
32. Yu, X.P.; Huang, F. *Research on Comprehensive Development Strategy of Jiangsu Coastal Area. Tidal Flat Volume. Evaluation and Reasonable Development and Utilization of Tidal Flat Resources in Jiangsu Coastal Area*; Jiangsu People's Publishing House: Nanjing, China, 2008. (In Chinese)
33. Pardo-Pascual, J.E.; Almonacid-Caballer, J.; Ruiz, L.A.; Palomar-Vazquez, J. Automatic extraction of shorelines from Landsat TM and ETM+ multi-temporal images with subpixel precision. *Remote Sens. Environ.* **2012**, *123*, 1–11. [CrossRef]
34. Daniels, R.C. Using ArcMap to extract shorelines from Landsat TM & ETM+ data. In Proceedings of the 32nd ESRI International Users Conference, San Diego, CA, USA, 23–27 July 2012; pp. 1–23.
35. Egbert, G.D.; Erofeeva, S.Y. Efficient inverse modeling of barotropic ocean tides. *J. Atmos. Ocean. Technol.* **2002**, *19*, 183–204. [CrossRef]
36. *2019 China Sea Level Bulletin*; Department of Marine Early Warning and Monitoring, Ministry of Natural Resources of the People's Republic of China: Beijing, China, 2020. (In Chinese)

remote sensing

Article

Fine-Resolution Mapping of Pan-Arctic Lake Ice-Off Phenology Based on Dense Sentinel-2 Time Series Data

Chong Liu [1,2], Huabing Huang [1,2], Fengming Hui [1,2], Ziqian Zhang [1] and Xiao Cheng [1,2,*]

[1] School of Geospatial Engineering and Science, Sun Yat-Sen University, Guangzhou 510275, China; liuc@mail.sysu.edu.cn (C.L.); huanghb55@mail.sysu.edu.cn (H.H.); huifm@mail.sysu.edu.cn (F.H.); zhangzq55@mail2.sysu.edu.cn (Z.Z.)

[2] Southern Marine Science and Engineering Guangdong Laboratory (Zhuhai), Zhuhai 519000, China

* Correspondence: chengxiao9@mail.sysu.edu.cn

Abstract: The timing of lake ice-off regulates biotic and abiotic processes in Arctic ecosystems. Due to the coarse spatial and temporal resolution of available satellite data, previous studies mainly focused on lake-scale investigations of melting/freezing, hindering the detection of subtle patterns within heterogeneous landscapes. To fill this knowledge gap, we developed a new approach for fine-resolution mapping of Pan-Arctic lake ice-off phenology. Using the Scene Classification Layer data derived from dense Sentinel-2 time series images, we estimated the pixel-by-pixel ice break-up end date information by seeking the transition time point when the pixel is completely free of ice. Applying this approach on the Google Earth Engine platform, we mapped the spatial distribution of the break-up end date for 45,532 lakes across the entire Arctic (except for Greenland) for the year 2019. The evaluation results suggested that our estimations matched well with both in situ measurements and an existing lake ice phenology product. Based on the generated map, we estimated that the average break-up end time of Pan-Arctic lakes is 172 ± 13.4 (measured in day of year) for the year 2019. The mapped lake ice-off phenology exhibits a latitudinal gradient, with a linear slope of 1.02 days per degree from 55°N onward. We also demonstrated the importance of lake and landscape characteristics in affecting spring lake ice melting. The proposed approach offers new possibilities for monitoring the seasonal Arctic lake ice freeze–thaw cycle, benefiting the ongoing efforts of combating and adapting to climate change.

Keywords: arctic lake; ice-off phenology; dense time series; Sentinel-2

Citation: Liu, C.; Huang, H.; Hui, F.; Zhang, Z.; Cheng, X. Fine-Resolution Mapping of Pan-Arctic Lake Ice-Off Phenology Based on Dense Sentinel-2 Time Series Data. *Remote Sens.* **2021**, *13*, 2742. https://doi.org/10.3390/rs13142742

Academic Editors: Adrian Ursu and Cristian Constantin Stoleriu

Received: 13 May 2021
Accepted: 7 July 2021
Published: 13 July 2021

Publisher's Note: MDPI stays neutral with regard to jurisdictional claims in published maps and institutional affiliations.

1. Introduction

Lakes, through their seasonal ice cover extent and duration, are a major landscape in the Arctic [1–5]. Governed by both geographical and morphological settings, the spring lake ice phenology (i.e., timing of ice-off) is a primary driver of Arctic hydrological processes [6], and has been identified as a sensitive metric to measure climate change [7–9]. Regional coherence or synchrony of lake break-up timing also regulates greenhouse gas emissions from fresh water into the atmosphere [9–12], affecting the land surface energy balance [13], which in turn generates feedbacks into the climate system. Given their wide-ranging importance, accurately estimating the spatial distribution of lake ice-off phenology is essential to understand Arctic ecosystem functions and project their responses to global environmental dynamics under various climate change scenarios [14,15].

Before the advent of remote sensing, lake ice-off phenology data could only be collected through in-field observations, which are nevertheless laborious and spatially sparse [16]. Satellite remote sensing has revolutionized our ability to monitor fresh water freeze–thaw cycling processes, especially at the continental to global scales [17]. Using high temporal frequency satellite image time series, many attempts have been made to extract lake ice-off timing information [18]. The microwave brightness temperature, with long available archives and the all-weather imaging capability, is perhaps the most widely used data source for lake ice phenology parameter identification [15,19–21]. However, most of the passive microwave sensors have coarse spatial resolutions (typically ranging from 6.25 km to 25 km), hence limiting their effectiveness within heterogeneous landscapes [15]. Therefore, other remote sensing data types with higher spatial resolutions, including synthetic aperture radar (SAR) [7,22], optical imagery [8,9,23,24], and UAV [25], have also been used for mapping lake ice-off phenology.

Despite previous efforts at estimating fresh water freeze–thaw cycling in many regions of the world, the extent and distribution of lake ice-off phenology have not yet been quantified across the entire Arctic, which vary with respect to ground thermal regime, topography, surface geology, and land cover [3]. Moreover, most of the current studies have been conducted by treating a lake as the smallest unit of interest, concealing the internal variability of lake ice-off timing distribution [26]. Technical challenges exist in almost every aspect of lake ice phenology mapping, including data acquisition, lake extent extraction, time series modeling, and ice melting/freezing identification. These challenges are even more critical for small lakes (typically less than 1000 ha), which account for approximately 99% of the total Arctic lake population [1], and which are more sensitive to climate change and land disturbance events [9,27].

This study aims to develop a spatially continuous map that first estimates Pan-Arctic lake ice-off phenology at a fine spatial resolution (i.e., 20 m). To achieve this goal, we design a new approach that identifies the per-pixel ice break-up date during the spring melting season based on dense time series stacks of Sentinel-2 data. We implement the algorithm on the Google Earth Engine (GEE) platform and apply it to lakes across the Pan-Arctic region (except for Greenland). To evaluate the mapping performance, we compare our results with those derived from daily freeze–thaw satellite products as well as in situ data for validation. In the remainder of this paper, we introduce the materials and methods (Section 2), report on the validation and mapping results of the Arctic lake ice-off phenology (Section 3), present discussions (Section 4), and draw some conclusions from the study (Section 5).

2. Materials and Methods

2.1. Study Area Boundaries and Lake Selection

The study area covers the entire Arctic, which is defined as the northernmost part of the Earth featured by tundra vegetation, an arctic climate, and arctic flora, with the tree line and the coastline joint-shaping its extent borders (Figure 1a) [27–29]. The lake selection in this study was based on HydroLAKES [30], which provides the shoreline polygons of global lakes with a surface area of no less than 10 ha. We filtered the HydroLAKES dataset by the extent of the study area, and excluded lakes that have a surface area less than 1 km^2 to avoid noise introduced by mixing pixels at the shore line [8]. We also masked out supraglacial lakes in Greenland because of their inter-annual fluctuations in extent and distribution [31]. After the abovementioned selection procedure, there remained 47,230 lakes, ranging from 1 km^2 to 4909 km^2, with their spatial concentration displayed in Figure 1b.

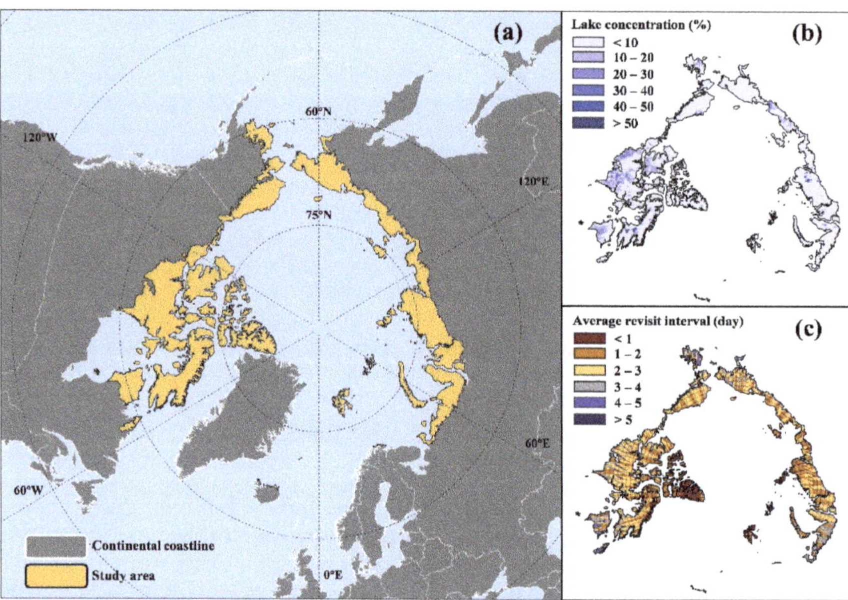

Figure 1. Overview of the study area. (**a**) Spatial coverage of the study area; (**b**) distribution of lake concentration, calculated as the percent area covered by lakes within a 25 km × 25 km grid; (**c**) distribution of the Sentinel-2 average revisit interval (day) within the study period (i.e., from 1 February to 1 September 2019).

2.2. Datasets

Our primary data source used was Sentinel-2 A/B satellite imagery, accessed from the European Space Agency (https://sentinel.esa.int/web/sentinel/user-guides/sentinel-2-msi, (accessed on 23 November 2020)) via the GEE platform [32]. Among the various Sentinel-2 products, we selected the category of Level 2A because it meets the formal quality of orthorectified Bottom-of-Atmosphere (BOA) reflectance, and thus multi-temporal images can be directly used to construct time series for further analysis [33]. Within the Sentinel-2 Level 2A data, we used mainly the Scene Classification Layer (termed SCL hereafter) band for lake ice-off phenology mapping at a 20 m resolution. For each Sentinel-2 image, the SCL band is a pixel classification map derived from four technical steps: cloud/snow detection, cirrus detection, cloud shadow detection, and classification map generation. A specific land cover scheme is designed for classifying each pixel into 11 different categories, including cloud, cloud shadows, vegetation, soils/deserts, water, snow/ice, etc. Readers can refer to the Sentinel-2 Level-2A Algorithm Overview (https://sentinels.copernicus.eu/web/sentinel/technical-guides/sentinel-2-msi/level-2a/algorithm, (accessed on 28 November 2020)) for more technical details of SCL generation. We processed all Sentinel-2 Level 2A data with less than 85% cloud obscuration acquired between 1 February and 1 September 2019, a collection of 1621 images with their average revisit interval pattern displayed in Figure 1c. In addition to satellite imagery, we also obtained ERA5 air temperature (ERA5 Ta hereafter), which is a reanalysis product that combines modeled data with observations into a globally gridded dataset, developed by the European Centre for Medium-Range Weather Forecasts (ECMWF). Compared with its predecessors, ERA5 Ta benefits from an enhanced spatial resolution and is becoming increasingly popular in lake/river ice applications [15,17]. Similar to the Sentinel-2 data collection, we accessed ERA5 Ta from the GEE's public data archive, and utilized the daily average air temperature at 2 m height for better identification of the ice-melting timing information.

2.3. Lake Ice-Off Phenology Mapping Algorithm

Many ice phenology variables have been used for characterizing the spring melting process at the level of a pixel or lake. Here, we mainly focus on break-up end (BUE), which is defined as the day of the year (DOY) when the pixel becomes totally ice free [20,24]. Figure 2 displays the proposed approach that aims to utilize pixel-level SCL time series information for BUE mapping on the GEE platform. Particularly, the framework includes three sequentially integrated parts. In the first part, we refined the lake extent derived from HydroLAKES. This procedure was implemented by referring to Sentinel-2 SCL class information (Section 2.3.1). In the second part, we again used time series of Sentinel-2 SCL and ERA5 Ta to generate a seamless, binary phenophase time series at the 20 m resolution and 5-day interval. Then, such a dataset was utilized as the input of the "max-seg-difference" algorithm, which identifies the per-pixel BUE by seeking the transition time point when the pixel is completely free of ice (Section 2.3.2). Finally, we performed an accuracy assessment (Section 2.3.3) and analyzed the spatial pattern of our estimated lake ice-off phenology result across the entire Arctic (Section 2.3.4).

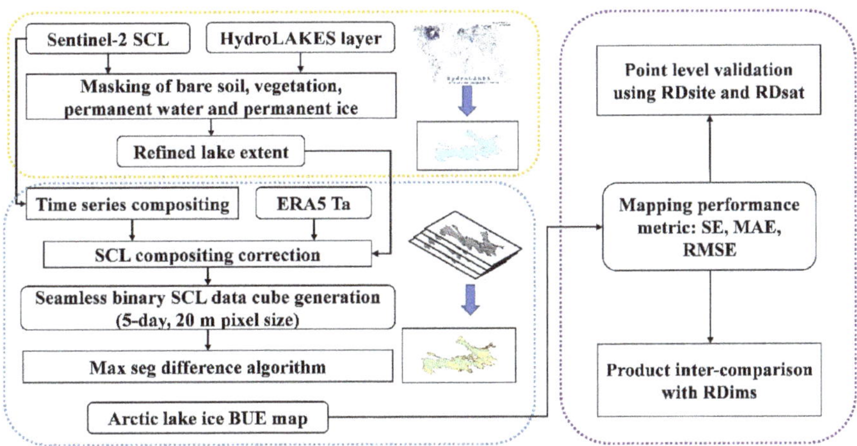

Figure 2. Flow chart of the lake ice BUE mapping proposed in this study.

2.3.1. Lake Extent Refinement

Despite overall high accuracy, the HydroLAKES shoreline layer still bears errors for lake ice-off phenology mapping. Some waterbodies are frozen or clear of ice all year round, neither of which should be included in our analysis. Moreover, several Arctic lakes may have changed in their extents [34] or seasonal fluctuation patterns [35] since the release of HydroLAKES. Therefore, a lake extent refinement is required in this study. We rasterized the original HydroLAKES shoreline layer into a binary image consisting of two types: lake and non-lake. Then, a frequency threshold-based method [36,37] was applied only to the lake pixels for selecting waterbodies with seasonal ice duration based on Sentinel-2 SCL data. In particular, for pixel i, only by satisfying the following criterion can it be identified as valid and used for lake ice BUE mapping.

$$P_i = \begin{cases} 1, & F_{bare} \leq 10\% \ and \ F_{veg} \leq 10\% \ and \ F_{ice} \geq 10\% \ and \ F_{water} \geq 10\% \\ 0, & otherwise \end{cases} \tag{1}$$

where P_i is the final lake extent refinement result, with "1" denoting valid and "0" invalid. F_{bare}, F_{veg}, F_{ice}, and F_{water} are percentages of the SCL time series classified as bare soil, vegetation, water, and ice, respectively, within the study period (from 1 February 2019 to 1 September 2019). The threshold of 10% was adopted due to the overall classification uncertainty of the SCL data [38]. To evaluate the refined lake extent result, 800 stratified random sample points were selected, with one half collected within the refined lake extent while the other half from the masked regions. After removing points that were unable to be identified, there remained 324 and 325 sample points, respectively. Three metrics, including commission error, omission error, and F1 score, were selected for refinement assessment [39]. Here, commission and omission represent the percentage of valid lake samples that were falsely masked, and the percentage of invalid lake samples that were not detected by the refinement procedure, respectively. The F1 score quantifies the level of balance between precision (commission) and recall (omission).

2.3.2. BUE Identification

- Creation of phenophase time series

At the pixel level, this study used Sentinel-2 SCL data to create a binary (ice/water) phenophase time series. Within the refined lake extent, we first masked poor quality or irrelevant observations, leaving only observations classified as water or ice by the SCL (labeled as 6 and 11, respectively). Second, an equitemporal phenophase composite was obtained by deriving the mode value of all valid observations during each temporal interval, which was set to 5 days as a balance of multiple factors, including the sensor revisit cycle, swath (the area imaged on Earth surface) overlap, and the temporal uncertainty control of the final BUE map. It should be noted that an extreme case may exist; that is, in a 5-day interval, there might be no valid data from any Sentinel-2 observations. In this case, a filling value was determined according to temporally adjacent (within 15 days) valid observations.

- Outlier removal

Several outliers remain in the created phenophase data cube because of the misclassification errors of the SCL. These errors on the temporal dimension could undermine or even lead to the failure of the BUE identification. In this study, we used ERA5 Ta data for outlier removal because the lake ice melting is highly correlated to the warming process of air temperature [9,25,40]. Due to the spatial resolution mismatch, the bicubic interpolation algorithm [41] was used for resizing the original ERA5 Ta data to 20 m, and the 28-day average air temperature (Ta^{28}) was adopted as the metric to correct unreasonable ice or water presence [9]. Two threshold values, $-5\ °C$ and $5\ °C$, were selected, corresponding to the upper and lower boundaries of Ta^{28} when the outlier removal takes effect. Specifically, two situations are involved. In the first situation, we assume no ice should melt if Ta^{28} is equal or lower than $-5\ °C$. By contrast, in the second situation, we assume no ice should exist when Ta^{28} is equal or higher than $5\ °C$. Such a threshold filter approach has been widely applied in previous studies [9,42], and proven effective for reducing the spurious detection results of ice melting.

- The "max-seg-difference" algorithm

Traditionally, the lake ice BUE is identified from satellite sensors by detecting a sharp change in surface reflectance or brightness temperature trajectory during the spring melting period [8,43]. This method is straightforward, but its usefulness could be affected by a number of factors, including sun illumination [8], ice thickness [25], and microwave dielectric properties [22]. As a consequence, additional intervention from the user is usually needed, hampering the application of BUE mapping over large areas. To deal with this issue, here we proposed a new BUE identification algorithm, aiming to be more self-adaptive at the per-pixel level through segmentation of the phenophase time series. Figure 3 graphically displays how the BUE date is determined for one pixel. First, we used the SCL trajectory to create a new Boolean time series profile, indicating the phenophase (water as 1, ice as 0) at each temporal interval. Then, for a given temporal interval DOY_i (highlighted with the blue circle), the entire phenophase profile can be divided into two parts: Seg^i_{prior} and Seg^i_{post}, with DOY_i as the boundary point. Since BUE represents the ending of ice cover and the beginning of open water, our core target here is to find a temporal interval DOY_{max} when the difference between Seg^i_{prior} (ice dominated) and Seg^i_{post} (water dominated) can be maximized. The objective function is thus written as

$$\mathrm{arg}max \left| \overline{Seg^i_{prior}} - \overline{Seg^i_{post}} \right|, \; i \in [1, \, N] \tag{2}$$

where $\overline{Seg^i_{prior}}$ and $\overline{Seg^i_{post}}$ are the average values of Seg^i_{prior} and Seg^i_{post}, respectively, and N is the length of the temporal domain (the total number of temporal intervals). In practice, we created a loop for each temporal interval and generated a "seg-diff" profile, through which BUE can be identified according to the maximum principle. A primary advantage of the "max-seg-difference" algorithm is that it can make full use of the time series information so that the impacts of "false" alarms (e.g., Case 1 and Case 3 in Figure 3) would be greatly reduced or even eliminated. We implemented the "max-seg-difference" algorithm within the refined lake extent using GEE, and the final Pan-Arctic BUE map were produced and exported at a 20 m spatial resolution.

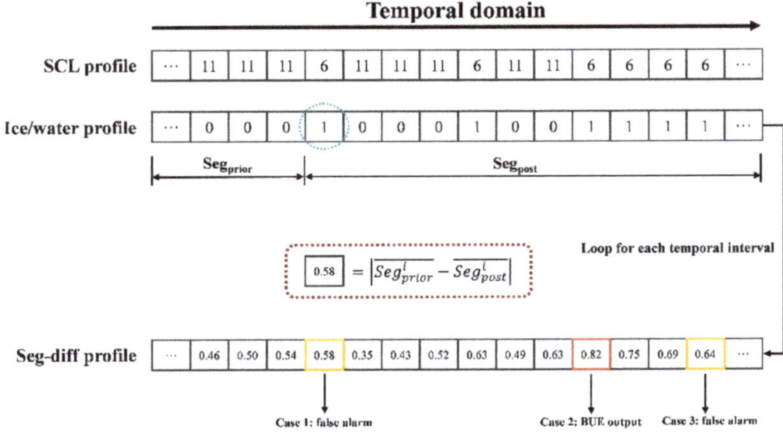

Figure 3. Illustration of the proposed "max-seg-difference" algorithm for BUE identification.

2.3.3. Evaluation of Lake Ice-Off Phenology Mapping

The performance evaluation of the Pan-Arctic lake ice BUE map was conducted in two ways. In the first evaluation method, we randomly created 400 sample points within the refined lake extent, and their BUE dates for the year 2019 were visually identified based on multi-temporal satellite images, including Sentinel-1, Sentinel-2, Landsat-8, and Planet, for the purpose of interpretation bias minimization. We kept only well-interpreted points with a high level of confidence, which eventually resulted in a total of 234 samples (termed RDsat hereafter). We also collected in situ ice phenology observations of 29 lakes provided by the Finnish Environmental Institute. It should be noted that these lakes are geographically beyond the Arctic extent. To make use of these valuable measurements (termed RDsite hereafter) for algorithm validation, we created a 20 m buffer for each lake ice observation site, and implemented our approach of lake ice BUE identification within each buffer. Previous studies suggested that using only a pixel-level evaluation may be insufficient to reflect the quality of a spatially continuous map [44]. Therefore, we implemented a second evaluation method by comparing our result with an independent lake ice BUE map derived from the 1 km Interactive Multisensor Snow and Ice Mapping System (IMS) snow and ice product for the year 2019 (termed RDims hereafter) [15]. We aggregated our 20 m × 20 m lake ice BUE map to 1 km × 1 km using the majority rule, and kept only those pixels whose surrounding area within the 1 km buffer is all lake, to reduce the uncertainty caused by sub-pixel heterogeneity. The evaluation metrics include mean error (*ME*), mean absolute error (*MAE*), and root mean square error (*RMSE*), where $\hat{f}_i$ and f_i are the estimated and reference BUE results for sample point i, respectively. M represents the sample number.

$$ME = \frac{1}{M}\sum_{i=1}^{M}\left(\hat{f}_i - f_i\right) \tag{3}$$

$$MAE = \frac{1}{M}\sum_{i=1}^{M}\left|\hat{f}_i - f_i\right| \tag{4}$$

$$RMSE = \sqrt{\frac{1}{M}\sum_{i=1}^{M}\left(\hat{f}_i - f_i\right)^2} \tag{5}$$

2.3.4. Analysis Method

To investigate the distribution patterns of lake ice-off phenology across the entire Arctic, we conducted our analysis through three levels of spatial aggregation. Based on our 20 m lake ice-off phenology mapping result, we first calculated the mean BUE value for each 5 km $\times$ 5 km grid. In this way, a coarse resolution map was derived, which was used for visual interpretation of the Pan-Arctic lake ice BUE distribution. We also divided the Arctic landmass into zones defined by administrative boundaries and calculated the country-level lake ice BUE estimates. To investigate the sensitivity of the lake ice BUE to the geographical environment, we examined the latitudinal trend of our estimated lake ice BUE map. For each country (except for Norway and Iceland, due to their limited lake numbers), a latitudinal profile of BUE was created with an aggregation interval of 500 m, and the trend was calculated as the slope of the linear regression outputs. The *t*-test was used to examine the significance level of the trend.

We further analyzed the lake-level relationship between the ice BUE and four morphology/landscape parameters: lake area, lake depth, residence time of the lake waters, and shoreline development, measured as the ratio between the shoreline length and the circumference of a circle with the same area. All these parameter records were derived from the HydroLAKES dataset. Similar to the latitudinal analysis, for each parameter–BUE pair, we derived the linear trend as well as its significance level.

3. Results

3.1. Assessment of the Refined Lake Extent

Table 1 shows the evaluation results of the lake extent refinement procedure. For the entire Pan-Arctic region, 45,532 lakes of 138,733 km^2 were identified as the valid lake extent within the study area. In general, we observed the refined Pan-Arctic lake extent exhibits high accuracies, with a commission error, omission error, and F1 score of 6.7%, 6.8%, and 93.2%, respectively. In other words, the refinement procedure is effective for masking out irrelevant land covers (bare soil, vegetation, stable ice, and stable water) that were included in the original HydroLAKES dataset. We also found substantial accuracy variations across countries. Referring to the F1 score, the best performances are observed in Iceland (97.3%) and Norway (95.3%), followed by the USA (93.9%), Russia (91.2%), and Canada (88.3%). Moreover, in the USA and Canada, the accuracies measured by commission are slightly lower than those by omission, while in other countries the opposite trend (higher omission than commission) is observed. Referring to the lake area, we found that the majority of Arctic lakes are located in three countries, including Canada (30,325 lakes, 100,938 km^2), Russia (11,495 lakes, 27,841 km^2) and the USA (3647 lakes, 9905 km^2), where the permafrost landscape is widely distributed. On the other hand, Norway and Iceland have tiny shares of the total lake area extent, primarily due to the smaller landmass of these two countries.

Table 1. Evaluation of the refined lake extent.

Region	Commission (%)	Omission (%)	F1 Score (%)	Refined Lake Number	Refined Lake Area (km²)
Canada	17.3	5.3	88.3	30,325	100,938
Iceland	1.3	4.0	97.3	8	6
Norway	2.7	6.7	95.3	57	43
Russia	5.3	12.0	91.2	11,495	27,841
USA	8.0	4.2	93.9	3647	9905
Pan-Arctic	6.7	6.8	93.2	45,532	138,733

The discrepancy between the original HydroLAKES product and the refined lake extents can be attributed to two primary reasons: misclassification errors of the HydroLAKES dataset and changes in lake extents. Figure 4 displays three typical subsets (200 × 200 pixels for each), representing the diversity of Arctic lake and landscape characteristics. The lake shoreline polygons before and after the refinement were compared and analyzed. Subset 1 is located in Siberia, Russia, while Subsets 2 and 3 are located in Alaska, USA. In Case 1, HydroLAKES mistakenly identifies a cluster of pixels as a lake, which is actually a shaded region covered by shrubs. Such an omission error, however, is not observed in our refined output. In Case 2, both the results with and without the refinement procedure capture the overall pattern of the lake boundaries. However, differences still existed following a further look. Specifically, we observed a slight shrinking trend in the largest lake in the middle, which is correctly reflected in the refined shoreline result. Case 3 further shows a more extreme situation in which the majority of lake areas has been lost and replaced by new plants. Therefore, the refined lake extent used in this study is more reasonable than the original HydroLAKES product for lake ice BUE mapping.

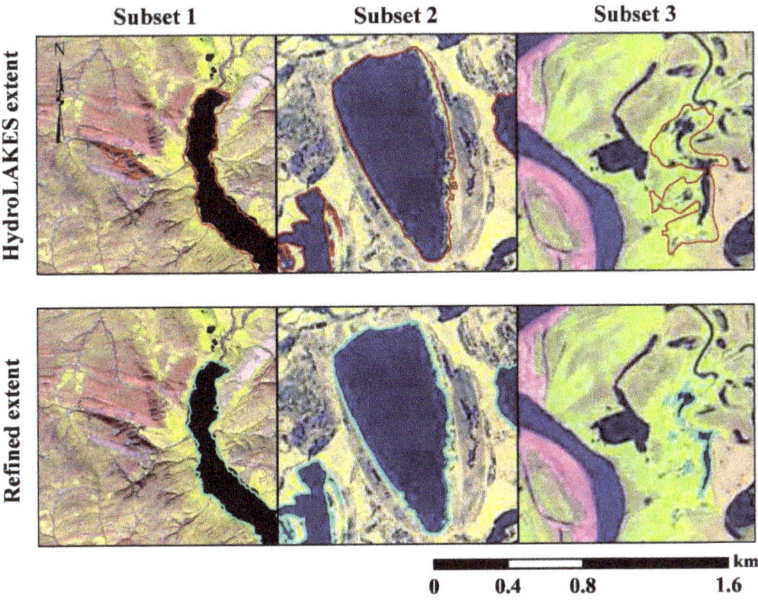

Figure 4. Three typical subsets showing lake shorelines before and after the refinement procedure. The background maps are Sentinel-2 images (shortwave infrared, near infrared, and red bands are displayed in as red, green, and blue layers) acquired on 18 July 2019 for Subset 1 and on 24 July 2019 for Subsets 2–3, respectively.

3.2. Evaluation of Lake Ice-Off Phenology Mapping

We evaluated the estimated lake ice BUE map based on the RDsite and RDsat datasets. We found that our predictions and the actual BUE dates are overall in good agreement (Figure 5). Table 2 further displays the quantitative accuracy metrics. We found that the predicted outputs are more subject to positive bias errors, with ME values of 0.34 and 6.14 days for RDsite and RDsat, respectively, meaning a higher probability of a delay in the lake ice BUE date detection. Comparatively, our BUE predictions show a slightly better performance in matching RDsite than RDsat by all measures.

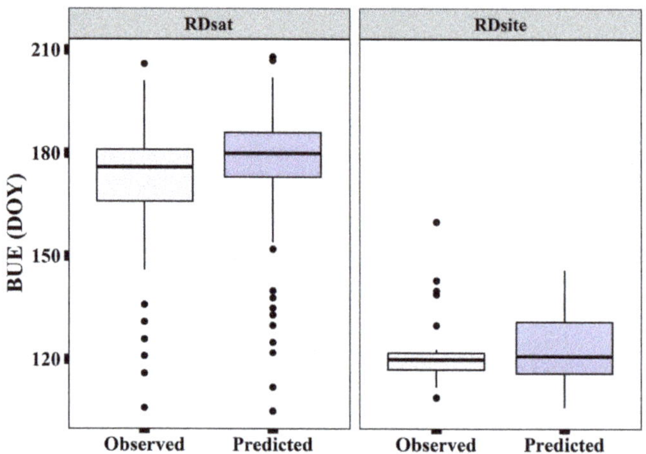

Figure 5. Box plots of the observed and predicted lake ice BUE dates using in situ observations (RDsite) and sample points from satellite visual interpretation (RDsat).

Table 2. Error metrics of the predicted BUE results using RDsite and RDsat (unit: day).

	Number of Observations	*ME*	*MAE*	*RMSE*
RDsite	29	0.34	6.55	7.40
RDsat	234	6.14	7.79	9.92

We further evaluated the performance of our proposed algorithm in estimating lake ice BUE using the RDims dataset. These results were summarized in the density plots using pixel and lake as the statistical unit, respectively (Figure 6). In general, we found that our predictions are generally consistent with those derived from the IMS time series, although the BUE difference is inherently variable within and among lakes. By forcing the intercept of the linear fit to zero [44], we found that our predictions show systematic underestimations (i.e., slope > 1.00 and $ME < 0$) with the IMS-based BUE. Specifically, at the pixel level our predictions explain 65% of the variations of RDims, while at the lake level 57% of the variation can be explained by the predicted BUE. Regarding MAE and $RMSE$, similar results were observed, indicating consistent performance across different spatial levels.

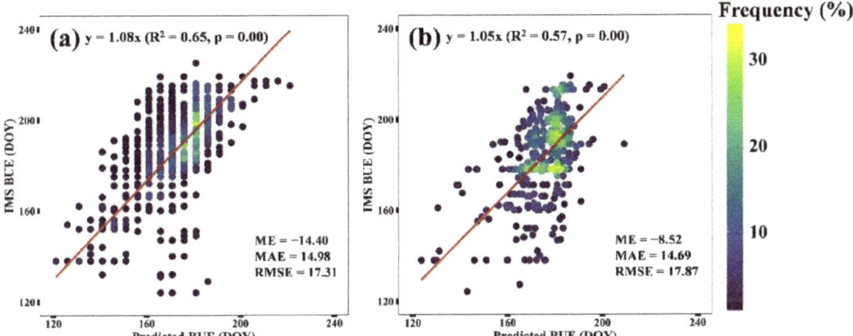

Figure 6. Density plots showing the relationships between the predicted and RDims BUE dates using pixel (**a**) and lake (**b**) as the statistical unit. The red line represents the linear fit with the intercept forced to 0.

Figure 7 further compares lake ice BUE maps from RDims and our predictions by selecting five Arctic lakes, each of which represents one typical landscape. In general, we found that both mapping results correctly reflect the BUE variation within and among lakes, with BUE dates ranging from 110 to 220. Spatially, our BUE maps can provide more details over space and exhibit more consistent BUE patterns. Taking lake Taymyr (A) and lake Hall (D) as examples, the BUE difference between the lake center and lake shore is better captured by our predictions than those of RDims. Such a mapping performance enhancement is primarily due to the improved spatial resolution of Sentinel-2, which allows us to reveal the hidden patterns of BUE heterogeneity that would otherwise not be detected using coarse resolution products (e.g., IMS). Apart from the improvement in mapping accuracy, our 20 m BUE map is also associated with reduced uncertainty caused by the noise introduced by mixing pixels at the shoreline (e.g., lake Ozero Yarato (C) and lake Nanvaranak (E) in Figure 7), and is capable of offering more spatially continuous lake ice BUE estimates across the entire Arctic.

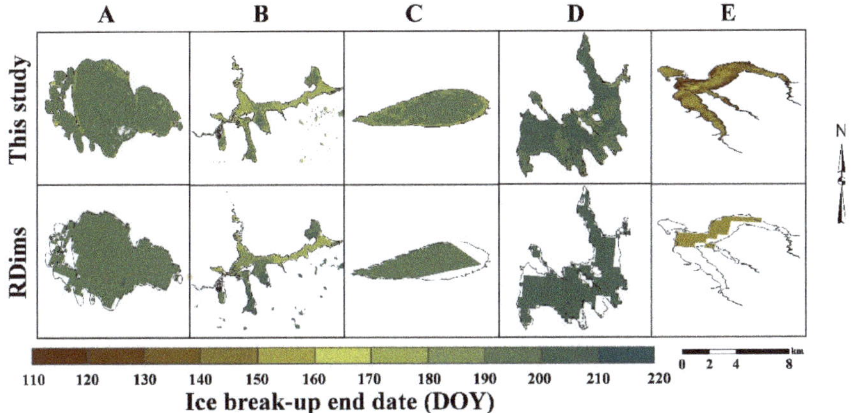

Figure 7. Comparison of BUE maps for five Arctic lakes. (**A–E**) represent lake Taymyr (Russia), lake Teshekpuk (U.S.), lake Ozero Yarato (Russia), lake Hall (Canada), and lake Nanvaranak (U.S.), respectively.

3.3. Spatial Pattern of Pan-Arctic Lake Ice BUE

Using the proposed approach, we calculated the Pan-Arctic lake ice BUE and mapped its distribution (Figure 8). Based on this map, we estimated that the average BUE of Pan-Arctic lakes for the year 2019 is 21 June (DOY = 172 ± 13.4 days, one standard deviation). Spatially, we found that the BUE distribution is strongly associated with latitudinal gradients. Earlier lake ice-off is observed in Iceland, Southwest Alaska, the Southampton Island of Canada, and some parts of Southern Siberia. Conversely, later clearing of lake ice is detected over relatively high latitudes, such as the Svalbard Archipelago of Norway, the Ellesmere Island of Canada, and the Severny Island of Russia. We also displayed finer-scale lake ice BUE mapping results by selecting four hydrological basins. In general, we found that small lakes have earlier ice BUE dates than large ones (see Basin 1 and Basin 3), though the magnitude of difference varies across basins. Similar with Figure 7, for some big lakes there exist discrepancies between the nearshore pixels and those located in the central lake areas.

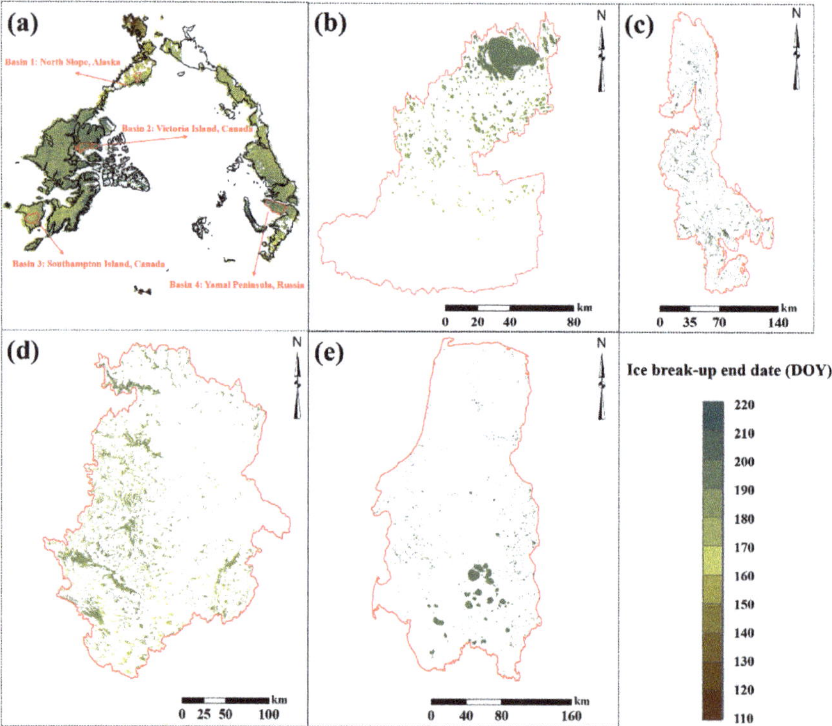

Figure 8. (**a**) Pan-Arctic lake ice BUE map for the year 2019 at a 25 km × 25 km tile scale. Four hydrological basins are highlighted for visual display at a 20 m spatial resolution, including (**b**) North Slope, (**c**) Victoria Island, (**d**) Southampton Island, and (**e**) Yamal Peninsula.

We further divided the Pan-Arctic lake ice BUE map into countries, and the results are shown in Figure 9a. Regarding the average value of the BUE, we found that Norway exhibits the latest BUE date (181), followed by Canada (175), Russia (169), and the USA (150), respectively. In contrast, an earlier BUE estimation by our output is observed in Iceland (112). Overall, countries with greater lake concentrations are less subject to internal variability, as reflected by the lower standard deviations (error bar in Figure 9a). In particular, for Russia, Canada, Iceland, the USA, and Norway, the BUE standard deviations are ±10.2, ±10.7, ±17.9, ±22.1, and ±27.6 days, respectively. Along latitude, all countries (except for Norway and Iceland) exhibit significant trends, together resulting in a linear slope of 1.02 days per degree for the Pan-Arctic (Figure 9b). Moreover, the latitudinal profile of Pan-Arctic lake ice BUE can be roughly divided into four phases by visual interpretation. There is a slight negative trend from 55°N to 62°N, followed by a dramatic increase from 62°N to 65°N, and a relatively stable sequence from 65°N to 78°N. Finally, the BUE is likely to increase from 78°N onward. At the country level, the greatest positive trend is detected in the USA, although its latitudinal range is limited (approximately from 59°N to 72°N). For Russia, relatively obvious increases are found in its southern and northern parts. The latitudinal profile of Canada exhibits the strongest agreement with the entire study area, indicating its prominent role in shaping the Pan-Arctic lake ice-off phenology.

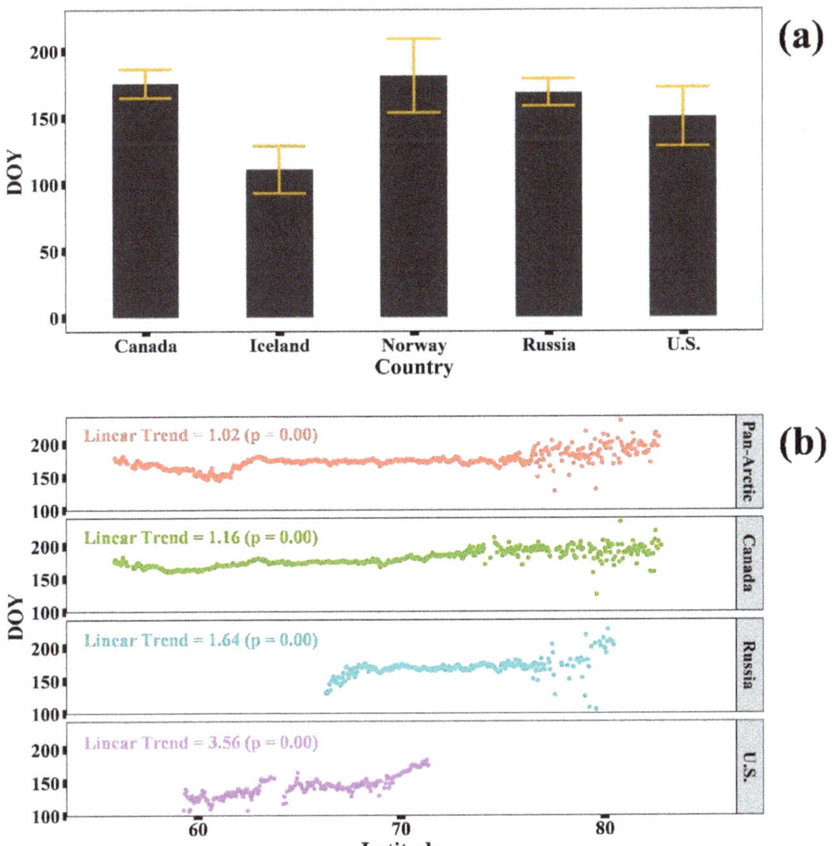

Figure 9. Geographical distribution of Pan-Arctic lake ice BUE. (**a**) Country-level statistics; (**b**) latitudinal profile, derived with an interval of 500 m.

To examine the sensitivity of the BUE to landscape parameters, we conducted an analysis of the influence of parameter change on BUE distribution at the lake level. These results are displayed and summarized in Figure 10. Except for shoreline development, we found that lake ice BUE is significantly impacted by the selected parameters, including lake area ($p = 0.00$), lake depth ($p = 0.05$), and residence time of the lake waters ($p = 0.00$). As suggested by previous studies [25,45], we demonstrated that bigger lakes are more likely to have later BUE dates and less internal variability than smaller lakes (Figure 10a). On the other hand, an overall later trend is also detected with lake depth increasing, with the most striking BUE change occurring at a lake depth of approximately 1 m (Figure 10b). Additionally, we found the BUE response to lake depth gradient is highly nonlinear, displaying a "saturation" phenomenon for deep lakes. Our analysis further indicates that the relationship between lake ice BUE and shoreline development is insignificant ($p = 0.82$, Figure 10c). Such a result, however, is different from that of several regional research studies, highlighting the impact of lake morphology on lake ice phenology [46,47]. Last but not least, we found a consistent and linear increase along the residence time profile (Figure 10d), suggesting that Arctic lakes with higher hydrologic connectivity tend to have earlier ice melting dates [48].

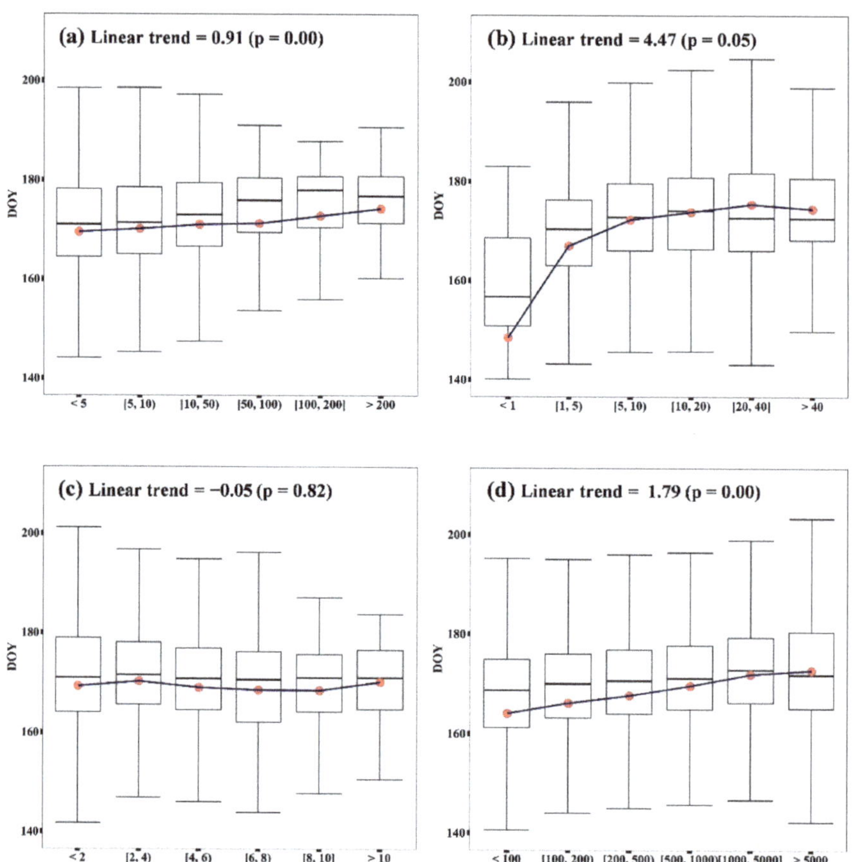

Figure 10. Characteristics of the Pan-Arctic lake ice BUE given (**a**) lake area (km^2); (**b**) lake depth (m); (**c**) shoreline development, measured as the ratio between the shoreline length and the circumference of a circle with the same area; (**d**) residence time of the lake waters (day). The linear trend was calculated using the group-aggregated parameter–BUE pair, and the red point denotes the average value of each data group along the *x*-axis.

4. Discussion

Regulating many aspects of Arctic ecosystems, the spatial distribution of lake ice-off phenology is important for us to understand the earliest detected impacts of climatic warming [40,43,49,50]. However, the diversity of freshwater lakes, landscape heterogeneity, and varying weather conditions make fine-resolution Pan-Arctic lake ice BUE mapping a challenging task. In this study, we proposed a novel approach for per-pixel BUE identification based on Sentinel-2 constellation data. We examined the lake ice BUE mapping performance and found that the estimated results matched well with in situ measurements and existing microwave satellite records (Table 2, Figures 5–7). The main advantages of the new approach are threefold. First, we confirmed the effectiveness of the SCL band of the Sentinel-2 Level 2A product for seasonal ice/water monitoring and incorporated it into our BUE identification approach. We showed that the refined lake extent can better reflect the shoreline boundaries than the original HydroLAKES dataset (Table 1, Figure 4); therefore, the errors caused by lake/land confusion can be greatly reduced or even eliminated. Second, we designed a "max-seg-difference" algorithm, which detects the abrupt change of the binary phenophase time series through temporal segmentation at the pixel level (Figure 3). Compared with commonly used threshold-based methods (e.g., [9,43]), the "max-seg-difference" algorithm is more robust and self-adaptive. As a result, it can be applied to various geographical environments without user intervention. Moreover, the use of phenophase information rather than initial satellite reflectance data has the potential of integrating more data types (e.g., SAR) for fine-resolution lake ice BUE mapping. Finally, we employed the GEE platform to run the full process under the proposed approach, which enables us to apply BUE mapping over a large spatial extent in an effective and efficient manner.

To the best of our knowledge, this study provides the first spatially continuous map of Pan-Arctic lake ice BUE at a 20 m resolution, which can facilitate ongoing efforts to understand the changing Arctic ecosystems. Finer spatial resolution BUE information is not only statistically beneficial for more accurate estimation, but also allows us to reveal the hidden patterns of lake ice melting that would otherwise not be detected using coarse resolution products [15,20]. For example, we found that the nearshore lake ice is likely to melt earlier than that distributed in the central lake area (Figures 7 and 8). Consistent with some previous studies [15,24], our lake ice BUE estimates exhibited an overall later trend with latitudinal gradients across the Pan-Arctic region. However, the magnitude of the latitudinal BUE shift is unevenly distributed (Figure 9), highlighting the complex suite of climatic and ecological forces that can play roles in spring lake ice melting. With an improved resolution, the new BUE map is particularly valuable for monitoring spring ice melting of small lakes, which constitutes the majority of total Arctic lake populations that are most vulnerable to climate change [27]. Theoretically, a small lake is fundamentally different from a big one in thermodynamic processes [25]. Our results confirmed this discrepancy by showing that the BUE date tends to occur later with increasing lake areas and lake depths (Figure 10a,b). Since small lakes are commonly ignored in previous Arctic lake ice phenology studies, reanalysis may be needed using finer-resolution data to clarify the existing findings of Arctic lake ice-off timing dynamics over space and time. We noted an insignificant change in lake ice BUE along the shoreline development profile (Figure 10c), reflecting the internal heterogeneity that obscures the expected effects of lake morphology. For example, within the lake extent, its shoreline topography can be very different. This may influence the wind speed and direction, both of which are drivers of initial ice breaking in early spring [25]. Additionally, our study also provides observational evidence of the hydrological drivers in joint-shaping BUE diversity through connectivity (Figure 10d). Water bodies with high hydrologic connectivity and short residence times are more subject to heat exchange, thus taking fewer days to become ice free in spring [25].

While presenting substantial confidence regarding the Pan-Arctic mapping of fine-resolution lake ice BUE, the proposed approach has limitations and uncertainties. In this study, we selected twin Sentinel-2 satellites as the base data source because of its relatively high spatial and temporal resolutions. With a temporal interval of 5 days, the reconstructed Sentinel-2 composite performed well in general for extracting Pan-Arctic BUE information. Nevertheless, cloud obscuration will lower the actual revisit frequency, making our approach less effective in tracking sub-seasonal variations in lake ice extent. In this sense, the increasing availability of miniature satellites in low Earth orbit, known as CubeSats (e.g., Planet and RapidEye), may offer added value for the timely and accurate estimation of lake ice phenology [5,49]. The derivation of accurate lake extent has always been a critical constraint on ice BUE mapping. In this study, the depicting of lake extent mainly relies on the HydroLAKES dataset, even after a refinement procedure was conducted. The errors of incorrect lake shoreline boundaries will be eventually propagated to the final BUE outputs, and make our estimates less informative. The "max-seg-difference" algorithm depends heavily on the accuracy of SCL, which inevitably contains classification errors. A recent study showed that there is room for improvement in the current scene classification algorithm by ESA [38]. Hence, a better BUE mapping performance of our approach can be expected once the new version of the SCL product is ready for use.

This study has both methodological and practical implications for future endeavors on Arctic-wide lake ice monitoring. By further including SAR time series data, the proposed approach can be extended to track the ice freeze-up process, which is equally important in understanding the Arctic lake system. Using this study as the baseline, the spatial-temporal dynamics of Arctic lake ice phenology can be quantified, making it possible to examine their drivers and project future changes. Moreover, the fine-resolution lake ice BUE map will benefit a variety of scientific research on Arctic ecosystems. For example, lakes are considered a major source of atmospheric methane (CH4) [11,12], and the timing of ice-off can affect the annual quantity of methane emitted from water to the atmosphere [17]. Finally, we emphasize the necessity of more comprehensive, spatially representative databases that can provide in situ ice phenology records to support model development and validation. So far, there are many lakes where the measurements are discontinued and outdated [8]. We expect this data gap can be filled by new and collaborative programs in the near future.

5. Conclusions

Accurately mapping lake ice-off phenology is essential for us to understand a changing Arctic. In this study, we developed a novel approach for fine-resolution BUE mapping with the use of dense Sentinel-2 time series data via the GEE platform. Experimental results in 45,532 lakes across the Arctic reveal the feasibility of this approach. From the estimated Pan-Arctic lake ice BUE map, four major conclusions can be drawn. First, the Sentinel-2 SCL time series dataset shows great potential in monitoring the sub-seasonal dynamics of ice/water coverage with high spatial (20 m) and temporal (5 day) resolutions, which can facilitate the fine-scale detection of lake ice melting. Second, the utilization of a binary phenophase profile rather than original spectral signatures can reduce data uncertainties, making the "max-seg-difference" algorithm self-adaptive for per-pixel BUE identification and efficient for large area application. Third, the mapped lake ice-off phenology exhibits gradients along the latitude profile and lake characteristics, indicating the joint control of climate and local landscapes in determining Arctic lake ice melting. Finally, we concluded that the new approach provided an improved estimate of Pan-Arctic lake ice BUE, which is expected to enlighten the continuing endeavor for achieving sustainable development goals. Future studies can consider environmental characteristics that were not explored in this study, such as lake ice quality, human footprint, and their impacts on Arctic lake ice BUE. Additionally, SAR satellite data hold potential in capturing ice melting/freezing information without the influence of unfavorable weather conditions. They can be thus combined with optical imagery for providing a refined characterization of Arctic lake ice phenology as well as its multi-decadal dynamics.

Author Contributions: Conceptualization, C.L., F.H. and X.C.; methodology, C.L. and H.H.; software, C.L. and Z.Z.; validation, C.L. and Z.Z.; formal analysis, C.L.; investigation, C.L. and Z.Z.; writing—original draft preparation, C.L.; writing—review and editing, C.L., H.H., F.H. and X.C.; visualization, C.L.; supervision, X.C. All authors have read and agreed to the published version of the manuscript.

Funding: This research was funded by the National Key R&D Program of China (No. 2019YFC1509104, No. 2018YFC1407103) and Innovation Group Project of Southern Marine Science and Engineering Guangdong Laboratory (Zhuhai) (No. 311021008).

Data Availability Statement: The Pan-Arctic lake ice-off phenology map for the year 2019 at 20 m resolution is publicly available from the corresponding author on reasonable request.

Acknowledgments: We are grateful to the European Space Agency for providing the Sentinel-2 data, European Centre for Medium-Range Weather Forecasts for providing the ERA5 data, World Wildlife Fund for providing the HydroLAKES dataset, and the Finnish Environmental Institute for providing in situ lake ice phenology measurements. This work would not have been possible without the computing support from the Google Earth Engine platform, which is a cloud-based platform for planetary-scale geospatial analysis. We would also like to thank Qi Zhang and Shiqi Tao for their assistance in manuscript editing.

Conflicts of Interest: The authors declare no conflict of interest.

References

1. Paltan, H.; Dash, J.; Edwards, M. A Refined Mapping of Arctic Lakes Using Landsat Imagery. *Int. J. Remote Sens.* **2015**, *36*, 5970–5982. [CrossRef]
2. Muster, S.; Roth, K.; Langer, M.; Lange, S.; Cresto Aleina, F.; Bartsch, A.; Morgenstern, A.; Grosse, G.; Jones, B.; Sannel, A.B.K.; et al. PeRL: A Circum-Arctic Permafrost Region Pond and Lake Database. *Earth Syst. Sci. Data* **2017**, *9*, 317–348. [CrossRef]
3. Nitze, I.; Grosse, G.; Jones, B.M.; Romanovsky, V.E.; Boike, J. Remote Sensing Quantifies Widespread Abundance of Permafrost Region Disturbances across the Arctic and Subarctic. *Nat. Commun.* **2018**, *9*, 5423. [CrossRef]
4. Olefeldt, D.; Hovemyr, M.; Kuhn, M.A.; Bastviken, D.; Bohn, T.J.; Connolly, J.; Crill, P.; Euskirchen, E.S.; Finkelstein, S.A.; Genet, H.; et al. The Boreal-Arctic Wetland and Lake Dataset (BAWLD). *Earth Syst. Sci. Data Discuss.* **2021**, *2021*, 1–40. [CrossRef]
5. Cooley, S.W.; Smith, L.C.; Ryan, J.C.; Pitcher, L.H.; Pavelsky, T.M. Arctic-Boreal Lake Dynamics Revealed Using CubeSat Imagery. *Geophys. Res. Lett.* **2019**, *46*, 2111–2120. [CrossRef]
6. Newton, A.M.W.; Mullan, D. Climate Change and Northern Hemisphere Lake and River Ice Phenology. *Cryosphere Discuss.* **2020**, *2020*, 1–60. [CrossRef]
7. Surdu, C.M.; Duguay, C.R.; Fernández Prieto, D. Evidence of Recent Changes in the Ice Regime of Lakes in the Canadian High Arctic from Spaceborne Satellite Observations. *Cryosphere* **2016**, *10*, 941–960. [CrossRef]
8. Weber, H.; Riffler, M.; Nõges, T.; Wunderle, S. Lake Ice Phenology from AVHRR Data for European Lakes: An Automated Two-Step Extraction Method. *Remote Sens. Environ.* **2016**, *174*, 329–340. [CrossRef]
9. Zhang, S.; Pavelsky, T.M. Remote Sensing of Lake Ice Phenology across a Range of Lakes Sizes, ME, USA. *Remote Sens.* **2019**, *11*, 1718. [CrossRef]
10. Cory, R.M.; Ward, C.P.; Crump, B.C.; Kling, G.W. Sunlight Controls Water Column Processing of Carbon in Arctic Fresh Waters. *Science* **2014**, *345*, 925. [CrossRef]
11. Engram, M.; Anthony, K.W.; Meyer, F.J.; Grosse, G. Synthetic Aperture Radar (SAR) Backscatter Response from Methane Ebullition Bubbles Trapped by Thermokarst Lake Ice. *Can. J. Remote Sens.* **2013**, *38*, 667–682. [CrossRef]
12. Engram, M.; Walter Anthony, K.M.; Sachs, T.; Kohnert, K.; Serafimovich, A.; Grosse, G.; Meyer, F.J. Remote Sensing Northern Lake Methane Ebullition. *Nat. Clim. Chang.* **2020**, *10*, 511–517. [CrossRef]
13. Rey, D.M.; Walvoord, M.; Minsley, B.; Rover, J.; Singha, K. Investigating Lake-Area Dynamics across a Permafrost-Thaw Spectrum Using Airborne Electromagnetic Surveys and Remote Sensing Time-Series Data in Yukon Flats, Alaska. *Environ. Res. Lett.* **2019**, *14*, 025001. [CrossRef]
14. Ruan, Y.; Zhang, X.; Xin, Q.; Qiu, Y.; Sun, Y. Prediction and Analysis of Lake Ice Phenology Dynamics Under Future Climate Scenarios Across the Inner Tibetan Plateau. *J. Geophys. Res. Atmos.* **2020**, *125*, e2020JD033082. [CrossRef]
15. Dauginis, A.A.; Brown, L.C. Recent Changes in Pan-Arctic Sea Ice, Lake Ice, and Snow on/off Timing. *Cryosphere Discuss.* **2021**, *2021*, 1–35. [CrossRef]
16. Brown, L.C.; Duguay, C.R. Modelling Lake Ice Phenology with an Examination of Satellite-Detected Subgrid Cell Variability. *Adv. Meteorol.* **2012**, *2012*, 529064. [CrossRef]
17. Yang, X.; Pavelsky, T.M.; Allen, G.H. The Past and Future of Global River Ice. *Nature* **2020**, *577*, 69–73. [CrossRef]
18. Duguay, C.R.; Bernier, M.; Gauthier, Y.; Kouraev, A. Remote sensing of lake and river ice. In *Remote Sensing of the Cryosphere*; John Wiley & Sons, Ltd.: Hoboken, NJ, USA, 2015; pp. 273–306. ISBN 978-1-118-36890-9.
19. Che, T.; Li, X.; Jin, R. Monitoring the Frozen Duration of Qinghai Lake Using Satellite Passive Microwave Remote Sensing Low Frequency Data. *Chin. Sci. Bull.* **2009**, *54*, 2294–2299. [CrossRef]
20. Kang, K.-K.; Duguay, C.R.; Howell, S.E.L. Estimating Ice Phenology on Large Northern Lakes from AMSR-E: Algorithm Development and Application to Great Bear Lake and Great Slave Lake, Canada. *Cryosphere* **2012**, *6*, 235–254. [CrossRef]
21. Xiong, C.; Lei, Y.; Qiu, Y. Contrasting Lake Ice Phenology Changes in the Qinghai-Tibet Plateau Revealed by Remote Sensing. *IEEE Geosci. Remote Sens. Lett.* **2020**, 1–5. [CrossRef]
22. Murfitt, J.; Duguay, C.R. Assessing the Performance of Methods for Monitoring Ice Phenology of the World's Largest High Arctic Lake Using High-Density Time Series Analysis of Sentinel-1 Data. *Remote Sens.* **2020**, *12*, 382. [CrossRef]
23. Yang, Q.; Song, K.; Wen, Z.; Hao, X.; Fang, C. Recent Trends of Ice Phenology for Eight Large Lakes Using MODIS Products in Northeast China. *Int. J. Remote Sens.* **2019**, *40*, 5388–5410. [CrossRef]
24. Šmejkalová, T.; Edwards, M.E.; Dash, J. Arctic Lakes Show Strong Decadal Trend in Earlier Spring Ice-Out. *Sci. Rep.* **2016**, *6*, 38449. [CrossRef] [PubMed]
25. Sharma, S.; Meyer, M.F.; Culpepper, J.; Yang, X.; Hampton, S.; Berger, S.A.; Brousil, M.R.; Fradkin, S.C.; Higgins, S.N.; Jankowski, K.J.; et al. Integrating Perspectives to Understand Lake Ice Dynamics in a Changing World. *J. Geophys. Res. Biogeosci.* **2020**, *125*, e2020JG005799. [CrossRef]
26. Scott, K.A.; Xu, L.; Pour, H.K. Retrieval of Ice/Water Observations from Synthetic Aperture Radar Imagery for Use in Lake Ice Data Assimilation. *J. Gt. Lakes Res.* **2020**, *46*, 1521–1532. [CrossRef]

27. Mishra, V.; Cherkauer, K.A.; Bowling, L.C.; Huber, M. Lake Ice Phenology of Small Lakes: Impacts of Climate Variability in the Great Lakes Region. *Glob. Planet. Chang.* **2011**, *76*, 166–185. [CrossRef]

28. Walker, D.A.; Reynolds, M.K.; Daniëls, F.J.A.; Einarsson, E.; Elvebakk, A.; Gould, W.A.; Katenin, A.E.; Kholod, S.S.; Markon, C.J.; Melnikov, E.S.; et al. The Circumpolar Arctic Vegetation Map. *J. Veg. Sci.* **2005**, *16*, 267–282. [CrossRef]

29. Raynolds, M.K.; Walker, D.A.; Balser, A.; Bay, C.; Campbell, M.; Cherosov, M.M.; Daniëls, F.J.A.; Eidesen, P.B.; Ermokhina, K.A.; Frost, G.V.; et al. A Raster Version of the Circumpolar Arctic Vegetation Map (CAVM). *Remote Sens. Environ.* **2019**, *232*, 111297. [CrossRef]

30. Messager, M.L.; Lehner, B.; Grill, G.; Nedeva, I.; Schmitt, O. Estimating the Volume and Age of Water Stored in Global Lakes Using a Geo-Statistical Approach. *Nat. Commun.* **2016**, *7*, 13603. [CrossRef]

31. Liang, Y.-L.; Colgan, W.; Lv, Q.; Steffen, K.; Abdalati, W.; Stroeve, J.; Gallaher, D.; Bayou, N. A Decadal Investigation of Supraglacial Lakes in West Greenland Using a Fully Automatic Detection and Tracking Algorithm. *Remote Sens. Environ.* **2012**, *123*, 127–138. [CrossRef]

32. Gorelick, N.; Hancher, M.; Dixon, M.; Ilyushchenko, S.; Thau, D.; Moore, R. Google Earth Engine: Planetary-Scale Geospatial Analysis for Everyone. *Big Remote. Sensed Data Tools Appl. Exp.* **2017**, *202*, 18–27. [CrossRef]

33. Louis, J.; Pflug, B.; Main-Knorn, M.; Debaecker, V.; Mueller-Wilm, U.; Iannone, R.Q.; Cadau, E.G.; Boccia, V.; Gascon, F. Sentinel-2 Global Surface Reflectance Level-2a Product Generated with Sen2Cor. In Proceedings of the IGARSS 2019—2019 IEEE International Geoscience and Remote Sensing Symposium, Yokohama, Japan, 28 July–2 August 2019; pp. 8522–8525.

34. Smith, L.C.; Sheng, Y.; MacDonald, G.M.; Hinzman, L.D. Disappearing Arctic Lakes. *Science* **2005**, *308*, 1429. [CrossRef]

35. Pickens, A.H.; Hansen, M.C.; Hancher, M.; Stehman, S.V.; Tyukavina, A.; Potapov, P.; Marroquin, B.; Sherani, Z. Mapping and Sampling to Characterize Global Inland Water Dynamics from 1999 to 2018 with Full Landsat Time-Series. *Remote Sens. Environ.* **2020**, *243*, 111792. [CrossRef]

36. Dong, J.; Xiao, X.; Menarguez, M.A.; Zhang, G.; Qin, Y.; Thau, D.; Biradar, C.; Moore, B., III. Mapping Paddy Rice Planting Area in Northeastern Asia with Landsat 8 Images, Phenology-Based Algorithm and Google Earth Engine. *Remote Sens. Environ.* **2016**, *185*, 142–154. [CrossRef]

37. Chen, B.; Xiao, X.; Li, X.; Pan, L.; Doughty, R.; Ma, J.; Dong, J.; Qin, Y.; Zhao, B.; Wu, Z.; et al. A Mangrove Forest Map of China in 2015: Analysis of Time Series Landsat 7/8 and Sentinel-1A Imagery in Google Earth Engine Cloud Computing Platform. *ISPRS J. Photogramm. Remote Sens.* **2017**, *131*, 104–120. [CrossRef]

38. Raiyani, K.; Gonçalves, T.; Rato, L.; Salgueiro, P.; Marques da Silva, J.R. Sentinel-2 Image Scene Classification: A Comparison between Sen2Cor and a Machine Learning Approach. *Remote Sens.* **2021**, *13*, 300. [CrossRef]

39. Zhu, Z.; Zhang, J.; Yang, Z.; Aljaddani, A.H.; Cohen, W.B.; Qiu, S.; Zhou, C. Continuous Monitoring of Land Disturbance Based on Landsat Time Series. *Time Ser. Anal. High Spat. Resolut. Imag.* **2020**, *238*, 111116. [CrossRef]

40. Arp, C.D.; Jones, B.M.; Grosse, G. Recent Lake Ice-out Phenology within and among Lake Districts of Alaska, U.S.A. *Limnol. Oceanogr.* **2013**, *58*, 2013–2028. [CrossRef]

41. Liu, C.; Zhang, Q.; Tao, S.; Qi, J.; Ding, M.; Guan, Q.; Wu, B.; Zhang, M.; Nabil, M.; Tian, F.; et al. A New Framework to Map Fine Resolution Cropping Intensity across the Globe: Algorithm, Validation, and Implication. *Remote Sens. Environ.* **2020**, *251*, 112095. [CrossRef]

42. Duguay, C.R.; Prowse, T.D.; Bonsal, B.R.; Brown, R.D.; Lacroix, M.P.; Ménard, P. Recent Trends in Canadian Lake Ice Cover. *Hydrol. Process.* **2006**, *20*, 781–801. [CrossRef]

43. Du, J.; Kimball, J.S.; Duguay, C.; Kim, Y.; Watts, J.D. Satellite Microwave Assessment of Northern Hemisphere Lake Ice Phenology from 2002 to 2015. *Cryosphere* **2017**, *11*, 47–63. [CrossRef]

44. Liu, C.; Zhang, Q.; Luo, H.; Qi, S.; Tao, S.; Xu, H.; Yao, Y. An Efficient Approach to Capture Continuous Impervious Surface Dynamics Using Spatial-Temporal Rules and Dense Landsat Time Series Stacks. *Remote Sens. Environ.* **2019**, *229*, 114–132. [CrossRef]

45. Kirillin, G.; Leppäranta, M.; Terzhevik, A.; Granin, N.; Bernhardt, J.; Engelhardt, C.; Efremova, T.; Golosov, S.; Palshin, N.; Sherstyankin, P.; et al. Physics of Seasonally Ice-Covered Lakes: A Review. *Aquat. Sci.* **2012**, *74*, 659–682. [CrossRef]

46. Williams, G.; Layman, K.L.; Stefan, H.G. Dependence of Lake Ice Covers on Climatic, Geographic and Bathymetric Variables. *Cold Reg. Sci. Technol.* **2004**, *40*, 145–164. [CrossRef]

47. Magee, M.R.; Wu, C.H. Effects of Changing Climate on Ice Cover in Three Morphometrically Different Lakes. *Hydrol. Process.* **2017**, *31*, 308–323. [CrossRef]

48. Beltaos, S.; Prowse, T. River-Ice Hydrology in a Shrinking Cryosphere. *Hydrol. Process.* **2009**, *23*, 122–144. [CrossRef]

49. Cooley, S.W.; Smith, L.C.; Stepan, L.; Mascaro, J. Tracking Dynamic Northern Surface Water Changes with High-Frequency Planet CubeSat Imagery. *Remote Sens.* **2017**, *9*, 1306. [CrossRef]

50. Sharma, S.; Blagrave, K.; Magnuson, J.J.; O'Reilly, C.M.; Oliver, S.; Batt, R.D.; Magee, M.R.; Straile, D.; Weyhenmeyer, G.A.; Winslow, L.; et al. Widespread loss of lake ice around the Northern Hemisphere in a warming world. *Nat. Clim. Chang.* **2019**, *9*, 227–231. [CrossRef]

remote sensing

Article

Drought Extent and Severity on Arable Lands in Romania Derived from Normalized Difference Drought Index (2001–2020)

Radu-Vlad Dobri [1], Lucian Sfîcă [1,*], Vlad-Alexandru Amihăesei [1,2], Liviu Apostol [1] and Simona Țîmpu [1]

[1] Department of Geography, Faculty of Geography and Geology, Alexandru Ioan Cuza University of Iași, 20A Carol I Blvd., 700505 Iasi, Romania; dobri.vlad@yahoo.com (R.-V.D.); vlad.amihaesei@meteoromania.ro (V.-A.A.); apostolliv@yahoo.com (L.A.); simona.timpu@yahoo.com (S.Ț.)
[2] Meteo Romania, National Meteorological Administration, 013686 Bucharest, Romania
* Correspondence: lucian.sfica@uaic.ro; Tel.: +407-242-87845

Abstract: The aim of this study was to evaluate the frequency and severity of drought over the arable lands of Romania using the Normalized Difference Drought Index (NDDI). This index was obtained from the Moderate Resolution Imaging Spectro-Radiometer (MODIS) sensor of the Terra satellite. The interval between March and September was investigated to study the drought occurrence from the early stage of crop growth to its harvest time. The study covered a long period (2001–2020), hence it is able to provide a sound climatological image of crop vegetation conditions. Corine Land Cover 2018 (CLC) was used to extract the arable land surfaces. According to this index, the driest year was 2003 with 25.6% of arable land affected by drought. On the contrary, the wettest year was 2016, with only 10.8% of arable land affected by drought. Regarding the multiannual average of the period 2001–2020, it can be seen that drought is not a phenomenon that occurs consistently each year, therefore only 11.7% of arable land was affected constantly by severe and extreme drought. The correlation between NDDI and precipitation amount was also investigated. Although the correlations at weekly or monthly levels are more complicated, the annual regional mean NDDI is overall negatively correlated with annual rainfall. Thus, from a climatic perspective, we consider that NDDI is a reliable and valuable tool for the assessment of droughts over the arable lands in Romania.

Keywords: drought monitoring; MODIS satellite images; arable lands; NDDI; Romania

Citation: Dobri, R.-V.; Sfîcă, L.; Amihăesei, V.-A.; Apostol, L.; Țîmpu, S. Drought Extent and Severity on Arable Lands in Romania Derived from Normalized Difference Drought Index (2001–2020). *Remote Sens.* **2021**, *13*, 1478. https://doi.org/10.3390/rs13081478

Academic Editor: Luca Brocca

Received: 7 March 2021
Accepted: 8 April 2021
Published: 12 April 2021

Publisher's Note: MDPI stays neutral with regard to jurisdictional claims in published maps and institutional affiliations.

1. Introduction

Drought is generally considered the most complex meteorological phenomenon [1], given that many factors contribute to its onset, such as precipitation amount, soil characteristics, terrestrial water accessible to plants, soil/air temperature and humidity or wind speed. Other factors that define the characteristics of the active surface, or the physiological peculiarities of the plants, in addition to the anthropogenic influence on the environment, are highly significant in its occurrence [1,2]. Despite the policies and efforts to reduce extreme weather effects, droughts will remain unavoidable.

As present, because regional development is considered to be one of the main factors contributing to economic growth, the better prepared a region to cope with adverse weather conditions, such as drought, the more the region can contribute to the development of the whole country [3,4]. Therefore, strategies and actions to mitigate drought impacts in less developed countries are essential, because these regions are among the most vulnerable, with the lowest financial and technical capacity to adapt or to mitigate the effect of this extreme phenomenon [5].

Scientific research has led to a wide variety of results and applications in the field of monitoring and control of drought effects [2]. However, several scientific problems and challenges remain, such as finding solutions to mitigate drought effects and improve living standards, which can be solved through a better understanding of meteorological extremes, with the aim of preventing possible future impacts [1,2].

The relevance of this topic derives, in particular, from the fact that in the context of climate change, the impact of this climatic phenomenon is expected to become more pronounced because periods of drought are expected to become longer in numerous regions of the world [5,6].

Classically, Wilhite and Glantz [7] defined four types of drought: meteorological (lack of precipitation), hydrological (lack of water supply), agricultural (crop water deficit) and socio-economic (combined effect of drought on human activities). Regardless of its type, the drought phenomenon has been investigated using a variety of indices which have been developed over time, such as the Palmer Drought Sensitivity Index (PDSI) [8], the Standardized Precipitation Index (SPI) [9,10], or the Standardized Precipitation Evaporation Index (SPEI) [11]. In addition, some indices based on the relationship between land surface temperature (LST) and land cover have been developed, such as the Temperature Condition Index (TCI) [12], Vegetation Temperature Condition Index (VTCI) [13] or Temperature Vegetation Dryness Index (TVDI) [14,15].

Taken into account the complicated definition of drought [7,16], vegetation indices appear to provide more appropriate estimations of drought occurrence and intensity and the associated agriculture impacts. In this regard, remote sensing products such as the Normalized Difference Vegetation Index (NDVI), Normalized Difference Water Index (NDWI), Leaf Area Index (LAI) and Fraction of Photosynthetically Active Radiation (FPAR) have also been developed [17–22]. Furthermore, other indices have been investigated, such as the Vegetation Condition Index (VCI) [12], Crop Water Stress Index (CWSI) [23], Vegetation Health Index (VHI) [12], Global Vegetation Moisture Index (GVMI) [24], Soil Water Index (SWI) [25] or Remote Sensing Drought Risk Index (RSDRI) [26,27]. Among these, one of the most flexible and useful indices to be applied for drought investigation on arable lands is the Normalized Difference Drought Index (NDDI) [20].

In Europe, many studies have focused on drought and its characteristics using various methods, indices and satellite products. For example, a study focusing on the north-east of the Iberian Peninsula developed by Vicente-Serrano [28] used vegetation indices derived from AVHRR (Advanced Very High Resolution Radiometer) images. In Spain and the Mediterranean region, Vicente-Serrano et al. [29] used AVHRR images and the NDVI index to study the drought impact over agricultural lands located in the Middle Ebro valley, one of the most arid regions in Europe. In addition, Gouveia et al. [30] analyzed the drought impacts on vegetation over the entire Mediterranean basin, using NDVI and SPEI indices, with the purpose of determining the stage at which vegetation is more impacted by drought. Drought was analyzed by Sepulcre-Canto et al. [31], who combined the SPI, the anomalies of soil moisture and the anomalies of the FPAR. Furthermore, Dalezios et al. [32] applied a number of drought indices based on NOAA (National Oceanic and Atmospheric Administration)-AVHRR and the Reconnaissance Drought Index (RDI) in Thessaly, central Greece, which is a drought-prone agricultural region characterized by vulnerable agriculture. The driest years, such as 2000, 2003 or 2008, were analyzed in many studies aiming to monitor the drought severity in Europe. For this, Sea-Viewing Wide Field-of-view Sensor (SeaWiFS) and Medium Resolution Imaging Spectrometer (MERIS) instruments [33] or MODIS vegetation indices (VIs), NDVI and enhanced vegetation index (EVI) [34] have been used. It should be noted that the assessment of drought at the continental scale identified the Carpathians region as subject to an increase in drought for 1950–2012 [35].

With the increase in the availability of remote sensing products and indices, their use has also become more frequent in Romania. These have been used to analyze and detect floods [36], land cover changes [37,38], landslide-prone hilly areas in Moldova and various other areas [39] and Saharan dust intrusion [40,41], and it is clear that they can also help to precisely monitor drought.

A comprehensive analysis of the correlation between NDVI and SPEI, aiming to evaluate the response of vegetation's photosynthetic activity to drought conditions from 1998 to 2014 over Romania and the Republic of Moldova, was undertaken by Páscoa et al. [11]. Changes in the forest ecosystems in south-western Romania, due to global climate change and anthropogenic impacts during the past three decades, and correlated with the evolution of aridization, were assessed by Prăvălie et al. [42] using the NDVI and the UNEP (United Nations Environmental Programme) aridity index. In addition, remote sensing data have been used in several studies to monitor the summer surface urban heat island of the city of Bucharest [43], Cluj-Napoca [44], and Galați [45]. Due to climate change, some studies have discussed the topic of desertification [46], which was studied in the south-east of Romania using LANDSAT TM images and LST.

Recently, Angearu et al. [47] analyzed drought severity using the Drought Severity Index (DSI) in Romania and its validation based on meteorological data, soil moisture content and agricultural production. In addition, drought assessment based on a multi-temporal analysis and trends of the DSI obtained from Terra MODIS satellite images was undertaken.

In this context, the main objective of our study was to enhance the knowledge on drought frequency and severity in Romania by providing a long-term (2001–2020) and comprehensive view of its impact over arable lands as reflected by the NDDI. Our results show that NDDI is a highly reliable and valuable tool for the assessment of droughts over the arable lands in Romania.

2. Materials and Methods

2.1. Study Area and Its Geographical Features

With a territory of about 238,500 km^2, Romania is the largest country in south-eastern Europe (Figure 1). The distribution of major landforms, with 31% mountainous area, 33% hills and sub-mountainous areas and 36% plains and meadows, provides its territory with significant climatic diversity [48].

First, regarding latitude, the average annual temperature decreases by 3 °C between the south and the north of Romania (from 11 to 8 °C), and regarding altitude it decreases by about 14 °C from the lowlands to the highest mountain peaks (from 11 to −3 °C). The amounts of precipitation are also strongly influenced by the topography. Compared to the average values in the plain areas from the west (about 600 mm) and east (about 400 mm) of the country, in the high mountain areas average precipitation rises on the slopes exposed to the advection of humid air masses at more than 1400 mm [49].

Second, the western and central regions of Romania are particularly impacted by the cyclones formed in the Atlantic and Mediterranean Seas (following a Pannonian track). These cyclones produce significantly more precipitation compared to the cyclones of Mediterranean origin, following a trans-Balkanian track, which produce more significant precipitation in the southern and eastern regions of the country [41,49–53].

It should be noted that the multiannual amount of precipitation at the country scale remained generally stable during the last interval [50], except for some areas in the north and north-west of Romania, with a positive trend, and from the east, south and south-east of the country, with a negative trend [54]. At the same time, there was an increase in evapotranspiration in Romania which led to increased aridity [46,50], which led some authors [46] to discuss a possible ongoing process of desertification, similarly to that in other regions of southern Europe, such as Spain [55], Italy [56] and Greece [57].

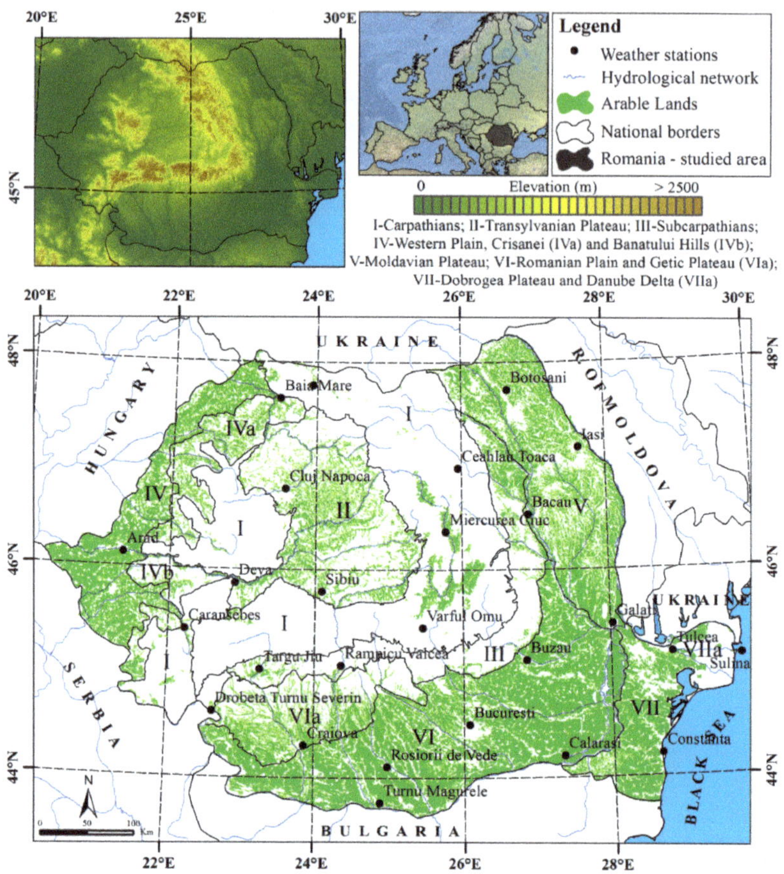

Figure 1. Geographical position of Romania and the extension of arable lands (source: Corine Land Cover (CLC) 2018).

The arable lands in Romania cover an area of about 11 million hectares, of which more than 50% are located in the plain regions of the south, west and extreme north-east of the country, where the arable lands represent the main type of land cover (Figure 1, Table 1).

Table 1. Territorial extension of arable lands according to CLC 2018 in the main regions of Romania (thousands of hectares).

Territorial Extension of Arable Lands in Romania (Source: CLC 2018)								
Carpathians	Transylvanian Plateau	Subcarpathians	Pannonian Plain	Crisana and Banat Hills	Moldavian Plateau	Romanian Plain	Getic Plateau	Dobrogea Plateau and Danube Delta
270	750	313	2600	363	1390	3.070	680	1680

Moreover, due to the large extent of arable lands, it should be noted that Romania represents one of the main producers of maize and wheat in Europe [50], but given the small ratio of irrigated land, representing less than 10% of all arable lands [58], the annual rank in cereal production at European level is highly variable depending on year-to-year weather conditions.

2.2. Data Used

2.2.1. Moderate Resolution Imaging Spectroradiometer (MODIS) Data

The MODIS Vegetation Indices (MOD13Q1) Version 6, based on a 16-day composite at 250 m spatial resolution [59–61], were used to calculate the NDDI. The MOD13Q1 product includes 12 layers, of which only three layers were used: band 1 (16 days Red reflectance—620 to 670 nm), band 2 (16 days Near Infrared reflectance—841 to 876 nm) and band 7 (16 days Middle Infrared reflectance—2105 to 2155 nm) (Tables 2 and 3). These MODIS products were extracted from the Land Processes Distributed Active Archive Center (LP DAAC), using Application for Extracting and Exploring Analysis Ready Samples (AppEEARS), ver. 2.42.1 [62]. Thus, 721 Surface Reflectance Bands for 16-day composite images (1, 2 and 7 bands) were used to calculate the NDDI.

Table 2. Composite periods of MOD13 used in the study [61].

Composite	Day of Year (No.)	Starting Day during Non-Leap Years	Starting Day during Leap Years *	Season
1	65	6-Mar	05-Mar	
2	81	22-Mar	21-Mar	
3	97	7-Apr	06-Apr	
4	113	23-Apr	22-Apr	Spring
5	129	9-May	08-May	
6	145	25-May	24-May	
7	161	10-Jun	09-Jun	
8	177	26-Jun	25-Jun	
9	193	12-Jul	11-Jul	
10	209	28-Jul	27-Jul	Summer
11	225	13-Aug	12-Aug	
12	241	29-Aug	28-Aug	

* Due to numerous corrections made to the acquired data, both in leap years and in non-leap years, 16-day Moderate Resolution Imaging Spectroradiometer (MODIS) composites have the same data range [61].

Table 3. Layers of MOD13Q1 product [59]; the bands used to calculate the Normalized Difference Drought Index (NDDI) (1, 2 and 7) are given in bold.

Layer Name	Description	Units	Data Type	Fill Value	No Data Value	Valid Range	Scale Factor
250 m 16 days NDVI	16 day NDVI	NDVI	16-bit signed integer	−3000	N/A	−2000 to 10,000	0.0001
250 m 16 days EVI	16 day EVI	EVI	16-bit signed integer	−3000	N/A	−2000 to 10,000	0.0001
250 m 16 days VI Quality	VI quality indicators	Bit Field	16-bit unsigned integer	65535	N/A	0 to 65534	N/A
250 m 16 days red reflectance	**Surface Reflectance Band 1**	**N/A**	**16-bit signed integer**	**−1000**	**N/A**	**0 to 10,000**	**0.0001**
250 m 16 days Near Infrared reflectance	**Surface Reflectance Band 2**	**N/A**	**16-bit signed integer**	**−1000**	**N/A**	**0 to 10,000**	**0.0001**
250 m 16 days blue reflectance	Surface Reflectance Band 3	N/A	16-bit signed integer	−1000	N/A	0 to 10,000	0.0001
250 m 16 days Middle Infrared reflectance	**Surface Reflectance Band 7**	**N/A**	**16-bit signed integer**	**−1000**	**N/A**	**0 to 10,000**	**0.0001**
250 m 16 days view zenith angle	View zenith angle of VI Pixel	Degree	16-bit signed integer	−10000	N/A	0 to 18,000	0.01
250 m 16 days sun zenith angle	Sun zenith angle of VI pixel	Degree	16-bit signed integer	−10000	N/A	0 to 18,000	0.01
250 m 16 days relative azimuth angle	Relative azimuth angle of VI pixel	Degree	16-bit signed integer	−4000	N/A	−18,000 to 18,000	0.01
250 m 16 days composite day of the year	Day of year VI pixel	Julian day	16-bit signed integer	−1	N/A	1 to 366	N/A
250 m 16 days pixel reliability	Quality reliability of VI pixel	Rank	8-bit signed integer	−1	N/A	0 to 3	N/A

The 16-day composite was chosen because it minimizes the errors induced by clouds. However, these errors generated by the clouds were still present and were eliminated as further explained in Section 2.3.1.

2.2.2. Corine Land Cover (CLC) Data for Arable Lands

The arable lands from Romania were extracted from CLC 2018. CLC 2018 is a dataset produced within the framework of the Copernicus Land Monitoring Service that refers to the land cover status of 2018 in Europe. The CLC service has a long heritage (formerly known as the "CORINE Land Cover Program"), and is coordinated by the European Environment Agency (EEA). It provides consistent and thematically detailed information on land cover and land cover changes in Europe. This project has a regular update time period of approximately six years [63]. In this classification, arable lands represent a distinct land cover category.

2.2.3. Precipitation Data

The relationship between atmospheric precipitation amount and NDDI was also analyzed, in order to understand how the crop vegetation responds to this major atmospheric driver of drought occurrence. The precipitation time series used in this study were taken from the ENSEMBLE project gridded data set E-OBS [64]. These data were obtained using a kriging interpolation procedure from the European Climate Assessment and Dataset (ECA&D) time series at meteorological stations [65]. The E-OBS version used in this study was version 21 (release date: May 2019) and covers Europe with a spatial resolution of $0.1°$ from 1950 to 2019 at a daily time step. The precipitation amount for the year 2020 was added from the ERA-5 land reanalysis hourly dataset of the Copernicus Climate Change Service C3S Climate Data Store (CDS) [66]. Reanalysis was conducted by combining the model data with observations across the world into a complete and consistent dataset. The spatial resolution of the ERA-5 land reanalysis is $0.1°$, thus no resampling was needed. The precipitation amount was calculated for each 16-day composite (12 composites between March and September) and for each year (20 years), similar to that of the NDDI.

2.3. Methodology
2.3.1. Gap-Filling of MODIS Images

The quality of remote sensing products is highly influenced by weather conditions. Among these factors, cloud cover can frequently induce gaps in the time series of the satellite optical imagery. To fill these gaps retrieved from MODIS over Romania, the Data Interpolating Empirical Orthogonal Functions (DINEOF) procedure was applied [67–69]. Recently, this method has been used in various remote sensing products such as for the reconstruction of total suspended matter [70] sea surface salinity [71], sea surface temperature derived from MODIS [68,72], MODIS-Aqua chlorophyll products [73] or LST over Bucharest [74]. More appropriate to our study, Filliponi et al. [75] applied the DINEOF algorithm to the reconstruction of the MODIS Fraction of Green Vegetation around the world.

The DINEOF procedure was run to gap-fill all MODIS NDDI composites and reconstruct the missing pixels. A full completeness (100% availability) composite (August 2012) was set as a profile mask, thus only pixels within the arable lands were filled. The DINEOF gap-filling method was applied using the rtsa R package version 0.3 for Raster time series Analysis (https://github.com/ffilipponi/rtsa/blob/master/DESCRIPTION (accessed on 15 February 2020) [75]. To evaluate the DINEOF method, artificial gaps were created for a full completeness composite and compared with the original. The difference between pixels of the original composite and DINEOF gap-filled pixels are shown in Table 4. For more details regarding this procedure see Figures S1–S3 in the Supplementary Materials.

Table 4. The statistical parameters of the original and gap-filled data pixels (//–not applicable).

Summary Statistical	Original Data	Gap-Filled Data	Original vs. Gap-Filled Data
Minimum	0.02	−0.08	//
1st Quartile	0.34	0.31	//
Median	0.46	0.44	//
Average	0.48	0.46	//
3rd Quartile	0.59	0.57	//
Maximum	1.75	1.72	//
Root-Mean-Square Error	//	//	0.03
Mean-Absolute-Error	//	//	0.09
R-squared	//	//	0.91

2.3.2. NDDI Calculation and Drought Assessment

Generally, no index can fully describe the complexity of drought at both temporal and spatial levels. Hence, it is recommended to combine several parameters, indicators or indices (including remote sensing data) in a single product for drought classification [20].

First, band 1 (B1), band 2 (B2) and band 7 (B7) from 16-day composite images at 250 m resolution of the MOD13Q1 product were extracted. Using these bands, the NDVI was calculated using B1 and B2 bands [20], by applying Equation (1), and NDWI was calculated from B2 and B7 [20,76], by applying Equation (2).

The NDVI is a classical index which measures the development and the density of vegetation and has values from −1.0 to 1.0. Negative values indicate clouds or water, whereas positive values indicate soil without vegetation (values near to zero), and dense green vegetation (values equal or higher than 0.6) [77,78]. The NDVI is widely used to evaluate the main parameters of vegetation, induced mainly by climate conditions, human activities and other anthropic or natural causes. NOAA/AVHRR, SPOT (French: Satellite Pour l'Observation de la Terre), MODIS or LANDSAT [9,10] imagery can be used to achieve these products. It should be noted that the NDVI is included in this product and can be downloaded already calculated. The NDVI was calculated using the following formula [20]:

$$NDVI_{Modis} = \frac{B2 - B1}{B2 + B1},\tag{1}$$

where $B1$ and $B2$ refer to MODIS band 1 and band 2, respectively.

Additionally, the NDWI is an index which measures the water content of leaves and is used for detecting and monitoring vegetation humidity. The NDWI is influenced by plant dehydration, and it is considered to be a better indicator for drought monitoring than the NDVI [77]. The NDWI also has values from −1.0 to 1.0. The common range for green vegetation is −0.1 to 0.4. This index increases with vegetation water content or from dry soil to free water [78,79]. Both the NDVI and NDWI have been used in different studies to observe their relationship with LAI for the study of the characteristics of vegetation that covers different regions, including arable lands [80–82] or FPAR [19,83]. The NDWI was calculated using the following formula [20]:

$$NDWI_{Modis} = \frac{B2 - B7}{B2 + B7},\tag{2}$$

where $B2$ and $B7$ refer to MODIS band 2 and band 7, respectively.

It has been found that the NDWI is more sensitive than the NDVI to drought conditions, providing information about the amount of water that enters the plant [20]. In addition, it has been found that the average in cases of drought is below 0.5/0.3 in the case of the NDVI/NDWI, whereas in periods without drought, the NDWI/NDVI has values above 0.4/0.6 [6].

Although meteorological drought has been and can be well studied further using the NDVI and NDWI, a new index combines the information provided by the NDVI and

NDWI. The NDDI was recently developed [20], and has been used to monitor the drought parameters in different regions of the world [6,84–87], representing a sensitive drought assessment tool for agriculture [88]. The first research to analyze the potential of this drought monitoring index was conceived with good results for the Flint Hills region of eastern Kansas and north-eastern Oklahoma [89].

Using the NDVI and NDWI results, the NDDI was calculated according to the equation below [20]:

$$NDDI = \frac{NDVI - NDWI}{NDVI + NDWI},$$

(3)

The resulting values of the NDDI range generally from 0 (no drought) to >1.0 (extreme drought).

The NDDI has a stronger response to summer drought conditions than a simple difference between the NDVI and NDWI, and is more sensitive indicator of drought in grasslands and arable lands than the NDVI alone. Because NDWI values decrease more than NDVI values during summers with severe drought, suggesting that the NDWI is more sensitive than the NDVI to drought conditions, the calculation of the NDDI is considered to be a more complex calculation compared to the NDWI and NDVI [20].

Finally, it was found that the NDDI combines well the information provided by the NDVI and NDWI, that it has a wider range of values than a simple NDVI–NDWI differentiation, and that it can be used, based on MODIS images at a good resolution, for the analysis of drought at local scales [90]. Thus, it has been used increasingly often in different regions to study the extent and severity of drought, especially during the vegetation period [6,22,86,91]. Because the NDDI is more sensitive and more accurate, drought-affected territories will be identified more often compared to using the NDVI or NDWI, with differences of up to 5% [6,90]. In Romania, this index has also been used based on MODIS images but to a small extent [83,84,92].

Using the final NDDI products for the 12 composites for each year between 2001 and 2020, a total of 240 16-day composites were derived. In these composite images, the absolute and relative frequency of NDDI values indicating drought (>0.5) were computed for each pixel. To assess the drought severity, the drought frequency for each pixel was considered, taking into account the NDDI classes [84,89] higher than 0.5, which indicate moderate drought (0.5–0.6), severe drought (0.6–1.0) and extreme drought (>1.0). Furthermore, these results were aggregated at annual and multiannual levels.

All of the cartographic presentations of the drought spatial distribution in this paper were constructed using ArcGIS software, version 10.3, produced by ESRI (Environmental Systems Research Institute).

2.3.3. Spearman's Correlation Analysis between NDDI and Precipitation Amount

To check the relationship between the NDDI and the precipitation amount—one of the major drivers of drought conditions—Spearman's correlation was applied. Using an iteration for each pixel, Spearman's correlation was first applied between each 16-day composite precipitation amount and the corresponding NDDI values for the same period, and then for the entire analyzed period (2001–2020). In this analysis a significance level of p-value <0.10 was used.

However, the relationship between atmospheric precipitation and the NDDI is far more complex. Therefore, in addition to a direct Spearman correlation between the two parameters, we determined the inertial effect of atmospheric precipitation on the state of crop vegetation. For this, a lagged correlation was applied between the composite n of atmospheric precipitation and the composite $n + 1$ of the NDDI (1lagged composite correlation). Similarly, we applied the correlation for 2composites lagged (n composite of atmospheric precipitation with $n + 2$ composite of NDDI) to fully cover the inertial response of the NDDI to precipitation input.

3. Results and Discussions

3.1. Drought Extent and Severity According to NDDI

The results of our analysis indicate firstly that, as a multiannual mean, 17.2% of arable land was affected by drought during the analyzed period (2001–2020), with a larger extent during the very dry years (25.6% in 2003, 24.1% in 2012, 23.0% in 2002 and 21.9% in 2020). By comparison, during very humid years, the arable lands were clearly less affected by drought (10.8% in 2016, 11.0% in 2014, 11.1% in 2018 and 11.2% in 2010). Thus, as a general feature, drought has constantly affected the territory of Romania, although only some regions have been severely impacted (Table 5 and Figure 2). Taking into account the fact that more than 70% of the arable lands in Romania are found in plain regions, we observe that these are the regions constantly facing the risk of drought occurrence and the associated impacts on agriculture production.

Additionally, the trend analysis applied to all drought types (moderate, severe, extreme) between 2001 and 2020 (Table 4) indicates a decreasing trend ($R^2 = 0.14$, p-value < 0.10), in contrast to other trends identified using other methods for drought assessment [46]. More details regarding the trend analysis are given in Figures S4 and S5 in the Supplementary Materials. Moreover, it can be observed that the first part of the analyzed period (2001–2010) recorded more years in terms of drought extent and severity than the second interval (2011–2020). Therefore, we can assume that the discussed increase in the frequency of drought events derived mainly from standardized precipitation index [93–95] is not firmly supported from the perspective of crop vegetation conditions, assessed using the NDDI.

Table 5. Relative frequency (%) of NDDI drought classes between 2001 and 2020 (ToD—Type of drought, Nd—No drought, Md—Moderate drought, Sd—Severe drought, Ed—Extreme drought and Ad—All types of drought) and the precipitation amount (mm) accumulated between 5 of March and 13 of September.

	Normal Difference Drought Index (NDDI)					
ToD	Nd	Md	Sd	Ed	Ad	Precipitation Amount (mm)
Range	<0.5	0.5–0.6	0.6–1	>1	>0.5	
2001	80.2	6.4	10.5	2.9	19.8	370.9
2002	77.0	7.2	12.3	3.5	23.0	310.0
2003	74.4	7.4	13.6	4.7	25.6	207.2
2004	81.3	5.5	9.8	3.5	18.7	319.8
2005	81.5	5.4	10.1	3.0	18.5	458.7
2006	83.0	4.9	9.8	2.3	17.0	375.3
2007	78.8	7.9	10.9	2.3	21.2	295.1
2008	87.0	4.5	6.9	1.5	13.0	289.9
2009	85.5	5.6	7.7	1.2	14.5	276.1
2010	88.8	4.0	6.1	1.1	11.2	381.0
2011	84.6	5.1	8.4	1.9	15.4	270.3
2012	75.9	7.2	13.4	3.5	24.1	267.3
2013	85.4	4.9	8.2	1.5	14.6	339.3
2014	89.0	3.8	6.0	1.3	11.0	405.5
2015	82.4	5.8	9.6	2.2	17.6	262.3
2016	89.2	3.7	5.9	1.2	10.8	329.4
2017	84.2	5.2	8.5	2.1	15.8	316.6
2018	88.9	4.2	5.9	1.0	11.1	340.9
2019	80.5	5.5	10.6	3.4	19.5	286.3
2020	78.1	7.2	11.8	2.9	21.9	360.1
2001–2020	82.8	5.6	9.3	2.4	17.2	323.1

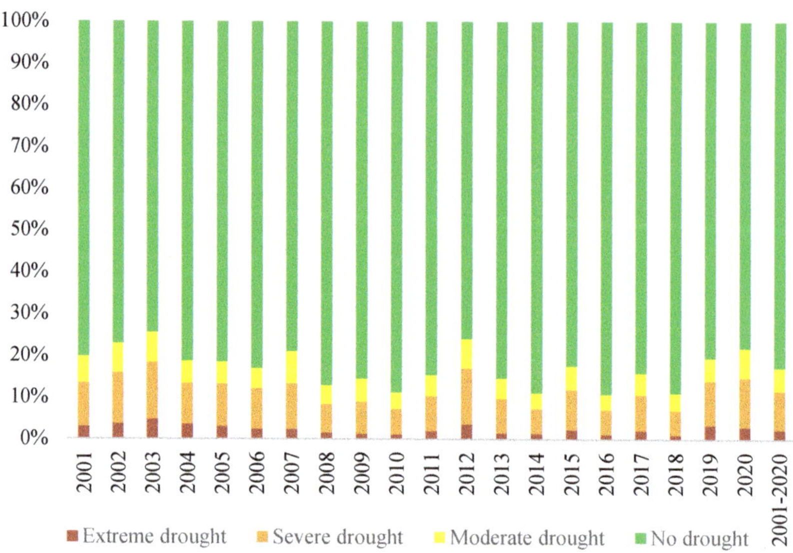

Figure 2. Relative frequency (%) of NDDI main drought classes between 2001 and 2020.

In terms of spatial distribution of drought for the 12 annual composites (Figure 3), and from year to year (Figure 4), we can see that the NDDI drought classes are present in all of the arable regions of Romania. As a main feature, a general country scale asymmetry, with the south-eastern regions more prone to drought than the western side of the country, can be clearly seen. This asymmetry is explained partially by the asymmetry in the field of atmospheric precipitation between these regions, but probably also by a plethora of other factors. For instance, it is known that the NDDI can also be influenced by the prevailing crop type (more wheat in the south-east of Romania), by some types of soils that easily lose the water reserve in the upper layers (such as those in the south and south-east of Romania developed on loess and sands), and by the level of underground water. For instance, analyzing the results for each composite between March and September (Figure 3), it can be seen that drought is less pronounced from April to June, an interval corresponding both to the peak in annual precipitation amount in Romania and to the maximum crop vegetation development. After the end of June, the drought increases in spatial frequency as a combined effect of the decrease in precipitation amount and the depletion of the arable lands by vegetation, due to the harvest of some important crops, such as wheat.

In general, the south-east of Romania was affected by drought even during the wettest years, such as in 2010, 2014, 2016 or 2018 (Figure 4). Normally, if we discuss the manifestation of the desertification process in Romania [46], one should expect this phenomenon in those regions that are affected by drought even during the wettest years. However, our analysis does not support the so-called theory of desertification in Romania, at least from the perspective of crop vegetation conditions.

To underline the variability of drought occurrence at the country scale, we identified, for each pixel on the map of arable lands in Romania, the year recording the highest value of the NDDI, indicating either severe or extreme drought conditions. The results were simplified by grouping them into four classes that indicate the 5 year interval in which these values were recorded (Figure 5).

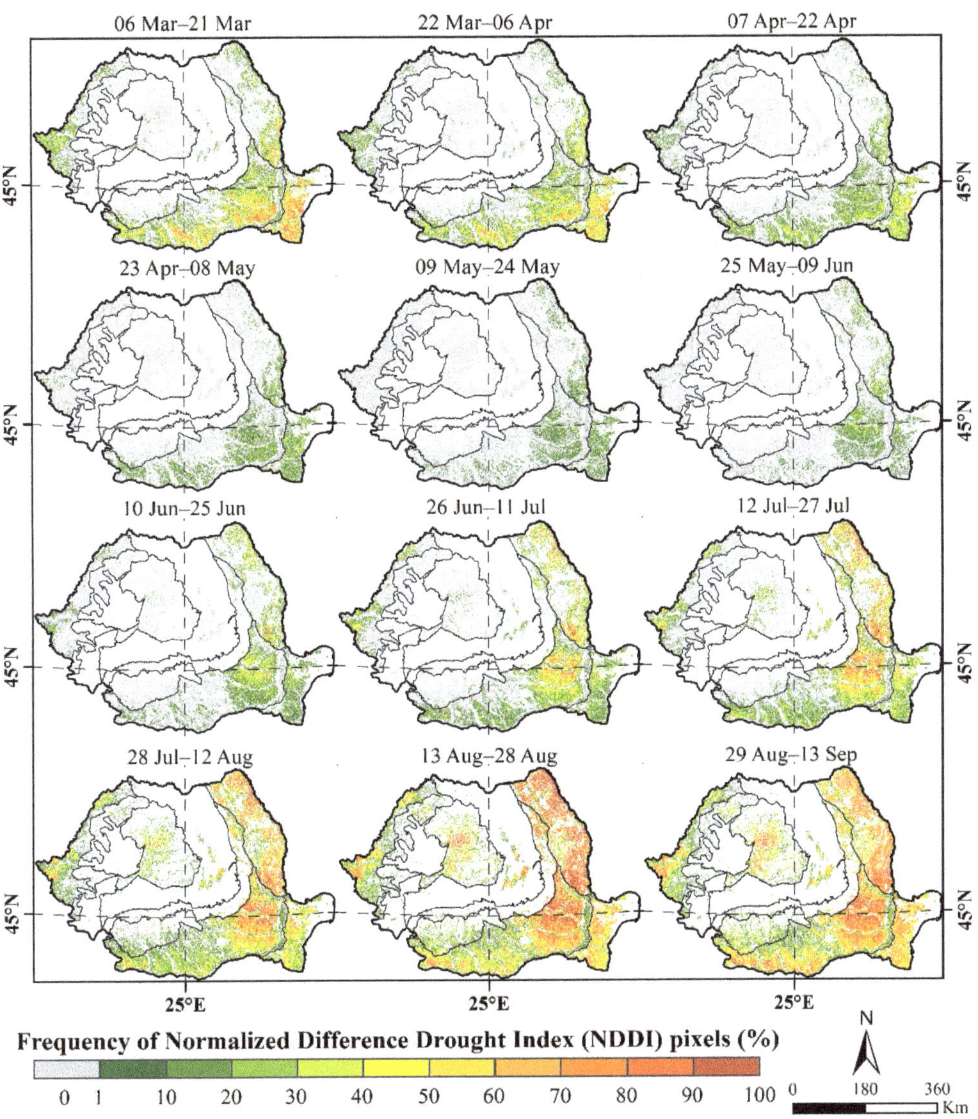

Figure 3. Spatial relative frequency of drought classes (NDDI >0.5) at the level of arable lands in Romania for the 16-day composite periods between March and September (2001–2020).

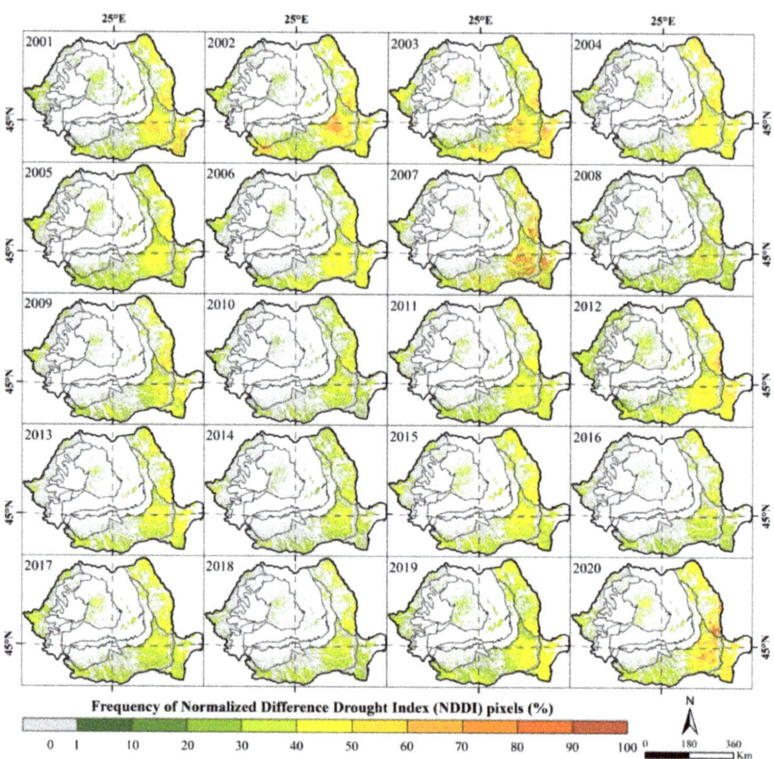

Figure 4. Spatial relative frequency of the drought classes (NDDI >0.5) at the annual level for the arable lands in Romania according to the NDDI between 2001 and 2020.

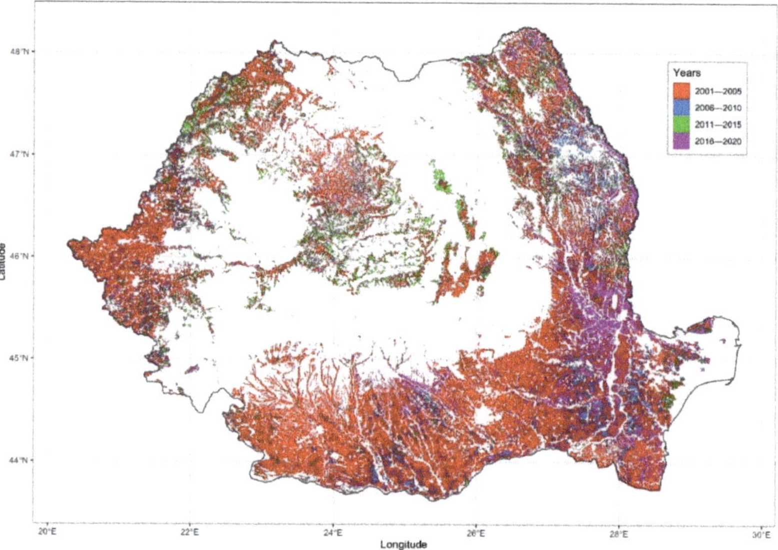

Figure 5. The year recording the highest value of the NDDI for each pixel in Romania for 2001–2020.

Although the southern and western parts of Romania recorded the highest values of the NDDI, mainly between 2001 and 2005 (due to 2003 and 2002 drought events), and secondly during 2006–2010 (as an effect of the 2007 drought episode), in the central and north-western parts of Romania, over large parts of the arable lands, the interval 2011–2015 had the highest drought impact (particularly due to the 2012 drought event). For the north-eastern part of Romania, even if 2001–2005 recorded most of the maximum values of the NDDI, no interval appears to be dominant for the most severe drought conditions as a whole: 2016–2020 prevailed over the southern part of the region together with the Bărăgan region, 2006–2010 prevailed over the central part of the region, and 2011–2020 was more present over the northern part, in particular (Figure 5).

Consequently, the most important region subject to drought in Romania (Figure 6), as derived from the NDDI values, extends over the southern part of the Moldavian Plateau and the eastern part of the Romanian Plain (the so-called Bărăgan Plain, one of the most important agricultural regions in Romania). However, within this region, drought was not pronounced along the valley of the rivers (Siret, Buzău, Ialomița). In addition, we can distinguish a compact strip of arable land oriented from north to south, located east of the Romanian Plain and west of the Dobrogea Plateau, along the Danube, between the cities of Galați and Călărași, which is not severely affected by drought. This region is represented by lands between the Danube branches, recording a high degree of soil humidity and benefiting from well-developed irrigation systems [47]. A second important region impacted by drought is located in the north-eastern part of Romania, within the Moldavian Plain. In addition to these two important regions, drought is also common in the Dobrogea region, in the southern part of the Romanian plain and in the western part of the country, in a region that is relatively distant from the extremity of the Pannonian Plain.

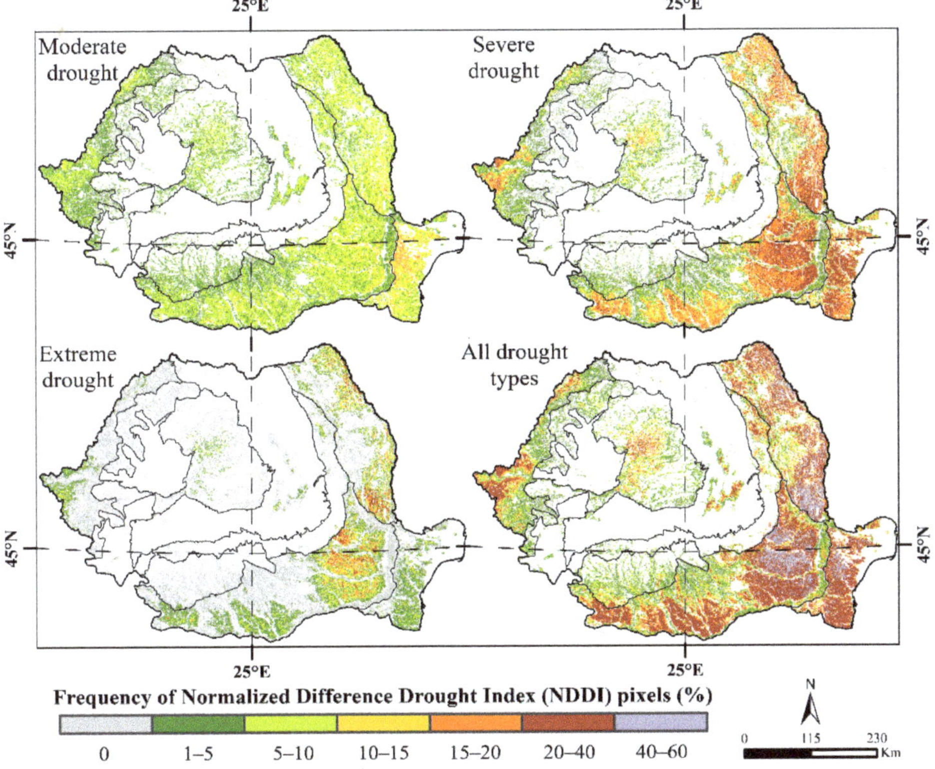

Figure 6. Spatial relative frequency of drought classes at the level of arable lands in Romania between 2001 and 2020.

It is important to underline that all of these regions that are affected by drought have at least three elements in common: the low amount of precipitation with a multiannual mean less than 300 mm between March and September [49], a very high level of groundwater vulnerability [96,97] and the prevalence of wheat crops in the arable lands.

For a more comprehensive view on drought, in addition to the frequency of drought classes for the entire period from 2001 to 2020, we selected the upper/lower third (that is, 7 years) of the driest/wettest years from 2001 to 2020, according to the results shown in Table 2. We thus present more synthetically the distribution of the frequency of drought classes of the NDDI, selecting the most relevant years for drought/humid conditions (Figures 7 and 8).

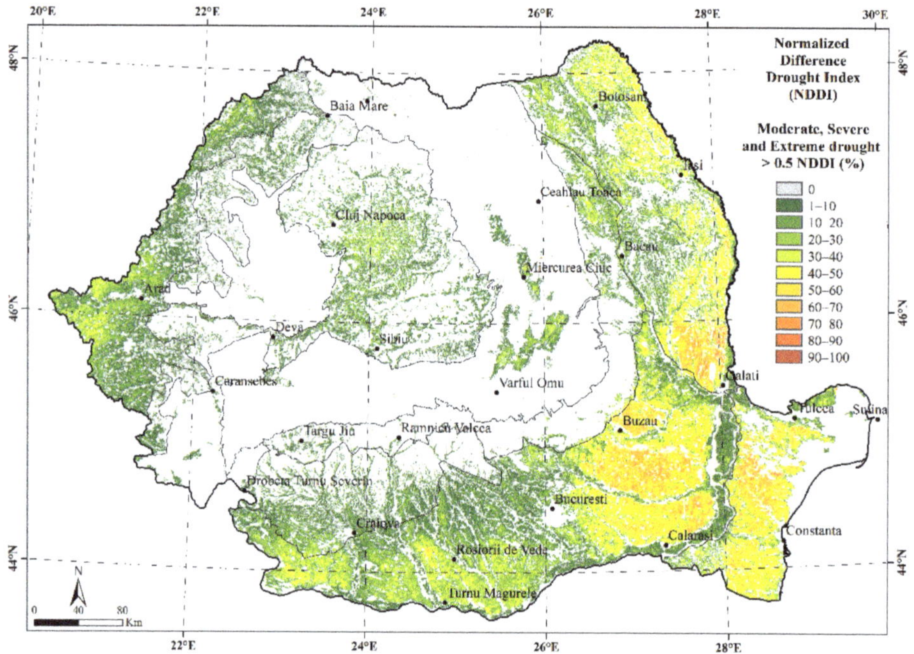

Figure 7. Spatial relative frequency of drought classes over arable lands in Romania for the seven driest years from 2001 to 2020.

The driest 7 years between 2001 and 2020, as shown by the NDDI values (Table 5), were by 2003, 2012, 2002, 2020, 2007, 2001 and 2019. Clearly 2003, with 7.4% of the arable land affected by moderate drought and 19.0% affected by severe and extreme drought, represents a record year from this point of view. The second driest year was 2012 with 7.2% of arable land affected by moderate drought, and 16.9% affected by severe and extreme drought.

The spatial extent of drought as shown by the NDDI values for these years presents a similar pattern as for multiannual mean, but extreme drought conditions are extended over the Dobrogea Plateau, the north-eastern part of the Moldavian Plateau and the southern central part of the Romanian Plain (Figure 7). For the western part of Romania during these years, the differences from the multiannual mean are not high, indicating a less pronounced impact of drought in this region.

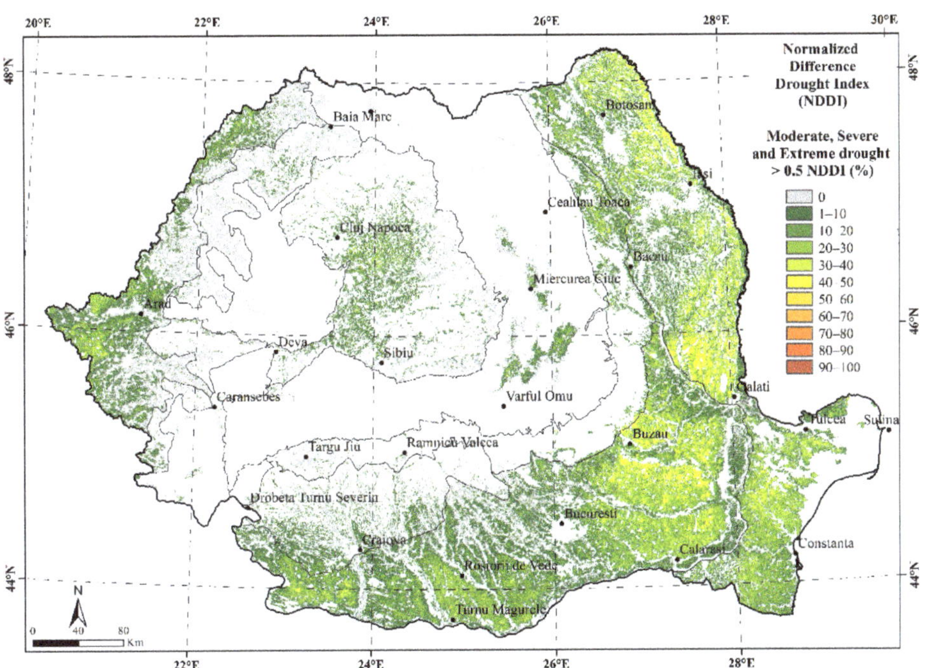

Figure 8. Spatial relative frequency of drought classes over arable lands in Romania for the seven wettest years from 2001 to 2020.

In contrast, the wettest 7 years between 2001 and 2020, as shown by the NDDI values, were 2016, 2014, 2018, 2010, 2008, 2009 and 2013. The wettest years were 2016 and 2014 with only 10.8% and 11%, respectively, of arable land affected by drought (Figure 8, Table 5). Regarding the territorial extension of the drought during these intervals, the most affected areas remained the eastern and southern parts of Romania. Certainly, during these years the drought manifested with a lower degree of severity. However, even in these general humid conditions the drought remained a threat in some regions pf eastern Romania, particularly in the northeast and southeast of the Moldavian Plateau, in the eastern part of the Romanian Plain and the Dobrogea Plateau. Because these are the regions in Romania that are constantly affected by drought, drought in these regions should not be considered an extreme meteorological event, but a common climate feature. That is, in these regions agriculture is not possible without intensive irrigation.

3.2. The Relationship between Atmospheric Precipitation and NDDI

The relationship between the atmospheric precipitation amount and the NDDI is governed by a logical inverse correlation (-0.37 Pearson coefficient, statistically significant for $p < 0.10$). In addition, the difference (70.4 mm) between the means of the precipitation amount for the five most dry/humid years, defined using the NDDI (Table 5), is statistically significant at $p < 0.10$.

Generally, we can observe that the 1-composite lagged correlation is prevalent for the entire interval with a spatial frequency of pixels with significant correlation reaching the maximum (ca. 30% of all the arable lands) from May to July (Figure 9). Moreover, for March and April (with a minimum between 23 April and 8 May) the direct correlation is very weak, as also observed by Potopová et al. [98] in the Republic of Moldova for the correlation between the NDVI and SPEI. The thermal increase in this period most likely causes a rapid increase in vegetal activity and so, although the precipitation amounts are not high, the plants develop rapidly (Figure 9a). In fact, the correlation is weak for this period, and also for the 1- and 2- composite lagged correlations, supporting the same explanation of the rapid increase in vegetal mass triggered by the temperature increase. Moreover, this is the period when the maize crop, which covers 25% of the arable lands in Romania [99] and is a highly demanding crop in terms of water supply, reaches its maximum vegetal activity. After this period, starting with the second 10-day period of May, the 1-composite lagged correlation increases in significance (more than 30% of the arable lands) with its peak at the beginning of June. This can represent the fact that the precipitation amount reaches its annual maximum during this period. The 2-composite lagged correlation was the most important towards the end of the study period, with its peak between 29 August and 13 of September. In brief, we can observe that the response of crop vegetation to atmospheric precipitation amount is fasterduring the maximum development phase of crops and early summer and slower at the end of crop development.

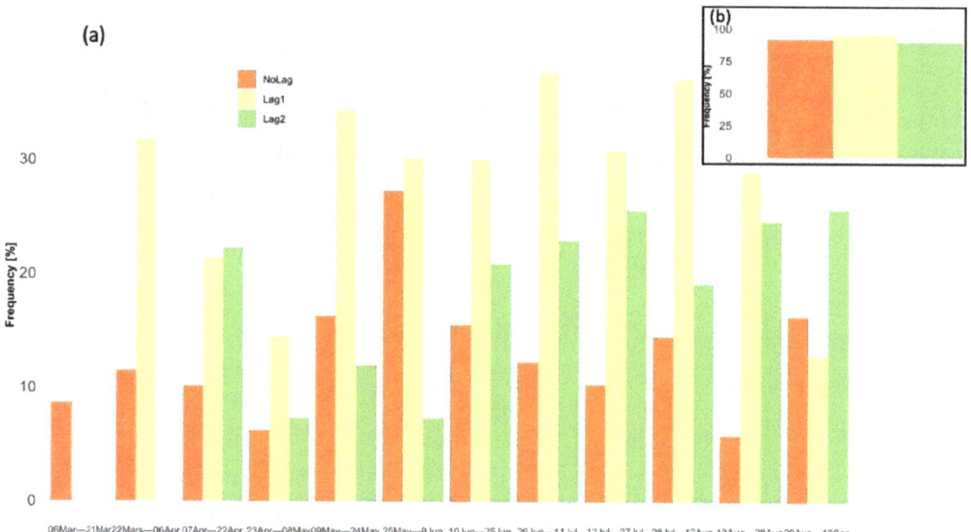

Figure 9. Spearman's correlation between precipitation amount and the NDDI with 1-composite lag and 2-composite lag for the 12 annual composites (**a**) and for annual level (**b**).

Overall, at the annual level the 1-composite lagged correlation reached the highest score of frequency of statistical significance on the arable lands, and was close to the direct correlation (Figure 9b). We note that for the annual level the correlation was aggregated for the p-value <0.05.

Therefore, we showed the 1-composite lagged correlation for each composite separately (Figure 10) and for annual conditions (Figure 11). For each composite separately, one can first observe that the significant correlations are mainly negative, reflecting the logical and expected relationship between low/high precipitation amount and drought/excessive humid conditions.

However, some pixels, particularly those located in the south-eastern part of Romania, present a positive significant correlation. This is mainly a result of the vegetation development despite the weak amount of atmospheric precipitation, especially when the plants benefit from a consistent reserve of soil humidity. In addition, the correlation is thoroughly influenced by the crop phenology and the year-to-year crop rotation.

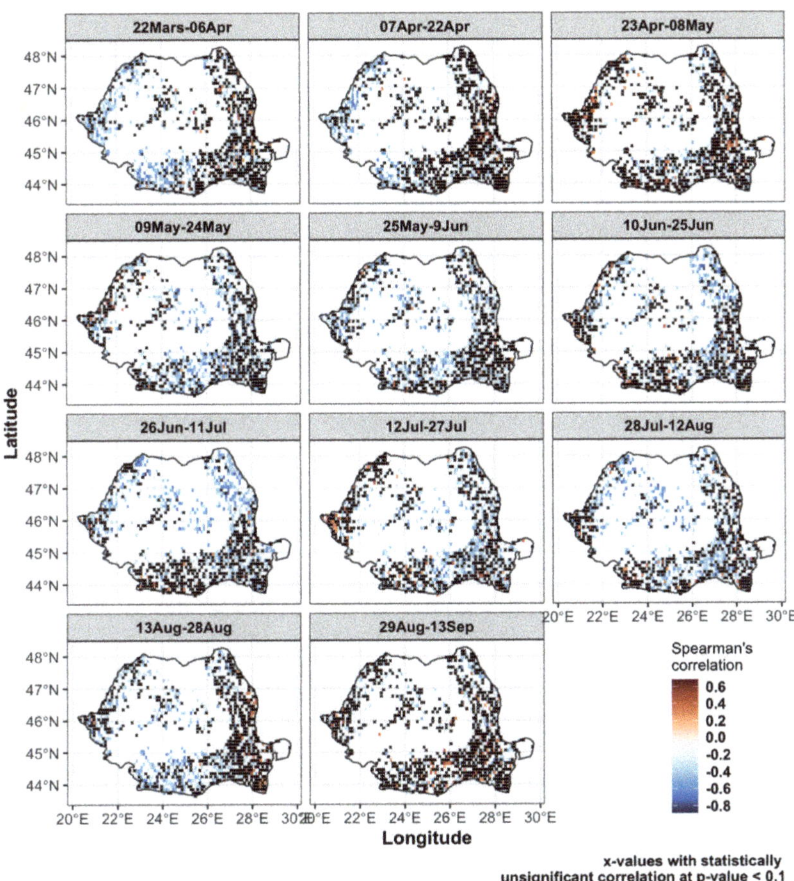

Figure 10. Spearman correlation between precipitation amount and the 1-composite lag of the NDDI.

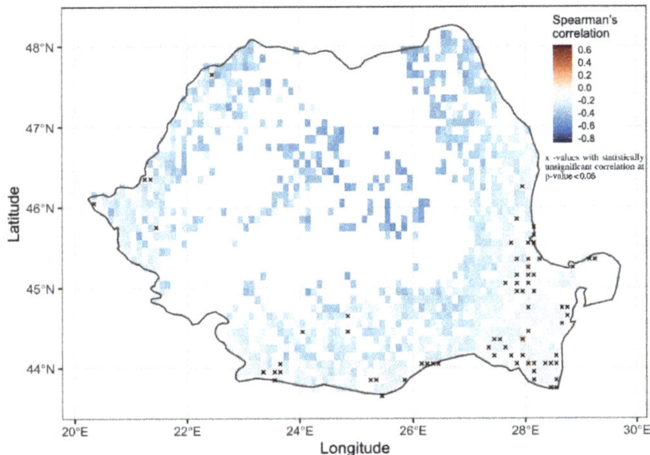

Figure 11. Spatial extent of Spearman's correlation values between precipitation amount and the 1-composite lag of the NDDI for March and September (2001–2020).

The annual conditions (Figure 11) resume the overall negative correlation between precipitation amount and NDDI values. Interestingly, this negative correlation is stronger over the arable lands in regions with higher precipitation, such as the regions in the northern half of the country. This is induced mainly by the prevalence in these regions of crops with a summer maximum in their development that corresponds with the annual peak in atmospheric precipitation, such as maize.

By comparison, the lack of statistical significance is specific for some spots in the south-eastern part of Romania, but also along the Danube and sparsely in the remainder of the country, and correspond mainly with areas with extensive irrigation.

4. Conclusions

Spatial and temporal characteristics of the NDDI were described in an attempt to analyze and monitor the frequency and severity of drought on the arable land in Romania for a relatively long period (2001–2020). The drought assessment was based on the calculation of the relative frequency of NDDI values indicating drought conditions (NDDI >0.5). The analysis focused on the March–September period to cover the crop development from the greening period of vegetation in early spring, to the stage of maximum vegetation and harvest in late summer and early autumn. In addition, the NDDI values were correlated with atmospheric precipitation, with the aim to determine the role of this important driver on drought frequency and severity.

The main conclusion of our study is that the NDDI represents a valuable tool to assess drought from both temporal and spatial perspectives. However, drought assessment using the NDDI should be interpreted with caution because the NDDI is a result of both meteorological and non-meteorological conditions, which are constantly changing over time. This aspect is underlined, in particular, by the weak significant correlation that we obtained between the NDDI and precipitation amount at the country scale. Moreover, when used on longer time period, as in our study, the NDDI offers a comprehensive view of the degree of aridity for the analyzed region.

Regarding the spatial distribution of drought using NDDI drought classes, it was observed that the most affected regions are the eastern and southern parts of the Romanian Plain, the entire Dobrogea region, and the north-eastern part of the Moldavian Plateau. In addition, most of the plain in the west, and the central and western parts of the Romanian Plain, are affected by moderate drought. Overall, we highlight that these are the arable regions in Romania for which drought should not be considered an extreme event, but a main climate feature, that can be handled by permanent irrigation.

Our results also do not indicate a clear trend regarding the multiannual frequency of drought, as expressed by NDDI values, during the analyzed period between 2001 and 2020. Therefore, based on our results, the discussion regarding a possible desertification manifesting in some parts of Romania is doubtful, at least from the point of view of vegetation conditions.

Supplementary Materials: The following are available online at https://www.mdpi.com/article/10.3390/rs13081478/s1, Figure S1: The maps of the original (top-left), artificially gaps (top-right) and dineof filling gaps (bellow left) of 241 -2012 composite, Figure S2: The density distribution of the original (orange) and gap-filled(gray) pixels, Figure S3: The boxjitterplot of the original (left) and gap-filled (right) pixels, Figure S4: Sen'slope trend for precipitation amount (6 of March–13 of September) between 2001 and 2020 (no significant values at $p < 0.10$), Figure S5: Figure S5 Sen'slope trend for drought frequency (6 of March–13 of September) derived from NDDI between 2001 and 2020 (sparse significant values, especially in the south-east of Romania).

Author Contributions: Conceptualization, R.-V.D., L.S.; methodology, R.-V.D., L.S., V.-A.A.; software, R.-V.D., S.Ț.; validation, R.-V.D., L.S., V.-A.A.; formal analysis, L.A., L.S.; investigation, R.-V.D., L.S., V.-A.A., L.A.; resources, R.-V.D.; data curation, L.S.; writing—original draft preparation, R.-V.D.; writing—review and editing, L.S., S.Ț.; visualization, S.Ț.; supervision, L.S., L.A.; funding acquisition, R.-V.D.; R.-V.D., L.S., V.-A.A. are the main authors of this research paper. All authors have read and agreed to the published version of the manuscript.

Funding: This work was financially supported by the Department of Geography, Faculty of Geography and Geology, the 'Alexandru Ioan Cuza' University of Iasi. Lucian Sfîcă was supported by a grant of the Romanian Ministry of Education and Research CNCS-UEFISCDI, project number PN-III-P1-1.1-TE-2019-0286, within PNCDI III, project leader Assoc. Prof. Ph.D Ionuț Minea.

Institutional Review Board Statement: Not applicable.

Informed Consent Statement: Not applicable.

Data Availability Statement: Not applicable.

Acknowledgments: This paper was co-financed from the European Social Fund, through the Human Capital Operational Program, Project Number POCU/380/6/13/123623 << Doctoral students and postdoctoral researchers prepared for the labor market!>>.

Conflicts of Interest: The authors declare no conflict of interest.

References

1. IPCC. *Climate Change 2013: The Physical Science Basis. Contribution of Working Group I to the Fifth Assessment Report of the Intergovernmental Panel on Climate Change*; Cambridge University Press: Cambridge, UK, 2013.
2. Trenberth, K.E.; Dai, A.; Van Der Schrier, G.; Jones, P.D.; Barichivich, J.; Briffa, K.R.; Sheffield, J. Global warming and changes in drought. *Nat. Clim. Chang.* **2014**, *4*, 17–22. [CrossRef]
3. Heim, R.R. A Review of Twentieth-Century Drought Indices Used in the United States. *Bull. Am. Meteorol. Soc.* **2002**, *83*, 1149–1166. [CrossRef]
4. Hao, Z.; Singh, V.P. Drought characterization from a multivariate perspective: A review. *J. Hydrol.* **2015**, *527*, 668–678. [CrossRef]
5. Rakonczai, J. Effects and Consequences of Global Climate Change in the Carpathian Basin. *Clim. Chang. Geophys. Found. Ecol. Eff.* **2011**, *12*, 297–322. [CrossRef]
6. Gulácsi, A.; Kovács, F. Drought Monitoring with Spectral Indices Calculated from Modis Satellite Images in Hungary. *J. Environ Geogr.* **2015**, *8*, 11–20. [CrossRef]
7. Wilhite, D.A.; Glantz, M.H. Understanding: The Drought Phenomenon: The Role of Definitions. *Water Int.* **1985**, *10*, 111–120. [CrossRef]
8. Palmer, W.C. *Meteorological Drought*; Research Paper No. 45; US Department of Commerce Weather Bureau: Washington, DC, USA, 1965; Volume 30.
9. McKee, T.B.; Doesken, N.J.; Kleist, J. The relationship of drought frequency and duration to time scales. In Proceedings of the 8th Conference on Applied Climatology, Anaheim, CA, USA, 17–22 January 1993; pp. 179–183.
10. Hayes, M.J.; Svoboda, M.D.; Wardlow, B.D.; Anderson, M.C.; Kogan, F. Drought monitoring: Historicaland current perspectives. In *Remote Sensing of Drought: Innovative Monitoring Approaches*; CRC Press: Boca Raton, FL, USA, 2012; pp. 1–19. ISBN 978143983560.
11. Páscoa, P.; Gouveia, C.; Russo, A.; Bojariu, R.; Vicente-Serrano, S.; Trigo, R. Drought Impacts on Vegetation in Southeastern Europe. *Remote Sens.* **2020**, *12*, 2156. [CrossRef]
12. Kogan, F. Application of vegetation index and brightness temperature for drought detection. *Adv. Space Res.* **1995**, *15*, 91–100. [CrossRef]
13. Wang, P.X.; Li, X.W.; Gong, J.Y.; Song, C. Vegetation temperature condition index and its application for drought monitoring. *Int. Geosci. Remote Sens.* **2001**, *1*, 141–143. [CrossRef]
14. Sandholt, I.; Rasmussen, K.; Andersen, J. A simple interpretation of the surface temperature/vegetation index space for assessment of surface moisture status. *Remote Sens. Environ.* **2002**, *79*, 213–224. [CrossRef]
15. Rahimzadeh-Bajgiran, P.; Omasa, K.; Shimizu, Y. Comparative evaluation of the Vegetation Dryness Index (VDI), the Temperature Vegetation Dryness Index (TVDI) and the improved TVDI (iTVDI) for water stress detection in semi-arid regions of Iran. *ISPRS J. Photogramm.* **2012**, *68*, 1–12. [CrossRef]
16. Lloyd-Hughes, B. The impracticality of a universal drought definition. *Appl. Clim.* **2014**, *117*, 607–611. [CrossRef]
17. Tucker, C.J.; Pinzon, J.E.; Brown, M.E.; Slayback, D.A.; Pak, E.W.; Mahoney, R.; Vermote, E.F.; El Saleous, N. An extended AVHRR 8-km NDVI dataset compatible with MODIS and SPOT vegetation NDVI data. *Int. J. Remote Sens.* **2005**, *26*, 4485–4498. [CrossRef]
18. Fensholt, R.; Rasmussen, K.; Nielsen, T.T.; Mbow, C. Evaluation of earth observation based long term vegetation trends—Intercomparing NDVI time series trend analysis consistency of Sahel from AVHRR GIMMS, Terra MODIS and SPOT VGT data. *Remote Sens. Environ.* **2009**, *113*, 1886–1898. [CrossRef]
19. Beck, P.S.; Atzberger, C.; Høgda, K.A.; Johansen, B.; Skidmore, A.K. Improved monitoring of vegetation dynamics at very high latitudes: A new method using MODIS NDVI. *Remote Sens. Environ.* **2006**, *100*, 321–334. [CrossRef]
20. Gu, Y.; Brown, J.F.; Verdin, J.P.; Wardlow, B. A five-year analysis of MODIS NDVI and NDWI for grassland drought assessment over the central Great Plains of the United States. *Geophys. Res. Lett.* **2007**, *34*. [CrossRef]
21. Fernandes, R.; Butson, C.; Leblanc, S.; Latifovic, R. Landsat-5 TM and Landsat-7 ETM+ based accuracy assessment of leaf area index products for Canada derived from SPOT-4 VEGETATION data. *Can. J. Remote Sens.* **2003**, *29*, 241–258. [CrossRef]

22. Gouveia, C.M.; Bastos, A.; Trigo, R.M.; Dacamara, C.C. Drought impacts on vegetation in the pre- and post-fire events over Iberian Peninsula. *Nat. Hazards Earth Syst. Sci.* **2012**, *12*, 3123–3137. [CrossRef]

23. Idso, S.; Jackson, R.; Pinter, P.; Reginato, R.; Hatfield, J. Normalizing the stress-degree-day parameter for environmental variability. *Agric. Meteorol.* **1981**, *24*, 45–55. [CrossRef]

24. Ceccato, P.; Gobron, N.; Flasse, S.; Pinty, B.; Tarantola, S. Designing a spectral index to estimate vegetation water content from remote sensing data: Part 1. *Remote Sens. Environ.* **2002**, *82*, 188–197. [CrossRef]

25. Wagner, W.; Scipal, K.; Pathe, C.; Gerten, D.; Lucht, W.; Rudolf, B. Evaluation of the agreement between the first global remotely sensed soil moisture data with model and precipitation data. *J. Geophys. Res. Space Phys.* **2003**, *108*. [CrossRef]

26. Liu, L.; Xiang, D.; Dong, X.; Zhou, Z. Improvement of the Drought Monitoring Model Based on the Cloud Parameters Methodand Remote Sensing Data. In Proceedings of the First International Workshop on Knowledge Discovery and Data Mining (WKDD 2008), Adelaide, Australia, 23–24 January 2008; pp. 293–296.

27. Belal, A.A.; El-Ramady, H.R.; Mohamed, E.S.; Saleh, A.M. Drought risk assessment using remote sensing and GIS techniques. *Arab. J. Geosci.* **2014**, *7*, 35–53. [CrossRef]

28. Vicente-Serrano, S.M. Evaluating the Impact of Drought Using Remote Sensing in a Mediterranean, Semi-arid Region. *Nat Hazards* **2007**, *40*, 173–208. [CrossRef]

29. Vicente-Serrano, S.M.; Cuadrat-Prats, J.M.; Romo, A. Early prediction of crop production using drought indices at different time-scales and remote sensing data: Application in the Ebro Valley (north-east Spain). *Int. J. Remote Sens.* **2006**, *27*, 511–518. [CrossRef]

30. Gouveia, C.; Trigo, R.; Beguería, S.; Vicente-Serrano, S. Drought impacts on vegetation activity in the Mediterranean region: An assessment using remote sensing data and multi-scale drought indicators. *Glob. Planet. Chang.* **2017**, *151*, 15–27. [CrossRef]

31. Sepulcre-Canto, G.; Horion, S.; Singleton, A.; Carrao, H.; Vogt, J. Development of a Combined Drought Indicator to detect agricultural drought in Europe. *Nat. Hazards Earth Syst. Sci.* **2012**, *12*, 3519–3531. [CrossRef]

32. Dalezios, N.R.; Blanta, A.; Spyropoulos, N.V. Assessment of remotely sensed drought features in vulnerable agriculture. *Nat. Hazards Earth Syst. Sci.* **2012**, *12*, 3139–3150. [CrossRef]

33. Gobron, N.; Pinty, B.; Mélin, F.; Taberner, M.; Verstraete, M.M.; Belward, A.; Lavergne, T.; Widlowski, J. The state of vegetation in Europe following the 2003 drought. *Int. J. Remote Sens.* **2005**, *26*, 2013–2020. [CrossRef]

34. Buras, A.; Rammig, A.; Zang, C.S. Quantifying impacts of the 2018 drought on European ecosystems in comparison to 2003. *Biogeosciences* **2020**, *17*, 1655–1672. [CrossRef]

35. Spinoni, J.; Naumann, G.; Vogt, J.; Barbosa, P. European drought climatologies and trends based on a multi-indicator approach. *Glob. Planet. Chang.* **2015**, *127*, 50–57. [CrossRef]

36. Enea, A.; Urzica, A.; Breabăn, I.G. Remote sensing, GIS and HEC-RAS techniques, applied for flood extentvalidation, based on Landsat imagery, LiDAR and hydrological data. Case study: Baseu River, Romania. *J. Environ. Prot. Ecol* **2018**, *19*, 1091–1101.

37. Rusu, A.; Ursu, A.; Stoleriu, C.C.; Groza, O.; Niacșu, L.; Sfîcă, L.; Minea, I.; Stoleriu, O.M. Structural Changes in the Romanian Economy Reflected through Corine Land Cover Datasets. *Remote Sens.* **2020**, *12*, 1323. [CrossRef]

38. Kuemmerle, T.; Müller, D.; Griffiths, P.; Rusu, M. Land use change in Southern Romania after the collapse of socialism. *Reg. Environ. Chang.* **2009**, *9*, 1–12. [CrossRef]

39. Mărgărint, M.C.; Niculiță, M. Landslide type and pattern in Moldavian Plateau, NE Romania. In *Landform Dynamics and Evolution in Romania*; Springer: Cham, The Netherlands, 2017; pp. 271–304. [CrossRef]

40. Mărmureanu, L.; Marin, C.A.; Andrei, S.; Antonescu, B.; Ene, D.; Boldeanu, M.; Vasilescu, J.; Vițelaru, C.; Cadar, O.; Levei, E. Orange Snow—A Saharan Dust Intrusion over Romania During Winter Conditions. *Remote Sens.* **2019**, *11*, 2466. [CrossRef]

41. Țîmpu, S.; Sfîcă, L.; Dobri, R.-V.; Cazacu, M.-M.; Nita, A.-I.; Birsan, M.-V. Tropospheric Dust and Associated Atmospheric Circulations over the Mediterranean Region with Focus on Romania's Territory. *Atmosphere* **2020**, *11*, 349. [CrossRef]

42. Prăvălie, R.; Sîrodoev, I.; Peptenatu, D. Changes in the forest ecosystems in areas impacted by aridization in south-western Romania. *J. Env. Health Sci. Eng.* **2014**, *12*, 2. [CrossRef]

43. Cheval, S.; Dumitrescu, A. The summer surface urban heat island of Bucharest (Romania) retrieved from MODIS images. *Appl. Clim.* **2015**, *121*, 631–640. [CrossRef]

44. Herbel, I.; Croitoru, A.-E.; Rus, A.V.; Roșca, C.F.; Harpa, G.V.; Ciupertea, A.-F.; Rus, I. The impact of heat waves on surface urban heat island and local economy in Cluj-Napoca city, Romania. *Appl. Clim.* **2018**, *133*, 681–695. [CrossRef]

45. Crețu, Ș.-C.; Ichim, P.; Sfîcă, L. Summer urban heat island of Galați city (Romania) detected using satellite products. *Present Environ. Sustain. Dev.* **2020**, *14*, 5–27. [CrossRef]

46. Vorovencii, I. Assessing and monitoring the risk of desertification in Dobrogea, Romania, using Landsat data and decision tree classifier. *Environ. Monit. Assess.* **2015**, *187*, 204. [CrossRef]

47. Angearu, C.-V.; Ontel, I.; Boldeanu, G.; Mihailescu, D.; Nertan, A.; Craciunescu, V.; Catana, S.; Irimescu, A. Multi-Temporal Analysis and Trends of the Drought based on MODIS Data in Agricultural Areas, Romania. *Remote Sens.* **2020**, *12*, 3940. [CrossRef]

48. Geografia României, I. *Geografia Fizică (Geography of Romania, I. Physical Geography)*; Romanian Academy Publishing: Bucharest, Romania, 1983; pp. 171–194. (In Romanian)

49. Sandu, I.; Pescaru, V.I.; Poiană, I.; Geicu, A.; Cândea, I.; Țâștea, D. *Clima României (Climate of Romania)*; Romanian Academy Publishing: Bucharest, Romania, 2008. (In Romanian)

50. Prăvălie, R.; Piticar, A.; Roșca, B.; Sfîcă, L.; Bandoc, G.; Tiscovschi, A.; Patriche, C. Spatio-temporal changes of the climatic water balance in Romania as a response to precipitation and reference evapotranspiration trends during 1961–2013. *Catena* **2019**, *172*, 295–312. [CrossRef]
51. Prăvalie, R.; Sîrodoev, I.; Peptenatu, D. Detecting climate change effects on forest ecosystems in Southwestern Romania using Landsat TM NDVI data. *J. Geogr. Sci.* **2014**, *24*, 815–832. [CrossRef]
52. Dobri, R.-V.; Sfîcă, L.; Ichim, P.; Harpa, G.-V. The Distribution of the Monthly 24-Hour Maximum Amount of Precipitation in Romania According to their Synoptic Causes. *Geogr. Tech.* **2017**, *12*, 62–72. [CrossRef]
53. Nita, I.A.; Sfîcă, L.; Apostol, L.; Radu, C.; Birsan, M.V.; Szep, R.; Keresztesi, A. Changes in cyclone intensity over Romania according to 12 tracking methods. *Rom. Rep. Phys.* **2020**, *72*, 706.
54. Croitoru, A.-E.; Piticar, A.; Dragotă, C.S.; Burada, D.C. Recent changes in reference evapotranspiration in Romania. *Glob. Planet. Chang.* **2013**, *111*, 127–136. [CrossRef]
55. Vicente-Serrano, S.M.; Azorin-Molina, C.; Sanchez-Lorenzo, A.; Revuelto, J.; López-Moreno, J.I.; González-Hidalgo, J.C.; Moran-Tejeda, E.; Espejo, F. Reference evapotranspiration variability and trends in Spain, 1961–2011. *Glob. Planet. Chang.* **2014**, *121*, 26–40. [CrossRef]
56. Colantoni, A.; Ferrara, C.; Perini, L.; Salvati, L. Assessing trends in climate aridity and vulnerability to soil degradation in Italy. *Ecol. Indic.* **2015**, *48*, 599–604. [CrossRef]
57. Nastos, P.T.; Politi, N.; Kapsomenakis, J. Spatial and temporal variability of the Aridity Index in Greece. *Atmos. Res.* **2013**, *119*, 140–152. [CrossRef]
58. Rusu, M.; Simion, G. Farm structure adjustments under the irrigation systems rehabilitation in the Southern plain of Romania: A first step towards sustainabile developments. *Carpathian J. Earth Environ. Sci.* **2015**, *10*, 91–100.
59. Didan, K. MOD13Q1 MODIS/Terra Vegetation Indices 16-Day L3 Global 250 m SIN grid V006. NASA EOSDIS Land Processes DAAC. 2015. Available online: https://doi.org/10.5067/MODIS/MOD13Q1.006 (accessed on 22 September 2020).
60. Du, T.L.T.; Du Bui, D.; Nguyen, M.D.; Lee, H. Satellite-Based, Multi-Indices for Evaluation of Agricultural Droughts in a Highly Dynamic Tropical Catchment, Central Vietnam. *Water* **2018**, *10*, 659. [CrossRef]
61. Testa, S.; Mondino, E.C.B.; Pedroli, C. Correcting MODIS 16-day composite NDVI time-series with actual acquisition dates. *Eur. J. Remote Sens.* **2014**, *47*, 285–305. [CrossRef]
62. Team, A. Application for Extracting and Exploring Analysis Ready Samples (AppEEARS). NASA EOSDIS Land Processes Distributed Active Archive Center (LP DAAC), USGS/Earth Resources Observation and Science (EROS) Center: Sioux Falls, SD, USA. 2019. Available online: https://lpdaacsvc.cr.usgs.gov/appeears/ (accessed on 22 September 2020).
63. Corine Land Cover, Copernicus Programme. Available online: https://land.copernicus.eu/pan-european/corine-land-cover/clc2018 (accessed on 7 January 2020).
64. Haylock, M.R.; Hofstra, N.; Tank, A.M.G.K.; Klok, E.J.; Jones, P.D.; New, M. A European daily high-resolution gridded data set of surface temperature and precipitation for 1950–2006. *J. Geophys. Res. Space Phys.* **2008**, *113*. [CrossRef]
65. Klok, E.J.; Tank, A.M.G.K. Updated and extended European dataset of daily climate observations. *Int. J. Clim.* **2009**, *29*, 1182–1191. [CrossRef]
66. Copernicus Climate Change Service (C3S). Climate Data Store (CDS). 2020. Available online: https://cds.climate.copernicus.eu/#!/home (accessed on 7 October 2020).
67. Beckers, J.-M.; Rixen, M. EOF calculations and data filling from incomplete oceanographic data sets. *J. Atmos. Oceanic Technol.* **2003**, *20*, 1839–1856. [CrossRef]
68. Alvera-Azcárate, A.; Barth, A.; Rixen, M.; Beckers, J. Reconstruction of incomplete oceanographic data sets using empirical orthogonal functions: Application to the Adriatic Sea surface temperature. *Ocean Model.* **2005**, *9*, 325–346. [CrossRef]
69. Alvera-Azcárate, A.; Barth, A.; Beckers, J.-M.; Weisberg, R.H. Multivariate reconstruction of missing data in sea surface temperature, chlorophyll, and wind satellite fields. *J. Geophys. Res. Space Phys.* **2007**, *112*. [CrossRef]
70. Sirjacobs, D.; Alvera-Azcárate, A.; Barth, A.; Lacroix, G.; Park, Y.; Nechad, B.; Ruddick, K.; Beckers, J.M. Cloud filling of ocean and sea surface temperature remote sensing products over the Southern North Sea by the Data Interpolating Empirical Orthogonal Functions methodology. *J. Sea Res.* **2011**, *65*, 114–130. [CrossRef]
71. Alvera-Azcárate, A.; Barth, A.; Parard, G.; Beckers, J.-M. Analysis of SMOS sea surface salinity data using DINEOF. *Remote Sens. Environ.* **2016**, *180*, 137–145. [CrossRef]
72. Beckers, J.-M.; Barth, A.; Alvera-Azcárate, A. DINEOF reconstruction of clouded images including error maps–application to the Sea-Surface Temperature around Corsican Island. *Ocean Sci.* **2006**, *2*, 183–199. [CrossRef]
73. Hilborn, A.; Costa, M. Applications of DINEOF to Satellite-Derived Chlorophyll-a from a Productive Coastal Region. *Remote Sens.* **2018**, *10*, 1449. [CrossRef]
74. Cheval, S.; Dumitrescu, A.; Amihaesei, V.-A. Exploratory Analysis of Urban Climate Using a Gap-Filled Landsat 8 Land Surface Temperature Data Set. *Sensors* **2020**, *20*, 5336. [CrossRef] [PubMed]
75. Filipponi, F.; Valentini, E.; Xuan, A.N.; Guerra, C.A.; Wolf, F.; Andrzejak, M.; Taramelli, A. Global MODIS Fraction of Green Vegetation Cover for Monitoring Abrupt and Gradual Vegetation Changes. *Remote Sens.* **2018**, *10*, 653. [CrossRef]
76. Tucker, C.J. Red and photographic infrared linear combinations for monitoring vegetation. *Remote Sens. Environ.* **1979**, *8*, 127–150. [CrossRef]
77. Gao, B.C. NDWI—A normalized difference water index for remote sensing of vegetation liquid water from space. *Remote Sens. Environ.* **1996**, *58*, 257–266. [CrossRef]

78. Chen, J.; Jönsson, P.; Tamura, M.; Gu, Z.; Matsushita, B.; Eklundh, L. A simple method for reconstructing a high-quality NDVI time-series data set based on the Savitzky–Golay filter. *Remote Sens. Environ.* **2004**, *91*, 332–344. [CrossRef]

79. Chen, J.; Quan, W.; Zhang, M.; Cui, T. A Simple Atmospheric Correction Algorithm for MODIS in Shallow Turbid Waters: A Case Study in Taihu Lake. *IEEE J. Sel. Top. Appl. Earth Obs. Remote Sens.* **2012**, *6*, 1825–1833. [CrossRef]

80. Wang, Q.; Adiku, S.; Tenhunen, J.; Granier, A. On the relationship of NDVI with leaf area index in a deciduous forest site. *Remote Sens. Environ.* **2005**, *94*, 244–255. [CrossRef]

81. Gamon, J.A.; Field, C.B.; Goulden, M.L.; Griffin, K.L.; Hartley, A.E.; Joel, G.; Penuelas, J.; Valentini, R. Relationships Between NDVI, Canopy Structure, and Photosynthesis in Three Californian Vegetation Types. *Ecol. Appl.* **1995**, *5*, 28–41. [CrossRef]

82. Myneni, R.; Williams, D. On the relationship between FAPAR and NDVI. *Remote Sens. Environ.* **1994**, *49*, 200–211. [CrossRef]

83. Angearu, C.-V.; Irimescu, A.; Mihailescu, D.; Virsta, A. Evaluation of Droughts and Fires in the Dobrogea Region, Using Modis Satellite Data. *Agric. Life. Life Agric. Conf. Proc.* **2018**, *1*, 336–345. [CrossRef]

84. Angearu, C.V. Analiza secetei asupra terenurilor arabile din România pe baza imaginilor satelitare. *Rev. Stiintifica A Adm. Natl. Meteorol.* **2018**, *1*, 61–76. (In Romanian)

85. Zhang, A.; Jia, G. Monitoring meteorological drought in semiarid regions using multi-sensor microwave remote sensing data. *Remote Sens. Environ.* **2013**, *134*, 12–23. [CrossRef]

86. Park, S.; Im, J.; Park, S. Probabilistic Drought Intensification Forecasts Using Temporal Patterns of Satellite-Derived Drought Indicators. EGU General Assembly Conference Abstracts. 2016; EPSC2016-11264. Available online: https://meetingorganizer. copernicus.org/EGU2016/EGU2016-11264-1.pdf (accessed on 10 October 2020).

87. Trinh, L.H.; Vu, D.T. Application of remote sensing technique for drought assessment based on normalized difference drought index, a case study of Bac Binh district, Binh Thuan province (Vietnam). *Russ. J. Earth Sci.* **2019**, *19*, 1–9. [CrossRef]

88. Gulácsi, A.; Kovács, F. Drought monitoring of forest vegetation using MODIS-based normalized difference drought index in Hungary. *Hung. Geogr. Bull.* **2018**, *67*, 29–42. [CrossRef]

89. Erdenetuya, M.; Bulgan, D.; Erdenetsetseg, B. Drought monitoring and assessment using multi satellite data in Mongolia. In Proceedings of the 32nd Asian Conference on Remote Sensing, Tapei, Taiwan, 3–7 October 2011.

90. Cheng-lin, L.; Jian-jun, W. Crop drought monitoring using MODIS NDDI over mid-territory of China. In Proceedings of the IEEE International Geoscience and Remote Sensing Symposium, IGARSS 2008, Boston, MA, USA, 7–11 July 2008.

91. Rhee, J.; Im, J.; Carbone, G.J. Monitoring agricultural drought for arid and humid regions using multi-sensor remote sensing data. *Remote Sens. Environ.* **2010**, *114*, 2875–2887. [CrossRef]

92. Stancalie, G.; Nertan, A.T.; Serban, F. Agricultural Drought Monitoring Using Satellite—Based Products in Romania. In Proceedings of the Third International Conference on Telecommunications and Remote Sensing, Luxembourg, 26–27 June 2014; ICTRS, SciTePress: Luxembourg, 2014; Volume 1, pp. 100–106, ISBN 978-989-758-033-8.

93. Cheval, S.; Busuioc, A.; Dumitrescu, A.; Birsan, M. Spatiotemporal variability of meteorological drought in Romania using the standardized precipitation index (SPI). *Clim. Res.* **2014**, *60*, 235–248. [CrossRef]

94. Ionita, M.; Scholz, P.; Chelcea, S. Assessment of droughts in Romania using the Standardized Precipitation Index. *Nat. Hazards* **2016**, *81*, 1483–1498. [CrossRef]

95. Paltineanu, C.; Mihailescu, I.F.; Prefac, Z.; Dragota, C.; Vasenciuc, F.; Claudia, N. Combining the standardized precipitation index and climatic water deficit in characterizing droughts: A case study in Romania. *Theor. Appl. Clim.* **2008**, *97*, 219–233. [CrossRef]

96. Prăvălie, R.; Patriche, C.; Săvulescu, I.; Sîrodoev, I.; Bandoc, G.; Sfîcă, L. Spatial assessment of land sensitivity to degradation across Romania. A quantitative approach based on the modified MEDALUS methodology. *Catena* **2020**, *187*, 104407. [CrossRef]

97. Minea, I.; Iosub, M.; Boicu, D. Groundwater Resources from Eastern Romania under Human and Climatic Pressure. *Sustainability* **2020**, *12*, 10341. [CrossRef]

98. Potopová, V.; Boroneant, C.; Boincean, B.; Soukup, J. Impact of agricultural drought on main crop yields in the Republic of Moldova. *Int. J. Climatol.* **2016**, *36*, 2063–2082. [CrossRef]

99. National Institute of Statistics in Romania (NISR). Available online: https://insse.ro/cms/en (accessed on 10 December 2020).

remote sensing

Article

Predicting Future Urban Flood Risk Using Land Change and Hydraulic Modeling in a River Watershed in the Central Province of Vietnam

Huu Duy Nguyen [1], Dennis Fox [2,*], Dinh Kha Dang [3], Le Tuan Pham [1], Quan Vu Viet Du [1], Thi Ha Thanh Nguyen [1], Thi Ngoc Dang [1], Van Truong Tran [1], Phuong Lan Vu [1], Quoc-Huy Nguyen [4], Tien Giang Nguyen [3], Quang-Thanh Bui [1,4] and Alexandru-Ionut Petrisor [5]

[1] Faculty of Geography, VNU University of Science, Vietnam National University, 334 Nguyen Trai, Thanh Xuan District, Hanoi 100000, Vietnam; nguyenhuuduy@hus.edu.vn (H.D.N.); phamletuan@hus.edu.vn (L.T.P.); duvuvietquan@hus.edu.vn (Q.V.V.D.); hathanh-geog@vnu.edu.vn (T.H.T.N.); dangthingoc@hus.edu.vn (T.N.D.); tranvantruong@hus.edu.vn (V.T.T.); vuphuonglan@hus.edu.vn (P.L.V.); thanhbq@vnu.edu.vn (Q.-T.B.)

[2] Department of Geography, Université Côte d'Azur, UMR ESPACE CNRS, 98 Blvd Edouard Herriot, 06204 Nice, France

[3] Faculty of Hydrology, Meteorology and Oceanography, VNU University of Science, Vietnam National University, 334 Nguyen Trai, Thanh Xuan District, Hanoi 100000, Vietnam; dangdinhkha@hus.edu.vn (D.K.D.); giangnt@vnu.edu.vn (T.G.N.)

[4] Centre for Applied Research in Remote Sensing and GIS, Faculty of Geography, VNU University of Science, Vietnam National University, 334 Nguyen Trai, Thanh Xuan District, Hanoi 100000, Vietnam; Huyquoc2311@hus.edu.vn

[5] Doctoral School of Urban Planning, Ion Mincu University of Architecture and Urbanism, 010014 Bucharest, Romania; alexandru.petrisor@uauim.ro

* Correspondence: Dennis.Fox@unice.fr

Citation: Nguyen, H.D.; Fox, D.; Dang, D.K.; Pham, L.T.; Viet Du, Q.V.; Nguyen, T.H.T.; Dang, T.N.; Tran, V.T.; Vu, P.L.; Nguyen, Q.-H.; et al. Predicting Future Urban Flood Risk Using Land Change and Hydraulic Modeling in a River Watershed in the Central Province of Vietnam. *Remote Sens.* 2021, 13, 262. https://doi.org/10.3390/rs13020262

Received: 13 December 2020
Accepted: 11 January 2021
Published: 13 January 2021

Publisher's Note: MDPI stays neutral with regard to jurisdictional clai-ms in published maps and institutio-nal affiliations.

Abstract: Flood risk is a significant challenge for sustainable spatial planning, particularly concerning climate change and urbanization. Phrasing suitable land planning strategies requires assessing future flood risk and predicting the impact of urban sprawl. This study aims to develop an innovative approach combining land use change and hydraulic models to explore future urban flood risk, aiming to reduce it under different vulnerability and exposure scenarios. SPOT-3 and Sentinel-2 images were processed and classified to create land cover maps for 1995 and 2019, and these were used to predict the 2040 land cover using the Land Change Modeler Module of Terrset. Flood risk was computed by combining hazard, exposure, and vulnerability using hydrodynamic modeling and the Analytic Hierarchy Process method. We have compared flood risk in 1995, 2019, and 2040. Although flood risk increases with urbanization, population density, and the number of hospitals in the flood plain, especially in the coastal region, the area exposed to high and very high risks decreases due to a reduction in poverty rate. This study can provide a theoretical framework supporting climate change related to risk assessment in other metropolitan regions. Methodologically, it underlines the importance of using satellite imagery and the continuity of data in the planning-related decision-making process.

Keywords: urban; risk flood; hazard; exposure; vulnerability; land cover change

1. Introduction

Floods are a frequent and damaging natural disaster that negatively impacts the socioeconomic development and lives of millions of people worldwide [1–5]. It is estimated that between 1990 and 2016, worldwide losses from flood damage amounted to USD 723 billion [6]. Due to population growth and climate change, urban areas in particular are expected to be seriously threatened by the effects of increased flood intensity and frequency [7–10]. Approximately 40% of the world's cities will be in areas under high flood risk by 2030, thereby affecting about 54 million people [11]. Furthermore, 82% of urban areas in Southeastern Asia will be in high-frequency flood zones by 2030 [12]. Understanding the complex relationships between urban growth and floods is essential in developing risk management strategies for sustainable land use planning [13].

The flood system is very complicated and has the characteristics of spatial-temporal dynamics, with uncertainties and integration of different challenges in a system generating complex phenomena [14]. Research is an essential tool in flood risk management since it plays an essential role in predicting flood hazard and improving societal comprehension of the complex environmental and socioeconomic components of flood risk [15]. The objective of this study is to develop a comprehensive approach to estimating the evolution in flood risk between 1995 and 2040 in a rapidly changing land cover context in Vietnam.

Flood risk is a combination of hazard, exposure, and vulnerability [16], so risk can be managed by reducing the flood intensity and/or damage [17,18]. Over the past few decades, with the development of science, the world has brought many changes in the approach to minimize flood effects. Traditional methods of flood control (i.e., structural measures such as dikes, embankments, etc.) are gradually being replaced by more comprehensive flood risk management models [19,20]. These approaches take into account the probabilities and potential consequences of flood events based on risk assessment studies where flood hazard and exposure/vulnerability factors are quantified [21]. Several methods have been applied to compute these indicators. The hazard approaches include: measurement-based, field surveys [22,23], hydrodynamic models [24,25], and GIS and Remote Sensing [26,27] in linear modeling of flood risk through overlaying component layers with associated analytical hierarchical process (AHP)-based computed weights. Land cover indicators can be grouped into two categories: (1) traditional terrestrial mapping [28]; and (2) land cover classification based mainly on satellite observations [29,30]. The emergence of satellite sensors like Landsat, Satellite Pour Observation de la Terre (SPOT), and Sentinel 2, among others, has greatly facilitated the rapid classification of land cover and its evolution. In addition, remote sensing has the advantage of rapid data acquisition with lower costs over field survey methods [31].

Flood risk assessment is an essential step in defining appropriate management strategies [32]. In recent decades, several studies have focused on developing flood risk assessment methods at different scales and with different objectives. Mishra et al. [33] developed a flood risk index for the Kosi River of India based on hazard (geomorphologic, distance to the active channel, slope, and rainfall) and socioeconomic vulnerability (population, household, and female densities; literacy rates; land cover and use changes; road–river intersections; road density). Chinh Luu et al. [34] studied the temporal changes of flood risk that integrated hazard, exposure, and vulnerability to better understand the evolving dynamics and formulate appropriate mitigation strategies. Dang et al. [35] outlined the essential roles in improving flood risk assessment methods to support decision-making processes. The authors classified flood risk indices into three components: social–economic, physical, and environmental. Kron [36] constructed flood risk indices based on the likelihood of flood occurrence, social–economic vulnerability, the environment, and flood consequences. Begun et al. [37] integrated floods-related damage with the probability of their occurrence. Several clues or methods of assessing flood risk have been developed in different areas. However, they are limited in a comprehensive framework that can support decision-makers to understand the aggravating risk causes. In addition, these studies focus on assessing the flood risk at a specific time while, according to Penning-Rowell et al. [38], flood risk reduction strategies are most effective when they are evaluated continuously. Jhong et al. [39] emphasize that understanding hazard, exposure, and vulnerability at different dates is an essential task in flood risk reduction that allows land managers to better see the spatial and temporal trends likely to arise in the future [40]. To fill the gaps identified, we integrate hydraulic modeling with land cover change analysis and prediction to assess flood risk at three dates: 1995, 2019, and 2040.

This research is different from the previous studies cited because it provides a novel and comprehensive approach to flood risk assessment based on state-of-the-art remote sensing and modeling techniques and assesses both historical and future trends. The initial hypothesis tested in this study is that flooding risk has increased substantially in the study area due to an increase in the population living in the flood plain. Flood hazard, exposure, and vulnerability were all accounted for in order to assess their potential interactions and compensatory effects. While this study explicitly examines flood risk in a catchment located in Vietnam, the findings are of importance to other rapidly evolving countries affected by floods and experiencing urban growth.

2. Study Area Characteristics

With an area of 5152.67 km^2, Quang Ngai province is located in Vietnam's central coastal region (Figure 1). In 2019, the total population was approximately 1.23 million, with a density of 237 people per km^2. The Tra Khuc River watershed within the city of Quang Ngai is central to the economy of the province; therefore, urbanization is growing rapidly. The population has increased from 1,218,600 in 2010 to 1,231,697 in 2019, with most of this growth occurring in the alluvial plains subject to floods. Apart from Ly Son Island, the Tra Khuc River watershed covers all Quang Ngai province districts. Its topography can be divided into three main segments: upslope, downslope, and smaller alluvial plains. Upslope, the catchment is dominated by steep slopes, narrow valleys, and gorges. Downstream, an extensive plain can be found at Quang Ngai city's location just before the channel's outlet into the East Sea. Between the upslope and downstream zones, smaller alluvial plains lie along the main river channels between the uplands.

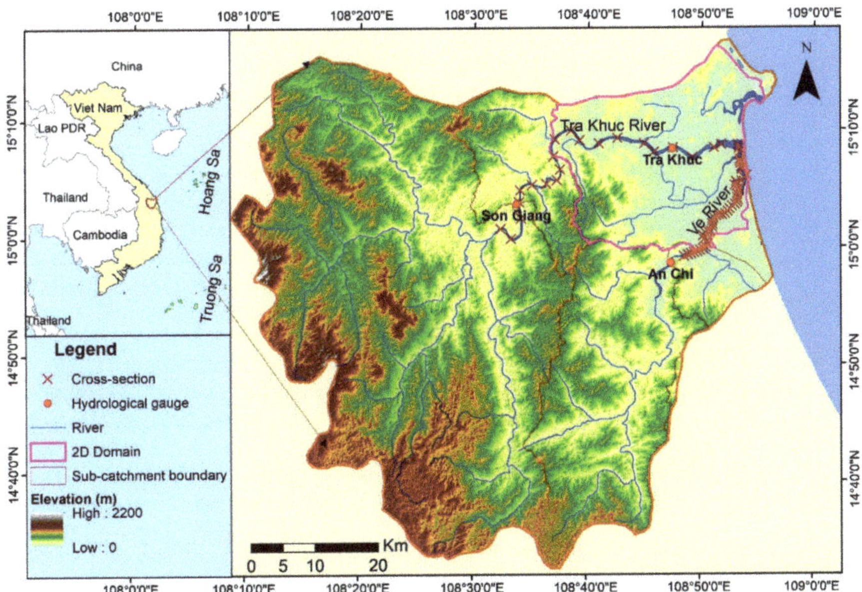

Figure 1. The Tra Khuc river watershed in Quang Ngai province.

The two main rivers forming the watersheds are the Tra Khuc River and the Ve River. Most rivers and streams originate from the Truong Son mountain range and flow into the East Sea with a common feature being shallow, narrow riverbeds and strong seasonal variations in channel width and discharge. While the Ve River has a length of 91 km, the Tra Khuc River is one of the largest rivers in central Vietnam and is nearly 135 km long.

A tropical monsoon regime influences the climate in this region, with distinct rainy and dry seasons. The dry season runs from April to August, and the rainy season runs from September to March; it is usually characterized by intense rainfall resulting from tropical depressions and storms. The average annual rainfall ranges from 2000 to 2500 mm with 80% of it occurring between September and December and heavy typhoon rains cause floods in the Tra Khuc watershed. The November 2013 flood caused 41 deaths and resulted in the evacuation of nearly 1700 households in addition to other substantial damage and is considered one of the most significant damage-causing events in Quang Ngai province's history. This event was selected as the case study for this research (Figure 1).

3. Data and Methods

In this study, flood risk is determined by a combination of flood hazard, exposure, and vulnerability. Hazard is a physical phenomenon—natural and uncontrollable with intense occurrence [41]. It is an elementary notion that expresses the probability of a situation, an event, or some causality—a flood in this case [36,42]. It is directly proportional to the intensity of the phenomenon's occurrence [43], since the combination of flood depth and flood peak velocity represents the capacity to destroy objects in the area through which the flood passes, directly affecting houses, buildings, and people's lives and health. Exposure is understood as values present at the location of a possible flood. These values can be goods, infrastructure, cultural heritage, people, or agricultural areas. This can be interpreted as the presence or availability of assets and people in flood risk areas. The level of exposure depends on the frequency of flood occurrence, the flood's intensity, and the value of the available properties and people [42,44]. In this study, land cover and population density were selected to analyze the level of exposure. The land cover categories are detailed below and include Cropland, Forest, Urban area, and Water. Potential flood impacts in urban areas are considered greater since the infrastructure is well developed and they concentrate more value than other land cover types; they also have the greatest population densities. Agricultural land follows urban areas as it provides the livelihood for many of the watershed inhabitants. Population density is a critical variable of flood exposure analysis; because it is linked directly with humans, it is proportional with the level of flood vulnerability [34]. Vulnerability represents the lack of individuals' or groups' ability or capacity to anticipate, counter, and resist floods [45]. The International Plant Protection Convention (IPPC) emphasizes that these capacities depend on the prevailing economic situation and social and political characteristics. Vulnerability increases in direct proportion to the flood level and decreases socioeconomic wellbeing during floods [46]. In this study, the poverty rates and number of hospitals were selected to build the vulnerability maps. The wealthy-to-poor ratio characterizes a community's resilience; the wealthy have sturdier homes, have the ability to access information about danger through modern media, and quick resumption of their normal lives is enhanced by their good economic standing [47]. Hospitals play a critical role during and after events. Patients place a heavy burden on communities and emergency management services in the event of an evacuation [48]. A flowchart of the methodological workflow is displayed in Figure 2. The quantification of hazards, vulnerability, and exposure is described in detail in the following sections.

3.1. Flood Hazard Estimation

Flow depth and velocity were used to estimate flood hazard. The hydraulic modeling tool MIKE FLOOD, which combines the MIKE 11 and MIKE 21 models, was used to build the depth and velocity map for the 2013 Quang Ngai historic flood.

3.1.1. Establishing the 1-Dimensional (1D) River Network

The 1D network is necessary for modeling water flow through the channels into the flood plains and was established for the Ve and Tra Khuc rivers with 46 cross-sections along 27.1 km of the Ve River (average distance about 0.59 km/sections), and 21 cross-sections along 51.6 km of the Tra Khuc (average distance about 2.6 km/sections). The cross-sections were measured directly in the field using total station (Figure 1). The discharge observation data at An Chi and Son Giang gauges were applied to the upstream boundary of Ve and Tra Khuc, respectively, and the tide level was applied to the downstream boundary using tidal sea level harmonic constants. MIKE NAM is the rainfall runoff model part of the MIKE 11 module developed by the Danish Hydraulic Institute (DHI), Denmark. This model provided discharge for sub-basins without streamflow data, and sub-catchment boundaries are presented in Figure 1. MIKE NAM was integrated into the MIKE 11 model and values were applied to the inflow boundaries for the MIKE 11 hydraulic model.

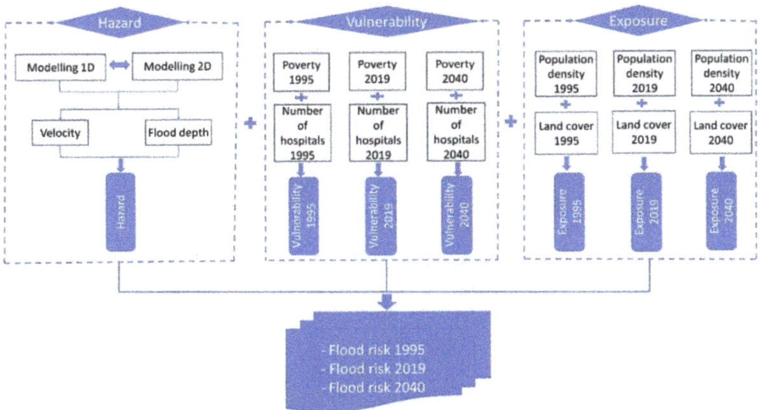

Figure 2. A flowchart with methodological workflow.

3.1.2. Establishing the River Network in the 2-Dimensional (2D) Hydraulic and FLOOD Models

The 2D channel network is necessary for estimating the spatial extent of the flooded areas when the river depth exceeds bank height. The area modeled covered an area of 507.2 km^2 based on the surface elevation data and observed flood inundation area defined by flood marks. The 1:10,000 topographic map published in 2010 by the Ministry of Water Resources and Environment was used for defining the 2D domain's elevation in MIKE-21. The 2D computational mesh was generated by discretizing the computational domain into 48,200 elements, with the size of elements ranging from 457 to 2500 m^2.

MIKE FLOOD is used to link MIKE 11 and MIKE 21 through the Lateral link which connects the tops of riverbanks in the 1D model to the 2D model's mesh elements. The water levels and time series of two flood events—one in November 2013 and the other in November 2017 at Tra Khuc gauge—served to calibrate and validate the model. Parameters for the 2D model were calibrated based on flood trace measurements taken in November 2013 (50 locations) and November 2017 (32 locations). Nash–Sutcliffe efficiency (NSE) [49], flood peak error, and coefficient of determination (R^2) were used to measure the model's reliability.

3.2. Indicators of Flood Exposure

Exposure was quantified based on property value and the population [22,24]; land-use categories and population density were selected to estimate these variables.

3.2.1. Land Cover Mapping in 1995, 2019, and 2040
Land Cover Mapping in 1995 and 2019

Land cover maps were created using satellite imagery—SPOT-3 (12/03/1995) with a 20 m spatial resolution and Sentinel-2 (27/02/2019) with a 10 m resolution, for 1995 and 2019, respectively. The data were projected to a common datum and coordinate system (WGS84/UTM Zone 48N). Both images were acquired with cloud cover less than 5%.

Images underwent radiometric and geometric corrections before re-sampling all bands to a 10 m spatial resolution using the bilinear interpolation sampling method.

The eCognition Developer of Trimble was used to perform object-oriented classification on the SPOT-3 and Sentinel-2 images; a sample output of the classification process is displayed in Figure 3. First, image objects were segmented by integrating similar pixels; then, each segment was assigned to land cover object layers [50,51]. In order to optimize the object-oriented classification, it was necessary to select the appropriate segmentation parameters as each segment must be homogeneous and separate from its neighbors [52].

Values for the 3 segmentation parameters were 400 for Scale, 0.2 for Shape, and 0.99 for Compactness. After segmentation, the objective variables were selected using random formation points, and the information related to the objects was extracted. Different values of objective characteristics were derived from spectral indices such as the Normalized Difference Vegetation Index (NDVI) and the Enhanced Vegetation Index (EVI). The samples and extracted objective information were then exported to a calculation table. The object-oriented classification was based on the training polygons with 5 land-use categories: Cropland, Developed (built area), Bare soil, Forest, and Water bodies.

The land cover map in 1995 was compared to panchromatic aerial photographs collected from the Ministry of Natural Resources and Environment (MONRE 1995), while the 2019 classification was compared to the 2020 land use map (MONRE 2020). Validation points were randomized with n = 205 for the map in 1995 and n = 262 for the map in 2019; the number of sampling points for each class depended on the class coverage.

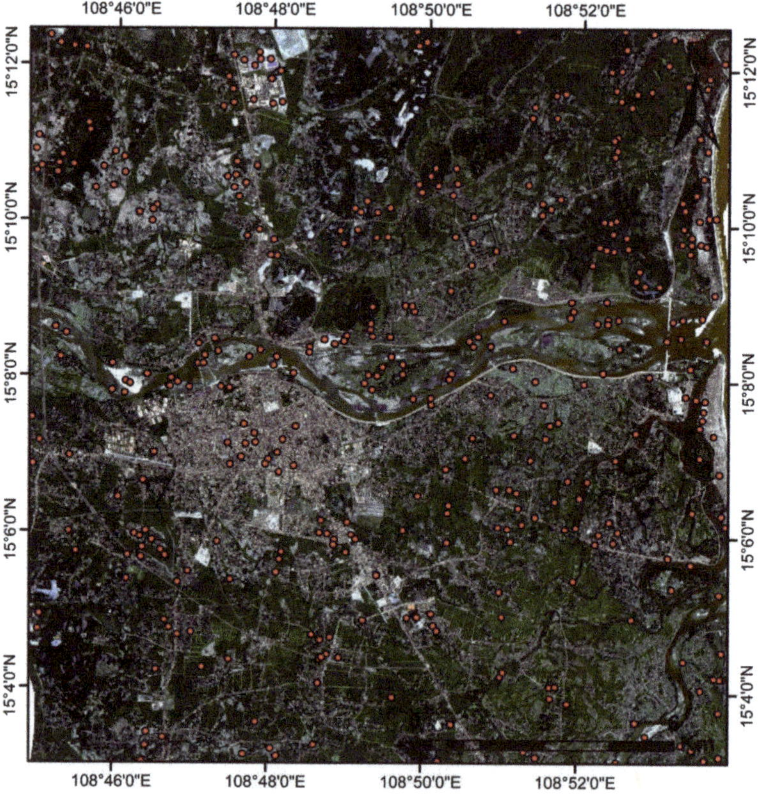

Figure 3. A sample of the classification of image satellite Sentinel 2A in the Quang Ngai city in 2019.

Change Prediction in 2040

The initial map had five categories, but Bare soil was mostly sand bars in the river or exposed beach areas along the coast, so it was integrated into the Water body category to facilitate land cover change modeling.

Future land cover change was modeled using the Land Change Modeler (LCM) module from Terrset [53] using the default Multi-Layer Perceptron (MLP) neural network option for predicting future change. LCM facilitates the quantification and mapping of historical land cover changes and predicts future changes based on past transition rates. Quantity of change from one land cover to another is based on Markov chain analysis, and spatial allocation depends on transition potentials derived from historical trends and explanatory variables such as topographic or distance variables [53,54]. Driver variables generally belong to one of three categories: accessibility, suitability, and zoning [55]. Accessibility is frequently modeled as the distance from roads or existing urban areas, and suitability for urban areas is strongly dependent on topography. There were no zoning constraints or initiatives in the study area. The initial explanatory variables tested were the following: altitude (10 m DEM), slope inclination (%), and distances from roads, developed area in 1995, and from water (river or sea). However, since slope inclination and altitude are closely correlated in the study area, only altitude was retained. The contribution of explanatory variables to predicting land cover change was estimated from Cramer's V values and from accuracy rates produced during the creation of potential transition layers [53]. Accuracy rates are calculated by running calibrating transitions on a sub-sample of cells and then evaluating the accuracy of predictions on the remaining cells. A land cover map for 2040 was produced from the transition potential maps.

3.2.2. Population Density

Population density is considered one of the critical indices for flood exposure. Population density by municipality was obtained from the municipalities for 1995 and 2019, and an estimate for 2040 by the municipalities, was obtained from the statistics and general offices of six districts in the study area (Quang Ngai, Son Tinh, Mo Duc, Tu Nghia, Nghia Hanh, and Binh Son).

3.3. Indicators of Flood Vulnerability

Poverty data by municipality and number of hospitals in the flood zone were obtained from six Departments of Statistics (Quang Ngai, Son Tinh, Mo Duc, Tu Nghia, Nghia Hanh, Binh Son) to build the vulnerability map as described below.

3.4. Assigning Normalized Weights for Hazard, Exposure and Vulnerability Using Analytical Hierarchical Process (AHP)

The components of flood risk are hazard, exposure, and vulnerability [36,56,57]. Each component was divided into sub-components and the AHP method was used to quantify the weight of each sub-component [35].

AHP is a quantitative method used to organize decision options and choose an option that satisfies a given criterion. Developed by Thomas L. Saaty in the 1970s, it is considered an efficient and flexible method for analyzing multi-criteria decisions [58]. It operates by setting priorities for multi-criteria rankings that are judged by experts involved in the decision-making process to derive the best decision possible. In this study, assessments were based on the authors' group experience and from previous similar studies [35,59–63]. The weights of flood risks in 4 steps:

(1) Constructing component and sub-component hierarchies: hazard, exposure, and vulnerability. Hazard is divided into two sub-components (depth and velocity), exposure into two sub-components (land cover and population density), and vulnerability into two sub-components (poverty rate and number of hospitals).

(2) Establishing priorities: The relative importance of the elements in each pair of sub-components was evaluated subjectively and assigned a value ranging from 1 to 9 as per the Saaty scale [64].

(3) After preparing a pairwise comparison matrix (Table 1), each column was divided by the corresponding sums to obtain the priority factors. This process is a normalized Eigenvector of the matrix (Table 2); the average values of the row were used as the priority of the subcomponents to calculate flood risk.

(4) Estimation of the Consistency Ratio (CR).

Table 1. Comparison matrix for flood hazard, exposure, and vulnerability indicators.

Hazard	Flood Depth	Velocity
Flood depth	1	3
Velocity	1/3	1
Column total	1.33	4
Exposure	**Population Density**	**Land Use**
Population density	1	2
Land use	1/2	1
Column total	1.5	3
Vulnerability	**Poverty Rates**	**Number of Hospitals**
Poverty rates	1	2
Number of hospitals	1/2	1
Column total	1.5	3

Table 2. Normalized matrix of flood hazard, exposure, and vulnerability.

Hazard	Flood Depth	Velocity	Sum	Average
Flood depth	0.75	0.75	1.5	0.75
Velocity	0.25	0.25	0.5	0.25
Exposure	**Population Density**	**Land use**	**Sum**	**Average**
Population density	0.66	0.66	1.32	0.66
Land use	0.34	0.34	0.68	0.34
Vulnerability	**Poverty Rates**	**Number of Hospitals**	**Sum**	**Average**
Poverty rates	0.66	0.66	1.32	0.66
Number of hospitals	0.34	0.34	0.68	0.34

The Consistency Ratio (CR) is used to ensure consistency in the experts' judgement throughout the application and is applied as follows:

CR = CI/RI, where Consistency Index (CI) can be calculated as per the equation $CI = (\lambda max - n)/(n - 1)$; λmax is the comparison matrix's eigenvalue obtained by multiplying each parameter's correlation matrix column's total summation by the normalized value of corresponding parameters; and n is the number of elements compared in pairs during a calculation.

RI must be defined in [64]

After calculating the relative importance and determining the weight of each factor in the hierarchy, flood hazard, exposure, and vulnerability were calculated using the following equations:

Flood hazard = 0.75 flood depth + 0.25 velocity

Flood exposure = 0.66 population density + 0.34 land cover

Vulnerability = 0.66 poverty rates + 0.34 number of hospitals

The output values were normalized, ranging from 0 to 1, and reclassified into five classes using the natural break method: Very low, Low, Moderate, High, and Very high.

4. Results

4.1. Flood Hazard Mapping

The differences between observed and simulated water level are illustrated in Figure 4a,b. Results from recorded flood events (Table 3) indicate a good match between observation and simulation; the flood peak error ranged from 3 to 9 cm (0.4–1% of peak flow depth) for both calibration and validation. The R^2 values were about 0.95-0.96, and the NSE value for water level ranged from 0.95 to 0.96.

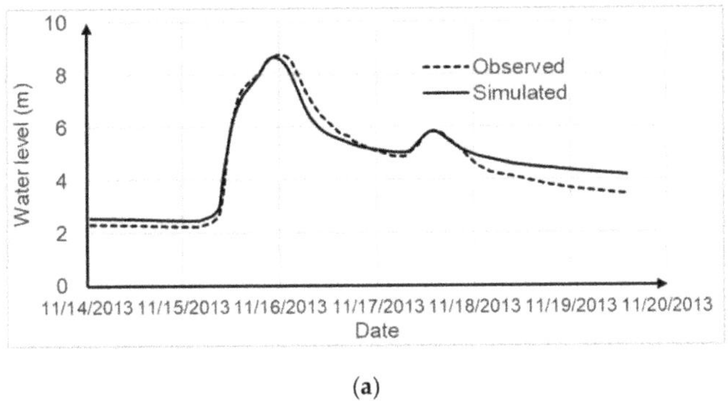

(**a**)

Figure 4. *Cont.*

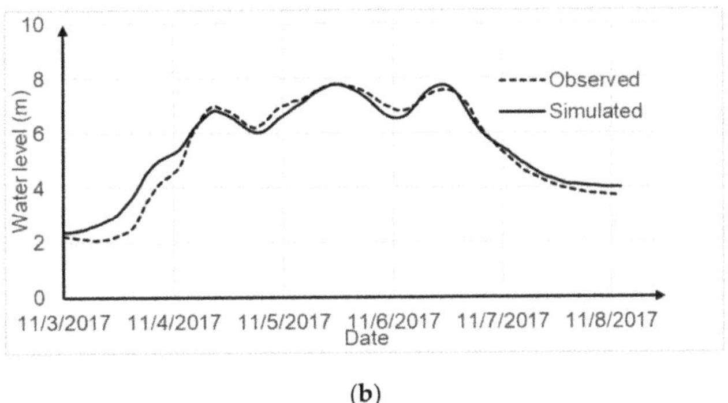

(**b**)

Figure 4. (**a**) The observed and simulated water level at Tra Khuc (calibration); (**b**) The observed and simulated water level at Tra Khuc (validation).

Table 3. Metric for model performance evaluation.

Metric	Flood Event in November 2013	Flood Event in November 2017
NSE	0.96	0.95
Error flood peak	9 cm (1%)	3 cm (0.4%)
R^2	0.95	0.96

Surveyed and simulated flood values are shown in Figure 5. R^2, the correlation coefficient between observed measurements and simulated values at these checkpoints, reaches 0.99 for calibration and validation. The model was therefore considered reliable for estimating flood hazard.

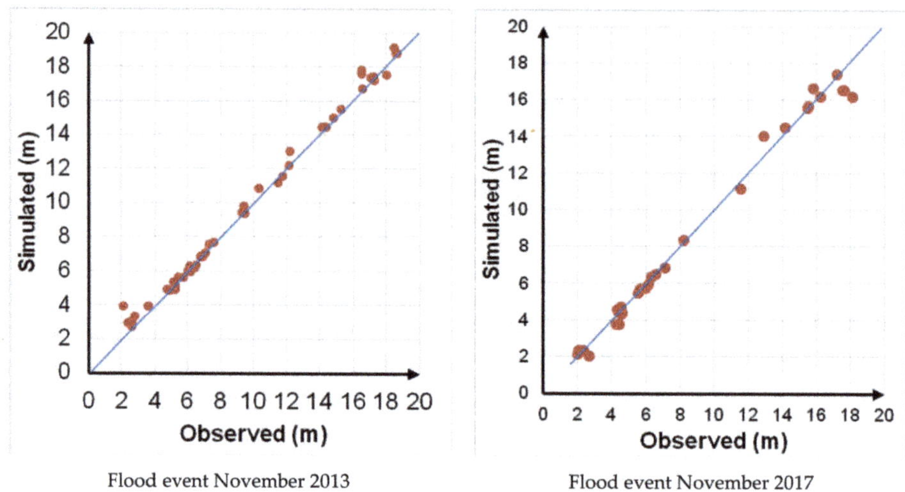

Figure 5. The observed and simulated flood marks.

From the hydrodynamic modeling, flood depth and velocity were mapped (Figure 6a). About 69.66 km^2 (29.3%) of the study area flooded to a depth of less than 0.5 m, 51.48 km^2 (21.7%) flooded to between 0.5 and 1 m, 70.33 km^2 (29.7%) flooded to between 1 and 2 m, 30.37 km^2 (12.7%) flooded to between 2 and 3 m, and 15.74 km^2 (6.67%) flooded to depths greater than 3 m.

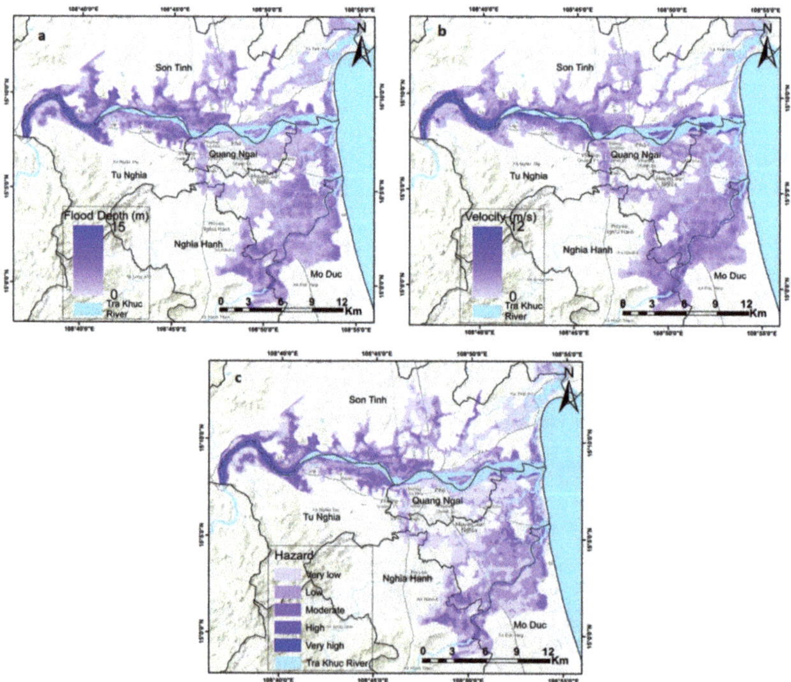

Figure 6. Flood (**a**), Velocity (**b**), and Hazard (**c**) mapping in the study area.

About 107.68 km^2 (45.3%) is affected by a velocity under 0.33 m/s, 96.51 km^2 (40.6%) by a velocity between 0.33 and 0.87 m/s, 22.85 km^2 (9.6%) by a velocity between 0.87 and 1.81 m/s, 7.64 km^2 (3.2%) by a velocity between 1.8 and 3.1 m/s, and 2.93 km^2 (1.2%) by a velocity over 3.1 m/s (Figure 6b).

As described above, the hazard map combines both depth and velocity; about 141.09 km^2 (59.3%) of the flood zone is located in the very low zone, 69.8 km^2 (29.3%) in the low zone, 11.06 km^2 (4.7%) in the moderate zone, 2.56 km^2 (1.07%) in the high zone, and 2.67 km^2 (1.1%) in the very high zone. The moderate, high, and very high zones are concentrated along the river in Son Tinh, Tu Nghia, and Quang Ngai city (Figure 6c).

4.2. Flood Exposure Mapping

4.2.1. Land Cover Change in 1995, 2019, and 2040

The classification accuracy for the land cover values shown in Table 4 is 78% in 1995 and 82% in 2019. In 1995, Cropland occupied more than half the landscape (54.8%) followed by Forest and Developed with similar areas at 19.9% and 18.5% of the landscape, respectively. Water and Bare soil combined account for less than 7% of the landscape. Forested areas tend to concentrate on steeper slopes at higher altitudes near the study area's northern and western zones (Figure 7a,b). Mean slope and altitude values for Forest are 8.2% and 19.3 m, respectively. Corresponding values for Cropland are 1.3% and 7.0 m, respectively, and values for Built-up area are 1.9% and 8.5 m, respectively; therefore, topography clearly influences land cover distribution and is a suitable land cover change driver.

Table 4. Land cover and land cover changes (km^2) for 1995 and 2019.

Land Cover	1995	2019	Change (km^2)	Change (%)
Cropland	285.8	190.5	−95.3	−33.3
Forest	103.9	101.1	−2.8	−2.7
Built-up area	96.3	189.1	92.8	96.4
Water	28.2	34.8	6.6	23.2
Bare soil	7.2	5.9	−1.3	−18.4
Total	521.5	521.5		

As shown in Table 4, between 1995 and 2019, Cropland lost 33.3% of its initial area (95.3 km^2), and Developed increased its area by 96.4%. The growth in Built-up area appears to spread out from existing Built-up area (Figure 7c); therefore, distance from Built-up area is also a reliable driver variable. Cross-tabulation analysis (Table 5) shows that the only substantial land cover transition is a shift from Cropland to Built-up area since changes in Forest and Water are negligible. The marginal transition from Built-up area to Cropland (1.2 km^2) in Table 5 suggests this is probably a classification error, as Developed lands rarely transition to agricultural or forested uses. Slope and altitude values for Forest, Cropland, and Built-up areas in 2019 are virtually unchanged from 1995, so Forest remains relatively untouched on steeper slopes (persistence of 91.5%) with minor swapping with water at lower altitudes. Built-up area is encroached in Cropland in the flatter, low-lying alluvial plain.

Cramer's V values are low for all explanatory variables (Table 5). According to Eastman (2016), values of 0.15 are useful, while values of 0.40 or greater are good. The explanatory variables used here all had values greater than 0.15, but none reached 0.40: Altitude (0.23), Distance from Roads (0.15), Distance from 1995—Developed (0.16), and Distance from Water (0.27). The accuracy rate of transition potential was 75.1%. However, Multi-Layer Perceptron (MLP) neural network results indicate that Distance from Developed is the most important variable in explaining Cropland to Developed transitions. It is followed by Distance to Water, Altitude, and Distance from Roads in that order.

Table 5. Cross-tabulation of land cover changes (km²) with 1995 values in columns and 2019 in rows. Bold values are aligned diagonally in the table show persistence values in km² and in % (in parentheses).

	Cropland	Forest	Built-Up Area	Water	Total 2019
Cropland	**189.2 (66.2)**	0.0	1.2	0.1	190.5
Forest	0.6	**95.1 (91.5)**	1.4	4.0	101.1
Built-up area	95.1	0.5	**93.3 (96.9)**	0.2	189.1
Water	0.9	8.3	0.3	**31.2 (88.0)**	40.7
Total 1995	285.8	103.9	96.3	35.5	521.5

The 2040 predicted land cover map (Figure 7d) shows a further decrease in Cropland area of 57.3 km² in 2040 to reach 133.2 km² and an equivalent increase in Built-up area to reach 246.4 km². These changes correspond to a loss of 30.1% in Cropland and gain of 30.3% in Built-up area for the 2019–2040 period. Since Cropland to Built-up area was the only transition modeled, both Forest and Water areas remained constant.

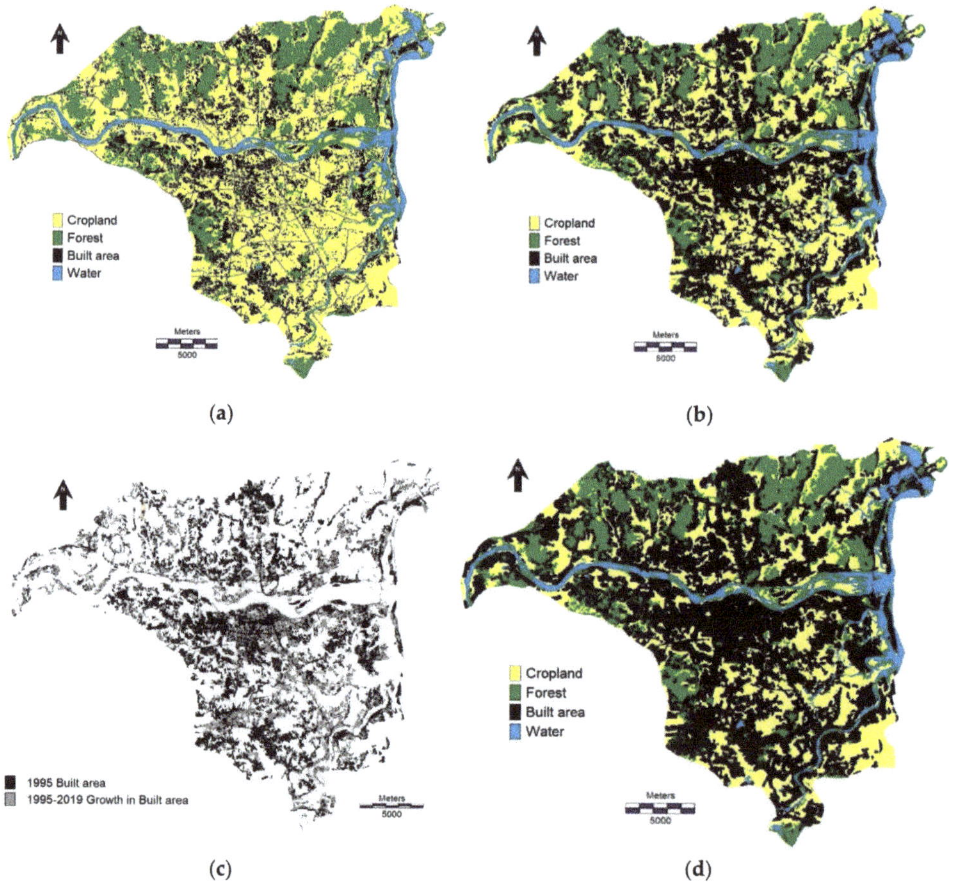

Figure 7. (**a**) Land cover map for 1995; (**b**) Land cover map for 2019; (**c**) Built-up area growth (grey) within and around 1995 developed (black) territory; (**d**) Predicted land cover map for 2040.

Urban areas have the greatest land values and population densities [65,66]. Agricultural land is the second most vulnerable land cover due to its subsistence and economic activities.

4.2.2. Population Density

Figure 8a–c show historical and future population densities within the flooded area. Population densities increased between 2019 and are expected to increase further by 2040.

For example, the values for the municipalities of Quang Ngai city are: Tran Phu (3799 in 1995 to 5461 in 2019, and 6762 in 2040), Quang Phu (2231 in 1995 to 2610 in 2019, and 3284 in 2040), and Le Hong Phong (2036 in 1995 to 2888 in 2019, and 3523 in 2040). Others were the Nghia An (5064 in 1995 to 5773 in 2019, and 6650 in 2040) and Song Ve (2713 in 1995 to 2801 in 2019, and 3118 in 2040) municipalities of Tu Nghia district, which increased (red color).

The great majority of municipalities witnessed an increase in population density, and the average values for the main cities are 3168.6 and 3906.6 inhabitants/km^2, for 1995 and 2019, respectively. Population density is expected to increase to 4667.4 inhabitants/km^2 in 2040, resulting in a net 47.3% increase between 1995 and 2040 despite marginal decreases in some municipalities.

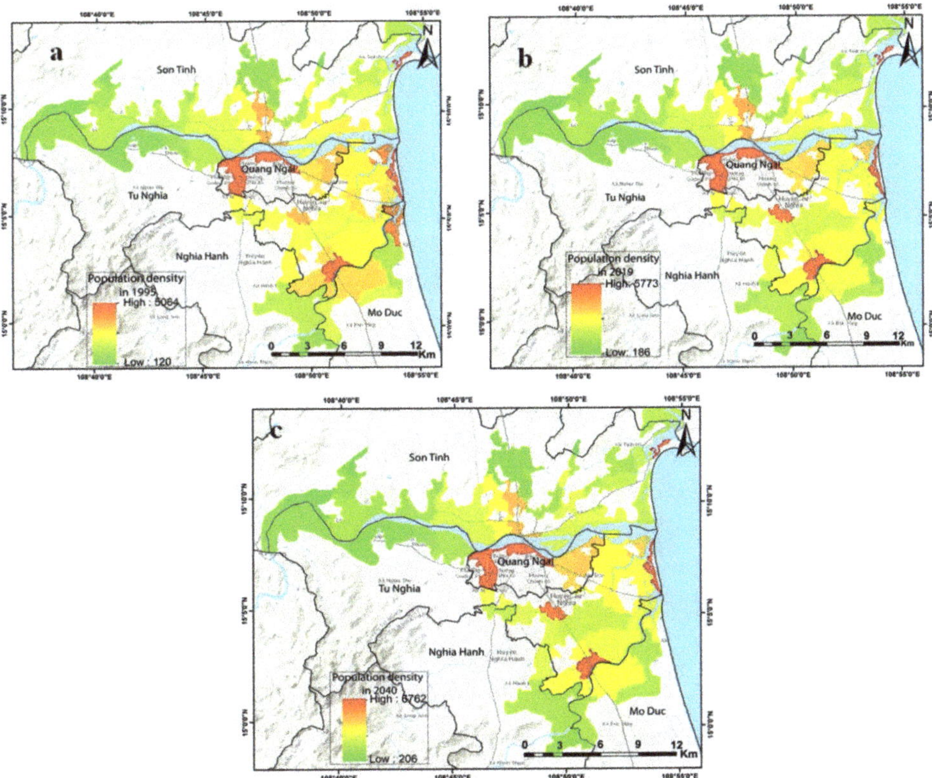

Figure 8. Population density in 1995, 2019, 2040 (**a–c**).

4.2.3. Flood Exposure

Flood exposure maps for 1995, 2019, and 2040 are shown in Figure 9a–c, respectively, with areas of High exposure having high population densities and high urban growth. Compared 1995 flood exposure, areas of low flood exposure increased by 2019, and will increase further by 2040, as shown in Table 6. Study areas of very high and high flood exposure categorization increased from approximately 18.8 in 1995 to 17.9 in 2019, and will increase to 30 km^2 by 2040.

Table 6. The distribution of flood exposure area in 1995, 2019, and 2040.

	Very Low (km^2)	Low (km^2)	Moderate (km^2)	High (km^2)	Very High (km^2)
1995	47.1	79.06	92.5	16.6	2.2
2019	47.1	81.2	91.3	13.5	4.4
2040	38.2	65.03	103.6	19.9	10.8

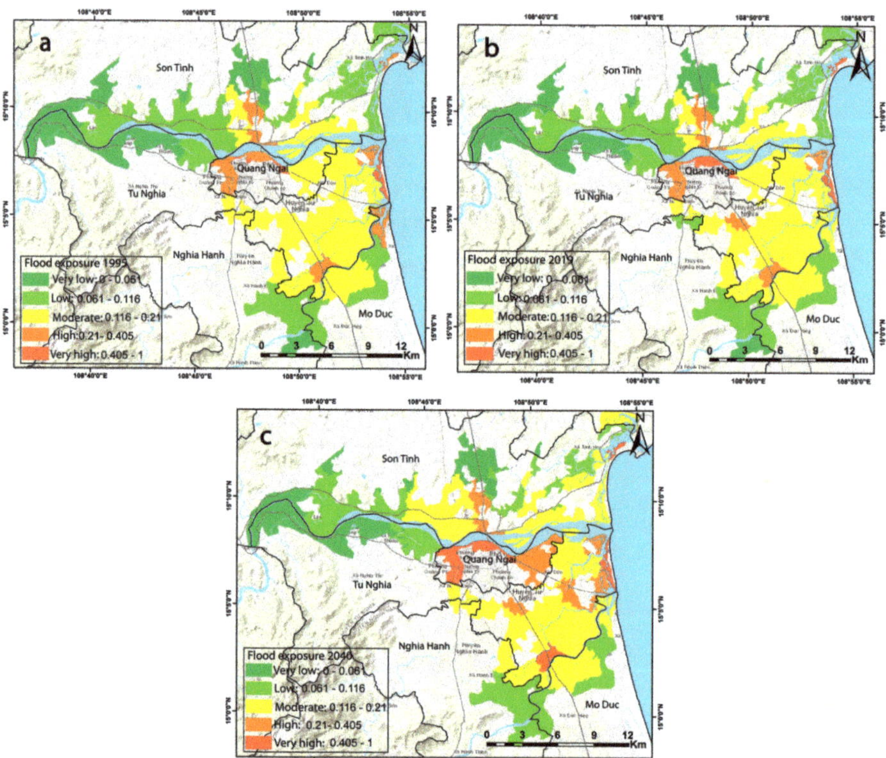

Figure 9. Flood exposure in 1995, 2019, and 2040 (**a–c**) in the study area.

4.3. Measuring Vulnerability to Flood

4.3.1. Poverty Rate and Number of Hospitals

Figure 10a–c show poverty rates for 1995, 2019, and 2040, respectively. The number of hospitals in the flood zone for corresponding years are shown in Figure 10d–f, respectively. These variables have undergone major changes over the study period. The maximum poverty rate was 72% in 1995, and it decreased to 18.3% in 2019; it is forecasted to continue decreasing to 4.9% by 2040.

In 1995, most municipalities had two hospitals, while some in Son Tinh and Mo Duc districts had none. Hospital numbers grew rapidly between 1995 and 2019 with Quang Ngai city and the eastern part of Tu Nghia district having a maximum of eight hospitals. This number is predicted to continue growing and should reach 12 by 2040.

4.3.2. Flood Vulnerability

In 1995, the entire flood plain of 240.1 km^2 was in a Very high risk zone; this value dropped by 62.0% to 91.2 km^2 and is expected to continue to decrease by a further 86.0% of its 2019 value in 2040 (Table 7 and Figure 11). Of the 240.1 km² in the Very high vulnerability category in 1995, about 38.7% will be in the Very low and Low categories in 2040; a further 30.0% will be in the Moderate vulnerability category. These trends show a major shift in flood vulnerability due to an improvement in the financial status of the population living in the flood plain. Only small patches, particularly in the city of Quang Ngai, will experience increased vulnerability due to a growth in the number of hospitals within the flood zone (Figure 11).

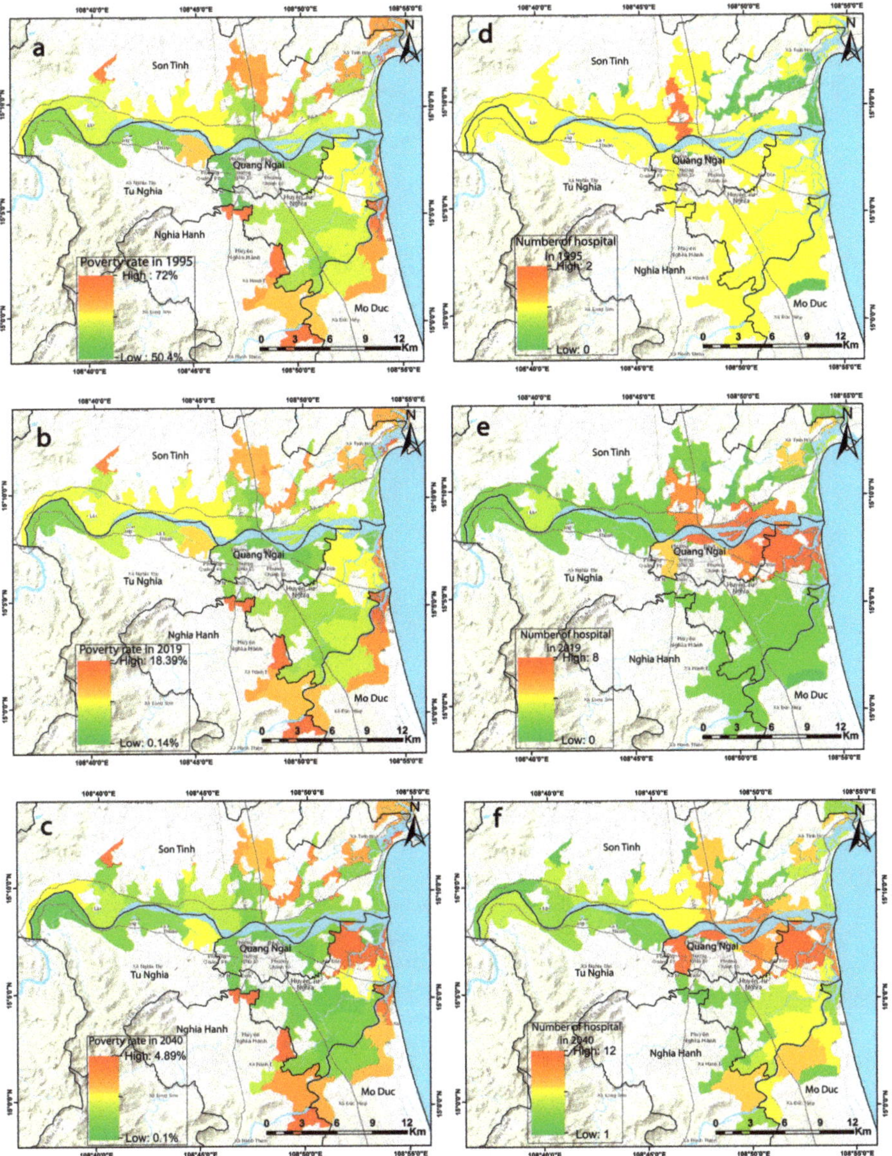

Figure 10. Poverty rate 1995, 2019, and 2040 (**a–c**) and number of hospitals in 1995, 2019, and 2040 (**d–f**) in the study area.

Table 7. The distribution of vulnerability area in 1995, 2019, and 2040.

	Very Low (km^2)	Low (km^2)	Moderate (km^2)	High (km^2)	Very High (km^2)
1995	0	0	0	0	240.1
2019	0.8	11.8	52.5	83.5	91.2
2040	17.02	75.9	72.02	62.2	12.8

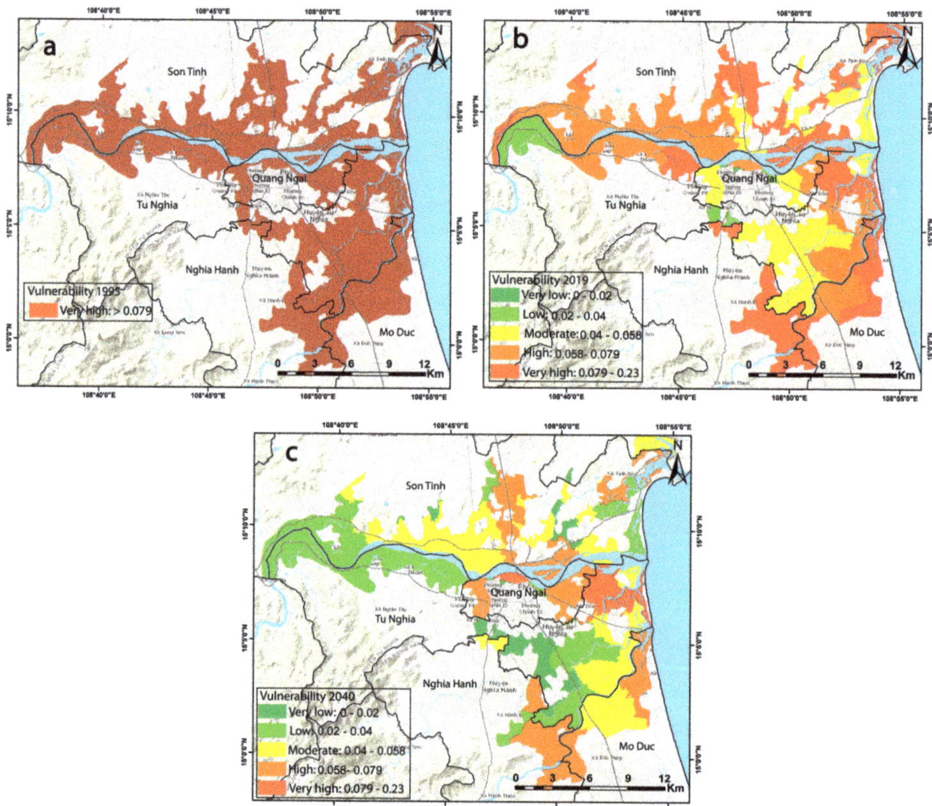

Figure 11. Vulnerability in 1995, 2019, 2040 (**a–c**) in the study area.

4.4. Flood Risk Mapping

The flood risk values shown in Table 8 and mapped in Figure 12 combine the results of flood hazard, exposure, and vulnerability described above. As for vulnerability, there is a shift from the High risk categories to Low risk categories over time. In 1995, about 82% of the flood plain was in the High or Very high risk categories. These categories are reduced about 1% in 2019 and 2040. Similarly, the Very Low and Low risk categories, which occupied less than 5% of the flood zone in 1995, attain more than 90% in 2019 and 2040. The major evolution occurs between 1995 and 2019 with little change afterwards.

Table 8. The distribution of flood risk area in 1995, 2019, and 2040.

	Very Low (km^2)	Low (km^2)	Moderate (km^2)	High (km^2)	Very High (km^2)
1995	0.05	8.2	33.8	74.8	120.6
2019	119.2	99.2	17.3	1.2	0.6
2040	145.7	73.3	15.3	2.6	0.5

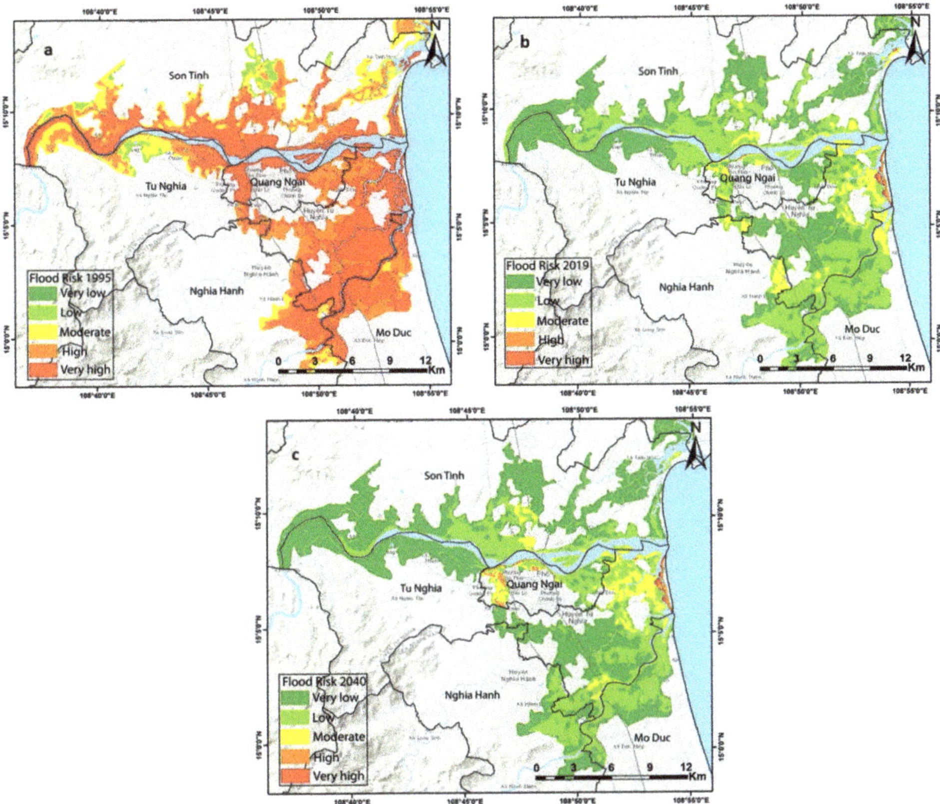

Figure 12. Flood risk in the Tra Khuc watershed in 1995 (**a**), 2019 (**b**), and 2040 (**c**).

5. Discussion

Flood risk analysis is a critical step towards sustainable economic development and protection against flood damage [67,68]. This study introduces a comprehensive flood risk assessment method that integrates hydraulic modeling, land cover change analysis and prediction, and socioeconomic trends. The flood risk maps of 1995, 2019, and 2020 have five risk levels (very low, low, moderate, high, and very high). These levels were classified by the natural break method, based on applying Jenk's optimization formula to minimize the variability of each category, having the advantage of automatically defining the final classes, underlining disparities in the best way possible [69,70]. Using different breakpoints would perhaps have divided the flood zone area slightly differently but the overall trend of diminishing risk over time would have remained the same.

The final results were unexpected as our initial hypothesis was that flood risk was increasing with increasing population density in the flood zone and with the transition from Cropland to Built-up area. This trend is typical of most cities undergoing rapid urbanization in a flood plain. Huu Duy et al., 2018 [71] reported that rapid urbanization growth, in addition to poor planning in the Gianh River watershed in Vietnam, has resulted in many populations becoming more vulnerable to floods. Areu-Rangel et al. [72] found urbanization was an important factor in increasing flood risk. The work by Waghwala [73] in the case of Surat City hammers in this fact—changes in LULC caused by urban expansion increase the flood risk. Bahrawi et al., 2020 [74] analyzed the flood risk in East-Ern Jeddah, Saudi Arabia. The authors pointed out that the rapid growth of urbanization is radically changing the characteristics of flooding and increasing the flood risk in this region. These results were confirmed by Zhang et al., 2018 [75]; according to the authors, the exacerbation of urbanization is not the only one to pose difficulties to the response to the floods, but also the total precipitation during the storms. Handayani et al., 2020 [76] reported that poorly planned urbanization has increased the surface area exposed to floods. Mustafa et al., 2018 [77] indicates that urbanization increases the risk of floods in the future due to the growth of population and infrastructure in the flood-prone area. This was confirmed by Neumann et al., 2015 [78]; the authors point out that population growth and urbanization increase vulnerability and risk. In 2060, Egypt, Nigeria, China, India, Bangladesh, Indonesia, and Vietnam are considered to be the countries most vulnerable to floods due to population growth. However, in this study, the reduction in poverty between 1995 and 2019 substantially increased the population's capacity to resist and bounce back from floods, therefore vulnerability, and consequently flood risk, diminished over time. This has been justified by research in other countries such as that by Rayhan et al., 2010 [79]. Tasnuva et al., 2020 [80] highlighted that poverty is considered one of the main factors of vulnerability. Thus, our study brings additional evidence for integrating socioeconomic status for flood risk assessment in the context of urbanization.

The trend identified here raises two major questions that must be addressed with regard to this specific study: variable weightings in the AHP method and the consequences of an extreme flood. Variable weightings were carried out based on the authors' experience and the scientific literature; the weightings have a subjective component that is difficult to evaluate. Flood hazard modeling gave particularly reliable results based on the calibration and validation event statistics; therefore, flow velocity and depth values are considered robust. However, the weightings attributed to these parameters in the AHP remain subjective, both for within hazard weighting (Table 1) and between risk component variables (Table 2). The same can be said for both exposure and vulnerability. In the end, the change in poverty level was the dominant factor accounting for an improvement in the flood risk levels mapped in Figure 11, and two further comments can be made about this. First, as cited above, a reduction in poverty has been shown to improve flood resilience in other studies, so this is not entirely surprising. However, the weights attributed to poverty level in Tables 1 and 2 are difficult to justify objectively since an increase in wealth is translated into a range of considerations that are beyond the limits of this study: home improvements that make houses more resistant to floods, insurance coverage that minimizes losses for personal homeowners and businesses, savings and other investments that allow homeowners and businesses to renovate or rebuild quickly, municipal wealth that can reestablish and/or rebuild municipal infrastructures quickly. These elements could not be quantified here and have been little explored in the scientific literature, particularly for Asian countries, but they represent a critical field of investigation in flood risk management, especially in countries where poverty rates are evolving quickly.

A second aspect related to the weight of poverty rate in determining the final flood risk map is related to the unexplored role of spatial scale in the data. Flood hazard and land cover were mapped at 10 m spatial resolutions. Population density and poverty levels were provided according to municipal boundaries where large areas were attributed identical values. The spatial limits of the municipal boundaries extend beyond the flood zone limitations; therefore, the results suggest that increases in population density were perhaps not as great in the flood zone as they were elsewhere in the study area. Hence, the important change in population density noted in the text above (overall increase of

47.3% between 1995 and 2040) is representative of municipalities within the watershed but not necessarily within the flood zone where population density may have changed less. A few municipalities even experienced minor decreases in population density, but it was not possible to find population density at a spatial scale compatible with the 10 m spatial resolution used in the study. Harmonizing socioeconomic data with physical (DEM) or remotely sensed raster data remains a permanent challenge in land cover change modeling.

In the case of an extreme event, like the November 2013 reference event, we could expect flood damage to be greater over time despite the decrease in flood risk. The flood zone is increasingly occupied by more people and land covers with greater economic value, so the damage from a given event would be expected to be of greater financial value. However, "flood risk" in our study is not limited to expected damage but to the overall capacity of the system to bounce back from that damage—its resilience. As noted by Fox et al. (2012) [81] in SE France, improvements in the river channel networks largely compensated the impacts of peri-urbanization on runoff in the catchment; there was a net increase in exposure, but the flood risk decreased. Despite this, in the case of an extreme event, more people and buildings would be affected. Therefore, although flood risk decreased in the Tra Khuc watershed thanks to improved socioeconomic conditions, flood prevention measures, largely ignored to date, must become a priority for the catchment.

The remaining question of the study is whether this novel approach may resolve the problems of previous studies related to sustainable land use planning. This question is very important because the "urban" status of the territory is the subject of an administrative decision, and other studies from countries experiencing economic transitions connect decisions with the socioeconomic dynamics [82–84]. Other studies have used approaches similar to ours to assess flood risk. Huu Xuan Nguyen et al., (2020) [85] used GIS-Based Fuzzy AHP–TOPSIS for assessing flood haz-ards along the South-Central Coast of Vietnam. Chinh Luu et al., in 2020 [59] used 300 flood marks in 2013 and a 5 m DEM to build a flood hazard map using the AHP method. Although these methods have some advantages, such as easy access to data, our study is more comprehensive compared to previous studies thanks to the use of hydrodynamic and land cover modeling to assess historical and future flood hazard and thanks to the integration of socioeconomic data such as poverty rate. Integrating socioeconomic changes to estimate flood risk was an essential part of this study, and it confirms that appropriate attention must be given to changes in resilience over time in estimating flood risk despite the uncertainties around AHP weights. Over the past 25 years, Vietnam has experienced rapid economic transition from an economy dependent on natural resources such as agriculture, fishing, and forestry to one based on industry and services [71]. The role of a continuous flood risk assessment from the past to the present is very important. This is a missing strategy in flood risk management in general and especially in developing countries like Vietnam.

The economic transition policy since 1986 has led to rapid urbanization. The coastal plains in Vietnam in general and in the study area are significantly affected by urbanization. This has been justified by several studies [71,86]. Zaninetti [86] emphasizes that people do not hesitate to occupy and prosper in areas afflicted by natural hazards such as floods and storms. Apprehension towards these dangers is offset by the daily usefulness of the land, fertility of the soil, presence of aquaculture and natural resources such as building materials, or simply the availability of flat, easy-to-build-on land. This transition will have serious consequences on people and assets, especially when no proper assessment strategy is in place. Therefore, the research presented in this paper is an appropriate flood mitigation strategy in the context of rapid urbanization. Our results are a useful contribution to the theoretical advancement of the field because they can assess flood risk at different times, including future probable trends. Based on this, policymakers can identify priority regions in need of spatial planning. Although this study was carried out in Vietnam, the results apply to other countries with flood challenges, particularly to regions with rapid socioeconomic changes and urbanization.

Finally, land cover change studies for flood risk assessment require accurate historical land cover maps [87]. A long-term, consistent database of satellite images offers researchers the opportunity to analyze phenomena from a historical perspective, and it is possible to assess long-term changes in local natural parameters. In addition, it provides a synoptic view of large areas [88–90]. Technically, the selection of an effective method to detect changes in land cover is considered an important step. Object-Based Image Analysis (OBIA) was selected to analyze the satellite images in this study. This method gives better results when compared to traditional pixel-based techniques for image classification, so it was used in this study. Pixel-based techniques often produce more "noise" (often referred to as a "salt-and-pepper" effect) than object-based approaches due to differences in reflection between adjacent pixels. The supervised OBIA technique helps researchers to be proactive about the size of the subjects they study (e.g., fine, medium-coarse), thereby choosing the appropriate level of segmentation for their research. In addition to multi-scale measures, OBIA can incorporate other thematic data (e.g., DEM, thematic maps, etc.) into the classification process to help filter out noise. Hence, object-based classification results will bring more accurate results.

This study is subject to the general limitations characteristic of the use of topography data in hydrodynamic modeling. Several previous studies have shown that the accuracy of hydrodynamic modeling depends on topography resolution. This study used the DEM extracted from the 1:10,000 topography map; however, future research would benefit more when more detailed information is used, e.g., LiDAR, because it presents information on land cover such as dike networks or vegetation to compute the friction coefficients [91,92]. In this study, several methods were used to limit these disadvantages, for example the artificial surface like dike networks or the main roads were measured and integrated into the 1D hydrodynamic modeling.

6. Conclusions

This study presents a new approach to flood risk assessment by integrating hydraulic modeling, population and land cover change analysis and prediction, and socioeconomic changes to better predict flood risk assessment in urban areas. Hydraulic modeling was applied to build a hazard map, and this was combined with exposure and vulnerability data to produce a comprehensive flood risk map. Risk assessment is critical in supporting local authorities' strategies and actions to mitigate flood risk. Results show that the area exposed to High and Very high risks decreased from 195 km^2 in 1995 to 1.85 km^2 in 2019 and to 3.2 km^2 by 2040. The decrease in risk occurred despite increases in land value and population density in the flood plain; these changes were compensated by a substantial decrease in poverty rate over time. The AHP method used identifies the need to better quantify and understand interactions between hazard, exposure, and vulnerability. Overall flood risk can decline thanks to improved socioeconomic conditions, but flood damage will nonetheless increase in extreme events; therefore, flood hazard mitigation must become or remain a priority despite an overall decrease in flood risk.

Author Contributions: Conceptualization, H.D.N., D.F.; methodology, H.D.N., D.F., Q.-T.B.; software, Q.V.V.D., Q.-H.N.; validation, H.D.N., V.T.T.; formal analysis, H.D.N., D.F., A.-I.P., T.H.T.N., D.K.D., T.G.N., P.L.V., Q.-H.N.; investigation, H.D.N., D.F.; resources, L.T.P., Q.-H.N.; data curation, H.D.N., D.F., Q.-T.B., T.N.D.; writing—Original draft preparation, H.D.N., D.K.D., T.H.T.N., V.T.T., Q.V.V.D., P.L.V., Q.-H.N., L.T.P., T.N.D.; writing—Review and editing, H.D.N., Q.-T.B., D.F., T.G.N., T.H.T.N., A.-I.P.; visualization, H.D.N., D.F., A.-I.P.; supervision, H.D.N., D.F., A.-I.P.; project administration, H.D.N., Q.-T.B., D.F. and A.-I.P.; funding acquisition, A.-I.P., H.D.N. All authors have read and agreed to the published version of the manuscript.

Funding: This research is funded by Vietnam National Foundation for Science and Technology Development (NAFOSTED) under grant number 105.07-2019.308.

Institutional Review Board Statement: Not applicable.

Informed Consent Statement: Not applicable.

Data Availability Statement: Restrictions apply to the availability of these data. Data was obtained within a project funded by Vietnam National Foundation for Science and Technology Development and is available from the funders by request.

Conflicts of Interest: The authors declare no conflict of interest.

Abbreviations

1D	1-dimensional
2D	2-dimensional
NSE	Nash-Sutcliffe Efficient
R^2	Coefficient of determination
NDVI	Normalized Difference Vegetation Index
EVI	Enhanced Vegetation Index
MONRE	Ministry of Natural Resources and Environment
LCM	Land Change Modeler
AHP	Analytical Hierarchical Process
CR	Consistency Ratio
CI	Consistency Index
MLP	Multi-Layer Perceptron
OBIA	Object-Based Image Analysis
DEM	Digital elevation model

References

1. Marchand, M.; Buurman, J.; Pribadi, A.; Kurniawan, A. Damage and casualties modelling as part of a vulnerability assessment for tsunami hazards: A case study from Aceh, Indonesia. *J. Flood Risk Manag.* **2009**, *2*, 120–131. [CrossRef]
2. Pradhan, B.; Youssef, D.A. A 100-year maximum flood susceptibility mapping using integrated hydrological and hydrodynamic models: Kelantan River Corridor, Malaysia. *J. Flood Risk Manag.* **2011**, *4*, 189–202. [CrossRef]
3. Dawod, G.M.; Mirza, M.N.; Al-Ghamdi, K.A. GIS-based estimation of flood hazard impacts on road network in Makkah city, Saudi Arabia. *Environ. Earth Sci.* **2012**, *67*, 2205–2215. [CrossRef]
4. Robi, M.A.; Abebe, A.; Pingale, S.M. Flood hazard mapping under a climate change scenario in a Ribb catchment of Blue Nile River basin, Ethiopia. *Appl. Geomat.* **2019**, *11*, 147–160. [CrossRef]
5. Francesch-Huidobro, M.; Dabrowski, M.; Tai, Y.; Chan, F.; Stead, D. Governance challenges of flood-prone delta cities: Integrating flood risk management and climate change in spatial planning. *Prog. Plann.* **2017**, *114*, 1–27. [CrossRef]
6. The Human Cost of Weather Related Disasters: 1995–2015. 2015. Available online: https://www.unisdr.org/files/46796_cop2 1weatherdisastersreport2015.pdf (accessed on 12 December 2020).
7. Carter, J.G. Urban climate change adaptation: Exploring the implications of future land cover scenarios. *Cities* **2018**, *77*, 73–80. [CrossRef]
8. Stoleriu, C.; Urzica, A.; Mihu-Pintilie, A. Improving flood risk map accuracy using high-density LiDAR data and the HEC-RAS river analysis system: A case study from North-Eastern Romania. *J. Flood Risk Manag.* **2019**, *13*, e12572. [CrossRef]
9. Iosub, M.; Minea, I.; Chelariu, O.E.; Ursu, A. Assessment of flash flood susceptibility potential in Moldavian plain (Romania). *J. Flood Risk Manag.* **2020**, *13*, e12588. [CrossRef]
10. Huțanu, E.; Mihu-Pintilie, A.; Urzica, A.; Paveluc, L.E.; Stoleriu, C.C.; Grozavu, A. Using 1D HEC-RAS modeling and LiDAR data to improve flood hazard maps' accuracy: A case study from Jijia floodplain (NE Romania). *Water* **2020**, *12*, 1624. [CrossRef]
11. Güneralp, B.; Seto, K. Futures of global urban expansion: Uncertainties and implications for biodiversity conservation. *Environ. Res. Lett.* **2013**, *8*, 014025. [CrossRef]
12. Güneralp, B.; Güneralp, I.; Liu, Y. Changing global patterns of urban exposure to flood and drought hazards. *Glob. Environ. Chang.* **2015**, *31*, 217–225. [CrossRef]
13. Ran, J.; Nedovic-Budic, Z. Integrating spatial planning and flood risk management: A new conceptual framework for the spatially integrated policy infrastructure. *Comput. Environ. Urban Syst.* **2016**, *57*, 68–79. [CrossRef]
14. Lin, K.; Chen, H.; Xu, C.-Y.; Yan, P.; Lan, T.; Liu, Z.; Dong, C. Assessment of flash flood risk based on improved analytic hierarchy process method and integrated maximum likelihood clustering algorithm. *J. Hydrol.* **2020**, *584*, 124696. [CrossRef]
15. Yang, W.; Xu, K.; Lian, J.; Bin, L.; Ma, C. Multiple flood vulnerability assessment approach based on fuzzy comprehensive evaluation method and coordinated development degree model. *J. Environ. Manag.* **2018**, *213*, 440–450. [CrossRef] [PubMed]
16. Koks, E.E.; Jongman, B.; Husby, T.G.; Botzen, W.J.W. Combining hazard, exposure and social vulnerability to provide lessons for flood risk management. *Environ. Sci. Policy* **2015**, *47*, 42–52. [CrossRef]
17. Vojtek, M.; Vojteková, J. Flood maps and their potential role in local spatial planning: A case study from Slovakia. *Water Policy* **2018**, *20*, 1042–1058. [CrossRef]
18. Meyer, V.; Scheuer, S.; Haase, D. A multicriteria approach for flood risk mapping exemplified at the Mulde river, Germany. *Nat. Hazards* **2009**, *48*, 17–39. [CrossRef]

19. Meyer, V.; Haase, D.; Scheuer, S. Flood risk assessment in European river basins-concept, methods, and challenges exemplified at the Mulde river. *Integr. Environ. Assess. Manag.* **2009**, *5*, 17–26. [CrossRef]

20. C, K.; Haase, D.; V, M.; Scheuer, S. Integrated urban flood risk assessment-Adapting a multicriteria approach to a city. *Nat. Hazards Earth Syst. Sci.* **2009**, *9*, 1881–1895.

21. Falter, D.; Dung, N.; Vorogushyn, S.; Schröter, K.; Hundecha, Y.; Kreibich, H.; Apel, H.; Theisselmann, F.; Merz, B. Continuous, large-scale simulation model for flood risk assessments: Proof-of-concept. *J. Flood Risk Manag.* **2016**, *9*, 3–21.

22. Luu, C.; Von Meding, J.; Kanjanabootra, S. Assessing flood hazard using flood marks and analytic hierarchy process approach: A case study for the 2013 flood event in Quang Nam, Vietnam. *Nat. Hazards* **2018**, *90*, 1031–1050. [CrossRef]

23. Chau, V.N.; Holland, J.; Cassells, S.; Tuohy, M. Using GIS to map impacts upon agriculture from extreme floods in Vietnam. *Appl. Geogr.* **2013**, *41*, 65–74. [CrossRef]

24. Teng, J.; Jakeman, A.J.; Vaze, J.; Croke, B.F.; Dutta, D.; Kim, S. Flood inundation modelling: A review of methods, recent advances and uncertainty analysis. *Environ. Model. Softw.* **2017**, *90*, 201–216. [CrossRef]

25. Fewtrell, T.J.; Duncan, A.; Sampson, C.C.; Neal, J.C.; Bates, P.D. Benchmarking urban flood models of varying complexity and scale using high resolution terrestrial LiDAR data. *Phys. Chem. Earth* **2011**, *36*, 281–291. [CrossRef]

26. Saha, A.K.; Agrawal, S. Mapping and assessment of flood risk in Prayagraj district, India: A GIS and remote sensing study. *Nanotechnol. Environ. Eng.* **2020**, *5*, 1–18.

27. Haq, M.; Akhtar, M.; Muhammad, S.; Paras, S.; Rahmatullah, J. Techniques of Remote Sensing and GIS for flood monitoring and damage assessment: A case study of Sindh province, Pakistan. *Egypt. J. Remote. Sens. Space Sci.* **2012**, *15*, 135–141.

28. Talukdar, S.; Singha, P.; Mahato, S.; Pal, S.; Liou, Y.-A.; Rahman, A. Land-Use Land-Cover Classification by Machine Learning Classifiers for Satellite Observations—A Review. *Remote Sens.* **2020**, *12*, 1135. [CrossRef]

29. Szuster, B.W.; Chen, Q.; Borger, M. A comparison of classification techniques to support land cover and land use analysis in tropical coastal zones. *Appl. Geogr.* **2011**, *31*, 525–532.

30. Akodéwou, A.; Oszwald, J.; Saïdi, S.; Gazull, L.; Akpavi, S.; Akpagana, K.; Gond, V. Land use and land cover dynamics analysis of the togodo protected area and its surroundings in Southeastern Togo, West Africa. *Sustainability* **2020**, *12*, 5439. [CrossRef]

31. Wittke, S.; Yu, X.; Karjalainen, M.; Hyyppä, J.; Puttonen, E. Comparison of two-dimensional multitemporal Sentinel-2 data with three-dimensional remote sensing data sources for forest inventory parameter estimation over a boreal forest. *Int. J. Appl. Earth Obs. Geoinf.* **2018**, *76*, 167–178. [CrossRef]

32. De Moel, H.; Jongman, B.; Kreibich, H.; Merz, B.; Penning-Rowsell, E.; Ward, P.J. Flood risk assessments at different spatial scales. *Mitig. Adapt. Strat. Glob. Chang.* **2015**, *20*, 865–890. [CrossRef] [PubMed]

33. Mishra, K.; Sinha, R. Flood risk assessment in the Kosi megafan using multi-criteria decision analysis: A hydro-geomorphic approach. *Geomorphology* **2020**, *350*, 106861. [CrossRef]

34. Luu, C.; Tran, H.X.; Pham, B.T.; Al-Ansari, N.; Tran, T.Q.; Duong, N.Q.; Dao, N.H.; Nguyen, L.P.; Nguyen, H.D.; Thu Ta, H. Framework of Spatial Flood Risk Assessment for a Case Study in Quang Binh Province, Vietnam. *Sustainability* **2020**, *12*, 3058. [CrossRef]

35. Dang, N.M.; Babel, M.S.; Luong, H.T. Evaluation of food risk parameters in the day river flood diversion area, Red River delta, Vietnam. *Nat. Hazards* **2011**, *56*, 169–194. [CrossRef]

36. Kron, W. Flood risk= hazard• values• vulnerability. *Water Int.* **2005**, *30*, 58–68. [CrossRef]

37. Begum, S.; Stive, M.J.; Hall, J.W. *Flood Risk Management in Europe: Innovation in Policy and Practice*; Springer: Dordrecht, The Netherlands, 2007.

38. Penning-Rowsell, E.; Yanyan, W.; Watkinson, A.; Jiang, J.; Thorne, C. Socioeconomic scenarios and flood damage assessment methodologies for the Taihu Basin, China. *J. Flood Risk Manag.* **2013**, *6*, 23–32. [CrossRef]

39. Chen, J.; Tachikawa, Y.; Tanaka, T.; Udmale, P.; Tung, C.-P. A generalized framework for assessing flood risk and suitable strategies under various vulnerability and adaptation scenarios: A case study for residents of Kyoto city in Japan. *Water* **2020**, *12*, 2508.

40. Mechler, R.; Bouwer, L. Understanding trends and projections of disaster losses and climate change: Is vulnerability the missing link? *Clim. Chang.* **2014**, *133*, 23–35. [CrossRef]

41. Sampson, C.C.; Smith, A.M.; Bates, P.D.; Neal, J.C.; Alfieri, L.; Freer, J.E. A high-resolution global flood hazard model. *Water Resour. Res.* **2015**, *51*, 7358–7381. [CrossRef]

42. Seneviratne, S.I.; Nicholls, N.; Easterling, D.; Goodess, C.M.; Kanae, S.; Kossin, J.; Luo, Y.; Marengo, J.; McInnes, K.; Rahimi, M.; et al. *Changes in Climate Extremes and Their Impacts on the Natural Physical Environment*; Cambridge University Press: Cambridge, UK, 2012.

43. Pham, B.; Tran, P.; Nguyen, H.; Qi, C.; Al-Ansari, N.; Amini, A.; Lanh, S.; Ho, L.S.; Tuyen, T.; Phan, H.; et al. A comparative study of Kernel Logistic Regression, Radial Basis Function Classifier, Multinomial Naïve Bayes, and Logistic Model Tree for flash flood susceptibility mapping. *Water* **2020**, *12*, 1–21.

44. Winsemius, H.; Van Beek, L.; Jongman, B.; Ward, P.; Bouwman, A. A framework for global river flood risk assessments. *Hydrol. Earth Syst. Sci.* **2013**, *17*, 1871–1892. [CrossRef]

45. Blaikie, P.; Cannon, T.; Davis, I.; Wisner, B. *At Risk: Natural Hazards, People Vulnerability and Disasters*, 1st ed.; Taylor & Francis: Abingdon-on-Thames, UK, 1994.

46. IPCC. *Climate Change 2014: Iimpacts, Adaptation, and Vulnerability*; Cambridge University Press: Cambridge, UK, 2014.

47. Kawasaki, A.; Kawamura, G.; Zin, W.W. A local level relationship between floods and poverty: A case in Myanmar. *Int. J. Disaster Risk Reduct.* **2020**, *42*, 101348. [CrossRef]
48. Fekete, A.; Tzavella, K.; Baumhauer, R. Spatial exposure aspects contributing to vulnerability and resilience assessments of urban critical infrastructure in a flood and blackout context. *Nat. Hazards* **2017**, *86*, 151–176. [CrossRef]
49. Nash, J.E.; Sutcliffe, J.V. River flow forecasting through conceptual models part I-A discussion of principles. *J. Hydrol.* **1970**, *10*, 282–290. [CrossRef]
50. Pu, R.; Landry, S.; Yu, Q. Object-based urban detailed land cover classification with high spatial resolution IKONOS imagery. *Int. J. Remote Sens.* **2011**, *32*, 3285–3308. [CrossRef]
51. Pham, V.-M.; Van Nghiem, S.; Bui, Q.-T.; Pham, T.M.; Van Pham, C. Quantitative assessment of urbanization and impacts in the complex of Huế Monuments, Vietnam. *Appl. Geogr.* **2019**, *112*, 102096. [CrossRef]
52. El-naggar, A.M. Determination of optimum segmentation parameter values for extracting building from remote sensing images. *Alex. Eng. J.* **2018**, *57*, 3089–3097. [CrossRef]
53. Eastman, J.R. *TerrSet Geospatial Monitoring and Modeling System*; Clark University: Worcester, MA, USA, 2016; pp. 345–389.
54. Gibson, L.; Münch, Z.; Palmer, A.; Mantel, S. Future land cover change scenarios in South African grasslands–implications of altered biophysical drivers on land management. *Heliyon* **2018**, *4*, e00693. [CrossRef] [PubMed]
55. García-Álvarez, D.; Camacho Olmedo, M.T.; Paegelow, M. Sensitivity of a common Land Use Cover Change (LUCC) model to the Minimum Mapping Unit (MMU) and Minimum Mapping Width (MMW) of input maps. *Comput. Environ. Urban Syst.* **2019**, *78*, 101389. [CrossRef]
56. Budiyono, Y.; Aerts, J.; Brinkman, J.; Marfai, M.A.; Ward, P. Flood risk assessment for delta mega-cities: A case study of Jakarta. *Nat. Hazards* **2014**, *75*, 389–413. [CrossRef]
57. Hill, C.; Dunn, F.; Haque, A.; Amoako-Johnson, F.; Nicholls, R.J.; Raju, P.V.; Addo, K.A. Hotspots of present and future risk within deltas: Hazards, Exposure and Vulnerability. In *Deltas in the Anthropocene*; Nicholls, R.J., Adger, W.N., Hutton, C.W., Hanson, S.E., Eds.; Springer Nature: Cham, Switzerland, 2020; pp. 127–151.
58. Saaty, T.L. What is the analytic hierarchy process. In *Mathematical Models for Decision Support*; Mitra, G., Greenberg, H.J., Lootsma, F.A., Rijkaert, M.J., Zimmermann, H.J., Eds.; Springer: Berlin, Germany, 1988; pp. 109–121.
59. Li, G.-F.; Xiang, X.-Y.; Tong, Y.-Y.; Wang, H.-M. Impact assessment of urbanization on flood risk in the Yangtze River Delta. *Stoch. Environ. Res. Risk Assess.* **2013**, *27*, 1683–1693. [CrossRef]
60. Gigović, L.; Pamučar, D.; Bajić, Z.; Drobnjak, S. Application of GIS-interval rough AHP methodology for flood hazard mapping in urban areas. *Water* **2017**, *9*, 360. [CrossRef]
61. Ali, S.A.; Khatun, R.; Ahmad, A.; Ahmad, S.N. Application of GIS-based analytic hierarchy process and frequency ratio model to flood vulnerable mapping and risk area estimation at Sundarban region, India. *Model. Earth Syst. Environ.* **2019**, *5*, 1083–1102. [CrossRef]
62. Tran, P.; Marincioni, F.; Shaw, R.; Sarti, M. Flood risk management in Central Viet Nam: Challenges and potentials. *Nat. Hazards* **2008**, *46*, 119–138. [CrossRef]
63. Dewan, A. *Floods in A Megacity: Geospatial Techniques in Assessing Hazards, Risk and Vulnerability*; Springer: Berlin, Germany, 2013.
64. Ghosh, A.; Kar, S.K. Application of analytical hierarchy process (AHP) for flood risk assessment: A case study in Malda district of West Bengal, India. *Nat. Hazards* **2018**, *94*, 349–368. [CrossRef]
65. Cammerer, H.; Thieken, A.H.; Verburg, P.H. Spatio-temporal dynamics in the flood exposure due to land use changes in the Alpine Lech Valley in Tyrol (Austria). *Nat. Hazards* **2013**, *68*, 1243–1270.
66. Papilloud, T.; Röthlisberger, V.; Loreti, S.; Keiler, M. Flood exposure analysis of road infrastructure-comparison of different methods at national level. *Int. J. Disaster Risk Reduct.* **2020**, *47*, 101548. [CrossRef]
67. Zou, Q.; Zhou, J.; Zhou, C.; Song, L.; Guo, J. Comprehensive flood risk assessment based on set pair analysis-variable fuzzy sets model and fuzzy AHP. *Stoch Environ. Res Risk Assess* **2013**, *27*, 525–546. [CrossRef]
68. Ogato, G.S.; Bantider, A.; Abebe, K.; Geneletti, D. Geographic information system (GIS)-Based multicriteria analysis of flooding hazard and risk in Ambo Town and its watershed, West shoa zone, oromia regional State, Ethiopia. *J. Hydrol. Reg. Stud.* **2020**, *27*, 100659.
69. Petrişor, A.-I.; Ianoş, I.; Tălângă, C. Land cover and use changes focused on the urbanization processes in Romania. *Environ. Eng. Manag. J.* **2010**, *9*, 765–771.
70. Ianoş, I.; Petrişor, A.-I.; Zamfir, D.; Cercleux, A.L.; Stoica, I.V. In search of a relevant indicator measuring territorial disparities in a transition country. Case study: Romania. *ERDE* **2013**, *144*, 69–81.
71. Nguyen, H.; Ardillier-Carras, F.; Touchart, L. Les paysages de rizières et leur évolution récente dans le delta du fleuve Gianh. *Cybergeo* **2018**, *876*. [CrossRef]
72. Areu-Rangel, O.S.; Cea, L.; Bonasia, R.; Espinosa-Echavarria, V.J. Impact of urban growth and changes in land use on river flood hazard in Villahermosa, Tabasco (Mexico). *Water* **2019**, *11*, 304. [CrossRef]
73. Waghwala, R.K.; Agnihotri, P. Flood risk assessment and resilience strategies for flood risk management: A case study of Surat City. *Int. J. Disaster Risk Reduct.* **2019**, *40*, 101155. [CrossRef]
74. Bahrawi, J.; Ewea, H.; Kamis, A.; Elhag, M. Potential flood risk due to urbanization expansion in arid environments, Saudi Arabia. *Nat. Hazards* **2020**, *104*, 795–809. [CrossRef]

75. Zhang, W.; Villarini, G.; Vecchi, G.A.; Smith, J.A. Urbanisation Exacerbated the Rainfall and Flooding Causedby Hurricane Harvey in Houston. *Nature* **2018**, *563*, 384–388. [CrossRef]
76. Handayani, W.; Chigbu, U.E.; Rudiarto, I.; Putri, I.H.S. Urbanization and Increasing Flood Risk in the Northern Coast of Central Java—Indonesia: An Assessment towards Better Land Use Policy and Flood Management. *Land* **2020**, *9*, 343. [CrossRef]
77. Mustafa, A.; Bruwier, M.; Archambeau, P.; Erpicum, S.; Pirotton, M.; Dewals, B.; Teller, J. Effects of spatial planning on future flood risks in urban environments. *J. Environ. Manag.* **2018**, *225*, 193–204. [CrossRef]
78. Neumann, B.; Vafeidis, A.T.; Zimmermann, J.; Nicholls, R.J. Future coastal population growth and exposure to sea-level rise and coastal flooding-a global assessment. *PLoS ONE* **2015**, *10*, e0118571. [CrossRef]
79. Rayhan, M.I. Assessing poverty, risk and vulnerability: A study on flooded households in rural Bangladesh. *J. Flood Risk Manag.* **2010**, *3*, 18–24. [CrossRef]
80. Tasnuva, A.; Hossain, M.R.; Salam, R.; Islam, A.R.M.T.; Patwary, M.M.; Ibrahim, S.M. Employing social vulnerability index to assess household social vulnerability of natural hazards: An evidence from southwest coastal Bangladesh. *Environ. Dev. Sustain.* **2020**, *111*, 1–23. [CrossRef]
81. Fox, D.M.; Witz, E.; Blanc, V.; Soulié, C.; Penalver-Navarro, M.; Dervieux, A. A case study of land cover change (1950–2003) and runoff in a Mediterranean catchment. *Appl. Geogr.* **2012**, *32*, 810–821. [CrossRef]
82. Rusu, A.; Ursu, A.; Stoleriu, C.C.; Groza, O.; Niacşu, L.; Sfîcă, L.; Minea, I.; Stoleriu, O.M. Structural changes in the Romanian economy feflected through Corine Land Cover datasets. *Remote Sens.* **2020**, *12*, 1323. [CrossRef]
83. Petrisor, A.-I.; Sîrodoev, I.; Ianoş, I. Trends in the national and regional transitional dynamics of land cover and use changes in Romania. *Remote Sens.* **2020**, *12*, 230. [CrossRef]
84. Petrisor, A.-I. Using Corine data to look at deforestation in Romania: Distribution & possible consequences. *Urban Arch. Constr.* **2015**, *6*, 83–90.
85. Nguyen, H.X.; Nguyen, A.T.; Ngo, A.T.; Phan, V.T.; Nguyen, T.D.; Do, V.T.; Dao, D.C.; Dang, D.T.; Nguyen, A.T.; Hens, L. A Hybrid Approach Using GIS-Based Fuzzy AHP–TOPSIS Assessing Flood Hazards along the South-Central Coast of Vietnam. *Appl. Sci.* **2020**, *10*, 7142. [CrossRef]
86. Zaninetti, J.-M.; Ngo, A.-T.; Grivel, S. La construction sociale de la vulnérabilité face au risque d'inondation au Viêt Nam. *Mappemonde* **2015**, *42*.
87. Armaş, I.; Toma-Danila, D.; Ionescu, R.; Gavriş, A. Vulnerability to earthquake hazard: Bucharest case study, Romania. *Int. J. Disaster Risk Sci.* **2017**, *8*, 182–195. [CrossRef]
88. Costache, R.-D.; Pham, Q.; Corodescu-Ros, E.; Cimpianu, C.; Hong, H.; Linh, N.; Chow, M.F.; Najah, A.-M.; Vojtek, M.; Pandhiani, S.; et al. Using GIS, Remote Sensing, and Machine Learning to highlight the correlation between the land-use/land-cover changes and flash-flood potential. *Remote Sens.* **2020**, *12*, 1422. [CrossRef]
89. Popa, M.C.; Peptenatu, D.; Drăghici, C.C.; Diaconu, D.C. Flood hazard mapping using the flood and flash-flood potential index in the Buzău River catchment, Romania. *Water* **2019**, *11*, 2116. [CrossRef]
90. Hoang, T.V.; Chou, T.-Y.; Fang, Y.-M.; Nguyen, N.; Nguyen, H.; Pham Xuan, C.; Dang Ngo, T.; Nguyen, X.L.; Meadows, M. Mapping forest fire risk and development of early warning system for NW Vietnam using AHP and MCA/GIS methods. *Appl. Sci.* **2020**, *10*, 4348. [CrossRef]
91. Bales, J.; Wagner, C. Sources of uncertainty in flood inundation maps. *J. Flood Risk Manag.* **2009**, *2*, 139–147. [CrossRef]
92. De Brito, M.M.; Evers, M.; Almoradie, A.D.S. Participatory flood vulnerability assessment: A multi-criteria approach. *Hydrol. Earth Syst. Sci.* **2018**, *22*, 373–390. [CrossRef]

Technical Note

Sensitivity of Spectral Indices on Burned Area Detection using Landsat Time Series in Savannas of Southern Burkina Faso

Jinxiu Liu [1,*], Eduardo Eiji Maeda [2], Du Wang [1] and Janne Heiskanen [2,3]

[1] School of Information Engineering, China University of Geosciences, Beijing 100083, China; 1004185227@cugb.edu.cn
[2] Department of Geosciences and Geography, University of Helsinki, P.O. Box 68, 00014 Helsinki, Finland; eduardo.maeda@helsinki.fi (E.E.M.); janne.heiskanen@helsinki.fi (J.H.)
[3] Institute for Atmospheric and Earth System Research, Faculty of Science, University of Helsinki, 00014 Helsinki, Finland
* Correspondence: jinxiuliu@cugb.edu.cn; Tel.: +86-10-8332-1744

Abstract: Accurate and efficient burned area mapping and monitoring are fundamental for environmental applications. Studies using Landsat time series for burned area mapping are increasing and popular. However, the performance of burned area mapping with different spectral indices and Landsat time series has not been evaluated and compared. This study compares eleven spectral indices for burned area detection in the savanna area of southern Burkina Faso using Landsat data ranging from October 2000 to April 2016. The same reference data are adopted to assess the performance of different spectral indices. The results indicate that Burned Area Index (BAI) is the most accurate index in burned area detection using our method based on harmonic model fitting and breakpoint identification. Among those tested, fire-related indices are more accurate than vegetation indices, and Char Soil Index (CSI) performed worst. Furthermore, we evaluate whether combining several different spectral indices can improve the accuracy of burned area detection. According to the results, only minor improvements in accuracy can be attained in the studied environment, and the performance depended on the number of selected spectral indices.

Keywords: burned area; spectral indices; Landsat time series; savanna

Citation: Liu, J.; Maeda, E.E.; Wang, D.; Heiskanen, J. Sensitivity of Spectral Indices on Burned Area Detection using Landsat Time Series in Savannas of Southern Burkina Faso. *Remote Sens.* **2021**, *13*, 2492. https://doi.org/10.3390/rs13132492

Academic Editors: Adrian Ursu and Cristian Constantin Stoleriu

Received: 19 May 2021
Accepted: 22 June 2021
Published: 25 June 2021

Publisher's Note: MDPI stays neutral with regard to jurisdictional claims in published maps and institutional affiliations.

1. Introduction

The African savanna frequently experiences extensive fires every year, as thousands of square kilometers are burned, making an important contribution to the total global burned area [1]. Fire is recognized as an important feature in savanna ecosystems, as it plays a key role in vegetation succession, carbon cycle, biodiversity, and land management [2]. Therefore, timely and accurate mapping of burned areas is essential for fire management, climate modeling, and environmental applications. The rapid development of remote sensing technology provides convenient and effective methods for burned area mapping from regional to global scale. Previously, coarse spatial resolution data from Advanced Very High Resolution Radiometer (AVHRR), Moderate Resolution Imaging Spectroradiometer (MODIS), or SPOT VEGETATION images have been widely used for monitoring burned areas [3]. However, the resolution of these sensors is too coarse to identify small burn patches and their dynamics on a regional scale. The increasing availability of medium spatial resolution satellite images, such as Landsat, is a valuable source of information for more accurate detection of the burned areas at local and regional scales.

A variety of methods have been developed for burned area monitoring and mapping using remote sensing data, including threshold-based method with spectral indices [4], supervised image classification, such as decision trees classification [5], artificial neural networks [6], logistic regression [2], principal component analysis [7], region growing [8], and spectral mixture analysis [9]. Many burned area mapping methods use spectral indices based on post-fire images or pre-fire and post-fire images to identify burned areas and separate those from other land cover categories and states, due to their simplicity and efficiency. Vegetation index and fire index are two kinds of spectral indices that are widely applied for burned area detection. Normalized Difference Vegetation Index (NDVI) is a popular vegetation index that is used to discriminate between the burned and unburned areas, due to its strong sensitivity to vegetation changes. There are numerous modifications of NDVI that have been developed to reduce atmospheric sensitivity and background variability and successfully applied in burned area detection, including the Global Environmental Monitoring Index (GEMI), the Soil-Adjusted Vegetation Index (SAVI), and the Enhanced Vegetation Index (EVI) [10]. Some fire indices are specifically designed and developed to be sensitive to post-fire spectral signals, such as the Normalized Burned Ratio (NBR), the Burned Area Index (BAI), and the Char Soil Index (CSI) [11].

The method based on image comparison by differencing the post-fire and pre-fire images with spectral indices is frequently applied for burned area detection, and achieved satisfactory results [12]. However, this method is constrained by the limited availability of cloud-free images, as well as challenges related to image-to-image normalization. With free and open access to the Landsat archive, Landsat time series data have been increasingly utilized for burned area detection and evaluation. Goodwin and Collett [13] proposed a method with Landsat time series data, including outlier identification caused by burned vegetation using near-infrared and mid-infrared spectral bands, region growing segmentation, and classification tree to distinguish burned area from other changes. Hawbaker et al. [14] used machine learning, thresholding, and region growing methods to identify burned areas with Landsat time series data using spectral band NBR. Liu et al. [15] developed an approach for mapping annual burned areas using a harmonic model fitting with BAI time series and breakpoint identification in Landsat time series. However, the sensitivity of different spectral indices on discriminating burned and unburned areas using Landsat time series has not been studied.

In this paper, our objective was to explore and evaluate the effect of using various spectral indices based on Landsat time series on the performance of burned area detection. To accomplish this goal, we used all available Landsat images between 2000 and 2016 in savannas of southern Burkina Faso for mapping burned areas. Furthermore, we tested whether fusion of burned area detections based on several spectral indices can improve accuracy.

2. Materials and Methods

2.1. Study Area and Remote Sensing Data

Our study area is located in southern Burkina Faso, belonging the West Sudanian savanna ecoregion (Figure 1). The mean annual precipitation was 827 mm, and the mean annual temperature was 27.5 °C [16]. Tropical dry forests and woodlands are surrounded by agroforestry parklands and agricultural fields. There was an increased conversion of forests and woodland into cropland during the last decade [17]. Fires, due to anthropogenic and natural reasons, occur regularly. Most of the fires take place during the dry season between November and February, even in early October, and late March and April. Burned vegetation typically recovers quickly without causing permanent land cover change.

We collected all available Landsat Surface Reflectance data with WRS-2 coordinates Path 195 Row 52 from the Earth Resources Observations and Science (EROS) Center Archive for the time period between October 2000 and April 2016, which was the same data in Liu et al. [15] for burned area detection. The full time series used in the study consisted of 281 images, including 40 Landsat 5 Thematic Mapper (TM) images, 185 Landsat 7 Enhanced Thematic Mapper Plus (ETM+) images, and 56 Landsat 8 Operational Land Imager (OLI) images. Table 1 shows the number of Landsat images from October in one year to September in the next year ranging from 2000 to 2015. In 2015–2016, images were from October 2015 to April 2016. All images have been atmospherically corrected, and clouds and shadows have been removed using the Fmask algorithm [18]. We selected surface reflectance in blue, green, red, near-infrared (NIR), and two shortwave infrared (SWIR1, SWIR2) bands for further analysis.

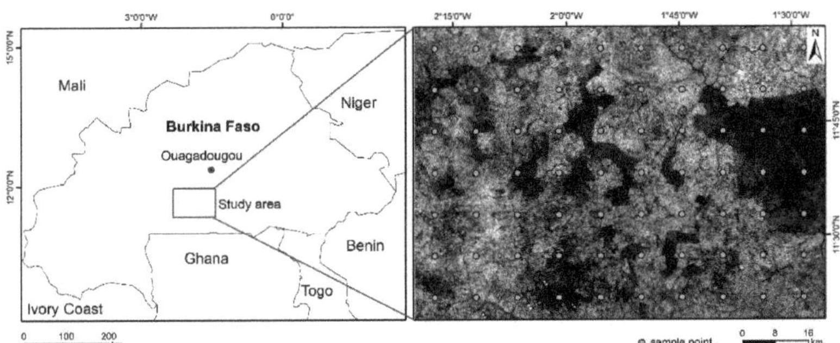

Figure 1. Study area and the distribution of random systematic sampling points based on Landsat 5 image acquired on 15 November 2014.

2.2. Burned Area Detection Algorithm Using Landsat Time Series

The algorithm for detecting burned areas with different spectral indices included four steps. First, the BFAST Monitor (Breaks for Additive Season and Trend Monitor) algorithm [19] was applied to detect land cover change using Landsat NDVI time series. Next, a harmonic model was fitted using different spectral indices for each stable period without land cover change. Then the outliers caused by fire events were detected by comparing model predictions with the observed values using an optimal threshold for different spectral indices. The optimal threshold was determined separately based on the performance of burned area detection with different spectral indices. The burned area pixels were combined from every single image into an annual Landsat burned area within fire season. Finally, we performed accuracy assessments using reference data interpreted from Landsat images and compared the performance of different spectral indices in burned area detection. Fusion of different spectral indices burned area detection results was also tested.

2.2.1. Breakpoint Detection Using Landsat Time Series

We applied the BFAST Monitor approach to detect deforestation in this study area described in more detail in Liu et al. [15]. The Landsat time series was separated into a historical period which is used as a baseline, and a monitoring period. We defined the baseline period between October 2000 and October 2002 as it is stable and has a minimum of nine observations for each pixel. The method fits a harmonic model to each pixel NDVI time series in a baseline period using ordinary least squares (OLS). By comparing the discrepancy between model predictions and observations in the monitoring period with a moving sums of residuals (MOSUM) approach, the breakpoint was detected when the deviation from zero was beyond a 95% significance boundary. The change magnitude was calculated by taking the median of all the residuals within the monitoring period. The disturbance pixel was determined by a breakpoint pixel with negative magnitude [19]. Each pixel-wise time series had its own stable period after detecting land cover change. The pixels without land cover change were regarded as stable from 2000 to 2016, and the pixels with breakpoints were separated into before and after land cover change periods, respectively.

Table 1. Frequency of all available Landsat images for the study area by year.

	2000–2001	2001–2002	2002–2003	2003–2004	2004–2005	2005–2006	2006–2007	2007–2008	2008–2009	2009–2010	2010–2011	2011–2012	2012–2013	2013–2014	2014–2015	2015–2016
TM	0	0	0	0	0	0	18	0	7	6	9	0	0	0	0	0
ETM+	10	7	7	15	12	14	14	12	11	10	12	9	13	12	15	12
OLI	0	0	0	0	0	0	0	0	0	0	0	0	8	21	22	5
Sum	10	7	7	15	12	14	32	12	18	16	21	9	21	33	37	17

2.2.2. Spectral Indices for Burned Area Detection

Following previous burned area mapping studies, several spectral indices were included and tested in our study based on their performance (Table 2). NDVI is sensitive to vegetation greenness and is widely used in burned area mapping studies. The Soil Adjusted Vegetation Index (SAVI) is selected as it suppresses the influence of soil brightness and improves the separability of burns from soil and water. The Enhanced Vegetation Index (EVI) does not saturate as quickly in high biomass regions by minimizing the effects of soil and atmosphere. The Global Environmental Monitoring Index (GEMI) is designed to minimize the influence of atmospheric effects, which are considered very important for dark surface detection. We also included the indices specifically developed for burn detection because they are sensitive to charcoal and ash deposition, such as NBR, NBR2, the Burned Area Index (BAI), the Burned Area Index Modified with longer SWIR band (BAIML), the Burned Area Index Modified with short SWIR band (BAIMs), and the Char Soil Index (CSI).

Table 2. Summary of spectral indices. R_{Blue}, R_{Red}, R_{NIR}, R_{SSWIR}, and R_{LSWIR} corresponds to surface reflectance in blue, red, near-infrared, short shortwave infrared, and long shortwave infrared spectral bands, respectively.

Index	Formula	Reference
Burned Area Index (BAI)	$\dfrac{1}{(0.06-R_{NIR})^2+(0.1-R_{Red})^2}$	Chuvieco et al. (2002) [20]
Burned Area Index Modified-SSWIR (BAIM$_S$)	$\dfrac{1}{(0.05-R_{NIR})^2+(0.2-R_{SSWIR})^2}$	Martín et al. (2006) [21]
Burned Area Index Modified-LSWIR (BAIML)	$\dfrac{1}{(0.05-R_{NIR})^2+(0.2-R_{LSWIR})^2}$	Martín et al. (2006) [21]
Mid-infrared burn Index (MIRBI)	$10R_{LSWIR} - 9.8R_{SSWIR} + 2$	Trigg and Flasse (2001) [22]
Char Soil Index (CSI)	R_{NIR}/R_{LSWIR}	Smith et al. (2005) [23]
Normalized Difference Vegetation Index (NDVI)	$(R_{NIR} - R_{Red})/(R_{NIR} + R_{Red})$	Tucker (1979) [24]
Normalized Burn Ratio (NBR)	$(R_{NIR} - R_{LSWIR})/(R_{NIR} + R_{LSWIR})$	Key and Benson (2003) [25]
Normalized Burn Ratio2 (NBR2)	$(R_{SSWIR} - R_{LSWIR})/(R_{SSWIR} + R_{LSWIR})$	Lutes et al. (2006) [26]
Global Environmental Monitoring Index (GEMI)	$n(1 - 0.25n) - (R_{Red} - 0.125)/(1 - R_{Red})$ with $n = \dfrac{2(R_{NIR}^2 - R_{Red}^2) + 1.5R_{NIR} + 0.5R_{Red}}{R_{NIR} + R_{Red} + 0.5}$	Pinty and Verstraete (1992) [27]
Soil-Adjusted Vegetation Index (SAVI)	$(1 + L)(R_{NIR} - R_{Red})/(R_{NIR} + R_{Red} + L)\ L = 0.5$	Huete (1988) [28]
Enhanced Vegetation Index (EVI)	$2.5(R_{NIR} - R_{Red})/(R_{NIR} + 6R_{Red} - 7.5R_{Blue} + 1)$	Huete et al. (2002) [29]

Large parts of the study area were burned during the dry season from November to March, and seasonal fires affected the spectral reflectance and indices, influencing the harmonic model fit and parameters. Therefore, the burned area detection method utilized the time series harmonic model with different spectral indices. We fitted a first order harmonic model considering the phenology in this study area which was driven by unimodal precipitation pattern. The following model was fit for each pixel:

$$y_t = a + b \times \sin\left(\frac{2\pi t}{T} + c\right) + e_t \tag{1}$$

where y_t is the dependent variable (spectral index), t is the independent variable (time as Julian date), T is the temporal frequency (365 days), a, b and c are the model parameters representing intercept, amplitude, and phase coefficients in the harmonic model, respectively, and et is the residual error. Parameters a, b and c are derived using ordinary least squares (OLS) linear regression for each pixel [19,30,31]. In addition to model coefficients, the root mean square error (RMSE) is also calculated.

The method was applied to different spectral indices time series within their stable period. The method consisted of three steps. First, we generated different spectral indices image stacks from the Landsat time series. Second, we fitted the time series harmonic model using the observations from different spectral indices within the stable period as the dependent variable. Third, a threshold was selected to detect burned pixels by comparing the observed and predicted values. We used the threshold to detect the burned pixels, because the clear observations, were always within the model predicted ranges ($\pm$threshold $\times$ RMSE) when the land cover was stable. By contrast, the burned pixels deviated from the ranges, and the threshold in our analysis was ranging from $0.5\times$RMSE to $3.5\times$RMSE depending on the selected spectral indices. We applied the process iteratively until all the burned pixels were detected. The burned pixels from separate images were combined into the annual Landsat burned area during fire season by checking the date from the MODIS burned area product (MCD45A1) as described in Liu et al. [15].

2.2.3. Optimal Threshold Selection and Accuracy Assessment

Different spectral indices had their own optimal threshold in detecting burned areas. To compare their performance, it is necessary to determine the optimal threshold for each spectral index. The basic idea of the optimal threshold approach is to select a threshold value from the training samples, assuming that the optimal threshold value leading to the maximum accuracy in extracting the burned area within the training samples is also optimal for the entire image. There were 70 points covering the whole study area based on systematic random sampling (Figure 1). The distances between the points in the west–east direction and the north–south direction were all 10 km. The starting point of the sampling was selected randomly. Therefore, we had 1120 points in total, considering the 16 years of observations. Reference data for burned area detection is difficult to acquire over a long period of time. Therefore, we visually interpreted each reference point using all the available observations in the Landsat time series for accuracy assessment, as it was demonstrated to be a practical method in previous studies [13]. We randomly selected 30% reference samples to determine the optimal threshold for each spectral index. We compared the overall accuracy and determined the optimal threshold when the highest overall accuracy occurred. After the optimal threshold was determined for each spectral index, the accuracy assessment was conducted using 70% of the reference samples for each spectral index with the corresponding optimal threshold.

2.2.4. Fusion of the Burned Area Results Based on Different Spectral Indices

To test the potential of the fusion of burned area detection results with different spectral indices, a majority vote rule was adopted. To be specific, the burned area detection results with different spectral indices were ranked based on the overall accuracy, and we selected the best 3, 5, 7, 9, and 11 burned area detection results corresponding to the spectral indices to complete the fusion process. In the fusion process, a pixel was determined as a burned pixel if the number of the burned pixel votes was greater than half of the votes, with the votes representing the pixel was burned according to one spectral index.

3. Results

3.1. Breakpoint Detection

The disturbance map generated using the BFAST monitor method with Landsat time series is shown in Figure 2. Conversion from forest and woodlands to cropland was the main cause of the land cover change from 2003 to 2016 in this study area. We observed that most land cover changes occurred in the south-eastern part of the study area. The example showed in Figure 2 demonstrated that land cover change identified with breakpoint time.

3.2. Optimal Threshold for Different Spectral Indices

The overall accuracy, producer's accuracy, and user's accuracy of burned area detection are demonstrated for a range of thresholds with different spectral indices (Figure 3). With an increasing threshold, the producer's accuracy showed a decreasing trend, while in contrast, the user's accuracy showed an increasing trend. The overall accuracy increased as the threshold increased, and then reached the peak value, but decreased as the threshold increased further. The optimal threshold is corresponding to the highest overall accuracy. Among all the tested indices, BAI achieved the best overall accuracy of 80.35% with a threshold of 2.9, followed by BAIML and BAIms with thresholds of 2.8 and 3.1, respectively.

NBR2 and MIRBI obtained overall accuracies around 73% and thresholds between 1.6 and 1.8. GEMI, NBR, SAVI, EVI, and NDVI had thresholds between 1.4 and 1.7. CSI provided an overall accuracy of 63.10% with a threshold of 0.8.

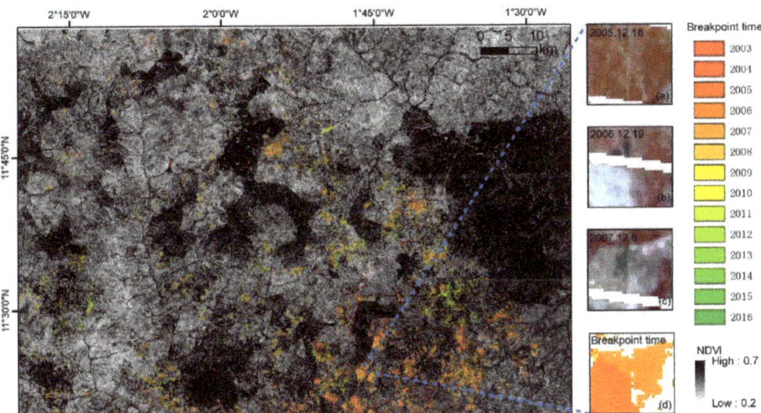

Figure 2. The disturbance map with breakpoint time (left panel), and an example of land cover change were identified (**a,b,c,d**) in the 2005–2007 period.

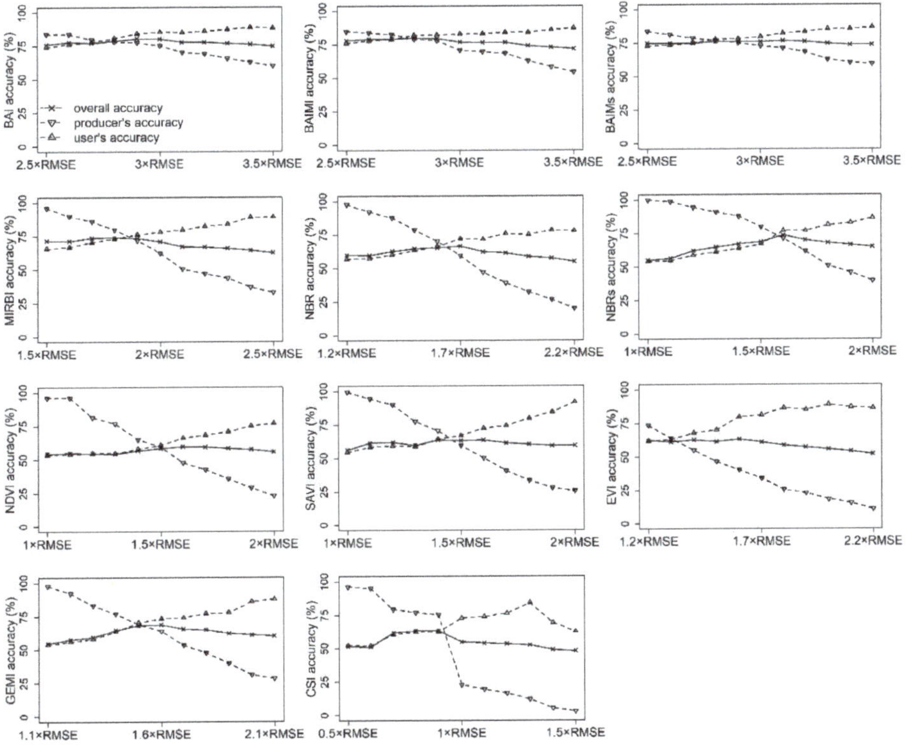

Figure 3. Overall accuracy and producer's and user's accuracies for different spectral indices as a function of threshold value.

3.3. Pixel-Wise Time Series for Burned Area Detection Using Different Spectral Indices

Figures 4 and 5 demonstrate the fit of the harmonic model before and after burned area detection for the same stable pixel time series with different spectral indices using an optimal threshold. Overall, the characteristics of burned pixels in the time series varied among different spectral indices time series. For BAI, BAIMs, BAIML, and MIRBI, the burned pixels had higher values compared to unburned pixels, by contrast, the burned pixels had lower values when compared to unburned pixels for NBR, NBR2, NDVI, SAVI, EVI, GEMI, and CSI. BAI time series achieved the best results by accurately detecting all the burned pixels, suggesting its sensitivity to burned signal. BAIMs and BAIML successfully captured most of the burned area in the time series with small omission and commission errors. More commission and omission errors were observed when using MIRBI, because there was only a subtle difference between burned and unburned values.

The NBR, NBR2, NDVI, SAVI, EVI, and GEMI time series behaved similarly and revealed poor performances with more commission and omission errors, demonstrating the burned observations in these time series data could not be correctly distinguished from unburned observations. In CSI time series, unburned pixels showed more fluctuations, leading to the most commission errors among all the indices.

3.4. Burned Area Detection Comparison and Accuracy Assessment

We evaluated the overall accuracy, producer's accuracy, and user's accuracy of burned area detection for each spectral index with the optimal threshold (Table 3). The overall accuracies of burned area detection varied by different spectral indices. Among the spectral indices tested, the most accurate was the BAI, with a good balance of commission and omission errors (overall accuracy 77.81%). The next best accuracies were achieved by BAIMs, BAIML, and MIRBI with similar performance (72.45% to 76.40%). NBR2 and NBR achieved overall accuracies of 70.79% and 66.71%, respectively. The overall accuracies for greenness-related spectral indices (GEMI, EVI, SAVI, NDVI) was between 63.52% and 67.99%, while GEMI outperformed EVI, SAVI, and NDVI, being the best vegetation index for burned area mapping among the four vegetation indices. The least successful spectral index was CSI, with an overall accuracy of 56.89%.

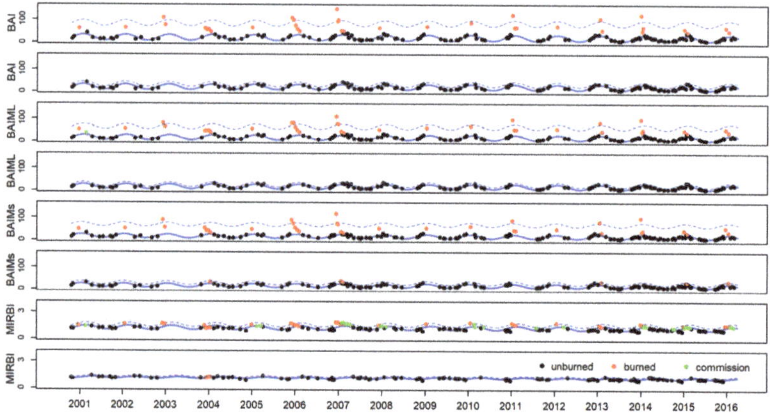

Figure 4. Demonstration of harmonic model fit before and after burned area detection for a single pixel, based on different spectral indices with burned pixels having higher values than unburned pixels. The black points are unburned pixels, the red points are burned pixels, the green points are commission pixels, the blue curve is the harmonic model, and the dashed blue line is the threshold for detecting burned pixels.

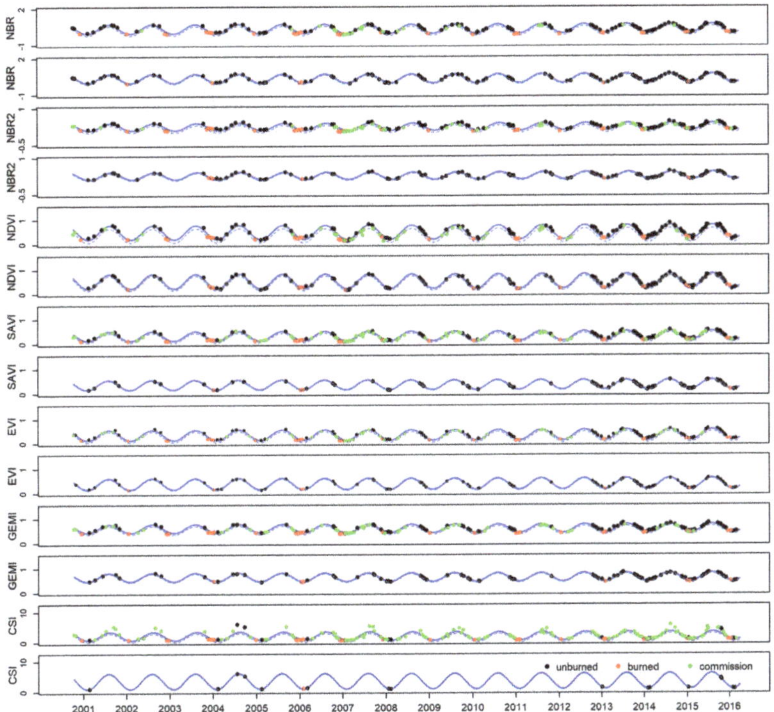

Figure 5. Demonstration of harmonic model fit before and after burned area detection for a single pixel, based on different spectral indices with burned pixels having lower values than unburned pixels. The black points are unburned pixels, the red points are burned pixels, the green points are commission pixels, the blue curve is the harmonic model, and the dashed blue line is the threshold for detecting burned pixels.

Table 3. Accuracy assessment for different classifications.

Input Features	OA	PA	UA
BAI	77.81	73.09	76.56
BAIMs	76.40	69.69	75.93
BAIML	76.28	73.94	73.52
MIRBI	72.45	76.77	66.91
NBR	66.71	55.81	65.23
NBR2	70.79	63.46	69.14
NDVI	63.52	53.26	60.84
SAVI	65.82	77.34	59.22
EVI	65.18	43.34	67.70
GEMI	67.99	63.46	64.74
CSI	56.89	81.87	51.33
Fusion of 3 spectral indices (BAI, BAIMs, BAIML)	77.30	73.09	75.66
Fusion of 5 spectral indices (BAI, BAIMs, BAIML, MIRBI, NBR2)	78.06	72.52	77.34
Fusion of 7 spectral indices (BAI, BAIMs, BAIML, MIRBI, NBR2, GEMI, NBR)	78.57	73.94	77.45
Fusion of 9 spectral indices (BAI, BAIMs, BAIML, MIRBI, NBR2, GEMI, NBR, SAVI, EVI)	78.57	72.81	78.12
Fusion of 11 spectral indices (BAI, BAIMs, BAIML, MIRBI, NBR2, GEMI, NBR, SAVI, EVI, NDVI, CSI)	77.42	71.96	76.51

OA = overall accuracy, PA = producer's accuracy, UA = user's accuracy.

The performance of fusion of different spectral indices with the majority vote method depended on the number of spectral indices used. In the fusion process, the overall accuracies increased as the number of spectral indices increased from 3 to 9, but decreased as the number of spectral indices increased further to 11. The best combination of spectral indices achieved an overall accuracy of 78.57% with a fusion of 7 or 9 spectral indices, with a slight increase compared to the best single spectral index BAI.

4. Discussion

Spectral indices are widely utilized in burned area monitoring, and different spectral indices are used in various environments to improve burned area detection accuracy. In this study, we tested different spectral indices in burned area detection over a savanna environment with Landsat time series.

We found that the most accurate spectral index tested in our study was BAI, followed by BAIMs and BAIML. This result was in line with previous studies demonstrating that BAI is effective in burned area detection, including the savanna environment [32]. Several studies confirm the potential of BAI in discriminating burned areas from other land cover types, due to its higher sensitivity to char signal in post-fire images in comparison to other spectral indices. MIRBI achieved a moderate accuracy among all the spectral indices, and it was found to discriminate burned area from soil and arable land, but not from shadows [33]. Our results also indicated that NBR and NBR2 are unsuitable for accurate burned area estimations, even though NBR is a common spectral index and widely applied in burned severity assessment [1]. The main reason is that NBR is more suitable for burned area detection for a short time after the fire, and hence, its use is limited by the temporal resolution of Landsat data [34]. Further assessments are necessary to evaluate the influence of the Landsat temporal resolution on the capability to detect and measure vegetation recovery, as well as how higher temporal resolution sensors would improve the detection of vegetation recovery dynamics after fire events. In addition, NBR has been proved to be less sensitive to burned areas in savanna environments, because of the general decrease in reflectance in all spectral bands in response to fire [13].

The vegetation indices, NDVI, SAVI, EVI, and GEMI, performed similarly, with limited capacity to identify a fire-affected area with other studies confirming these findings [11]. The main reason is that they were designed to determine vegetation properties, not intended for burned area detection. The low accuracies are mainly caused by confusions between the burned area with dark surfaces, such as soil, shadow, and water, and especially in the dry season, the characteristic of the vegetation can lead to errors and confusion with burned areas [20]. CSI was the least successful in burned area estimation among all the spectral indices, and our findings on the poor performance of CSI in burned area detection are similar to those by Schepers et al. [35], who pointed that CSI easily confused char signal with soil.

The performance of burned area detection using a fusion of different spectral indices depended on the number of selected spectral indices, suggesting careful selection is necessary when fusing burned areas detection based on several spectral indices. However, the small improvement in the overall accuracy in comparison to the best index (BAI) suggests that there are only minor complementary benefits among the best indices in the studied environment, although greater benefits might be attained elsewhere. Considering that selecting the best indices for fusion can vary from one region to another, burned area detection based on BAI remains a logical first choice for burned area monitoring. However, the method based on fuzzy set theory could be a more effective way to combine indices as indicated by a previous study [36]; hence, it deserves to be tested in future studies.

5. Conclusions

In this study, we explored the performance and sensitivity of eleven different spectral indices on burned area detection with Landsat time series in the savanna area of southern Burkina Faso. The burned area detection method was based on breakpoint identification and harmonic model fitting with Landsat data from 2000 to 2016. The result demonstrated that BAI was the most accurate index for burned area mapping among all the tested spectral indices. Fire-related spectral indices outperformed vegetation indices, and CSI performed worst in burned area detection. Fusing different spectral indices only achieved minor improvement in accuracy in our study area, and the number of selected spectral indices influenced the burned area mapping performance.

Author Contributions: Conceptualization, J.L., J.H.; methodology, J.L., E.E.M. and J.H.; software, J.L. and D.W.; validation, J.L.; formal analysis, J.L.; investigation, J.L., E.E.M. and J.H.; resources, J.L., E.E.M. and J.H.; data curation, J.L., E.E.M. and J.H.; writing—original draft preparation, J.L. and D.W.; writing—review and editing, J.L., D.W., E.E.M. and J.H.; visualization, J.L. and D.W.; supervision, E.E.M. and J.H. All authors have read and agreed to the published version of the manuscript.

Funding: This research was supported by the Fundamental Research Funds for the Central Universities (grant number 2652018077), Building Bio-Carbon and Rural Development in West Africa (BIODEV) program funded by the Ministry for Foreign Affairs of Finland, and Academy of Finland (decision numbers 318252 and 319905).

Data Availability Statement: The Landsat data is downloaded from Earth Explorer (http://earthexplorer. usgs.gov/ accessed on 17 May 2021). MODIS burned area product (MCD45A1) is downloaded from http://modis-fire.umd.edu/ (accessed on 17 May 2021).

Acknowledgments: The authors are grateful to the editors and the anonymous reviewers for their suggestions for improving the manuscript. We also thank some data agencies who contributed to the datasets used in this study.

Conflicts of Interest: The authors declare no conflict of interest.

References

1. Musyimi, Z.; Said, M.Y.; Zida, D.; Rosenstock, T.S.; Udelhoven, T.; Savadogo, P.; de Leeuw, J.; Aynekulu, E. Evaluating Fire Severity in Sudanian Ecosystems of Burkina Faso Using Landsat 8 Satellite Images. *J. Arid. Environ.* **2017**, *139*, 95–109. [CrossRef]
2. Siljander, M. Predictive Fire Occurrence Modelling to Improve Burned Area Estimation at a Regional Scale: A Case Study in East Caprivi, Namibia. *Int. J. Appl. Earth Obs. Geoinf.* **2009**, *11*, 380–393. [CrossRef]
3. Mouillot, F.; Schultz, M.G.; Yue, C.; Cadule, P.; Tansey, K.; Ciais, P.; Chuvieco, E. Ten Years of Global Burned Area Products from Spaceborne Remote Sensing-A Review: Analysis of User Needs and Recommendations for Future Developments. *Int. J. Appl. Earth Obs. Geoinf.* **2014**, *26*, 64–79. [CrossRef]
4. Koutsias, N.; Pleniou, M.; Mallinis, G.; Nioti, F.; Sifakis, N.I. A Rule-Based Semi-Automatic Method to Map Burned Areas: Exploring the USGS Historical Landsat Archives to Reconstruct Recent Fire History. *Int. J. Remote Sens.* **2013**, *34*, 7049–7068. [CrossRef]
5. Kontoes, C.C.; Poilve, H.; Florsch, G.; Keramitsoglou, I.; Paralikidis, S. A Comparative Analysis of a Fixed Thresholding vs. a Classification Tree Approach for Operational Burn Scar Detection and Mapping. *Int. J. Appl. Earth Obs. Geoinf.* **2009**, *11*, 299–316. [CrossRef]
6. Maeda, E.E.; Formaggio, A.R.; Shimabukuro, Y.E.; Balue Arcoverde, G.F.; Hansen, M.C. Predicting Forest Fire in the Brazilian Amazon Using MODIS Imagery and Artificial Neural Networks. *Int. J. Appl. Earth Obs. Geoinf.* **2009**, *11*, 265–272. [CrossRef]
7. Koutsias, N.; Mallinis, G.; Karteris, M. A Forward/Backward Principal Component Analysis of Landsat-7 ETM+ Data to Enhance the Spectral Signal of Burnt Surfaces. *ISPRS-J. Photogramm. Remote Sens.* **2009**, *64*, 37–46. [CrossRef]
8. Hardtke, L.A.; Blanco, P.D.; del Valle, H.F.; Metternicht, G.I.; Sione, W.F. Semi-Automated Mapping of Burned Areas in Semi-Arid Ecosystems Using MODIS Time-Series Imagery. *Int. J. Appl. Earth Obs. Geoinf.* **2015**, *38*, 25–35. [CrossRef]
9. Smith, A.M.S.; Drake, N.A.; Wooster, M.J.; Hudak, A.T.; Holden, Z.A.; Gibbons, C.J. Production of Landsat ETM plus Reference Imagery of Burned Areas within Southern African Savannahs: Comparison of Methods and Application to MODIS. *Int. J. Remote Sens.* **2007**, *28*, 2753–2775. [CrossRef]
10. Veraverbeke, S.; Harris, S.; Hook, S. Evaluating Spectral Indices for Burned Area Discrimination Using MODIS/ASTER (MASTER) Airborne Simulator Data. *Remote Sens. Environ.* **2011**, *115*, 2702–2709. [CrossRef]
11. Stroppiana, D.; Boschetti, M.; Zaffaroni, P.; Brivio, P.A. Analysis and Interpretation of Spectral Indices for Soft Multicriteria Burned-Area Mapping in Mediterranean Regions. *IEEE Geosci. Remote Sens. Lett.* **2009**, *6*, 499–503. [CrossRef]
12. Liu, S.; Zheng, Y.; Dalponte, M.; Tong, X. A Novel Fire Index-Based Burned Area Change Detection Approach Using Landsat-8 OLI Data. *Eur. J. Remote Sens.* **2020**, *53*, 104–112. [CrossRef]
13. Goodwin, N.R.; Collett, L.J. Development of an Automated Method for Mapping Fire History Captured in Landsat TM and ETM plus Time Series across Queensland, Australia. *Remote Sens. Environ.* **2014**, *148*, 206–221. [CrossRef]

14. Hawbaker, T.J.; Vanderhoof, M.K.; Beal, Y.-J.; Takacs, J.D.; Schmidt, G.L.; Falgout, J.T.; Williams, B.; Fairaux, N.M.; Caldwell, M.K.; Picotte, J.J.; et al. Mapping Burned Areas Using Dense Time-Series of Landsat Data. *Remote Sens. Environ.* **2017**, *198*, 504–522. [CrossRef]

15. Liu, J.; Heiskanen, J.; Maeda, E.E.; Pellikka, P.K.E. Burned Area Detection Based on Landsat Time Series in Savannas of Southern Burkina Faso. *Int. J. Appl. Earth Obs. Geoinf.* **2018**, *64*, 210–220. [CrossRef]

16. Hijmans, R.J.; Cameron, S.E.; Parra, J.L.; Jones, P.G.; Jarvis, A. Very High Resolution Interpolated Climate Surfaces for Global Land Areas. *Int. J. Climatol.* **2005**, *25*, 1965–1978. [CrossRef]

17. Gessner, U.; Knauer, K.; Kuenzer, C.; Dech, S. Land Surface Phenology in a West African Savanna: Impact of Land Use, Land Cover and Fire. In *Remote Sensing Time Series*; Kuenzer, C., Dech, S., Wagner, W., Eds.; Remote Sensing and Digital Image Processing; Springer International Publishing: Dordrecht, The Netherlands, 2015; Volume 22, pp. 203–223. ISBN 978-3-319-15967-6.

18. Zhu, Z.; Woodcock, C.E. Object-Based Cloud and Cloud Shadow Detection in Landsat Imagery. *Remote Sens. Environ.* **2012**, *118*, 83–94. [CrossRef]

19. DeVries, B.; Verbesselt, J.; Kooistra, L.; Herold, M. Robust Monitoring of Small-Scale Forest Disturbances in a Tropical Montane Forest Using Landsat Time Series. *Remote Sens. Environ.* **2015**, *161*, 107–121. [CrossRef]

20. Chuvieco, E.; Martin, M.P.; Palacios, A. Assessment of Different Spectral Indices in the Red-near-Infrared Spectral Domain for Burned Land Discrimination. *Int. J. Remote Sens.* **2002**, *23*, 5103–5110. [CrossRef]

21. Martín, M.P.; Gómez, I.; Chuvieco, E. Burnt Area Index (BAIM) for Burned Area Discrimination at Regional Scale Using MODIS Data. *For. Ecol. Manag.* **2006**, *234*, S221. [CrossRef]

22. Trigg, S.; Flasse, S. An Evaluation of Different Bi-Spectral Spaces for Discriminating Burned Shrub-Savannah. *Int. J. Remote Sens.* **2001**, *22*, 2641–2647. [CrossRef]

23. Smith, A.M.S.; Wooster, M.J.; Drake, N.A.; Dipotso, F.M.; Falkowski, M.J.; Hudak, A.T. Testing the Potential of Multi-Spectral Remote Sensing for Retrospectively Estimating Fire Severity in African Savannahs. *Remote Sens. Environ.* **2005**, *97*, 92–115. [CrossRef]

24. Tucker, C.J. Red and Photographic Infrared Linear Combinations for Monitoring Vegetation. *Remote Sens. Environ.* **1979**, *8*, 127–150. [CrossRef]

25. Key, C.H.; Benson, N.C. *The Normalized Burn Ratio (NBR): A Landsat TM Radiometric Measure of Burn Severity*; United States Geological Survey, Northern Rocky Mountain Science Center: Bozeman, MT, USA, 1999.

26. Lutes, D.C.; Keane, R.E.; Caratti, J.F.; Key, C.H.; Benson, N.C.; Sutherland, S.; Gangi, L.J. *FIREMON: Fire Effects Monitoring and Inventory System*; US Department of Agriculture, Forest Service, Rocky Mountain Research Station: Fort Collins, CO, USA, 2006; Gen. Tech. Rep. RMRS-GTR-164. 1 CD.

27. Pinty, B.; Verstraete, M. Gemi—a Nonlinear Index to Monitor Global Vegetation from Satellites. *Vegetatio* **1992**, *101*, 15–20. [CrossRef]

28. Huete, A. A Soil-Adjusted Vegetation Index (Savi). *Remote Sens. Environ.* **1988**, *25*, 295–309. [CrossRef]

29. Huete, A.; Didan, K.; Miura, T.; Rodriguez, E.P.; Gao, X.; Ferreira, L.G. Overview of the Radiometric and Biophysical Performance of the MODIS Vegetation Indices. *Remote Sens. Environ.* **2002**, *83*, 195–213. [CrossRef]

30. Verbesselt, J.; Zeileis, A.; Herold, M. Near Real-Time Disturbance Detection Using Satellite Image Time Series. *Remote Sens. Environ.* **2012**, *123*, 98–108. [CrossRef]

31. Zhu, Z.; Woodcock, C.E. Continuous Change Detection and Classification of Land Cover Using All Available Landsat Data. *Remote Sens. Environ.* **2014**, *144*, 152–171. [CrossRef]

32. Disney, M.I.; Lewis, P.; Gomez-Dans, J.; Roy, D.; Wooster, M.J.; Lajas, D. 3D Radiative Transfer Modelling of Fire Impacts on a Two-Layer Savanna System. *Remote Sens. Environ.* **2011**, *115*, 1866–1881. [CrossRef]

33. Bastarrika, A.; Chuvieco, E.; Pilar Martin, M. Mapping Burned Areas from Landsat TM/ETM plus Data with a Two-Phase Algorithm: Balancing Omission and Commission Errors. *Remote Sens. Environ.* **2011**, *115*, 1003–1012. [CrossRef]

34. Fornacca, D.; Ren, G.; Xiao, W. Evaluating the Best Spectral Indices for the Detection of Burn Scars at Several Post-Fire Dates in a Mountainous Region of Northwest Yunnan, China. *Remote Sens.* **2018**, *10*, 1196. [CrossRef]

35. Schepers, L.; Haest, B.; Veraverbeke, S.; Spanhove, T.; Vanden Borre, J.; Goossens, R. Burned Area Detection and Burn Severity Assessment of a Heathland Fire in Belgium Using Airborne Imaging Spectroscopy (APEX). *Remote Sens.* **2014**, *6*, 1803–1826. [CrossRef]

36. Stroppiana, D.; Bordogna, G.; Boschetti, M.; Carrara, P.; Boschetti, L.; Brivio, P.A. Positive and Negative Information for Assessing and Revising Scores of Burn Evidence. *IEEE Geosci. Remote Sens. Lett.* **2012**, *9*, 363–367. [CrossRef]

MDPI AG
Grosspeteranlage 5
4052 Basel
Switzerland
Tel.: +41 61 683 77 34

Remote Sensing Editorial Office
E-mail: remotesensing@mdpi.com
www.mdpi.com/journal/remotesensing